Molecular Biology Intelligence Unit

Molecular Mechanisms of Epithelial Cell Junctions: From Development to Disease

Sandra Citi, M.D., Ph.D.

Università degli Studi di Padova
Padova, Italy

R.G. Landes Company
Austin

Molecular Biology Intelligence Unit

MOLECULAR MECHANISMS OF EPITHELIAL CELL JUNCTIONS:
FROM DEVELOPMENT TO DISEASE

R.G. LANDES COMPANY
Austin

CRC Press is the exclusive worldwide distributor of publications of the
Molecular Biology Intelligence Unit.
CRC Press, 2000 Corporate Blvd., NW, Boca Raton, FL 33431. Phone: 407/994-0555.

Submitted: March 1994
Published: July 1994

Production Manager: Deborah Molsberry
Copy Editor: Constance Kerkaporta

Printed on recycled paper by Armstrong Printing Co., Austin, Texas, U.S.A.

Please address all inquiries to the Publisher:
R.G. Landes Company, 909 Pine Street, Georgetown, TX 78626
or
P.O. Box 4858, Austin, TX 78765
Phone: 512/ 863 7762; FAX: 512/ 863 0081

ISBN 1-57059-091-5 CATALOG # LN9091

To Our Readers

R.G. Landes Company publishes four book series: *Medical Intelligence Unit, Molecular Biology Intelligence Unit, Neuroscience Intelligence Unit,* and *Biotechnology Intelligence Unit.* Our goal is to publish the most recent information in biomedical science for sophisticated researchers and physicians.

To achieve this goal we have accelerated our publishing program to conform to the fast pace in which information grows in biomedical science. The book you have in hand, like all titles in our series, was published *within 90 days of receipt of the manuscript.*

As you might expect, this causes a few problems for us; sometimes it makes our office more like a big city newspaper than a scholarly publisher. So as you look through this book you may see something that isn't just right. Please let us know. Or if you have an idea for improving our books, we'd like very much to hear from you. If the problem you describe hasn't already been discovered or if your idea is provocative enough to promote a discussion here, we'll give you a free book of your choice. Just list three titles in order of preference and we'll send you one, based on availability—with our thanks. Our address is printed on the copyright page of each of our books.

Elinda McKenna
Director of Operations
R.G. Landes Company

CONTENTS

CONTRIBUTORS

Sandra Citi, M.D., Ph.D., *Editor*
Dipartimento di Biologia
Università degli Studi di Padova
Via Trieste 75
35121 Padova, Italy
Chapter 5

Masayuki Amagai, M.D., Ph.D.
Department of Dermatology
Keio University School of Medicine
35 Shinanomachi, Shinjuku-ku
Tokyo 160, Japan
Chapter 14

Jürgen Behrens, Ph.D.
Max Delbruck Centrum für Molekulare Medizin
Robert Rössle Strasse 10
13122 Berlin, Germany
Chapter 13

Walter Birchmeier, Ph.D.
Max Delbruck Centrum für Molekulare Medizin
Robert Rössle Strasse 10
13122 Berlin, Germany
Chapter 13

Peter J. Bryant, D. Phil.
Developmental Biology Center
University of California, Irvine
Irvine, California 92717
Chapter 1

Paolo Burighel, Ph.D.
Dipartimento di Biologia
Università degli Studi di Padova
Via Trieste 75
35121 Padova, Italy
Chapter 2

Marcelino Cereijido, M.D., Ph.D.
Centro de Investigacion y Estudios Avanzados
Apartado Postal 14-740
Mexico, D.F.
07000 Mexico
Chapter 6

Ruben G. Contreras, Ph.D.
Centro de Investigacion y Estudios Avanzados
Apartado Postal 14-740
Mexico, D.F.
07000 Mexico
Chapter 6

Pierre A. Coulombe, Ph.D.
Department of Biological Chemistry
Johns Hopkins University School of Medicine
725 N. Wolfe Street
Baltimore, Maryland 21205
Chapter 15

Pamela Cowin, Ph.D.
Department of Cell Biology
Ronald O. Perelman Department of Dermatology and Kaplan Cancer Center
New York University Medical Center
550 First Avenue
New York, NY 10016
Chapter 8

CONTRIBUTORS

Romano Dallai, Ph.D.
Dipartimento di Biologia Evolutiva
Università degli Studi di Siena
Via Mattioli 4
Siena, Italy
Chapter 2

Michele De Luca, M.D.
Istituto Nazionale per la Ricerca sul Cancro
16132 Genova, Italy
Chapter 16

Tom P. Fleming, Ph.D.
Department of Biology
University of Southampton
Bassett Crescent East
Southampton SO9 3TU
United Kingdom
Chapter 4

Elaine Fuchs, Ph.D.
Department of Molecular Genetics and Cell Biology
and Howard Hughes Medical Institute
The University of Chicago
Chicago, Illinois 60637
Chapter 15

Maria del Refugio Garcia-Villegas, Ph.D.
Centro de Investigacion y Estudios Avanzados
Apartado Postal 14-740
Mexico, D.F.
07000 Mexico
Chapter 6

Kathleen J. Green, Ph.D.
Departments of Pathology and Dermatology
Northwestern University Medical School
303 East Chicago Avenue
Chicago, Illinois 60611
Chapter 9

Lorenza Gonzalez-Mariscal, Ph.D.
Centro de Investigacion y Estudios Avanzados
Apartado Postal 14-740
Mexico, D.F.
07000 Mexico
Chapter 6

Nancy J. Lane, D. Phil., Sc.D.
Department of Zoology
University of Cambridge
Downing Street
Cambridge CB2 3EJ, United Kingdom
Chapter 2

Catherine Leclerc, Ph.D.
Centre de Biologie du Développement UMR9925
Université Paul Sabatier
118 route de Narbonne
31062 Toulouse cedex, France
Chapter 3

Robert D. Lynch, D. Sc.
Department of Biological Sciences
University of Massachusetts Lowell
Lowell, Massachusetts
Chapter 7

Pier Carlo Marchisio, M.D.
Dipartimento di Ricerca Biologica e Tecnologica
Istituto Scientifico S. Raffaele
Via Olgettina 58
20132 Milano, Italy
Chapter 16

Giambruno Martinucci, Ph.D.
Dipartimento di Biologia
Università degli Studi di Padova
Via Trieste 75
35121 Padova, Italy
Chapter 2

CONTRIBUTORS

Susan Mechanic, Ph.D.
Department of Cell Biology
Ronald O. Perelman Department of Dermatology
and Kaplan Cancer Center
New York University Medical Center
550 First Avenue
New York, NY 10016
Chapter 8

Linda S. Musil, Ph.D.
Department of Cell Biology
Harvard Medical School
220 Longwood Avenue
Boston, Massachusetts 02115
Chapter 10

Akira Nagafuchi, Ph.D.
Department of Information Physiology
National Institute for Physiological Sciences
Myodaiji-cho, Okazaki
Aichi 444, Japan
Chapter 12

Eveline E. Schneeberger, M.D.
Department of Pathology
Cox 5
Massachusetts General Hospital
Boston, Massachusetts 02114
Chapter 7

David C. Spray, Ph.D.
Departments of Neuroscience and Medicine
Albert Einstein College of Medicine
1300 Morris Park Avenue
Bronx, New York 10461
Chapter 11

Thaddeus S. Stappenbeck, Ph.D.
Departments of Pathology *and* Dermatology
Northwestern University Medical School
303 East Chicago Avenue
Chicago, Illinois 60611
Chapter 9

Sachiko Tsukita, Ph.D.
Department of Information Physiology
National Institute for Physiological Sciences
Myodaiji-cho, Okazaki
Aichi 444, Japan
Chapter 12

Shoichiro Tsukita, M.D., Ph. D.
Department of Information Physiology
National Institute for Physiological Sciences
Myodaiji-cho, Okazaki
Aichi 444, Japan
and Department of Medical Chemistry
Faculty of Medicine, Kyoto University
Kyoto, Japan
Chapter 12

Jesus Valdes, Ph.D.
Centro de Investigacion y Estudios Avanzados
Apartado Postal 14-740
Mexico, D.F.
07000 Mexico
Chapter 6

Shigenobu Yonemura, Ph.D.
Department of Information Physiology
National Institute for Physiological Sciences
Myodaiji-cho, Okazaki
Aichi 444, Japan
Chapter 12

PREFACE

In multicellular Eukaryotes, cells interact and exchange information through a variety of mechanisms. One is based on the direct contact between adjacent cells, which occurs at highly specialized sites called junctions. Recent years have witnessed remarkable progress in characterization of the composition, function and regulation of junctions. The molecular basis of several human diseases has been investigated, and in some cases understood in depth, due to the elucidation of the structure and organization of junction-associated proteins.

This monograph includes chapters by leading investigators in the field and focuses on cell-cell junctions (tight junction, *zonula adhaerens*, desmosome, gap junction), cell-substratum junctions, and associated cytoskeletal components. The aim of the book is to review the most important recent advances in clarifying the organization of these structures, and their roles in evolution, development and pathogenesis of human disease. Such progress has been made possible by a combination of morphological, biochemical, physiological and molecular genetic approaches.

My sincerest thanks to all contributors for their participation, and my loving gratitude to my husband David Shore for his help and support.

This book is dedicated to our newborn daughter, Carolina.

Sandra Citi

CHAPTER 1

Molecular Structure and Function of Invertebrate Junctions

Peter J. Bryant

1. INTRODUCTION

Until recently, the molecular structure of invertebrate epithelial cell junctions was largely unknown. There were several brave attempts to investigate these structures by subcellular fractionation and biochemical analysis, leading to a preliminary understanding of the proteins associated with junctions. However, over the past few years new molecular genetic approaches using *Drosophila* have led to striking advances in our knowledge of the proteins associated with adherens, septate and basal junctions and the investigation of mutant phenotypes has opened up new ways to probe the functions of these molecules, especially with regard to their roles in cell-cell signaling. In this review I will concentrate on the results from these new approaches.*

2. ULTRASTRUCTURE OF INVERTEBRATE JUNCTIONS

The lateral cell membranes of typical invertebrate epithelial cells show the set of junctions illustrated in Figure 1.1. The apical-most structure resembles the vertebrate adherens junction and new molecular evidence indicates that a similar set of proteins is present. Just basal to the adherens junction is the septate junction, characterized by regularly spaced septa traversing the intercellular space. As seen in freeze-fracture replicas, the septa in ectodermal cells are made up of distinct particles ("pleated septate

* This chapter will follow the convention of using italics for gene names and roman text with a capital first letter for the corresponding protein.

Molecular Mechanisms of Epithelial Cell Junctions: From Development to Disease
Edited by Sandra Citi, M.D., Ph.D.;

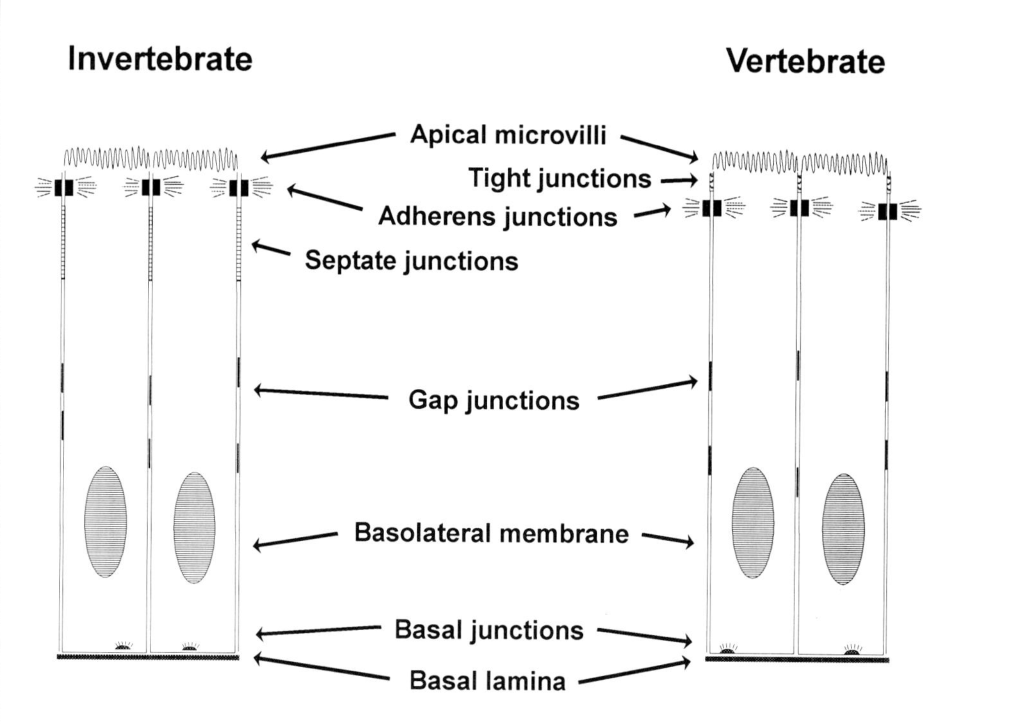

Fig. 1.1. Schematic view of the arrangement of specialized junctions in invertebrate and vertebrate epithelial cells.

junctions") whereas in endodermal cells they have a smoothly curved profile ("smooth septate junctions"); this is discussed in chapter 2 in this volume. Based on the fact that septate junctions appear to replace the tight junctions of vertebrates and that they have at least some of the same properties, it has been suggested that they are analogous structures.[1] This idea is now supported by the finding of homologous proteins in tight and septate junctions.[2] Basal to the septate junctions (and sometimes interspersed within them) are gap junctions that resemble but show some morphometric differences with those of vertebrates. Connecting the basal surface of epithelial cells to the basal lamina are specialized junctions that we will call basal junctions. For the general ultrastructure of junctions in *Drosophila* see ref. 3 and for an account of the development of junctions in the *Drosophila* embryo see ref. 4.

In the vertebrate literature the term adherens junction has been applied to a variety of structures including both cell-substrate junctions and cell-cell junctions. Cell-substrate adherens junctions include focal adhesions of cultured cells, myotendinous junctions, dense plaques of smooth muscle cells and basal adhesions to basal lamina.[5] Cell-cell adherens junctions include zonulae adherentes ("intermediate junctions" or "belt desmosomes") of epithelial cells, endothelial junctions, fasciae adherentes of cardiac myocytes and spot adhesions in other cell types. These junctions all show association with actin microfilaments and some of the proteins they have in common are localized at the interface between the junctional membrane and the actin microfilament network.[5] In invertebrates there are clear ultrastructural and molecular differences between the cell-cell adherens junction localized at the apical margin of the lateral cell membrane of epithelial cells and the basal junctions that connect the basal surface of these cells to the basal lamina. We therefore deal with these junctions separately.

3. MOLECULAR STRUCTURE OF ADHERENS JUNCTIONS

3.1 CADHERINS

A critical component of vertebrate adherens junctions is a member of the cadherin family of calcium-dependent cell adhesion molecules,[6] which appear to act in a homophilic manner by direct interactions between the extracellular domains of cadherin molecules on adjacent cells.[7] Efforts to identify *Drosophila* homologs of typical vertebrate cadherins based on homology were not successful. However, M. Takeichi and co-workers (personal communication) have now identified a 150 kD protein (*D*E-cadherin; Fig. 1.2) that co-immunoprecipitates with *Drosophila* α-catenin and is expressed in the epidermis and gut, but not in the nervous system or mesoderm. In well-differentiated epithelia, the protein is localized at the position of adherens junctions. cDNA sequencing indicates that this protein contains an extracellular domain with six cadherin repeats, a cysteine-rich region and a laminin A-globular domain and a cytoplasmic domain with homology to potential catenin-binding sequences. The first cadherin repeat (CR0) is probably cleaved off, since the sequence corresponding to the N-terminus of vertebrate cadherins is located between this domain and CR1. Each cadherin repeat shows more homology to the repeat in the corresponding position in vertebrate cadherins than to other *Drosophila* repeats, indicating the variation between the repeats is highly conserved.

3.2

A second molecule related to cadherins was identified as the product of the *fat* tumor suppressor gene.[8] The presumptive extracellular domain is much larger than that of typical cadherins, containing 34 repeats of a sequence showing homology to the four extracellular repeats of typical cadherins, as well as four EGF-like repeats. The presumptive cytoplasmic domain of the predicted *fat* protein has only very limited homology to the cytoplasmic domain of typical cadherins and the homology is separated into three disjunct domains. Therefore, it seems likely that the cytoplasmic domain of the *fat* protein has evolved some novel functions as compared to typical cadherins. It is not yet known whether the *fat* product is localized to cell junctions.

3.3 α-CATENIN

In vertebrate cells, cytoplasmic proteins called α-, β- and γ-catenins, are associated with the cytoplasmic tail of cadherins in adherens junctions.[6,7] Since α-catenin-negative cell lines show loss of adhesive properties,[9] it seems likely that the catenins are necessary for cadherin adhesive function. The gene

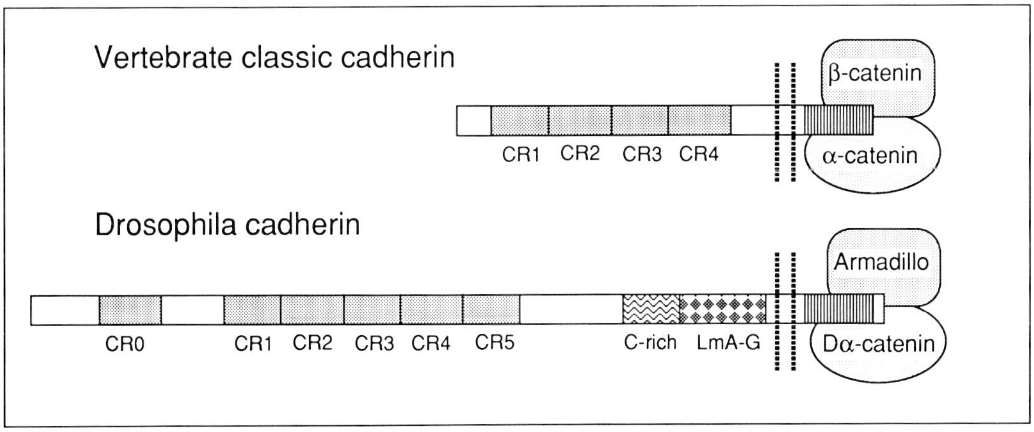

Fig. 1.2. Modular structure of Drosophila *cadherin, compared to a classic vertebrate cadherin, showing interaction with catenins (courtesy of M. Takeichi). CR, cadherin repeat; C-rich, cysteine-rich domain; LmA-G, laminin A globular domain.*

encoding the *Drosophila* homolog of α-catenin has been cloned by virtue of its homology with vertebrate α-catenin.[10] The protein product is localized in cell junctions in both embryos and imaginal discs and the amino acid sequence predicted from cDNA sequence shows 60% identity with mouse αE-catenin.[10]

3.4. Armadillo (β-catenin)

The *armadillo* gene product Arm of *Drosophila* shows 71% identity to the vertebrate adherens junction protein β-catenin[11] and 63% identity with the desmosome-associated protein plakoglobin.[12] Arm is enriched at adherens junctions[12,13] and forms a complex with *Drosophila* α-catenin and *D*E-cadherin, as indicated by co-immunoprecipitation;[13] (M. Takeichi, personal communication). The complex runs at about 8S on sucrose gradients and is therefore similar in size to the vertebrate complex of cadherin, α-catenin and β-catenin.[13,14] Thus the molecular complex at the adherens junction in *Drosophila* may be quite similar to the vertebrate structure (Fig. 1.2).

4. Apical Signalling Pathways

Genetic studies are providing several examples of cell-cell signaling pathways in which some of the critical components occur at or near adherens junctions (Fig. 1.3; see ref. 15 for review). These results are adding considerably to the list of adherens junction proteins and they are suggesting that adherens junctions are much more dynamic than previously recognized. In fact, it seems likely that these junctions are the most important channels for communication between epithelial cells undergoing coordinated developmental activities.

4.1. The Wingless/Armadillo pathway

Soon after cellularization in the *Drosophila* embryo, the anterior-posterior pattern of differentiation within each body segment is elaborated by interactions between the newly formed cells. These interactions involve the expression of genes in the segment polarity class, notably *wingless* (*wg*) which encodes a secreted protein Wg,[16,17] and *engrailed* (*en*) which encodes a homeodomain-

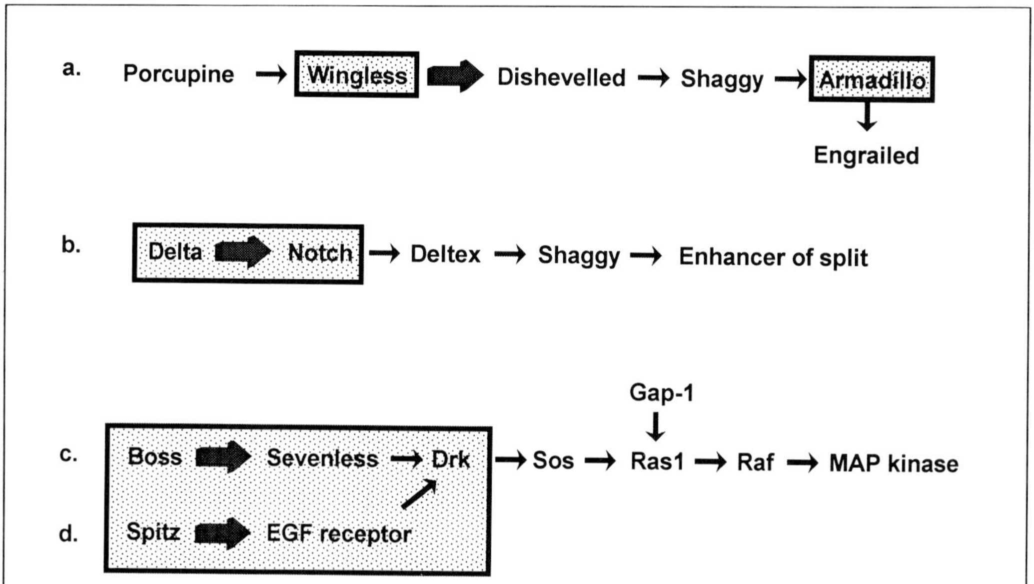

Fig. 1.3. Schematic view of putative signaling pathways that are mediated through proteins localized at the apical ends and/or at the adherens junctions between epithelial cells in Drosophila. *Proteins localized at the apical surface and/or adherens junctions are boxed and shaded. a, the Wingless/Armadillo pathway. b, the Delta/Notch pathway. c, the Boss/Sev pathway. d, the Spitz/EGF receptor pathway. References and description in text.*

containing transcription factor.[18] Lethal mutations in these genes cause the replacement of structures normally found in the posterior parts of body segments (bare cuticle) by structures typical of anterior regions (denticles), leading to an embryo covered in denticles. *wg* is expressed in the anterior part of each segment and this maintains expression of *en* in the posteriorly adjacent cells[19] as shown by the premature loss of *en* expression in *wg* embryos[20] and by the expanded domain of *en* expression produced by ectopic expression of *wg*.[21] Immunocytochemical studies of Wg show that the protein is transferred from the expressing cell into the responding cell across the anterior/posterior compartment boundary,[22] and that the protein can be found two or three cell diameters away from the expressing cell.[23] It has been suggested that Wg may be responsible for patterning the segment, if cells at different anterior/posterior levels within the segment select different fates depending on the local level of Wg protein.[23] The predicted Wg protein contains a signal sequence but no transmembrane domain, consistent with it being a secreted protein.[16,17] Wg is always found in association with membranes and the transfer of Wg occurs in small vesicles that are sometimes found close to "fascia adherens" a term that presumably refers to adherens junctions.[23]

In developing imaginal discs, *wg* is also expressed in specific patterns and has an important role in patterning. Certain allelic combinations lead to loss of ventral and distal structures in legs and wings and their replacement by dorsal structures.[24] Studies of genetic mosaics show that *wg* behaves in a non-autonomous fashion;[25] that is, a clone of *wg* mutant cells in a *wg* requiring area of the imaginal disc can develop normally, presumably because surrounding non-mutant cells provide Wg protein to the clone. This result is consistent with the idea that the protein acts as an intercellular signal. In the leg disc, *wg* is expressed in a ventral sector,[24] and when expressed ectopically in the dorsal part of the disc it can influence the behavior of surrounding, non-expressing cells so that they participate in forming pattern duplications.[26] Thus the Wg protein acts in cell-cell signaling in the imaginal discs as well as in the embryo.

Several genes have been identified whose functions appear to be required for operation of the Wg pathway (Fig. 1.3). Studies of epistasis and of autonomy in genetic mosaics indicate that the *porcupine* gene product is required in the signaling cell for the secretion of Wg while the products of *armadillo*, *disheveled* and *zeste-white 3* are required for transduction of the signal.[27,28] Of these gene products, the subcellular location is known only for Arm and this protein is, at least in part, junction-associated. Mutations in *armadillo* produce a similar phenotype to that of *wingless* in developing embryos—production of denticles all along the embryonic segment instead of only in the anterior half.[29] However, this effect is autonomous in genetic mosaics,[30] suggesting that Arm may function in the responding cell rather than in the signaling cell. Recent studies indicate that the membrane localization of Arm may be regulated as part of the signal transduction mechanism.[12,31] In the stage 9 embryo, about three-fourths of the Arm protein is membrane-associated and probably localized in adherens junctions and the remainder is cytoplasmic.[13] However, the distribution between membrane and cytoplasm varies along the segment. In a stripe at about the middle of each segment (the Arm stripe), centered on the Wg-expressing region, Arm staining is more intense and much of the protein is cytoplasmic although it is still enriched at the membrane. Between the stripes (interstripe regions) there is noticeably less Arm and nearly all of it is membrane-associated. From these observations, it has been suggested that Arm protein is released from the membrane into the cytoplasm upon receipt of the Wg signal. This is the earliest known response to Wg signal,[31] and is presumed to result from interaction between the Wg protein and its still unidentified receptor. Consistent with this idea, in embryos lacking *wg* function the entire segment shows predominantly membrane-associated Arm like that of the interstripe in wild type. In contrast in *zw3* (=*shaggy*) mutants, which show the opposite transformation to *wg* (transformation of anterior to posterior) the entire segment shows the stripe pattern with cytoplasmic Arm.[31] *zw3* encodes a serine/threonine kinase[32,33] that appears to be required for Arm phosphorylation, leading to speculation that phosphorylation could control junctional localization of Arm.[31] In *disheveled* mutant embryos, like those mutant for *wg*, the entire embryo shows the interstripe pattern with Arm enriched at the membrane,[34] indicating that the product of *disheveled* is required for the release of Arm from the cell membrane. Presumably the release of Arm from the membrane is an important step in the signal transduction mechanism that ultimately leads to activation of *en* expression and other downstream events.

4.2 The Delta/Notch pathway

Shortly after blastoderm formation, cell-cell interactions within the ventral "neurogenic" region of the *Drosophila* embryo result in the specification of neuroblasts in a predictable pattern.[35] About 25% of the cells acquire the neuroblast fate as a result of a two-step process. First, small groups of contiguous cells acquire the ability to become neuroblasts and second, inhibitory interactions within each group lead to the selection of one neuroblast that proliferates to form neurons, the remainder of the cells becoming epidermis.[36] These inhibitory interactions depend upon the products of the *Notch* and *Delta* genes.[35,37] In the absence of either function, all of the cells in the neurogenic region become neuroblasts. The two genes are also required for cell interactions controlling neurogenesis in the peripheral nervous system[35] the formation of bristle patterns and for several different cell interactions controlling the fate of cells in the developing retina including the Boss/Sev interaction (see below).[38–40] Notch is a transmembrane glycoprotein with 36 sequences homologous to epidermal growth factor (EGF) in its extracellular domain,[41,42] and six copies of the cdc10/ankyrin repeat in the intracellular domain. Delta is also a transmembrane protein containing nine EGF repeats in its extracellular domain.[43–45]

Notch is localized at adherens junctions in various epithelia,[37] whereas Delta shows a more general cell-surface localization.[39,46] Delta and Notch proteins expressed on different cells can mediate calcium-dependent adhesion between them,[47] they co-localize in membrane patches when both are expressed on the same cell, they co-immunoprecipitate from embryo extracts[47] and genetic studies provide evidence for direct interactions between the two proteins in vivo.[48] Single amino acid substitutions in the EGF-like domains of Delta interact in an allele-specific way with single amino acid substitutions in specific EGF-like domains of Notch,[49] indicating direct interactions between these domains. An attractive hypothesis is therefore that these two proteins act as a signal-receptor pair for cell-cell interactions operating at adherens junctions to determine cell fates. Experiments using genetic mosaics to study the role of these genes in bristle pattern formation indicate that cells mutant for *Notch* autonomously adopt the mutant phenotype of excess bristle production, whereas cells mutant for *Delta* can non-autonomously show the wild-type phenotype if they are adjacent to wild-type cells.[50] These results argue in favor of a model in which Delta is the signal and Notch is the receptor for the signaling events that specify bristle sites and presumably also for the events that specify neuroblast positions in the embryo. The product of the *Serrate* gene, which is also a transmembrane protein with extracellular EGF-like repeats[51] when expressed on cultured cells can also mediate adhesion to Notch-expressing cells and is therefore another candidate ligand.[52]

Studies of the phenotypes produced in vivo by various truncated derivatives of Notch have begun to elucidate the functions of different domains. Expression of a protein consisting of only the intracellular domain causes a dominant "antineurogenic" phenotype in which elements of the embryonic nervous system are underproduced[53,54]—the opposite abnormality from the loss-of-function *Notch* phenotype. This is interpreted as resulting from constitutive activation of the intracellular domain leading to unregulated downstream signaling, implying that the function of the extracellular domain is to regulate the signaling activity of the intracellular domain.

We only have an incomplete picture of the signal transduction pathways upstream or downstream of the Delta/Notch interaction and interpretation is difficult because the pathway may not be linear (Fig. 1.3[35,55]). Genetic epistasis tests indicate that *mastermind*, *bigbrain* and *neuralized* act upstream of Notch,[56,57] but it is not known whether they act upstream of Delta. The *zw3*/shaggy ser/thr kinase[58] and the nuclear protein encoded by *Enhancer of split*[57] act downstream of Notch; in fact, *Enhancer of split* appears to be downstream of all other known elements of the pathway. The product of the *deltex* gene binds to the cdc10/ankyrin repeats in the cytoplasmic domain of Notch,[40] and deletion studies show that this region of Notch is required for its signaling function and for its interaction with other proteins.[59] The interaction with Deltex therefore seems likely to be an important early step in signal transduction. The isolated intracellular domain of Notch gets translocated to the nucleus,[54,57] suggesting that this part of the intact, normal protein may interact with a protein that normally moves to the nucleus. Based on all of these data, the tentative pathway shown in Figure 1.3 is proposed.

Both Delta and Notch are down-regulated following cell fate specification and then Delta or, less abundantly, both Notch and Delta, appear in apparent endocytic vesicles in the interacting cells.[39] In mixtures of Notch-expressing and Delta-expressing cells in culture, vesicles containing both Notch and Delta are found in the responding, Notch-expressing cells, implying that the ligand and receptor are co-internalized in a manner resembling the case of Boss and Sev[39] (see below). The internalization in vivo presumably reflects clearance from the cell surface upon completion of the signaling function.

4.3. The Boss/Sevenless pathway

The induction of the R7 photoreceptor cell by the R8 cell is an important event in the patterning of ommatidia during eye development.[60] This interaction requires the products of two genes, *bride-of-sevenless* (*boss*;[61,62]) and *sevenless* (*sev*;[63,64]). The two proteins (Boss and Sev) have become well known as two of the key players in a cell-cell signaling event in the eye imaginal disc that is required for the specification of the R7 photoreceptor cell in each ommatidium. Genetic mosaic analysis indicates that Boss is required in the R8 precursor[65] and Sev is required in the R7 precursor[66] for a successful inductive event. Predicted amino acid sequences indicate that both Boss and Sev are transmembrane proteins: Boss is a seven-pass transmembrane protein[67] and Sev is a transmembrane protein tyrosine kinase.[68–71] This leads to the hypothesis that the Boss protein expressed on the surface of the R8 cell interacts directly with the Sev protein expressed on the surface of the R7 cell to specify R7 fate. Both proteins are localized at apical microvilli and adherens junctions.[64,72] The Drk protein, an SH3-SH2-SH3 adaptor, couples the Sev receptor to the Sos guanine nucleotide exchange factor, possibly creating a signaling complex that activates Ras in the Sev signaling pathway.[73,74] Drk is localized to the apical plasma membrane of epithelial cells in the eye imaginal disc,[74] making it likely that Sos has a similar localization, at least transiently. Many other components of this pathway are known (Fig. 1.3),[60,75] but their subcellular localization in epithelial cells has not been reported so their relationship to junctions is unknown. The entire Boss protein seems to be transferred from the R8 cell into the apical part of the R7 cell during R7 induction and it is contained in a prominent multivesicular body similar to a late endosome.[72] Whether this internalization is important for the signaling process, or simply reflects clearing of the proteins from the cell surface following the signaling event, is not known.

4.4. The Spitz/EGF receptor pathway

The *Drosophila* homolog of the EGF receptor (DER) is localized at the apical microvillar surface of imaginal disc cells[76] but it is not clear whether its distribution extends to adherens junctions. It is required for cell proliferation in imaginal discs in general and for ommatidial patterning in the eye disc.[77] In the follicle cells of the ovary DER is required for the formation of dorsal pattern elements[78] and in the embryo it is required for determining the identity of ventral ectoderm.[79] A candidate for the DER ligand is the product of the *spitz* gene, which is also required for ventral ectoderm identity and shows homology to TGF-α,[80] a ligand for the mammalian EGF receptor. The Spitz protein can stimulate autophosphorylation of the DER protein expressed on cultured cells, supporting the possibility of a direct interaction in vivo.[79] The events downstream of DER are not known, but probably involve elements in common with the Sev pathway (Fig. 1.3). The DER protein has not been shown directly to interact with Drk, but the *Caenorhabditis* homologs of these two proteins do bind to each other in vitro and the *Drk* gene can rescue *sem-5* mutants of *Caenorhabditis*, which are defective in the homolog of Drk.[81] It therefore seems highly likely that DER and Drk interact with one another in signal transduction.

4.5 Protein Tyrosine Kinases

In epithelial cells of vertebrates, most of the phosphotyrosine detectable by immunolocalization is localized at adherens junctions.[82–84] This has important functional significance, since increasing the level of phosphotyrosine (by inhibition of phosphatases) leads to rapid deterioration of adherens junctions and this effect can be reversed by inhibitors of tyrosine kinases.[85] Immunolocalization studies show that most of the phosphotyrosine in *Drosophila* epithelial cells is also present at adherens junctions (J.-W. Wu, D. F. Woods and PJB, unpublished). In mammals, the proteins phosphorylated on tyrosine at adherens junctions include those of the cadherin/catenin complex[86–88] but this has not yet been demonstrated in invertebrates.

The kinases responsible for the high level of phosphorylation at adherens junctions could include receptor protein tyrosine kinases (PTKs) and/or non-receptor PTKs. As discussed above, the Sev receptor in *Drosophila* is clearly localized to adherens junctions and the EGF receptor is also localized apically. In mammals, the non-receptor PTKs $pp62^{c-yes}$ and $pp60^{c-src}$ are highly enriched in the adherens junctions of hepatocytes, kidney epithelial cells and keratinocytes.[83] In *Drosophila*, the non-receptor PTK Dabl is enriched at regions of cell contact in the cellular blastoderm embryo and in the apical cortical region (the region containing adherens junctions) of imaginal disc cells.[89] The src-related protein product of the *Dsrc*28C locus is localized at the cell periphery in early embryos and larval salivary glands.[90]

4.6. Other Proteins Showing Apical Localization

Several interesting proteins have been reported as showing localization to the apical region of *Drosophila* epithelial cells, without clear evidence as to localization in adherens junctions. The product of the *crumbs* gene, a transmembrane protein with EGF-like repeats in its extracellular domain[91] that is required to maintain epithelial integrity in the early embryo, is localized apically in embryonic epithelia.[92] The *expanded* gene, in which mutations cause overgrowth of some parts of imaginal discs but loss of other regions, encodes a homolog of the human neurofibromatosis type 2 (NF2) gene and other members of the band 4.1 family and is localized apically.[93,94]

5. Molecular Structure of Septate Junctions

Lane and Dilworth[95] reported on the protein composition of membrane fractions that were highly enriched in septate junctions from tissues of the moth *Manduca sexta*, the cockroach *Periplaneta americana* and the lobster *Nephrops norvegicus*. They found major proteins of 18 and 22 kD from pleated septate junctions and 17 and 24 kD from smooth septate junctions. Since these proteins could not be extracted with high concentrations of urea, they were thought to be integral membrane components. However, further molecular studies of these proteins have not been reported.

5.1. The discs large-1 tumor suppressor gene product (Dlg)

One protein associated with septate junctions is the product of the *discs large-1* tumor suppressor gene in *Drosophila*.[96] This gene was identified by the phenotype of imaginal disc overgrowth caused by recessive lethal mutations. The imaginal discs lose their single-layered epithelial structure either entirely or over large regions and the disc cells become rounded instead of highly polarized.[96] They also lose the ability to differentiate, even after transplantation into genetically normal hosts. Mutant embryos are normal if they are derived from heterozygous mothers, but when produced from mutant germ-line clones and fertilized by mutant sperm they show loss of epithelial structure and excess cell proliferation, dying as late embryos. Therefore, the gene product is required for epithelial structure and cell proliferation control in the embryo as well as in imaginal discs. At the ultrastructural level, strong *dlg* alleles cause complete loss of septate junctions from imaginal discs and almost complete loss from epithelial tissues, such as salivary gland, which have grown during larval development by cell enlargement rather than proliferation (D. F. Woods and PJB, unpublished). The small amount of septate junction present in the latter tissues may indicate the persistence of a small amount of maternal Dlg protein or protein produced on maternal transcript.

Antibodies against a bacterial fusion gene product containing about the C-terminal half of Dlg recognize a membrane-associated protein as early as the cellular blastoderm stage, before any septate junctions are present. Later in embryogenesis a strong signal is detected in all epithelial tissues, on the subapical part of the lateral cell membrane coinciding with the position of septate junctions.[96] The protein shows a similar distribution in late larval and adult epithelial tissues. It is also present in the larval central nervous system, but its subcellular localization in this tissue is not known. Among the tissues in the larva, the brain and imaginal discs are the only ones to show a neoplastic phenotype, possibly because these tissues undergo extensive cell proliferation and therefore dilute out maternal gene product during larval development.

The *dlg* gene gives rise to several transcripts, of which we have most information on the one represented by cDNA DlgA. Sequencing of cDNA indicates that the DlgA protein contains 960 amino acids (aa) (Fig. 1.4). It has no signal sequence or transmembrane domain, so its membrane association is presumably a result of binding to other localized molecules. The C-terminal 179 amino acids shows clear homology to the cytosolic guanylate kinase of both yeast[97] and pig.[98] This enzyme catalyzes the conversion of GMP to GDP, using ATP as the phosphate donor. All of the amino acids that, in the yeast protein, are known to interact with the GMP substrate or the Mg^{2+} cofactor, are either conserved or replaced by conservative substitution in the predicted DlgA protein. However, the ATP binding site, which is highly conserved in a large number of kinases, shows a three amino acid deficiency in DlgA. This might mean that the ATP-binding site of DlgA is modified in an unusual but still functional way, or that this domain has evolved a new function requiring GMP binding. It might have a novel catalytic function, converting GMP to a product other than GDP, or it may bind GMP in a regulatory rather than a catalytic mode.

The predicted DlgA protein also contains a copy of the 71-amino acid SH3 domain,[99,100] which is also found in non-receptor PTKs and other signal transduction proteins and mediates binding to other proteins. SH3 domains in different proteins show some specificity for different binding proteins,[101] but this is not well enough understood to form the basis for predictions in the case of DlgA. The N-terminal half of DlgA contains three copies of a newly recognized ≈91-amino acid domain of unknown function called DHR (Discs-large Homologous Region,[102] previously called GLGF[103]). The DlgA protein also contains a region encoded by OPA repeats[104] and a PEST motif,[105] both of which are thought to control protein synthesis and turnover rates.

At least four mammalian proteins have been found with a similar modular structure

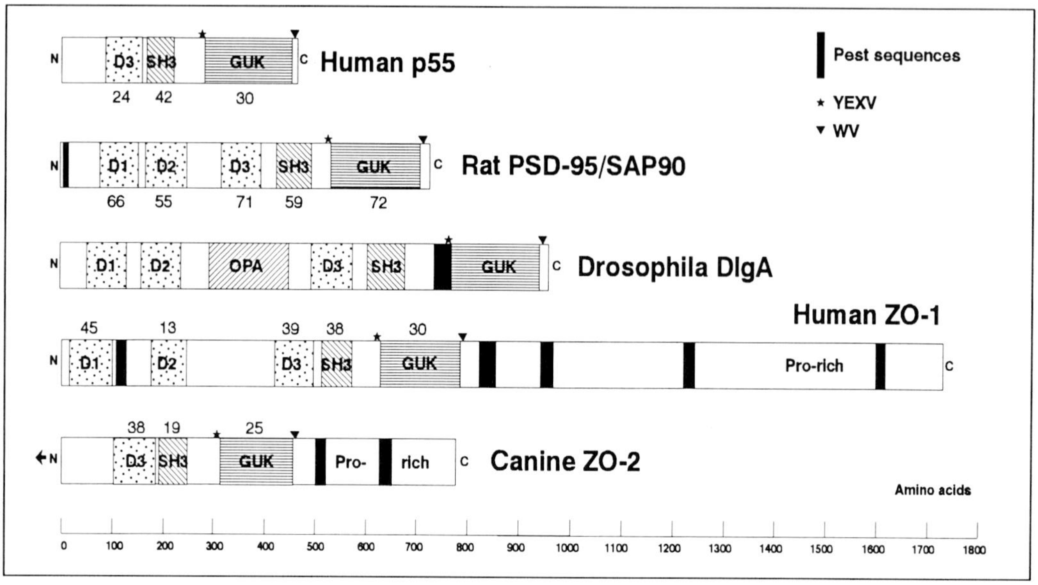

Fig. 1.4. Modular structures of human p55,[106] rat PSD-95/SAP90,[103,107] Drosophila *DlgA*,[96] human ZO-1[108] and canine ZO-2.[109] D1, D2, D3, discs-large homologous regions (DHRs); SH3, src-homology region 3;[99] GUK, guanylate kinase-homologous region; OPA, repetitive sequence.[104] Domain boundaries and sequence identities were determined using the BLAST algorithm[149] on line at the National Center for Biotechnology Information and the multiple alignment program MACAW.[150] Numbers outside the boxes indicate percent amino acid identity compared with the corresponding domain of DlgA. Regions shown in black are PEST sequences, scoring >6.0 on the PESTFIND program.[105] Arrow indicates incomplete sequence.

and significance sequence similarity to DlgA (Fig. 1.4). All are membrane-associated and three of them are components of cell junctions. p55, the major palmitoylated protein of the erythrocyte membrane cytoskeleton, has the GUK, SH3 and one of the DHR domains.[106] PSD-95/SAP90,[103,107] which is localized in synaptic junctions and ZO-1[108] and ZO-2[109] which are localized in tight junctions, have the GUK, SH3 and all three DHR domains. The finding of homologous proteins in tight and septate junctions reinforces the argument that these structures are analogous.[1] Both PSD-95/SAP90 and ZO-1 show the same deficiency in the ATP-binding site that is seen in DlgA, suggesting that this is not simply an inactivating alteration, but that it represents a highly conserved modification of function in comparison to cytosolic guanylate kinases. ZO-1 also shows loss of some of the amino acids involved in GMP binding in guanylate kinase, making it difficult to predict what catalytic or regulatory role might be served by this domain.

One copy of the DHR domain is found in the *disheveled* gene product and several other mammalian proteins.[110] Because of the membrane localization of many of the DHR-containing proteins, we speculate that this domain is involved in controlling subcellular localization through binding to other proteins.

5.2. Protein 4.1 (Coracle)

Fehon and coworkers[111] have cloned cDNAs representing transcripts encoding the *Drosophila* homolog of the human and *Xenopus* protein 4.1, a major component of the erythrocyte membrane cytoskeleton that is thought to link the spectrin-actin cytoskeleton to the membrane, thus stabilizing cell shape[112] (Fig. 1.5). The predicted *Drosophila* protein (D4.1) contains 1698 amino acids and shows 54% identity with the vertebrate sequences over a stretch of 350 aa in the N-terminal region. In this region it shows significant but lower similarity to other members of the protein 4.1 family including ezrin, talin, moesin A

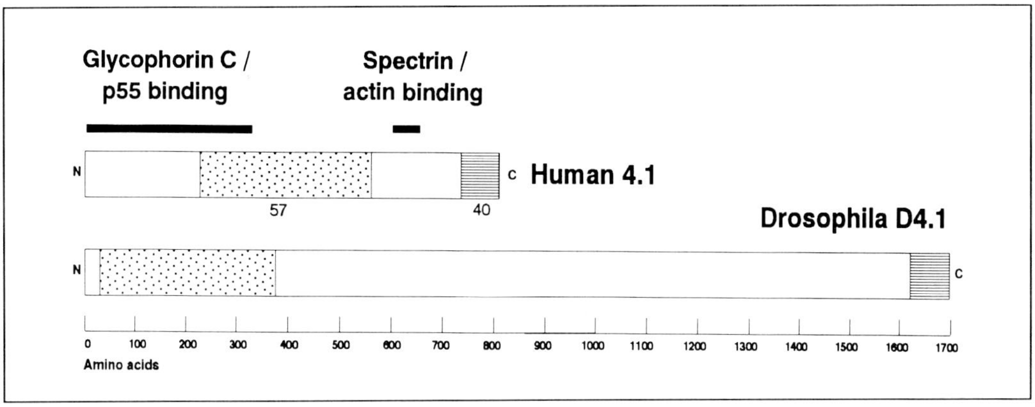

Fig. 1.5. Comparison of the predicted amino acid sequence of Drosophila *D4.1 with human protein 4.1. Domains showing significant similarity are shaded and numbers outside the boxes indicate percent amino acid identity compared with the corresponding domain of D4.1. From data in ref. 111.*

and B, NF2 and two tyrosine phosphatases. A region at the C-terminus shows similarity to protein 4.1 but not to other members of the protein 4.1 family, indicating that D4.1 is a true homolog of protein 4.1 rather than of one of the other family members. Compared to vertebrate protein 4.1, D4.1 has an insert of 1200 aa with no sequence similarity to known proteins. The *Drosophila* gene appears to give rise to one abundant and five less abundant transcripts and cDNA sequencing indicates that three of the transcripts are formed by alternative splicing. The human gene also shows multiple alternative splicing giving rise to tissue-specific isoforms, but the patterns do not correspond to those seen in *Drosophila*.

The D4.1 protein first becomes detectable by immunolocalization at embryonic stage 12, the start of germ-band shortening. At that time it is seen in epidermis, hindgut and foregut and at a later stage of germ band shortening it also appears in the tracheae and salivary glands. It is not expressed in the central nervous system. During larval development it is abundantly expressed in the imaginal discs. In all these tissues the protein is localized in a subapical belt of the lateral cell membrane, at the position of septate junctions; in fact, its subcellular localization is almost identical to that of the Dlg protein. Septate junctions develop soon after the beginning of D4.1 expression, at the end of germ band shortening. Thus the expression pattern of D4.1 is distinct from that of Dlg which appears much earlier in development and also exists in tissues (central nervous system) that do not have septate junctions.

The gene encoding D4.1 is named *coracle*, after the phenotype caused by homozygous mutations. Embryos homozygous for these mutations fail to complete dorsal closure, with the time of onset of abnormalities corresponding to the period of abundant D4.1 expression. The resulting embryos show an open-dorsal phenotype. The existing *coracle* alleles also show dominant suppression of the rough-eye phenotype caused by the *ellipse* mutation, a dominant alteration in the gene encoding the EGF receptor which is localized at the apical cell surface.[76] This suggests an intriguing functional interaction between events occurring at the apical surface and those occurring at septate junctions.

5.3. Fasciclin III

The transmembrane protein fasciclin III[113,114] (also reported as DENS antigen and as the antigen detected by the monoclonal antibodies DA.1B6[115] and 6D6[114,116] is also localized at septate junctions in epithelial cells (D. F. Woods, personal communication) but unlike Dlg it is expressed in stripes in the embryo.[113] It is also expressed on specific axon bundles and other neurons and accessory cells of both central and peripheral nervous systems.[113] The predicted amino acid sequence features three divergent immunoglobulin repeats in an extracellular domain, an apparent transmembrane domain and a cytoplasmic domain.[117] It can mediate cell aggregation in vitro in a homophilic manner,[114] suggesting a possible role in cell adhesion in vivo. Null mutations in the fasciclin III gene produce a surprisingly mild phenotype, involving subtle effects on adult morphology and behavior.[117]

5.4. Possible Interactions between Septate Junction Proteins

Interactions between septate junction proteins have not yet been studied directly, but the work on mammalian homologs of these proteins suggests some possibilities. The Dlg homolog p55, purified from human erythrocytes, binds both to the N-terminal 30 kD fragment of protein 4.1 and to the transmembrane protein glycophorin C.[118] This raises the intriguing possibility that Dlg may bind to homologous or similar proteins in *Drosophila*. D4.1 and fasciclin III are obvious candidates, given their location in septate junctions.

Sequences near the C-terminus of mammalian protein 4.1 interact with the cytoskeletal proteins spectrin and actin, stabilizing their interactions with each other and with the cell membrane.[119] However, the *Drosophila* D4.1 isoforms apparently lack these sequences.[111] Furthermore, both actin and spectrin are localized in the apical domain of the cell as well as in the basolateral region and are not exclusively colocalized with D4.1.[111] Thus D4.1 does not appear to interact with the cytoskeleton in the manner proposed for band 4.1 in the erythrocyte.

6. Molecular Structure of Gap Junctions

Gap junctions are specialized structures, found in all multicellular animals, that allow direct transfer of ions and small molecules between adjacent cells. The functional significance of this potential transfer is obscure, although studies using antisense RNA or antibodies to interfere with gap junction formation or function have indicated that gap junctional communication is important for a variety of developmental patterning events (for reviews see ref. 120 and chapters 3, 10 and 11 in this volume). These results might suggest that gap-junctional communication is an important route of cell-cell signaling, but they are also consistent with gap junctions playing a more general role in cell physiology.

In transmission electron microscopy, gap junctions are recognized as regions where the plasma membranes of adjacent cells are closely apposed to one another, whereas in freeze-fracture replicas they appear as plaques made up of structures called connexons. Each connexon has a hexagonal or star shape with a central pore.[121,122] The dimensions of insect gap junctions are slightly different from those seen in vertebrates;[120] the connexon pore diameter is 2-3 nm compared with 1.5-2 nm for vertebrates; connexon diameter is 12-14 nm compared with 6-10 nm for vertebrates; transverse profile width is 15-18 nm compared with 13-16 nm for vertebrates; and the extracellular gap width is 2-5 nm compared with 1.5-3 nm. Insect gap junctions also differ from vertebrate gap junctions in the behavior of connexons during freeze-fracture.[120]

Studies of the molecular structure of invertebrate gap junctions have involved subcellular fractionation and protein separation followed by immunochemical and molecular studies of candidate proteins.[120,121] The hepatopancreas of crayfish or lobster has often been used as the starting point for purification, because of its large size and high content of gap junctions. A 16-18 kD candidate gap junction protein identified from crustacean hepatopancreas shows homology to a protein ("ductin") associated with mammalian gap junctions that is also the major component of the proton channel of the vacuolar H^+-ATPase.[123] Protease digestion studies and secondary structure predictions suggest that the molecule is arranged in a bundle of four α helices that are mostly buried in the membrane. Ductin is highly conserved between fungi, plants and animals and the gene encoding the *Drosophila* homolog has been sequenced.[124]

Fractions enriched in "collapsed" gap junctions have been isolated from *Drosophila* whole larvae.[125] An 18 kD protein from these preparations showed no homology to ductins or any other known proteins.[120] Antibodies to this protein stained the non-native gap junctions in tissue fractions and sections as well as the 18 kD band on immunoblots,[125,126] but they failed to stain gap junctions in "native" configuration defined by appropriate profile width.

Gap junction preparations have recently been made from the tobacco budworm *Heliothis virescens*[127] using NaOH extraction and these show a more normal ("native") gap junction morphology in the electron microscope than previously reported preparations. A 40 kDa protein from these preparations has been identified immunologically and partially sequenced. It does not match any known sequences.

Vertebrate gap junctional channels are generally believed to consist mainly of proteins called connexins (reviewed in ref. 123, 128, 129, see also chapters 10 and 11 in this volume). These proteins range in size from 26 to 50 kD and are required for active cell coupling although they can apparently be removed from gap junctional preparations by treatment with detergent, urea and protease without destroying the basic junctional morphology of hexagonally packed channels. This suggests that other proteins, presumably ductin, may be important structural components of the gap junction.[123] So far, no invertebrate homologs of connexins have been reported.

7. MOLECULAR STRUCTURE OF BASAL JUNCTIONS

At several times in development specialized junctions are formed between epithelial cells and other cells basal to the epithelial layer. These junctions have been called basal junctions[130] or anchoring junctions,[131] and they involve transmembrane proteins of the integrin family[132] that probably interact with components of the extracellular matrix. One of the best studied examples occurs during the prepupal and pupal period of *Drosophila* metamorphosis, when imaginal discs undergo dramatic morphogenetic changes that convert them into adult structures. In the case of the wing disc, the epithelium undergoes a folding process that brings the presumptive dorsal and ventral wing surfaces into basal/basal apposition, after which the cells secrete the wing cuticle over their apical microvillar surfaces. The developmental process is complicated in that the two surfaces fuse and separate during prepupal development, then fuse and separate again during pupal development.[130,133] The molecular events have been correlated with ultrastructural changes occurring during pupal development. During this stage, cytoplasmic processes extend basally from each of the two epithelial layers until they contact processes extending from the opposite layer. Integrin-containing junctions are then formed where the dorsal and ventral processes meet and these junctions capture the "plus" ends of microtubules that nucleate at the opposite end of the cell, at hemidesmosome-like structures on the apical, cuticle-secreting surface.[134] Together with actin filaments, these microtubules form the "transalar cytoskeleton" that connects apical hemidesmosomes on one wing surface, via the basal junctions to apical hemidesmosomes on the other wing surface. The basal junctions show electron-dense material in the 30-50 nm extracellular space and a fringe of fibers on the cytoplasmic face, similar to adherens junctions. Later in the pupal period the wing cells flatten and the dorsal and ventral cell layers are then connected by a broad, flat area of contact with about 3-4 basal junctions per pair of connected cells. Eventually, after secreting the adult cuticle the epithelial cells are lost, leaving a wing blade that consists of two fused layers of cuticle, except at the wing veins where cells persist.[130]

The integrins important for dorsal/ventral wing surface fusion were initially identified as position-specific antigens on larval imaginal discs and are therefore known as PS integrins.[131,135] Like vertebrate integrins, they function as α/β dimers. The 110 kd β_{PS} subunit is common to the PS1 and PS2 antigens, whereas the α subunits α_{PS1} and α_{PS2} are distinct. The α_{PS2} subunit is cleaved on the extracellular side of the transmembrane domain to form heavy (140-145 kDa) and light (~20 kD) chains[136] which are probably disulfide linked. The α_{PS1} subunit also appears to be cleaved.[137] Furthermore, two isoforms of α_{PS2} arise from alternative splicing.[138] The α_{PS1} subunit is expressed on the presumptive dorsal wing while α_{PS2} is expressed on the presumptive ventral wing, both mainly on the basal cell surfaces.[139] This pattern exists in the third larval instar, long before the apposition of the two wing surfaces.[139] The common β_{PS} subunit is expressed on the basal surface of cells over most of the disc. As the two wing layers fuse during pupal development, β_{PS} is present on the apposed basal surfaces, but does not appear to be localized exclusively to the basal junctions.[130] The α_{PS1} subunit is encoded by the *multiple edematous wings* locus (D. Brower, personal communication), the α_{PS2} subunit is encoded by the *inflated* locus,[140] and β_{PS} is encoded by the *myospheroid* locus.[141,142] Mutations in each of these genes lead, among other phenotypes, to failure in the dorsal/ventral fusion process, producing a blistered wing[137,140,143–145] (D. Brower, personal communication). These effects may not be independent, since, at least in the embryo, absence of β_{PS} leads to the disappearance of both α_{PS1} and α_{PS2} from the cell surface[142] accompanied by the accumulation of uncleaved α_{PS2} precursor.[142] Mosaic analysis shows that tissue mutant for *multiple edematous wings* causes wing blistering only when it is present on the dorsal wing surface (D. Brower, personal communication), whereas tissue mutant for *inflated* has the effect only when it occupies ventral territory,[145] consistent with the expression patterns of α_{PS1} and α_{PS2} respectively.

The PS integrins also appear to be important in interactions between cell layers during embryogenesis. β_{PS} is found between the ectodermal and mesodermal cell layers soon after they are formed at the extended germ band stage, at which time spot adherens junctions are found between the same layers.[4] Later in development, muscles differentiate and attach to the epidermis and to the gut epithelium by specialized junctions. As in the case of fusion between the wing epidermal layers, in the muscle attachment sites there is a complementary α subunit distribution, with α_{PS1}/β_{PS} expressed on the epidermis and the gut epithelium and α_{PS2}/β_{PS} on the muscle cells.[136,142] In mutants lacking β_{PS} (*myospheroid*) or α_{PS2} (*inflated*), the muscles detach from their attachment sites and round up;[145,146] in fact this phenotype is what suggested the name *myospheroid*.

The extracellular ligands for the PS integrins have yet to be identified. The size of the intercellular space between fusing wing layers is consistent with the possibility of a heterophilic interaction between the integrin molecules themselves, but based on the properties of vertebrate integrins it seems more likely that the PS integrins bind to the extracellular matrix, possibly to different sites on a shared ligand.[130] In vertebrates, many integrins bind to specific sites (RGD) on extracellular components including fibronectin, vitronectin, laminin and collagen.[131,132] *Drosophila* cells will spread on a surface coated with vitronectin and this can be inhibited with an antibody against PS integrins.[147] *Drosophila* cells expressing transfected α_{PS2} and β_{PS} genes will spread on both vertebrate vitronectin and fibronectin and this spreading is inhibited by soluble RGD peptides.[148] Recently, a large RGD-containing *Drosophila* protein called tiggrin has been shown to be a ligand for the PS2 integrin (D. Brower and J. Fessler, personal communication).

8. CONCLUSIONS

It is too early to be able to suggest detailed models of the molecular structure of any invertebrate junctions. However, many proteins that are important in cell-cell signaling, as well as epithelial integrity, have been localized to the junctions and the details of some of the pathways of signal transduction are being worked out. These studies are providing some interesting new insights, among which are the following. First, most of the developmentally significant cell-cell signaling events that have been identified in epithelia appear to occur at or near the adherens junctions at the apical end of the lateral cell membrane.[15] Second, several of these cell-cell signaling events appear to involve intercellular movement of proteins during signaling (Boss, Wg, Delta) or intracellular movement of proteins as part of the signal transduction mechanism (Arm, Notch?). Third, at least one protein (Dlg) appears to combine a structural role in junctions with a signaling role involving a putative catalytic domain. Lastly, a review of this area reveals the enormous power of genetic methods for identifying the components of junctions. Cloning of genes identified by their mutant phenotypes has already led to the identification of many new junction components. But it is now clear that genetic methods can also be used to identify additional interacting proteins by the use of screens for interacting mutations. The increasing use of this methodology is likely to lead to the identification of not only more structural components of junctions, but also the upstream and downstream components of the signaling pathways that are mediated through cell junctions.

Acknowledgment

The author's research is supported by grants from the National Institutes of Health and the National Science Foundation.

References

1. Green CR, Bergquist PR. Phylogenetic relationships within the Invertebrata in relation to the structure of septate junctions and the development of 'occluding' junctional types. J Cell Sci 1982; 53:279-305.
2. Woods DF, Bryant PJ. ZO–1, DlgA and PSD–95/SAP90: homologous proteins in tight, septate and synaptic cell junctions. Mech Develop 1993; 44:85-89.
3. Poodry CA. Epidermis: Morphology and Development. In: Ashburner M, Wright TRF, eds. The Genetics and Biology of *Drosophila*. London, NY, San Franscisco, Toronto, Sydney, Academic Press, 1980:443-497.
4. Tepass U, Hartenstein V. The development of cellular junctions in the *Drosophila* embryo. Develop Biol 1994; 161:563-596.
5. Geiger B, Ginsberg D. The cytoplasmic domain of adherens–type junctions. Cell Motil Cytoskeleton 1991; 20:1-6.
6. Geiger B, Ayalon O. Cadherins. Annu Rev Cell Biol 1992; 8:307-332.
7. Magee AI, Buxton RS. Transmembrane molecular assemblies regulated by the greater cadherin family. Curr Opin Cell Biol 1991; 3:854-861.
8. Mahoney PA, Weber U, Onofrechuk P, Biessmann H, Bryant PJ, Goodman CS. The *fat* tumor suppressor gene in *Drosophila* encodes a novel member of the cadherin gene superfamily. Cell 1991; 67:853-868.
9. Hirano S, Kimoto N, Shimoyama Y, Hirohashi S, Takeichi M. Identification of a neural α–catenin as a key regulator of cadherin function and multicellular organization. Cell 1992; 70:293-301.
10. Oda H, Uemura T, Shiomi K, Nagafuchi A, Tsukita S, Takeichi M. Identification of a *Drosophila* homolog of α–catenin and its association with the *armadillo* protein. J Cell Biol 1993; 121:1133-1140.
11. Peifer M, McCrea PD, Green KJ, Wieschaus E, Gumbiner BM. The vertebrate adhesive junction proteins β–catenin and plakoglobin and the *Drosophila* segment polarity gene *armadillo* form a multigene family with similar properties. J Cell Biol 1992; 118:681-691.
12. Peifer M, Wieschaus E. The segment polarity gene *armadillo* encodes a functionally modular protein that is the *Drosophila* homolog of human plakoglobin. Cell 1990; 63:1167-1178.
13. Peifer M. The product of the *Drosophila* segment polarity gene *armadillo* is part of a multi–protein complex resembling the vertebrate adherens junction. J Cell Sci 1993; 105:993-1000.
14. Ozawa M, Kemler R. Molecular organization of the uvomorulin–catenin complex. J Cell Biol 1992; 116:989-996.
15. Woods DF, Bryant PJ. Apical cell junctions and cell signaling in epithelia. J Cell Sci 1993; Suppl. 17:171-181.
16. Cabrera CV, Alonso MC, Johnston P, Phillips RG, Lawrence PA. Phenocopies induced with antisense RNA identify the *wingless* gene. Cell 1987; 50:659-663.
17. Rijsewijk F, Schuermann M, Wagenaar E, Parren P, Weigel D, Nusse R. The *Drosophila* homolog of the mouse mammary oncogene *int–1* is identical to the segment polarity gene *wingless*. Cell 1987; 50:649-657.

18. Poole SJ, Kauvar LM, Drees B, Kornberg T. The *engrailed* locus of *Drosophila*: structural analysis of an embryonic transcript. Cell 1985; 40:37-43.
19. DiNardo S, Sher E, Heemskerk–Jongens J, Kassis JA, O'Farrell PH. Two–tiered regulation of spatially patterned *engrailed* gene expression during *Drosophila* embryogenesis. Nature 1988; 332:604-609.
20. Heemskerk J, DiNardo S, Kostriken R, O'Farrell PH. Multiple modes of *engrailed* regulation in the progression towards cell fate determination. Nature 1991; 352:404-410.
21. Noordermeer J, Johnston P, Rijsewijk F, Nusse R, Lawrence PA. The consequences of ubiquitous expression of the *wingless* gene in the *Drosophila* embryo. Development 1992; 116:711-719.
22. Van den Heuvel M, Nusse R, Johnston P, Lawrence PA. Distribution of the *wingless* gene product in *Drosophila* embryos: a protein involved in cell–cell communication. Cell 1989; 59:739-749.
23. González F, Swales L, Bejsovec A, Skaer H, Martinez-Arias A. Secretion and movement of *wingless* protein in the epidermis of the *Drosophila* embryo. Mech Dev 1991; 35:43-54.
24. Couso JP, Bate M, Martínez–Arias A. A *wingless*–dependent polar coordinate system in *Drosophila* imaginal discs. Science 1993; 259:484-489.
25. Morata G, Lawrence PA. The development of *wingless*, a homeotic mutation of *Drosophila*. Develop Biol 1977; 56:227-240.
26. Struhl G, Basler K. Organizing activity of wingless protein in *Drosophila*. Cell 1993; 72:527-540.
27. Blair S. A role for the segment polarity gene *shaggy–zeste white 3* in the specification of regional identity in the developing wing of *Drosophila*. Develop Biol 1994; 162: 229-244.
28. Peifer M, Orsulic S, Pai L–M, et al.: A model system for cell adhesion and signal transduction in *Drosophila*. Development (suppl) 1993; (in press)
29. Peifer M, Rauskolb C, Williams M, Riggleman B, Wieschaus E. The segment polarity gene *armadillo* interacts with the *wingless* signaling pathway in both embryonic and adult pattern formation. Development 1991; 111:1029-1043.
30. Wieschaus E, Riggleman R. Autonomous requirements for the segment polarity gene *armadillo* during *Drosophila* embryogenesis. Cell 1987; 49:177-184.
31. Peifer M, Sweeton D, Casey M, Wieschaus E. *wingless* signal and Zeste–white 3 kinase trigger opposing changes in the intracellular distribution of Armadillo. Development 1994; 120:369-380.
32. Siegfried E, Perkins LA, Capaci TM, Perrimon N. Putative protein kinase product of the *Drosophila* segment–polarity gene *zeste–white3*. Nature 1990; 345:825-829.
33. Siegfried E, Chou T–B, Perrimon N. *wingless* signaling acts through *zeste–white 3*, the *Drosophila* homolog of *glycogen synthase kinase–3*, to regulate *engrailed* and establish cell fate. Cell 1992; 71:1167-1179.
34. Riggleman B, Schedl P, Wieschaus E. Spatial expression of the *Drosophila* segment polarity gene *armadillo* is posttranscriptionally regulated by *wingless*. Cell 1990; 63:549-560.
35. Artavanis–Tsakonas S, Delidakis C, Fehon RG. The *Notch* locus and the cell biology of neuroblast segregation. Annu Rev Cell Biol 1991; 7:427-452.
36. Doe CQ, Goodman CS. Early events in insect neurogenesis. II. the role of cell interactions and cell lineage in the determination of neuronal precursor cells. Develop Biol 1985; 111:206-219.
37. Fehon RG, Johansen K, Rebay I, Artavanis-Tsakonas S. Complex cellular and subcellular regulation of *Notch* expression during embryonic and imaginal development of *Drosophila*: Implications for *Notch* function. J Cell Biol 1991; 113:657-669.
38. Cagan RL, Ready DF. *Notch* is required for successive cell decisions in the developing *Drosophila* retina. Genes Develop 1989; 3:1099-1112.
39. Kooh PJ, Fehon RG, Muskavitch MAT. Implications of dynamic patterns of Delta and Notch expression for cellular interactions during *Drosophila* development. Development 1993; 117:493-507.
40. Fortini ME, Artavanis–Tsakonas S. *Notch*: Neurogenesis is only part of the picture. Cell 1993; 75:1245-1247.

41. Wharton KA, Johansen KM, Xu T, Artavanis–Tsakonas S. Nucleotide sequence from the neurogenic locus *Notch* implies a gene product that shares homology with proteins containing EGF–like repeats. Cell 1985; 437:567-581.
42. Kidd S, Kelley MR, Young MW. Sequence of the *Notch* locus of *Drosophila melanogaster*: relationship of the encoded protein to mammalian clotting and growth factors. Molec Cell Biol 1986; 6:3094-3108.
43. Vässin H, Bremer KA, Knust E, Campos-Ortega JA. The neurogenic gene *Delta* of *Drosophila melanogaster* is expressed in neurogenic territories and encodes a putative transmembrane protein with EGF–like repeats. EMBO J 1987; 6:3431-3440.
44. Kopczynski CC, Alton AK, Fechtel K, Kooh PJ, Muskavitch MA. *Delta*, a *Drosophila* neurogenic gene, is transcriptionally complex and encodes a protein related to blood coagulation factors and epidermal growth factor of vertebrates. Genes Develop 1988; 2:1723-1735.
45. Johansen KM, Fehon RG, Artavanis–Tsakonas S. The *Notch* gene product is a glycoprotein expressed on the cell surface of both epidermal and neuronal precursor cells during *Drosophila* development. J Cell Biol 1989; 109:2427-2440.
46. Bender LB, Kooh PJ, Muskavitch MAT: Complex function and expression of *Delta* during *Drosophila* oogenesis. Genetics 1993; 133:967-978.
47. Fehon RG, Kooh PJ, Rebay I, Regan CL, Xu T, Muskavitch MA, Artavanis-Tsakonas S. Molecular interactions between the protein products of the neurogenic loci *Notch* and *Delta*, two EGF–homologous genes in *Drosophila*. Cell 1990; 61:523-534.
48. Xu T, Rebay I, Fleming RJ, Scottgale TN, Artavanis-Tsakonas S. The *Notch* locus and the genetic circuitry involved in early *Drosophila* neurogenesis. Genes Dev 1990; 4:464-475.
49. Lieber T, Wesley CS, Alcamo E, Hassel B, Krane JF, Campos-Ortega JA, Young MW. Single amino acid substitutions in EGF–like elements of Notch and Delta modify *Drosophila* development and affect cell adhesion in vitro. Neuron 1992; 9:847-859.
50. Heitzler P, Simpson P. The choice of cell fate in the epidermis of *Drosophila*. Cell 1991; 64:1083-1092.
51. Fleming RJ, Scottgale TN, Diederich RJ, Artavanis–Tsakonas S. The gene *Serrate* encodes a putative EGF–like transmembrane protein essential for proper ectodermal development in *Drosophila melanogaster*. Genes Develop 1990; 4:2188-2201.
52. Rebay I, Fleming RJ, Fehon RG, Cherhas L, Cherhas P, Artavanis–Tsakonas S. Specific EGF repeats of Notch mediate interactions with Delta and Serrate: implications for Notch as a multifunctional receptor. Cell 1991; 67:687-699.
53. Struhl G, Fitzgerald K, Greenwald I. Intrinsic activity of the Lin–12 and Notch intracellular domains in vivo. Cell 1993; 74:331-345.
54. Lieber T, Kidd S, Alcamo E, Corbin V, Young MW. Antineurogenic phenotypes induced by truncated Notch proteins indicate a role in signal transduction and may point to a novel function for Notch in nuclei. Genes Dev 1993; 7:1949-1965.
55. Campos–Ortega JA. Mechanisms of early neurogenesis in *Drosophila melanogaster*. J Neurobiol 1993; 24:1305-1327.
56. Concha A, Dietrich U, Weigel D, Campos-Ortega JA. Functional interactions of neurogenic genes of *Drosophila melanogaster*. Genetics 1988; 118:499-508.
57. Fortini ME, Rebay I, Caron LA, Artavanis–Tsakonas S. An activated Notch receptor blocks cell–fate commitment in the developing *Drosophila* eye. Nature 1993; 365:555-557.
58. Ruel L, Bourouis M, Heitzler P, Pantesco V, Simpson P. *Drosophila shaggy* kinase and rat glycogen synthase kinase–3 have conserved activities and act downstream of *Notch*. Nature 1993; 362:557-560.
59. Rebay I, Fehon RG, Artavanis–Tsakonas S. Specific truncations of *Drosophila* Notch define dominant activated and dominant negative forms of the receptor. Cell 1993; 74:319-329.
60. Venkatesh T. Neuronal development in the *Drosophila* retina. J Neurobiol 1993; 24:740-756.

61. Krämer H, Cagan RL, Zipursky SL. Interaction of *bride of sevenless* membrane–bound ligand and the *sevenless* tyrosine–kinase receptor. Nature 1991; 352:207-212.
62. Cagan RL, Ready DF. The emergence of order in the *Drosophila* pupal retina. Develop Biol 1989; 136:346-362.
63. Banerjee U, Renfranz PJ, Hinton DR, Rabin BA, Benzer S. The *sevenless*+ protein is expressed apically in cell membranes of developing *Drosophila* retina; it is not restricted to cell R7. Cell 1987; 51:151-158.
64. Tomlinson A, Bowtell DD, Hafen E, Rubin GM. Localization of the *sevenless* protein, a putative receptor for positional information, in the eye imaginal disc of *Drosophila*. Cell 1987; 51:143-150.
65. Reinke R, Zipursky SL. Cell–cell interaction in the *Drosophila* retina: the *bride of sevenless* gene is required in photoreceptor cell R8 for R7 cell development. Cell 1988; 55:321-330.
66. Tomlinson A, Ready DF. Cell fate in the *Drosophila* ommatidium. Develop Biol 1987; 123:264-275.
67. Hart AC, Krämer H, Van Vactor DL, Jr., Paidhungat M, Zipursky SL. Induction of cell fate in the *Drosophila* retina: The *bride of sevenless* protein is predicted to contain a large extracellular domain and seven transmembrane segments. Genes Develop 1990; 4:1835-1847.
68. Hafen E, Basler K, Edstroem JE, Rubin GM. *Sevenless*, a cell–specific homeotic gene of *Drosophila*, encodes a putative transmembrane receptor with a tyrosine kinase domain. Science 1987; 236:55-63.
69. Basler K, Hafen E. Control of photoreceptor cell fate by the *sevenless* protein requires a functional tyrosine kinase domain. Cell 1988; 54:299-311.
70. Bowtell DD, Simon MA, Rubin GM. Nucleotide sequence and structure of the *sevenless* gene of *Drosophila melanogaster*. Genes Develop 1988; 2:620-634.
71. Simon MA, Bowtell DD, Rubin GM. Structure and activity of the *sevenless* protein: a protein tyrosine kinase receptor required for photoreceptor development in *Drosophila*. Proc Natl Acad Sci USA 1989; 86:8333-8337.
72. Cagan RL, Krämer H, Hart AC, Zipursky SL. The *bride of sevenless* and *sevenless* interaction: internalization of a transmembrane ligand. Cell 1992; 69:393-399.
73. Simon MA, Dodson GS, Rubin GM. An SH3–SH2–SH3 protein is required for $p21^{Ras1}$ activation and binds to sevenless and Sos proteins in vitro. Cell 1993; 73:169-177.
74. Olivier JP, Raabe T, Henkemeyer M, Dickson B, Mbamalu G, Margolis B, Schlessinger J, Hafen E, Pawson T. A *Drosophila* SH2–SH3 adaptor protein implicated in coupling the *sevenless* tyrosine kinase to an activator of *Ras* guanine nucleotide exchange, *Sos*. Cell 1993; 73:179-191.
75. Fortini ME, Simon MA, Rubin GM. Signalling by the *sevenless* protein tyrosine kinase is mimicked by *Ras1* activation. Nature 1992; 355:559-561.
76. Zak NB, Shilo BZ. Localization of DER and the pattern of cell divisions in wild–type and *Ellipse* eye imaginal discs. Develop Biol 1992; 149:448-456.
77. Xu T, Rubin GM. Analysis of genetic mosaics in developing and adult *Drosophila* tissues. Development 1993; 117:1223-1237.
78. Shilo B–Z: Roles of receptor tyrosine kinases in *Drosophila* development. FASEB J 1992; 6:2915-2922.
79. Raz E, Shilo B–Z: Establishment of ventral cell fates in the *Drosophila* embryonic ectoderm requires DER, the EGF receptor homolog. Genes Dev 1993; 7:1937-1948.
80. Rutledge BJ, Zhang K, Bier E, Jan YN, Perrimon N. The *Drosophila spitz* gene encodes a putative EGF–like growth factor involved in dorsal–ventral axis formation and neurogenesis. Genes Dev 1992; 6:1503-1517.
81. Stern MJ, Marengere LEM, Daly RJ, et al. The human GRB2 and *Drosophila Drk* genes can functionally replace the *Caenorhabditis elegans* cell signaling gene *sem–5*. Mol Biol Cell 1993; 4:1175-1188.
82. Maher PA, Pasquale EB, Wang JYJ, Singer SJ. Phosphotyrosine–containing proteins are concentrated in focal adhesions and intercellular junctions in normal cells. Proc Natl Acad Sci USA 1985; 82:6576-6580.

83. Tsukita S, Oishi K, Akiyama T, Yamanashi Y, Yamamoto T, Tsukita S. Specific proto–oncogenic tyrosine kinases of src family are enriched in cell–to–cell adherens junctions where the level of tyrosine phosphorylation is elevated. J Cell Biol 1991; 113:867-879.
84. Volberg T, Geiger B, Dror R, Zick Y. Modulation of intercellular adherens–type junctions and tyrosine phosphorylation of their components in RSV–transformed cultured chick lens cells. Cell Regul 1991; 2:105-120.
85. Volberg T, Zick Y, Dror R, Sabanay I, Gilon C, Levitzki A, Geiger B. The effect of tyrosine–specific protein phosphorylation on the assembly of adherens–type junctions. EMBO J 1992; 11:1733-1742.
86. Matsuyoshi N, Hamaguchi M, Taniguchi S, Nagafuchi A, Tsukita S, Takeichi M. Cadherin–mediated cell–cell adhesion is perturbed by v–*src* tyrosine phosphorylation in metastatic fibroblasts. J Cell Biol 1992; 118:703-714.
87. Behrens J, Vakaet L, Friis R, Winterhager E, Van Roy F, Mareel MM, Birchimeier W. Loss of epithelial differentiation and gain of invasiveness correlates with tyrosine phosphorylation of the E–cadherin/β–catenin complex in cells transformed with a temperature–sensitive v–*src* gene. J Cell Biol 1993; 120:757-766.
88. Hamaguchi M, Matsuyoshi N, Ohnishi Y, Gotoh B, Takeichi M, Nagai Y. $p60^{v\text{–}src}$ causes tyrosine phosphorylation and inactivation of the N–cadherin–catenin cell adhesion system. EMBO J 1993; 12:307-314.
89. Bennett RL, Hoffmann FM. Increased levels of the *Drosophila* Abelson tyrosine kinase in nerves and muscles: Subcellular localization and mutant phenotypes imply a role in cell–cell interactions. Development 1992; 116:953-966.
90. Vincent WS, 3d., Gregory RJ, Wadsworth SC. Embryonic expression of a *Drosophila src* gene: alternate forms of the protein are expressed in segmental stripes and in the nervous system. Genes Develop 1989; 3:334-347.
91. Tepass U, Theres C, Knust E. *crumbs* encodes an EGF–like protein expressed on apical membranes of *Drosophila* epithelial cells and required for organization of epithelia. Cell 1990; 61:787-799.
92. Tepass U, Knust E. Phenotypic and developmental analysis of mutations at the *crumbs* locus, a gene required for the development of epithelia in *Drosophila melanogaster*. Roux's Arch Develop Biol 1990; 199:189-206.
93. Boedigheimer M, Laughon A. *expanded*: a gene involved in the control of cell proliferation in *Drosophila* imaginal discs. Development 1993; 118:1291-1301.
94. Boedigheimer M, Bryant PJ, Laughon A. *expanded*, a negative regulator of cell proliferation in *Drosophila*, shows homology to the NF2 tumor suppressor. Mech Dev 1993; 44:83-84.
95. Lane NJ, Dilworth S. Isolation and biochemical characterization of septate junctions: differences between the proteins in smooth and pleated varieties. J Cell Sci 1989; 93:123-131.
96. Woods DF, Bryant PJ. The *discs–large* tumor suppressor gene of *Drosophila* encodes a guanylate kinase homolog localized at septate junctions. Cell 1991; 66:451-464.
97. Stehle T, Schulz GE. Refined structure of the complex between guanylate kinase and its substrate GMP at 2.0 Å resolution. J Molec Biol 1992; 224:1127-1141.
98. Zschocke PD, Schiltz E, Schulz GE. Purification and sequence determination of guanylate kinase from pig brain. Eur J Biochem 1993; 213:263-269.
99. Musacchio A, Gibson T, Lehto V–P, Saraste M. SH3—An abundant protein domain in search of a function. FEBS Lett 1992; 307:55-61.
100. Musacchio A, Noble M, Pauptit R, Wierenga R, Saraste M. Crystal structure of a Src–homology 3 (SH3) domain. Nature 1992; 359:851-855.
101. Gout I, Dhand R, Hiles ID, Fry MJ, Panayotou G, Das P, Truong O, Totty NF, Hsuan J, Booker GW et al.: The GTPase dynamin binds to and is activated by a subset of SH3 domains. Cell 1993; 75:25-36.
102. Bryant PJ, Watson KL, Justice RW, et al. Tumor suppressor genes encoding proteins required for cell interactions and signal transduction in *Drosophila*. Development 1993; (in press)
103. Cho K–O, Hunt CA, Kennedy MB. The rat brain postsynaptic density fraction contains a homolog of the *Drosophila discs–large* tumor suppressor protein. Neuron 1992; 9:929-942.

104. Wharton KA, Yedvobnick B, Finnerty VG, Artavanis-Tsakonas S. *opa*: A novel family of transcribed repeats shared by the *Notch* locus and other developmentally regulated loci in *D. melanogaster*. Cell 1985; 40:55-62.
105. Rogers S, Wells R, Rechsteiner M. Amino acid sequences common to rapidly degraded proteins: the PEST hypothesis. Science 1986; 234:364-368.
106. Ruff P, Speicher DW, Husain–Chishti A. Molecular identification of a major palmitoylated erythrocyte membrane protein containing the *src* homology 3 motif. Proc Natl Acad Sci USA 1991; 88:6595-6599.
107. Kistner U, Wenzel BM, Veh RW, Cases-Langhoff, Garner AM, Appeltauer U, Voss B, Gundelfinger ED, Garner CC. SAP90, a rat presynaptic protein related to the product of the *Drosophila* tumor suppressor gene *dlg*–A. J Biol Chem 1993; 268:4580-4583.
108. Willott E, Balda MS, Fanning AS, Jameson B, Van Itallie C, Anderson JM. The tight junction protein ZO–1 is homologous to the *Drosophila* discs–large tumor suppressor protein of septate junctions. Proc Natl Acad Sci USA 1993; 90:7834-7838.
109. Jesaitis LA, Goodenough DA. Molecular characterization and tissue distribution of ZO–2, a tight junction protein homologous to ZO–1 and the *Drosophila* discs–large tumor suppressor protein. J Cell Biol 1994; 124: 949-961.
110. Theisen H, Purcell J, Bennett M, et al. *dishevelled* is required during *wingless* signaling to establish both cell polarity and cell identity. Development 1994; 120: 347-360.
111. Fehon RG, Dawson IA, Artavanis–Tsakonas S. A *Drosophila* homolog of membrane–skeleton protein 4.1 is associated with septate junctions and is encoded by the *coracle* gene. Development 1994; 120:545-557.
112. Marchesi VT. Stabilizing infrastructure of cell membranes. Ann.Rev.Cell Biol. 1985; 1:531-561.
113. Patel NH, Snow PM, Goodman CS. Characterization and cloning of fasciclin III: a glycoprotein expressed on a subset of neurons and axon pathways in *Drosophila*. Cell 1987; 48:975-988.
114. Snow PM, Bieber AJ, Goodman CS. Fasciclin III: a novel homophilic adhesion molecule in *Drosophila*. Cell 1989; 59:313-323.
115. Brower DL, Smith RJ, Wilcox M. A monoclonal antibody specific for diploid epithelial cells in *Drosophila*. Nature 1980; 285:403-405.
116. Zipursky SL, Venkatesh TR, Teplow DB, Benzer S. Neuronal development in the *Drosophila* retina: monoclonal antibodies as molecular probes. Cell 1984; 36:15-26.
117. Grenningloh G, Bieber AJ, Rehm EJ, Snow PM, Traquina ZR, Hortsch M, Patel NH, Goodman CS. Molecular genetics of neuronal recognition in *Drosophila*: Evolution and function of immunoglobulin superfamily cell adhesion molecules. Cold Spring Harbor Symp Quant Biol 1990; 55:327-340.
118. Marfatia SM, Lue RA, Branton D, et al. In vitro binding studies suggest a membrane–associated complex between erythroid p55, protein 4.1 and glycophorin C. J Chem Biol 1994; 269:8631-8634.
119. Bennett V. The spectrin–actin junction of erythrocyte membrane skeletons. Biochim Biophys Acta 1989; 988:107-121.
120. Ryerse J. Gap junctions. In: Harrison FW, Locke M, eds. Microscopic Anatomy of Invertebrates, Volume 11. New York, Wiley–Liss, 1994,
121. Berdan RC, Gilula NB. The arthropod gap junction and pseudo–gap junction: isolation and preliminary biochemical analysis. Cell Tissue Res 1988; 251:257-274.
122. Holzenburg A, Jones PC, Franklin T, Pali T, Heimburg T, Marsh D, Findlay JB, Finbow ME. Evidence for a common structure for a class of membrane channels. Eur J Biochem 1993; 213:21-30.
123. Finbow ME, Pitts JD. Is the gap junction channel - *the connexon* - made of connexin or ductin?. J Cell Sci 1993; 106:463-472.
124. Meagher L, McLean P, Finbow ME. Sequence of a cDNA from *Drosophila* coding for the 16 kD proteolipid component of the vacuolar H^+–ATPase. Nucleic Acids Res 1990; 18:6712.
125. Ryerse JS. Isolation and characterization of gap junctions from *Drosophila* melanogaster. Cell Tissue Res 1989; 256:7-16.
126. Ryerse JS. Electron microscope immunolocation of gap junctions in *Drosophila*. Tissue Cell 1989; 21:835-839.

127. Ryerse JS. Structural, immunocytochemical and initial biochemical characterization of NaOH–extracted gap junctions from an insect, *Heliothis virescens*. Cell Tissue Res 1993; 274:393-403.
128. Pitts JD, Finbow ME, Kam E. Junctional communication and cellular differentiation. Br J Cancer Suppl 1988; 9:52-57.
129. Haefliger JA, Bruzzone R, Jenkins NA, Gilbert DJ, Copeland NJ. Four novel members of the connexin family of gap junction proteins. Molecular cloning, expression and chromosome mapping. J Biol Chem 1992; 267:2057-2064.
130. Fristrom D, Wilcox M, Fristrom J. The distribution of PS integrins, laminin A and F–actin during key stages in *Drosophila* wing development. Development 1993; 117:509-523.
131. Brown NH. Integrins hold *Drosophila* together. BioEssays 1993; 15:383-390.
132. Hynes RO. Integrins: versatility, modulation and signaling in cell adhesion. Cell 1992; 69:11-25.
133. Waddington CH. The genetic control of wing development in *Drosophila*. J Genet 1940; 41:75-139.
134. Mogensen MM, Tucker JB, Stebbings H. Microtubule polarities indicate that nucleation and capture of microtubules occurs at cell surfaces in *Drosophila*. J Cell Biol 1989; 108:1445-1452.
135. Brower DL, Wilcox M, Piovant M, Smith RJ, Reger LA. Related cell–surface antigens expressed with positional specificity in *Drosophila* imaginal discs. Proc Natl Acad Sci U S A 1984; 81:7485-7489.
136. Bogaert T, Brown N, Wilcox M. The *Drosophila* PS2 antigen is an invertebrate integrin that, like the fibronectin receptor, becomes localized to muscle attachments. Cell 1987; 51:929-940.
137. Wehrli M, DiAntonio A, Fearnley IM, Smith RJ, Wilcox M. Cloning and characterization of α_{PS1}, a novel *Drosophila melanogaster* integrin. Mech Dev 1993; 43:21-36.
138. Brown NH, King DL, Wilcox M, Kafatos FC. Developmentally regulated alternative splicing of *Drosophila* integrin PS2 alpha transcripts. Cell 1989; 59:185-195.
139. Brower DL, Piovant M, Reger LA. Developmental analysis of *Drosophila* position–specific antigens. Develop Biol 1985; 108:120-130.
140. Wilcox M, DiAntonio A, Leptin M. The function of PS integrins in *Drosophila* wing morphogenesis. Development 1989; 107:891-897.
141. MacKrell AJ, Blumberg B, Haynes SR, Fessler JH. The *lethal myospheroid* gene of *Drosophila* encodes a membrane protein homologous to vertebrate integrin beta subunits. Proc Natl Acad Sci USA 1988; 85:2633-2637.
142. Leptin M, Bogaert T, Lehmann R, Wilcox M. The function of PS integrins during *Drosophila* embryogenesis. Cell 1989; 56:401-408.
143. Brower DL, Jaffe SM. Requirement for integrins during *Drosophila* wing development. Nature 1989; 342:285-287.
144. Zusman S, Patel–King RS, Ffrench–Constant C, Hynes RO. Requirements for integrins during *Drosophila* development. Development 1990; 108:391-402.
145. Brabant MC, Brower DL. PS2 integrin requirements in *Drosophila* embryo and wing morphogenesis. Dev Biol 1993; 157:49-59.
146. Newman SM, Jr., Wright TRF: Histological and ultrastructural analysis of developmental defects produced by the mutation, *lethal(1)myospheroid*, in *Drosophila melanogaster*. Develop Biol 1981; 86:393-402.
147. Hirano S, Ui K, Miyake T, Uemura T, Takeichi M. *Drosophila* PS integrins recognize vertebrate vitronectin and function as cell–substratum adhesion receptors in vitro. Development 1991; 113:1007-1016.
148. Bunch TA, Brower DL. *Drosophila* PS2 integrin mediates RGD–dependent cell–matrix interactions. Dev Biol 1992; 116:239-247.
149. Altschul SF, Gish W, Miller W, Myers EW, Lipman DJ. Basic local alignment search tool. J Molec Biol 1990; 215:403-410.
150. Schuler GD, Altschul SF, Lipman DJ. A workbench for multiple alignment construction and analysis. Proteins 1991; 9:180-190.

CHAPTER 2

Electron Microscopic Structure and Evolution of Epithelial Junctions

Nancy J. Lane
Giambruno Martinucci
Romano Dallai
Paolo Burighel

1. HISTORICAL INTRODUCTION TO THE FIELD: THE CELL JUNCTION AS A MODEL FOR PHYLOGENETIC CONSIDERATIONS

Intercellular junctions are membrane specializations localized at the regions of contact between cells. It is reasonable to assume that they were initially formed for mechanical adhesion with neighboring cells, possibly after cell-to-cell recognition and that this occurred when cells came into close vicinity with one another up to a critical distance: this distance would be determined by the size of the protein chains projecting from the external surface of the plasma membrane. The transient focal adhesions between adjacent blastomeres during early embryonic development would seem to support this view.[1]

More specific functions at the junctional level, either for mechanical or physiological purposes, might be developed secondarily when adjacent cells have formed permanent intercellular junctional structures.

The two gamonts of the protozoan *Gregarina polymorpha* at mating are connected by a kind of septate junction.[2] Similarly, the sperm cells of the firebrat insect *Thermobia domestica* are usually coupled with a single junction.[3] These two examples indicate that the feasibility of organizing cell junctions appeared very early in evolution and this was undoubtedly of great importance in the organization of the first metazoans. It would seem that only those animals, that were able to organize specific sites for cell junctions on their membranes, have had an evolutionary success. Porifera (sponges),

Molecular Mechanisms of Epithelial Cell Junctions: From Development to Disease
Edited by Sandra Citi, M.D., Ph.D.;

for instance, which do not express well defined junctional structures, remain at the lower end of the evolutionary ladder.

For some time it has been generally considered that the two main epithelial cell junctions with adhesive functions, other than desmosomes, are the *tight junctions* (*zonula occludens*), which apically seal the intercellular space and which were thought originally to be present only in vertebrates and the *septate junctions*, found only in the invertebrates.

This view, however, is not entirely accurate: tight junctions are also found in tunicates, which are primitive chordates,[4] a simple type of tight junction has been described in insect tissues and more complex tight junctions occur in arachnids (chelicerates).[5] On the other hand, septate-like junctions have been occasionally described in vertebrates, particularly in pathological conditions and at axo-glial paranodal junctions. Moreover, under the term septate junction, different morphological variations have been described in the animal groups examined so far and often a careful study of these junctions has revealed conspicuous differences between those that are found in the different invertebrate phyla.

It has been debated whether these two kinds of junctions play analogous functions due to their similar apical position between epithelial cells; it is generally accepted that septate junctions mainly serve as adhesive devices, but may also represent a partial barrier to paracellular flow.[6] Tight junctions, however, consistently occur where highly occluding permeability barriers are required. Some data, such as those which demonstrate a limit to lipophylic markers moving from one side of the septate junctional belt to the other, seem to confirm certain similarities between the two types of junction[7] as this has been also shown to occur in tight junctions.[8] Moreover, despite the clear morphological differences between tight and septate junctions, they have recently been shown to share some protein constituents.[9] This suggests that a protein family exists which is specialized for signal transduction, on the cytoplasmic surface of certain intercellular junctions. This could also indicate that the final morphological appearance of a junction may depend on a "membrane spacing factor" controlling the size of the intercellular cleft.[10] In this context the mutual interaction between the cytoskeletal components and the cell membrane with its associated glycocalyx, is undoubtedly of considerable importance.

In the evaluation of the existence of certain kinds of junction between any given cells it is important to take into consideration the capability of the cell in question to express the proteins characteristic of that type of junction. For example, if it is true, as claimed, that arthropod cells do not possess intermediate filaments, this would explain the fact that they lack adhesive junctions such as *maculae adhaerentes* (spot desmosomes), which are known to be in close association with cytokeratin filaments.

With the beginning of studies on cell interactions, attention was initially focused on the structure and then on the putative mechanical role of the different junctional complexes (desmosomes, tight junctions, septate junctions). Further, when the occurrence of intercellular communication was acknowledged, it was recognized that gap junctional specializations allowed the exchange of ions and small molecules between neighboring cells.[11] These regions were established as electrotonic or gap junctions with a reduced intercellular cleft of only 2-3 nm. With but a few exceptions, gap junctions seem to be present between most animal cells with relatively similar morphological features.

The available observations on intercellular junctions can be used for phylogenetic purposes to establish relationships. The comparison of junctional structures usually should be between homologous tissues, i.e., those with the same embryological origin. The risk of establishing false relationships between groups owing to convergences or parallelisms is, of course, always possible.

Primitive Cnidaria exhibit several kinds of septate junctions, Echaracterized by relatively straight septa with lateral projections,[10] a variation not yet found in other organisms. Platyhelminthes (flatworms) have pleated septate junctions and this kind of junction, with a number of variations, occurs in many phyla belonging to the proterostome and deuterostome lineages (see Table 2.1). Interestingly, the primitive onychophorans and the more sophisticated arthropods share the same type of smooth septate junction at the midgut level, giving further support to the view held for many years of a close relationship between these taxa. In this context it is surprising to have found a smooth septate junction in the aschelmint nematodes, but Eernisse et al.[12] have recently reconsidered the position of this group; in a strict consensus phylogram they have suggested the existence of a large clade including the Arthropoda, Onychophora, Kinorhyncha and Nematoda. One cannot exclude, however, a common junctional structure existing in these groups which has arisen merely due to convergence.

Chaetognatha (arrow-worms) have a paired septate junction as do both the hemichordates and the echinoderms (star fish, sea-urchin, etc.) and this suggests a relationship of some sort between these groups. In addition, the anastomosing septate junctions of the echinoderms and hemichordates are also very similar in many aspects.[10]

Table 2.1. Phylogenetic relationships of the animal taxa quoted in the text

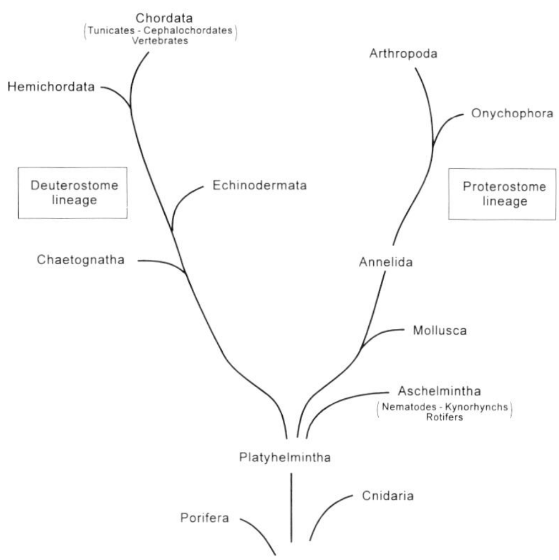

From the studies on cell junctions the two phyletic lines, proterostomes and deuterostomes, are confirmed (see Table 2.2). Within these two lineages some relationships are evident: Onychophora-Arthropoda; Chaetognatha-Echinodermata-Hemichordata; Tunicata-Vertebrata. Small variations on a type of cell junction within a group, if well documented, may indicate a clear relationship between the taxa exhibiting that particular junctional model. Among arthropods this is the case, for instance, for sea-spiders, horse-shoe crabs and apterygote insects, which show the same type of variation in their smooth septate junctions.[13]

2. CHANGES IN SEPTATE AND SCALARIFORM JUNCTIONS THROUGH THE PHYLA

Intercellular septate junctions are the most common type of cell junction in the invertebrates. In their more typical appearance they are characterized by a constant intercellular cleft of 15 to 20 nm, crossed by regularly spaced bridges or septa (Fig. 2.2). The glycoprotein material present in the cleft, however, often obscures the septa. The first description of a septate junction was reported in *Hydra* and since that time variations on the theme have been observed in a range of tissues from different organisms.[5,10,14]

Septate junctions usually form a belt at the apical region of epithelial cells and are often associated with other kinds of junctions such as *zonulae adhaerentes* and gap junctions. Cross sectioned tissues show a rather uniform appearance of the septate junction.

Table 2.2. Junctional types found in Metazoans

	Lower Invertebr.			Proterostomes									Deuterostomes					
				Mollusca					Arthropoda							Chordata		
	Cnidaria	Platyhelmintha	Nematoda	*Gastropods*	*Bivalves*	*Cephalopods*	Annelida	Onychophora	*Crustaceans*	*Chelicerates*	*Myriapods*	*Insects*	Chaetognatha	Echinodermata	Hemichordata	*Tunicates*	*Cephalochordates*	*Vertebrates*
TJ	0	0	0	0	0	0	0	0	0	+	0	+	0	0	0	+	0	+
GJ	+	+	+	+	+	+	+	0	+	+	+	+	+	+	0	+	0	+
SJ	+	+	+	+	+	+	+	+	+	+	+	+	+	+	+	0	0	0
ScJ	0	0	-	-	-	-	-	+	+	+	+	+	0	0	0	+	0	0
ZA	+	+	+	+	+	+	+	+	+	+	+	+	+	-	-	+	+	+
MA	?	?	?	-	-	-	+	?	-	-	-	?	?	-	-	?	+	+
LJ	0	0	0	-	-	+	0	0	-	-	+	+	0	-	-	0	0	*

TJ = tight junctions; GJ = gap junctions; SJ = septate junctions; ScJ = scalariform junctions; ZA = zonulae adherentes; MA = maculae adherentes; LJ = linker junctions; + = present; 0 = apparently absent; – = no data; ? = stated in literature to be present; * = embryos only

Differences are noticeable, however, when tracer-infiltration and freeze fracture replicas are used to characterize these junctions; variations are to be found in the structure and disposition of the intercellular septa and the intramembranous organization at the junctional contact.

The following types of septate junction have been described:

2.1 Straight Septate Junctions

The septate junction of Cnidaria, after lanthanum treatment, appears as double straight lined structures with lateral projections in gastrodermal tissue, while the epithelial septate junctions are single structures. The intramembrane organization at freeze fracture shows grooves or pits on the protoplasmic face of the membrane (P-face), while intramembranous particles (IMPs) are associated with the external face of the membrane (E-face).

2.2 Pleated Septate Junctions (Fig. 2.9)

Platyhelmintes, Aschelmintes, Mollusca, Annelida, Arthropoda, Chaetognatha and Echinodermata exhibit distinct undulating septa in oblique sections of lanthanum treated material. These electron-lucent zig-zag strands are visible against a dense background of lanthanum tracer (Figs. 2.1, 2.3). In freeze cleaved replicas rows of 8-10 nm IMPs are visible. These are separated from one another, but are associated proximally with the P-face of the membrane and exhibit complementary E-face grooves (Fig. 2.4). The degree of order of the septate junctions, in freeze-fractured replicas, varies widely according to the tissue and organism examined, but the pattern appears not to be influenced by the fixation of material. Cytoplasmic fibrils have been seen to insert into the membrane at the junctional level and these have been shown to bind the S_1 subunit of rabbit myosin; thus they appear to be actin-like in

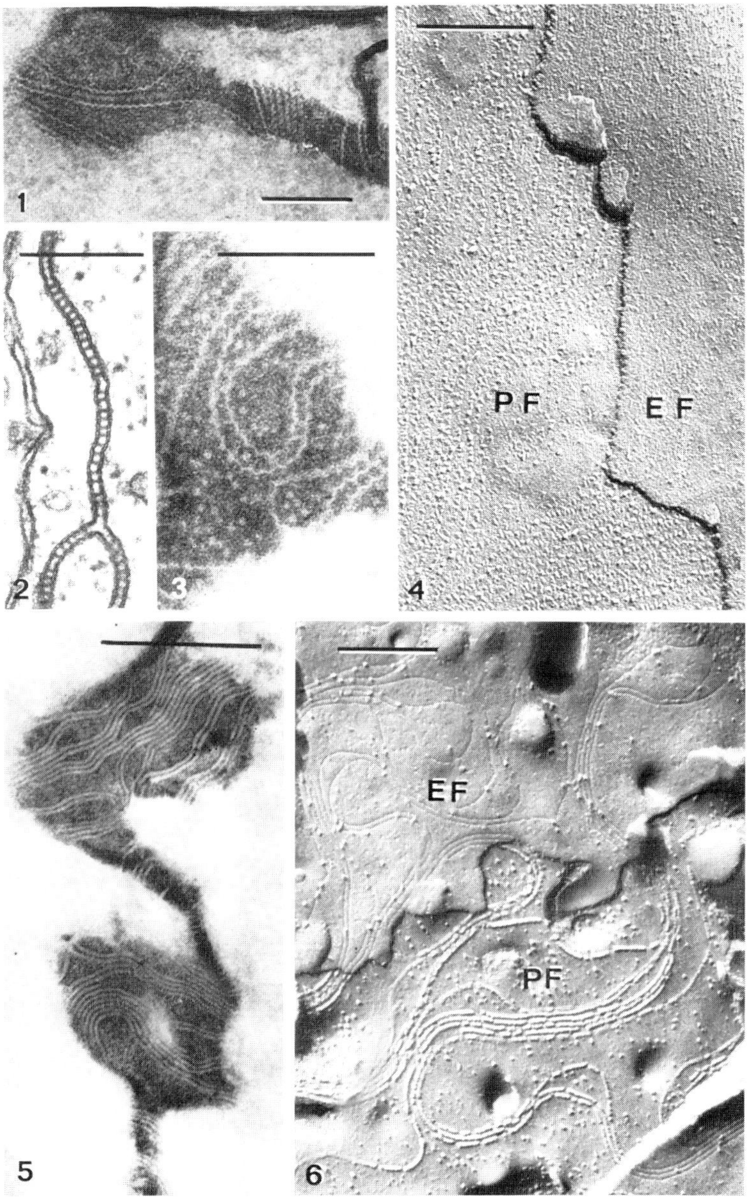

Fig. 2.1. Cross-section through cells of the foregut of the cricket Achaeta sp. *(Insecta) in lanthanum-treated material. The pleated septate junctions appear as zig-zagging electron-lucent ribbons. Scale bar = 0.25 µm.*

Fig. 2.2. Cross-section through the epithelial cells from the locust (Insecta) eye. The typical appearance of a pleated septate junction is apparent with septa straddling a uniformly wide intercellular space. Scale bar = 0.25 mm.

Fig. 2.3. Pleated septate junction in lanthanum-treated epithelium from the digitiform glands of Cepaea sp. *(Mollusca). Scale bar = 0.25µm.*

Fig. 2.4. Freeze-fracture replica of Peripatopsis moseleyi *(Onychophora) mucous glands showing the pleated septate junctions. Rows of IMPs are present on the P-face (PF) and pits and furrows on the E-face (EF) of the plasma membrane. Scale bar = 0.25 µm.*

Fig. 2.5. Smooth septate junction from lanthanum-treated midgut of the onychophoran Peripatopsis moseleyi. *Continuous parallel electron-transparent ribbons are visible. Scale bar = 0.25 µm.*

Fig. 2.6. Replica of the medfly Ceratitis capitata *(Insecta) midgut. The smooth septate junction appears as continuous rows of fused IMPs on the P-face (PF) and complementary furrows on the E-face (EF) of the plasma membrane. Scale bar = 0.25 µm.*

nature.[15] In some cases the actin is tethered onto microtubules running closely to the junction.[16] Recently an antibody against fusion peptides has shown that the product of the tumor suppressor gene of *Drosophila* is localized at the septate junctional level.[17]

2.3 Paired Septate Junctions

These are typically found in Chaetognatha.[18] In tangential view they are characterized by two parallel undulating strands in tight or loose formation, running through the intercellular space at the junctional level of lanthanum-treated material. This kind of junction is present at the apex of the intestinal cells of *Sagitta*; below these junctions a typical pleated septate junction occurs. In freeze-fracture replicas a single row of IMPs, around 10 nm in diameter, closely associated with the P-face of the membrane, runs

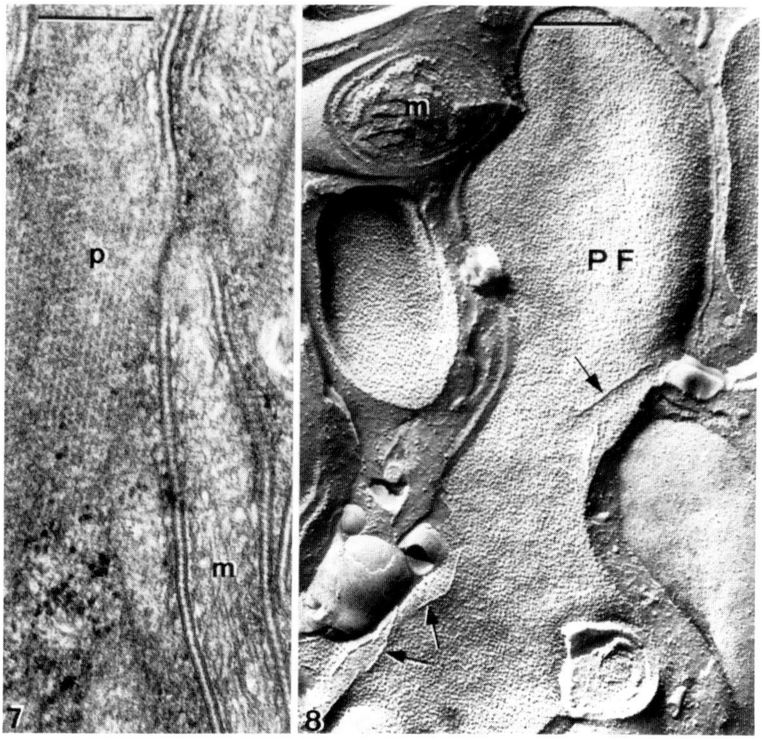

Fig. 2.7. Tangential section through the rectal papilla of the thrip Frankliniella occidentalis *(Insecta) showing the scalariform junctions. The intercellular space is crossed by bridges which, in tangential view, appear as an orderly array of pillars (p). m, mitochondrion. Scale bar = 0.25 μm.*

Fig. 2.8. Freeze-fracture replica of the rectal pad from the caddisfly Stenophylax permistus *(Insecta). At the scalariform junctional level a uniform array of IMPs is associated with the P-face (PF) of the plasma membrane. A few linear aligments of IMPs of the reticular junction are also visible (arrows). m, mitochondrion. Scale bar = 0.25 μm.*

to the apex of ridges. Complementary furrows to the rows of particles, are on the E-face. Also the hemichordate *Balanoglossus* and the echinoderms exhibit the same type of double-septum septate junction in their epidermal tissues.

2.4 Anastomosing septate junctions

This variation on the septate junctional theme was first described in the endothelial cells of echinoderms. Later, this junction was also found in the endodermal tissues of the hemichordates.[19] This junction is characterized, in tangential sections of lanthanum-treated material, by a network of electron-lucent strands forming an anastomosing pattern of circles. Freeze-fracture replicas show rows of 7-8 nm particles forming a network on the P-face of the membrane.

2.5 Smooth septate junctions

Also known as continuous junctions, smooth septate junctions are found in the midgut of arthropods, onychophorans and nematodes, which have an endodermal embryological origin. This junctional type has also been described in the Malpighian tubules of insects alongside a pleated septate junction.[20,21] These tubules have a mixed endo-ectodermic embryological origin. This coexistence of junctions has led to the possibility of determining the embryological origin of tissues based on the type of septate junction present. Smooth septate junctions are characterized by sinuous electron-transparent strands in tangential sections of lanthanum-treated tissue (Fig. 2.5). Between these strands translucent small dots are often visible; these correspond to columns cut in cross section. In freeze fracture replicas of fixed material, rows of particles fused into ridges are associated with the P-face of the membrane, while on the complementary E-face grooves are evident (Fig. 2.6); in unfixed material this pattern of particle distribution is reversed. Subtle variations on the smooth septate junction are found in different groups. In horse-shoe crabs, sea spiders and some apterygotan insects the strands in lanthanum-treated material and the rows of particles in freeze-fracture replicas appear as double structures.

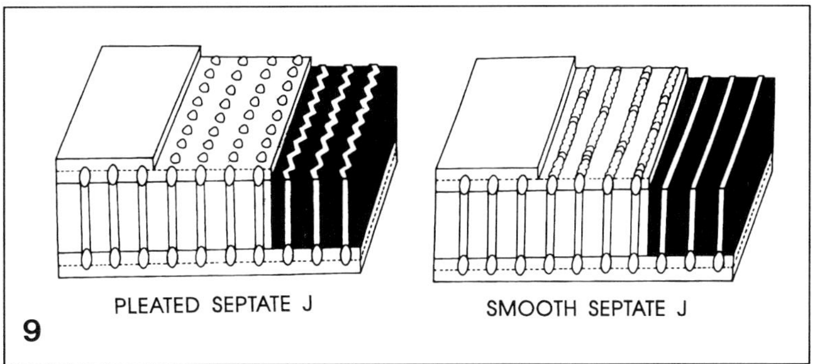

Fig. 2.9. Schematic reconstruction of the pleated septate and the smooth septate junctions.

Cytoskeletal elements are associated with these junctions, such as actin, α-actinin, myosin and vinculin: these are localized close to the specialized plasma membrane.[15,22,23] A protein of 23-24 kD molecular weight has been found, associated possibly with the integral membrane component of the smooth septate junction. This protein seems to be somewhat larger than that constituting the component of the pleated septate junction (22-23 kD).[24]

2.6 Scalariform junctions

Scalariform junctions are defined by a wide intercellular cleft crossed by very fine striations giving a faint ladder-like appearance in cross section.[25] Often their plasma membrane is in close association with mitochondria. They are not true septate junctions as they lack septate ribbons, apparently instead possessing intercellular columns as shown in oblique sections (p in Fig. 2.7). After freeze-fracture, a layer of 9 nm IMPs is visible on the P-face of the membranes, while the complementary E-face does not show any particular differentiation (Fig. 2.8). This kind of junction was found originally in the excretory organs of primitive insects, but it is now found to be widely present in invertebrate tissues involved in the active transport of fluids and solutes against a concentration gradient. It is particularly well developed in the rectum of insects,[26] but it has also been observed in a number of arthropod and related groups as well as in tunicates.[27]

2.7 Reticular septate junctions

This kind of junction is often associated with scalariform junctions in the rectal tissues of some insects, particularly the Diptera.[26] Its presence is defined by isolated cross-striations or columns in the intercellular cleft after thin sections and by aligned particles in randomly oriented ridges on the P-face after freeze-fracturing (Fig. 2.8). This junction seems to act to maintain the plasma membranes spacing and to prevent any distortions of the stacked membranes involved in ion pumping and transport, which are very active functions in this district of the insect gut.

3. Changes in Tight Junctions through the Phyla

The presence of tight junctions (TJs, *zonulae occludentes*) in polarized epithelia is commonly demonstrated by using one or all of the following methods: electrophysiological measurement of resistance between the opposite sides of the tissue, ultrastructural observations in thin sections or freeze-fracture replicas and immunolocalization of proteins known to be associated with TJs.

In sections of vertebrate tissues prepared for electron microscopy, TJs are morphologically detectable owing to the obliteration of the intercellular space that occurs at the discrete points of apposition between the outer leaflets of the plasma membranes of the adjacent cells. However, in many cases the presence of TJs cannot unequivocally be ascertained in such thin sections. The study of replicas made by freeze-fracture is the tool of choice to identify the presence and the three-dimensional morphology of TJs, especially when the membranes are exposed in *en face* view. In such preparations, TJs are shown as anastomosing rows of fibrils or aligned IMPs on the P-face and complementary grooves and pits on the E-face. It is assumed that the observed IMPs or fibrils correspond to the putative points of membrane apposition seen in the thin sections.

TJs are commonly located at the juncture between the basolateral and apical plasmalemmas to form a circumferential belt, measuring somewhere between 0.1 and 0.8 μm in width. Circumscribing the apex of all the cells making up an epithelial layer, TJs form a gasket-like structure which seals and controls the paracellular permeability between the luminal and the basolateral fluid compartments.

TJs are found in most tissues of vertebrates with epithelial organization, irrespective of their embryological origin. In fact, they are common in tissues with polarized epithelial organization differentiated from not only the ectoderm (e.g., epidermis), but also the endoderm (e.g., intestinal mucosa) and mesoderm (e.g., endothelia).

The most important roles attributed to TJs are the creation of a permeability barrier between the apical and the basolateral compartments and the fence function to separate apical and basolateral plasma membrane domains (see also chapters 6 and 7 in this volume). However, an adhesive, mechanical function has also been attributed to them by many investigators.[28,29] As tight junctions participate in the control of the passive diffusion of ions through the paracellular pathway, they become responsible for the transepithelial resistance. However, all TJs do not form a comparable, rigid permeability barrier. They can be either 'tight' or 'leaky' and initially a correlation between the number of junctional strands, their overall depth and their permeability properties was postulated.[30,31] Subsequently other factors, relating to the packing of their IMPs into strands, have been deemed important in TJ leakiness.[32,33]

The organization of vertebrate TJs is highly variable in different tissues both with regard to the number of the rows constituting the TJs and the extent of their branching and anastomoses. For example, in renal proximal tubule, TJs may be represented by a single continuous strand, but in the toad bladder, TJs often form an elaborate pattern of more than eight strands.

As above mentioned TJs were initially considered to be exclusive to vertebrates, but beginning in the 1970s, they were also found to be present in two phylogenetically distant groups of invertebrates: the arthropods (among them, insects and arachnids)[5,34,35] and the tunicates.[36,37] In both arthropods and tunicates they form a permeability barrier as shown by electrophysiology and by the use of tracers such as ionic lanthanum. TJs appear to be the morphological basis of the blood-eye, blood-testis and blood-brain barrier exhibited by insects. Moreover, in ascidians, TJs create the barrier between the restricted pericardial cavity and the blood; in the myocardium they separate the extraluminal from the luminal membrane, which appears to be the unique site for conducting the excitation stimulus for muscle contraction.[38]

In many epithelia in arthropods, the lateral borders of adjacent cells are enormously interdigitated; the TJs are at the basal surface of the highly folded perineurial cell layer in insects and nearer the outer surface in spiders. While in the arthropods, TJ occurrence in forming barriers is limited to particularly important organs and potentially vulnerable to uncontrolled fluctuation in ionic concentration of the hemolymph, in tunicates, TJs are present in all the epithelial tissues. In this respect and by virtue of the fact that TJs are commonly located in the apical cell regions, as well as in their ontogeny, the tunicates resemble vertebrates far more than do the arthropods. Nevertheless, in thin sections of both arthropod and tunicate material, the features of the TJs are essentially similar to those of vertebrates. In all cases the junctional membranes of the adjacent cells are fused together in a number of punctate appositions so that the intercellular cleft is completely obliterated at these points (Figs. 2.10, 2.12) and an undulating, mirror-image scalloped appearance along the lateral borders of the two participating cells is created. When tissues are incubated with ionic lanthanum, the penetration of this electron opaque tracer is restricted and cannot penetrate beyond the level of the apposition points (Fig. 2.13).[35,37]

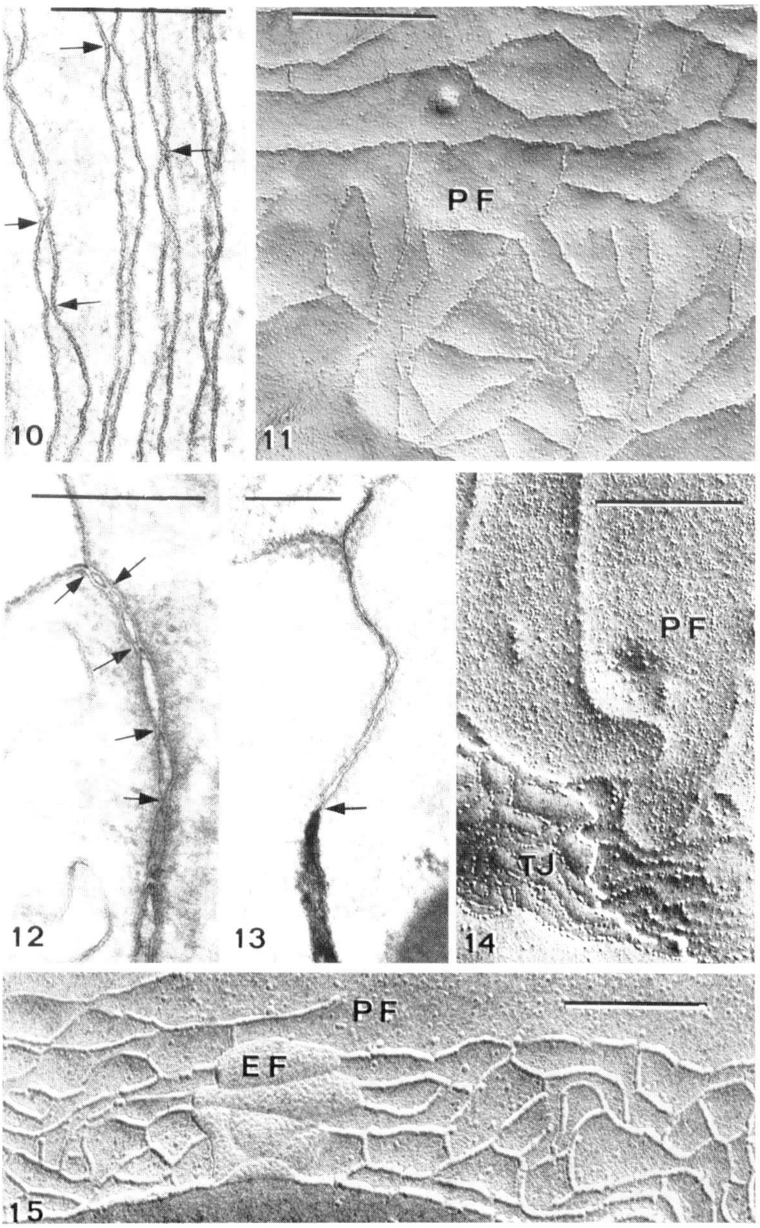

Fig. 2.10. Cross-section through the perineurial sheath of the CNS of a spider (chelicerate, arachnid). The close membrane appositions forming the tight junctions (arrows) are visible. Scale bar = 0.25 mm.

Fig. 2.11. Freeze-fracture replica of the perineurial sheath of a spider showing rows of differently oriented IMPs, making up the tight junctional ridges on the P-face (PF). Scale bar = 0.5 mm.

Fig. 2.12. Section through the pharynx of the ascidian Botryllus schlosseri (tunicate) in the region of the cells of the branchial slits. Note the punctate cell-to-cell appositions forming the tight junctions (arrows). At these points, fusion of the outer half membrane leaflets of the adjacent cells can be seen. Scale bar = 0.25 mm

Fig. 2.13. Lanthanum infiltration between mucous cells of the stomach of Botryllus schlosseri. Note that the progress of the tracer is halted at a punctate tight-junctional apposition (arrow), indicating that these junctions are able to produce a permeability barrier. Scale bar = 0.25 mm.

Fig. 2.14. Replica from stomach of the ascidian Molgula socialis (tunicate). Note the tight junctional network (TJ) from which rows of IMPs run for considerable distance along the lateral border of the P-fractured face of the plasma membrane (PF). Scale bar = 0.5 mm.

Fig. 2.15. Replica of the gut cells of the ascidian Botrylloides leachi (tunicate) showing the tight-junctional network. Rows of fused IMPs are visible on the P-face (PF) and complementary furrows on the E-face (EF) of the membrane. Scale bar = 0.25 mm.

The intramembranous TJ particles range around 8-10 nm in arthropods and 8-12 nm in tunicates. Frequently they are aligned in ridges on the P-face and have complementary grooves on the E-face, but there is not always a complete coincidence between the two. Indeed, when the fracture plane is in transition from EF to PF, the PF ridges are found to be slightly offset with respect to the EF grooves. Such an offset two-fibril model of TJs is consistent with data from both arthropod and vertebrate material.[39] The ridges are built up by linear arrays of the IMPs which are fused laterally and they show great variations in the degree of complexity of their arrangements (Figs. 2.14, 2.15).

Commonly in insects, TJs are characterized by single P-face ridges or E-face grooves which are often discontinuous and variable in length; in some cases two or three of them may lie in parallel with one another. In spiders and scorpions, more complex TJs are found, composed of a network of moniliform ridges on the P-face (Fig. 2.11) and a complementary interconnecting array of grooves on the E-face.[35]

Similar variations in the degree of complexity of the arrangement of TJ ridges were also encountered in tunicates. For example, a single row of junctional IMPs on the PF may be observed above the thick *zonula adhaerens* of the epidermis of larvaceans,[40] a few parallel ridges are encountered in the appropriate zones of the branchia,[41] but more often in tunicates, the TJs are formed by networks of a number of anastomosing ridges (Fig. 2.15). In several cases, especially in the gut of ascidians, there is a tendency for the basal component of the TJ network to be composed of extensive individual ridges, running in wavy alignments parallel to the longitudinal axis of the cells, separately, with few anastomoses (Fig. 2.14). This may be related to junctional turnover and to the capacity of TJs to grow, assemble and disassemble during many natural processes and by varying environmental conditions.[42,43]

In arthropods, the preferential fracture on the P-face (Fig. 2.11), is irrespective of the pretreatment of the tissue, that is whether the material is frozen without fixing or after chemical fixation. However, in many cases, single particles or ridges of IMPs fracture onto the EF. This has occurred routinely in fixed cryoprotected tissues of scorpions[35] and several tunicates,[4,10,37] but unfixed cryoprotected material of the former reveals the conventional P-face ridges. In this respect, the invertebrate tissues which possess EF ridges resemble the tight junctional system in some vertebrate tissues, notably that in the blood-testis barrier[44] and the endothelium of small blood vessels.[45]

In many cases TJs are seen to coexist with other intercellular junctions. In the insect nervous system, TJs lie beneath or between extensive septate junctions, along with gap junctions and adherens-like junctions.[34] In tunicates, which lack septate junctions and in arachnids, TJs are intimately associated with gap junctions; in the branchial ciliated epithelium of tunicates, TJs, gap junctions and *zonulae adhaerentes* are associated in a characteristic sequence from lumen to the region of insertion of the striated rootlets of the cilia.[29,41]

Commonly, in TJs both of vertebrates, and invertebrates, a fibrous material is associated with their cytoplasmic face, in close proximity to the cell-cell apposition points. In arthropods and tunicates, as in vertebrates immunocytochemistry and/or treatment with the S_1 subfragment of heavy meromyosin[28,46] demonstrated that this associated material consisted of actin fibrils. It is possible that these microfilaments play a part in positioning the tight junctional strands. It is not yet clear if cingulin and ZO-1 proteins are also associated with invertebrate TJs. As in vertebrates, the cytoskeletal complex of invertebrates, acting on the TJs, could participate in regulation of the paracellular permeability of epithelia. So, both the structural features and apparent association with cytoskeletal elements suggest that there may be a fundamental similarity in physiology between invertebrate and vertebrate TJs. As yet, as in vertebrates, we have no evidence as to the chemical composition of the intramembranous components of the TJs in invertebrates. Models have been proposed based on rapid freezing, suggesting a lipoprotein moiety in the membranes, linking up with the cytoplasmic cytoskeletal molecules[39] in TJs of both vertebrates and invertebrates.

In both tight[8] and septate[7] junctions, fluorescent dyes, introduced into the apical membrane domain and incorporated into the inner half membrane leaflet, cannot diffuse into the basolateral membrane surface. In the case of TJs, this has been interpreted to represent an intramembrane 'fence', with TJ strands preventing lateral diffusion of proteins and lipids, in contrast to the intercellular 'gate' of the outer membrane half leaflet, which is fused between adjacent cells minimizing the cleft and increasing transepithelial resistance. This 'gate' decreases nonelectrolyte permeability and gives rise to the occlusion of the intercellular space between cell membranes;[47] hence a permeability barrier is established with different concentrations of solutions on either side of the junction. Studies involving the depletion of ATP in cultured MDCK cells induced a loss of transepithelial electrical resistance, thereby abolishing the 'gate', while leaving the 'fence' function intact, as indicated by the maintenance of lipid polarity;[48] this has been interpreted as an *uncoupling* of the molecular 'fence' from the paracellular 'gate' function in the tight junctions of epithelia and is considered to demonstrate the previously merely postulated separate nature of the two functions.

4. CHANGES IN ADHERENS JUNCTIONS (MACULAE AND ZONULAE ADHAERENTES) THROUGH THE PHYLA

In epithelial tissues the function of maintaining cell-to-cell adhesion and providing mechanical support is particularly attributed to two types of intercellular adherens junctions: the desmosomes (spot desmosomes) or *maculae adhaerentes* and the *zonulae adhaerentes* (ZA, belt desmosomes), the so-called intermediate junctions. These types of adherens junctions are widely distributed in different tissues of both vertebrate and invertebrate organisms. They are usually easy to discern although in some invertebrate tissues they appear to be somewhat simplified and not immediately recognizable. Desmosomes differ from ZA in both morphology and biochemical composition (see also chapters 8 and 9 in this volume).

The structure of vertebrate desmosomes, as revealed by electron microscopy, has been extensively reported in review articles.[49,50] Desmosomes are usually disc-shaped membrane differentiations, 0.2-0.5 µm in diameter and the adjacent membranes lie parallel, separated by an intercellular space of 22-35 nm. In thin sections, the interspace appears filled with a filamentous desmoglea which frequently forms a central stratum in the midline. This filamentous material crosses the membranes and holds them together, structurally connecting one cell to the other. In addition dense intracellular plaques are closely associated with the cytoplasmic surfaces of the junctional membranes. The plaques provide support for the attachment of looping bundles of cytoskeletal intermediate filaments (IFs) or tonofilaments, 10 nm in diameter (see also chapter 9 in this volume). Therefore, desmosomes appear symmetrical about the intercellular midline, the two halves of the junction being furnished by the adjacent cells. Studies of freeze-fracture replicas of these desmosomes, unlike septate, tight and gap junctions, have added very little information. The junctional area may often be recognizable as patches of irregularly sized particles on each of the membrane faces in vertebrate tissues. In contrast, epithelia of invertebrates do not exhibit this freeze-fracture profile and any differentiation into patches of intramembrane particles is rarely seen. Typical desmosomes function as button-like devices connecting several cells together via the IFs, thus allowing the passive distribution of mechanical stress between cells. Their role seems to be to limit the distensibility and to enhance the resistance of the tissue in which they are located.

The basic structure of desmosomes is similar in a wide range of animal species. Variations in size, form and frequency of desmosomes, however, occur from tissue to tissue of the same organism and from species to species. In some invertebrates, such as annelids, the extracellular space between the junctional membranes is considerably wider and contains a denser desmoglea than that of vertebrates.[51] Desmosomes connected with cytoskeletal IFs are very frequent in annelid tissues and numerous desmosomes are also found in the epidermal tissues of flatworms and arrow-worms. In arthropods, the desmosomal plaques seem to be associated with underlying cytoplasmic microtubules, not with IFs.[52,53] This situation does not generally occur in vertebrate tissues, although recent evidence shows that a microtubule-associated protein mediates the association between microtubules and desmosomes in MDCK cell monolayers.[54] In arthropods it is rare to find desmosomes with typical features; sometimes their macular nature is not unequivocal and often the dimensions of their associated filaments are smaller than that of the intermediate size (10 nm). In other invertebrate phyla, if the fine structure of a junction does not correspond exactly to that of vertebrate desmosomes, investigators often prefer to refer to them as "desmosome-like" structures.

ZA, the belt desmosomes, are often confused with spot desmosomes as their morphology superficially appears similar. Nevertheless, they exhibit several important differences. The ZA of vertebrates are continuous junctions which form a belt around the apico-lateral borders of epithelial cells immediately below the tight junctions and occupy a position intermediate to the subjacent desmosomes. In thin sections, the ZA can be seen to occupy a specialized region in which the two adjacent membranes lie parallel to one another and are separated by an intercellular space, 15-25 nm in width. This space is filled with filamentous material but it is narrower than the desmosome interspace and no central stratum is generally visible in the midline. On the cytoplasmic sides, dense mats of filaments, approximately 7 nm in diameter, are closely apposed to the membranes. These microfilaments have the dimensions of fibrous actin and do not form the compact plaques characteristic of desmosomes. In freeze-fracture replicas ZA are unnoticeable; the only reported evidence of difference is an occasional decrease in the density of particles on both E- and P-faces of the junctional membrane in comparison to adjacent non-junctional regions.[49] This description appears valid for ZA of polarized epithelia. The group of adherens junctions, in which actin microfilaments are associated with membrane specializations, also includes the *fasciae* and *puncta adhaerentes*. The first of these are discontinuous strip-like junctions that occur in cardiac muscle and exhibit a quite different topology from the epithelial ZA. The second of these are small cell-cell contact areas of either epithelial or non-epithelial cells. The morphological features of the actin-associated junctions show a great variability among different tissues and between species of vertebrates.[55] In contrast to desmosomes, the function of the ZA and *fasciae adhaerentes* is to transmit active forces between cells of a tissue, based on the contractile actin cytoskeleton.

In invertebrates, membrane-filament contact zones with diverse structures but apparently similar function of specialized cell-to-cell adhesions, are widely distributed through animal phyla from the turbellarians (Platyhelminthes) to the tunicates. In invertebrates, it is not always easy to classify ZA or *fasciae adhaerentes* on the basis of morphological features alone. As septate junctions, which have an adhesive function too, have a widespread occurrence in invertebrates, ZA are often limited to particular tissues or organs. When present, the ZA are usually the outermost junction, occurring above the septate junctions at the luminal surface of epithelial layers, such as in the molluscan midgut (Fig. 2.16), insect accessory glands (Fig. 2.17) or Malpighian tubules (Fig. 2.18). In tunicates, the ZA are found underneath tight junctions in the typical vertebrate position. In the epithelial ciliated cells of ascidians, complexes of different junctional types coexist: tight junctions, ZA and gap junctions are all present. The ZA are belt-like in distribution; their parallel junctional membranes are separated by an interspace 25 nm wide, which is filled with filamentous material denser in the midline (Figs. 2.19, 2.20). On both cytoplasmic sides, dense mats anchor bundles of ciliary rootlets (Fig. 2.19) and no actin can be detected.[29,46] The fine structure

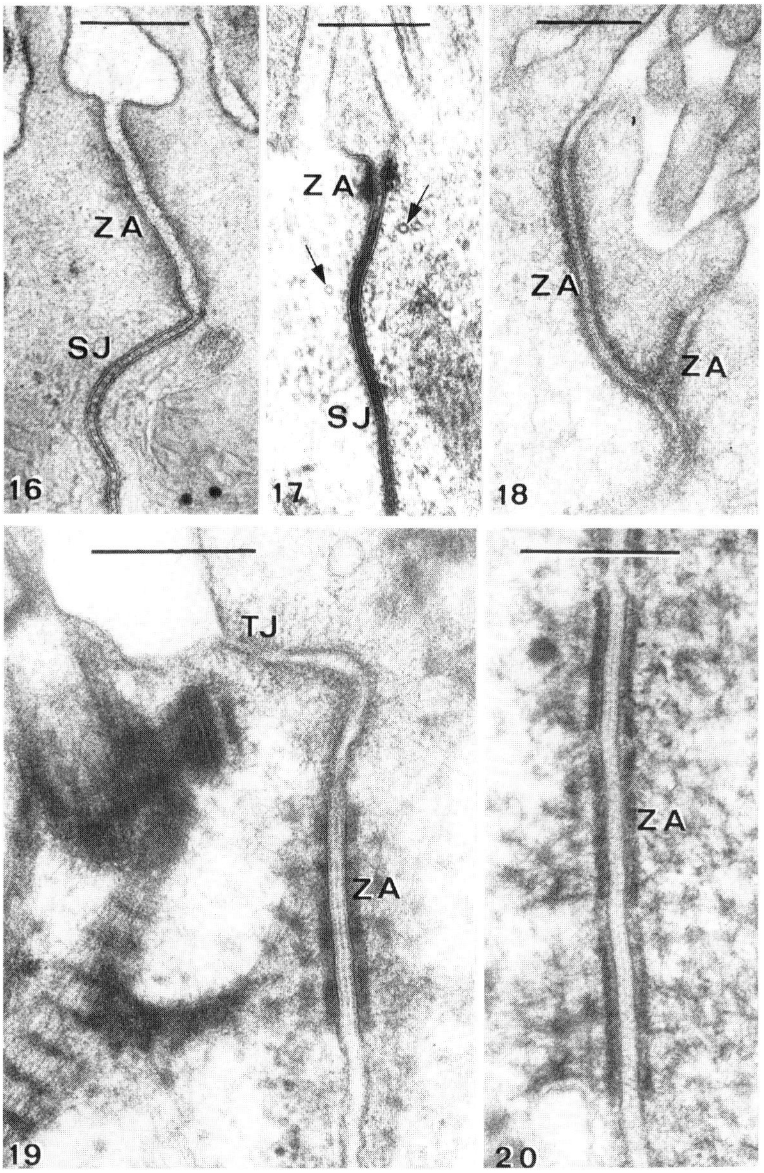

Fig. 2.16. Cross-section through the midgut of Cepaea sp. *(Mollusca). A* zonula adherens *(ZA) is present between the luminal region of the epithelial cells. Beneath this belt desmosome a septate junction (SJ) is visible. Scale bar = 0.25 µm.*

Fig. 2.17. Cross-section through a deferens duct of the caddisfly Rhyacophila italica *(Insecta). A short* zonula adherens *(ZA) is present at the apex of the epithelial cells. Microtubules (arrows) and microfilaments are visible close to the septate junction (SJ). Scale bar = 0.25 µm.*

Fig. 2.18. Cross-section through the Malpighian tubules of the cockroach Periplaneta americana *(Insecta). A* zonula adherens *(ZA) is visible between the apical region of the epithelial cells. Scale bar = 0.25 µm.*

Fig. 2.19.-2.20. Cross-section through the ciliated epithelium of the branchial slits in the ascidian Diplosoma listerianum *(Fig. 2.19) and* Ascidiella aspersa *(Fig. 2.20) (tunicates) showing the atypical dense zonulae adhaerentes (ZA) with ciliary rootlets inserting on them. TJ, tight junction. Scale bar = 0.25 µm.*

of these zonular adherens junctions is somewhat ambiguous, because they have some features in common with desmosomes. In the cephalochordate *Amphioxus*, surprisingly, ZA are the only type of junctional specialization found in the ciliated epithelia.[56]

Ultrastructural investigations are insufficient to understand the formation, nature and function of adherens junctions, so immunocytochemistry and the analysis of their biochemical composition can be illuminating. In vertebrates, many papers have revealed the molecular components of intercellular junctions. Both ZA junctions and desmosomes are multiprotein complexes organized around transmembrane cell adhesion molecules (see also chapters 1, 8, 9, 12 and 14 in this volume). They are quite distinct biochemically and each junctional complex contains at least one junction-specific component. In

ZA, the protein α-catenin may be able, even if indirectly, to mediate anchoring of actin filaments.[57] In desmosomes, the protein desmoplakin appears to be involved in anchoring the intermediate filaments, mainly cytokeratins.[58] Unfortunately, the adherens junctions in invertebrate phyla have received much less attention and efforts to identify their molecular components are more difficult owing to the paucity of material. A survey of representative species of nearly all invertebrate phyla for the screening of intermediate filaments has been performed, by immunofluorescence, using IF antibody.[59] The presence of intermediate filaments has been ascertained in nearly all invertebrates with the exception of arthropods. The suggestion is that, where IFs are present, there the epithelial cell membranes have the potential to differentiate into desmosome-type junctions. If it is true that arthropods do not possess IFs, a fundamental cytoplasmic component of desmosomes is lacking and this would explain the unusual morphological features of these junctions.[28] There is a long way to go before we achieve a full understanding of the molecular components of invertebrate adherens junctions. Recently, a promising new direction has been opened with the study on the segment polarity gene *armadillo* of *Drosophila* (see chapters 1 and 8 in this volume). The gene product seems to be a key component of the intercellular adherens junctions of *Drosophila*, which appear structurally more analogous to the actin-associated adherens-type junctions than to the desmosomes of vertebrates.[60]

5. Gap Junctions and Linker Junctions: Evolutionary Considerations

5.1 Gap Junctions

Epithelial cells are nearly always associated with one another by gap junctions, one exception being the Onychophora, which appears to lack them entirely.[61,62] These are intercellular associations where the cleft between the adjacent cells is reduced from the conventional 15-20 nm to 2-3 nm (Fig. 2.21). These close appositions often show a striated appearance across the space and in en face view, as is evident in tangential sections, subunits of the component connexons can sometimes be seen. These connexons straddle the intercellular cleft and form a transmembrane hemichannel which, when juxtaposed to and aligned with, a similar hemichannel in the membrane of the adjacent cell, form a channel or pore between the two cells. The channels of such connexons, arranged in clusters or plaques, form the morphological basis of the 'coupling' that is observed to occur in cells linked by gap junctions; exchange of ions or small molecules, less than about 1.2 kD ± in molecular weight, can take place between cells, leading to ionic or metabolic communication. The structure of gap junctions looks roughly similar in thin sections of most epithelial cells except in some crustacean preparations[63,64] where the intercellular space may be as wide as 4 nm or more. In freeze-fracture replicas there are certain differences, notably with respect to the size of the individual transmembrane connexons making up to the gap junctional aggregate and as regards the membrane face to which connexons adhere after the cleaving of the frozen epithelial cells in replicas. In all arthropods, the connexons are 13 nm in diameter and the preferential plane of cleavage is on the E-face (Fig. 2.22), in contrast with that of the vertebrates and other invertebrate groups, where smaller particles are observed, which cleave on the P-face (Fig. 2.23).

Each connexon is thought to consist of six subunits surrounding a central hydrophilic channel; each of these subunits is composed of a single polypeptide which spans the plasma membrane four times with the hydrophilic lining of the connexon channel formed by an amphiphilic α-helix. Biochemically, gap junctional polypeptides have been shown to belong to a family of connexin (CX) proteins[65] which exhibit homologous sequences in their extracellular and transmembrane domains (see also chapters 3, 10 and 11 in this volume). On the cytoplasmic side are found both the C and N terminals of the polypeptide chain; variations in their length are responsible for the differing molecular weights that have been reported for gap

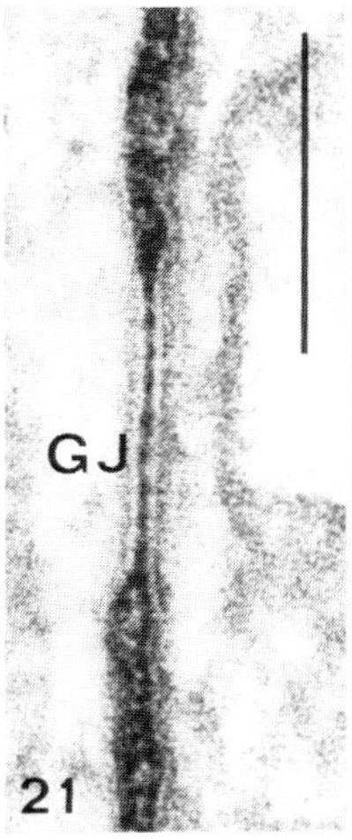

Fig. 2.21. Gap junction (GJ) in the intestine of the ascidian Botryllus schlosseri, *after lanthanum tracer impregnation. Note the reduced intercellular space at the junctional level freely crossed by tracer. Scale bar = 0.25 μm.*

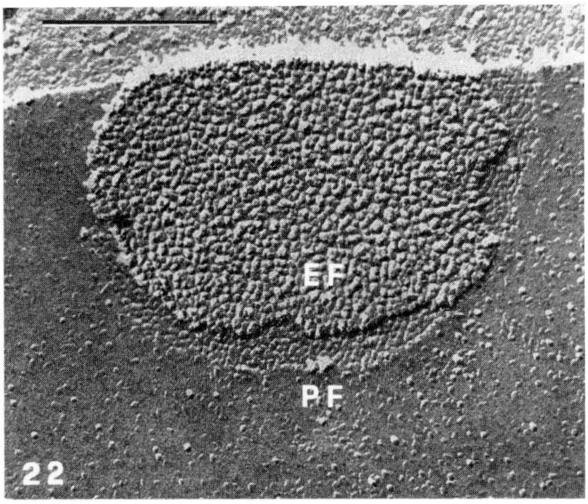

Fig. 2.22. Freeze-fracture replica of an inverted gap junction from the milliped Pachyjulus *sp. (Arthropoda). IMPs are associated with the E-face (EF) of the plasma membrane, while complimentary pits are visible on the P-fracture face (PF). Scale bar = 0.25 μm.*

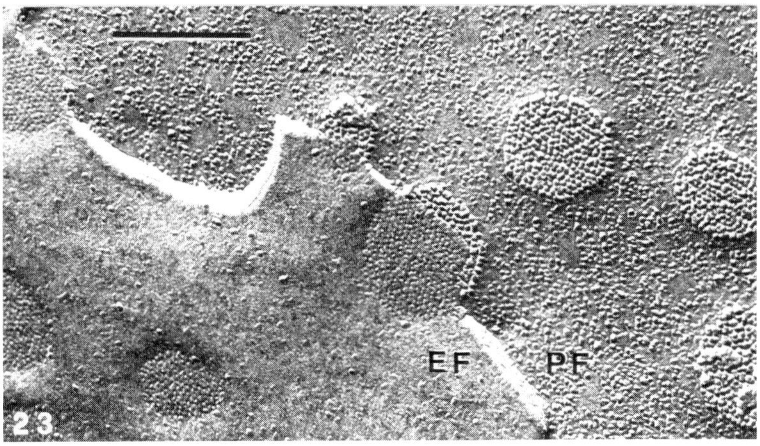

Fig. 2.23. Freeze-fracture replica of a gap junction from the stomach of the ascidian Clavelina lepadiformis. *IMPs are associated with the P-face (PF) and complementary pits are on the E-face (EF) of the plasma membrane. Scale bar = 0.25 μm.*

junctional proteins from different, or even the same, tissues, for example heart, where these are CX37, CX40, CX42, CX43, CX46 and MP70. In liver, two connexins of different molecular weights (21 and 27 kD) have also been found to co-exist; individual connexons may contain both proteins, or different connexons may exist, each consisting of only type of subunit.[66] Another, protein, 16 to 18 kD in molecular weight, highly conserved, has also been found to be present in gap junctions isolated from many species.[67,68] This 16 to 18 kD protein was originally found in arthropod tissues as an 18 kD molecular weight protein[69] but has now been found in tissues of vertebrates as well, as a 16 kD protein.[70] Since the 27 kD connexin protein can be removed from liver by proteases, while the 16 to 18 kD protein is present in junctional preparations regardless of the method of purification, the latter may turn out to be the major structural protein. The vertebrate 18 kD and the invertebrate 16 kD proteins are homologous with one another as well as with both the 16 kD subunit of the vacuolar H^+-ATPase and the F_0 proton channel of the F_0 F_1 ATP synthetase of mitochondria and chloroplasts.[71] This 16-18 kD protein, now called ductin,[71] may line the gap junction channels and in its other form, originally found in endomembrane systems (such as the lysosomes and Golgi), it is now implicated in epithelial ion transport in both vertebrates and arthropods. Antibodies to ductin have recently been found to inhibit gap junction-based cell-cell communication in *Drosophila* oocyte-cumulus complexes.[71] Clearly the evolutionary trend here would seem to have been to utilize a primitive protein as a channel component of both endomembrane proton pumps and plasma membrane gap junctions. The 16 kD ductin protein is a tandem repeat of two 8-kD domains, suggesting evolution from the 8 kD F_0 subunit by intragenic duplication.[72] Each domain has homology with the 8 kD subunit C of the F_0 proton channel from the F_0 F_1 ATP synthetase of mitochondria.[73]

5.2 "Linker" Junctions (Fig. 2.24)

This kind of intercellular junction has been observed in only a few situations, mainly between epithelial-like or endothelial cells. They have been found between cells in invertebrates, in the ensheathing glia of the CNS of insects,[5] myriapods[74] and cephalopods[75] as well as between embryonic cells in the brains of certain vertebrates[76] and in placental and myocardial microvessels.[77,78]

The linker junctions consist of areas where there are electron-dense cross-striations between cells, bridging the gap between adjacent plasma membranes (Fig. 2.24). This intercellular gap may be reduced in these structures as in myriapods[74] or widened, as in insect glial cells[5] and in the endothelia of microvessels where it may be wide as 18 nm.[76] There are no available data regarding their biochemical or molecular composition. It has been postulated that linker junctions may form the basis of a primitive intercellular 'barrier' to certain charged molecules which may become enmeshed in the intercellular glycocalyx formed by the intercellular cross-striations. In the cells of the cephalopod brain, certainly, these junctions preclude the entry of exogenous molecules such as horseradish peroxidase.[79]

The existence of these junctions in the embryos of vertebrates and the fact that they seem not to be present in the adult vertebrate brain have led investigators to postulate that they are primitive structures.[80] They could have been lost to the adult form of vertebrates when their barrier function is taken on by the more effective occluding tight junctions. They may be examples of ontogeny recapitulating phylogeny, giving way, in the vertebrates, to the tight junctions found in the adult organism. Linker junctions have been maintained by certain invertebrate groups some of which lack tight junctions. Their relationship to the more ubiquitous septate junction of invertebrates, which also may possess a partial barrier function, is unclear, owing to the dearth of any biochemical data. The recent immunohistochemical evidence on brain microvessels of the rat suggests that adherens junction components may be associated with these linker junctional regions;[78] this could lead to interesting experiments using vertebrate antibodies against invertebrate junctional structures.

6. Evolutionary Trends and Future Perspectives

In any analysis of the evolution of intercellular junctions from the invertebrates through to the vertebrates the major problem is that structure has to be the basis for comparison as so little biochemical data is available from invertebrates. Immunological cross-reactivity is another avenue of approach to this problem, using antibodies raised against vertebrate junctional proteins in immunocytochemical preparations of invertebrate tissues. One of problems here may be that of loss of antigenicity after

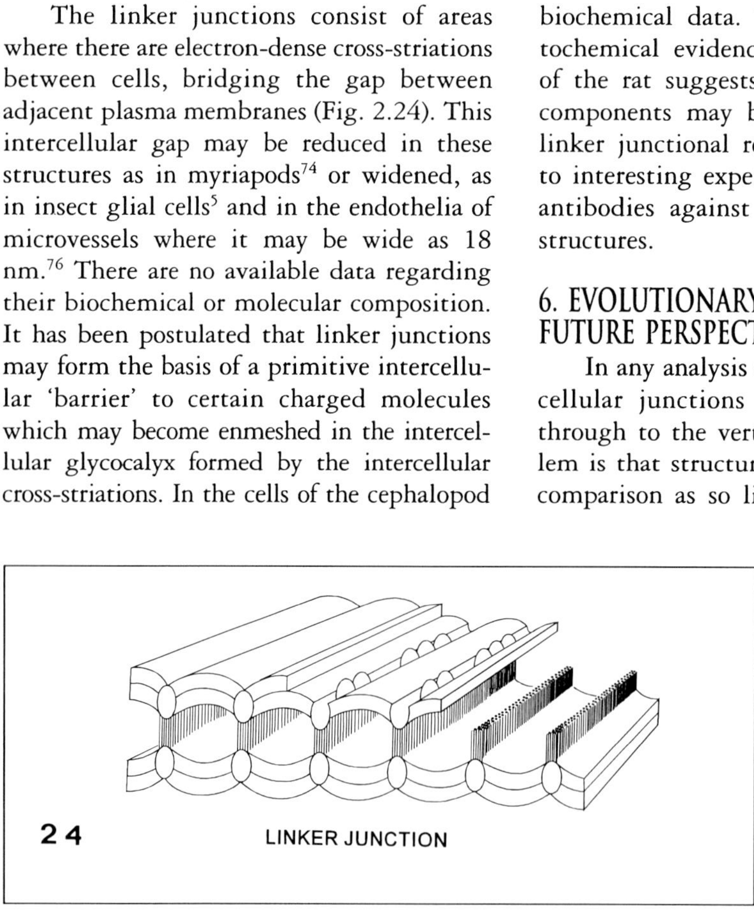

2 24

Fig. 2.24. Schematic reconstruction of the linker junction

fixation and other preparative procedures; more use might therefore be made of tissue extracts from invertebrate sources and cross-reacting them with antibodies to specific vertebrate junctional components. Using *Drosophila* as a genetic tool, progress may be made along the lines investigated by Bryant and his colleagues[81] regarding the septate junctions and the newly identified associated tight junction-like proteins.[9] (see also chapters 1 and 12 in this volume) The range and variation within the evolutionary tree under consideration (Table 2.1) indicates the enormous scope of the problem in establishing evolutionary links between the junctions of so many diverse groups. Our knowledge of junctional types that exist in these different groups of organisms can be seen in Table 2.2; this information is based primarily on morphological EM observations and not on biochemical data. Why some groups lack certain junctional types is not always clear; the absence of some kinds of junctions may be in relation to the primitive position in the evolutionary tree of the taxon, or may indicate the capability of that taxon to resolve similar functions through different devices.

There is no doubt that the emergence of the importance of the cellular adhering molecules (CAMs) in stabilizing and maintaining intercellular contact, prior to the more definite establishment of junctions, is one of the most important developments in this field. Speculation centers on whether the CAMs are also basic as intercellular components of the septate and linker junctions, as they have been shown to be in the adhering junctions; a further query is whether CAMs initiating the events leading to cells becoming loosely associated with one another be fore adherens and occludens junction formation, is also the preface to the more intimate and close association observed between cell membranes at the gap junctions. The role of the cytoskeleton is also of enormous importance in the establishing of contacts between cells and the subsequent differentiation of specific junctional structures. In this capacity, the lack of observed cytoskeletal attachments to the connexons makes gap junctional assembly and development the more mysterious. Recent developments in this regard show that the assembly of connexin43 into individual connexons occurs after exit from the endoplasmic reticulum and distinguishes this process from that of other integral plasma membrane proteins.[82]

The convergence to similar functions between junctions with different structures is another of the interesting trends to be observed. For example, barrier function appears to be associated in some way with a range of junctional types, starting in more primitive systems with linker and septate junctions and moving on to the tight junctions. It appears that the more sophisticated an organism's physiology, particularly with regard to, for example, its nervous system, the more need there is to have an effective junctional barrier. Hence the blood-brain barrier in arachnids is based on very complex tight junctions, while that in cephalopod molluscs is based on linker junctions, with no occlusion of the intercellular cleft, but elaboration of the intercellular matrix materials and their organization. Clearly the latter is very effective in the context of cephalopod brain anatomy,[79] but the more highly organized tight junctions, found in all the vertebrates, have presumably been more effective ultimately, in that they have been conserved throughout the vertebrates.

Recently, there have been a number of interesting developments with regard to junctions, as indicated by Gumbiner[83] in his article on "breaking through the tight junction barrier". The idea is emerging that certain cell adhesion molecules may function as tumor suppressors,[84] whereby the expression of their products results in the suppression of dedifferentiation, invasion and metastasis. These cell adhesion molecules are intimately associated with the cytoskeleton at specialized junctional regions and may also play a role in signal transduction from the adhesion receptors or junctional components (see chapters 5, 8, 12, 13 in this volume). ZO-1 is a member of this family of proteins[9] as is the discs-large (dlg) tumor suppressor protein of septate junctions[17] and submembrane junctional plaque proteins.[84] Reviews considering these developments[43,85,86] discuss the relationships between ZO-1, the 160 kD protein from ZO-2, the 130 kD protein (ZO-3?), unrelated to cingulin (140 kD), the 155 kD protein called the 7H6 antigen and occludin[87] (see also chapter 5 in this volume). It is interesting that a 4.1 protein,[88] encoded by the *coracle* gene, has been found to be a component of *Drosophila* septate junctions, where it can be involved in signal transduction (see chapter 1 in this volume).

The colinear distribution of homologous domains within ZO-1 and the dlg tumor suppressor gene product suggests a common evolutionary origin and a similarity between the tight junctions of vertebrates and the septate junctions of the invertebrates. Their possible role in the transduction of membrane signals may be related to the regulation of paracellular permeability as well as intercellular junction assembly (see also chapters 1, 5, 6 and 12 in this volume). The extent to which similar homologies may be found among other junctional proteins has yet to be established but present findings will provide a direction for the field for some time to come, leading to fresh insights into junctional function.

References

1. Eichenberger-Glinz S. Intercellular junctions during development and in tissue cultures of *Drosophila melanogaster*: an electron-microscopy study. Wilhelm Roux's Arch 1979; 186:333-349.
2. Dallai R, Vegni Talluri M. Evidence for septate junctions in the syzygy of the protozoon *Gregarina polymorpha* (Protozoa, Apicomplexa). J Cell Sci 1988; 89:217-224.
3. Dallai R, Afzelius BA. Paired spermatozoa in *Thermobia* (Insecta, Thysanura). J Ultrastruct Res 1984; 86:67-74.
4. Georges D. Gap and tight junctions in tunicates. Study in conventional and freeze-fracture techniques. Tissue Cell 1979; 11:781-92.
5. Lane NJ. Tight junctions in arthropod tissues. Int Rev Cytol 1981; 73:243-318.
6. Skaer H leB, Maddrell SHP, Harrison JB. The permeability properties of septate junctions in Malpighian tubules of *Rhodnius*. J Cell Sci 1987; 88: 251-265.
7. Wood RL. The septate junction limits mobility of lipophilic markers in plasma membranes of *Hydra vulgaris (attenuata)*. Cell Tissue Res 1990; 259:61-66.
8. van Meer G, Simons K. The function of tight junctions in maintaining differences in lipid composition between the apical and basolateral cell surface domains of MDCK cells. EMBO J 1986; 5:1455-464.
9. Willot E, Balda MS, Fanning AS, Jameson B, van Itallie C, Anderson JM. The tight junction protein ZO-1 is homologous to the *Drosophila* discs-large tumor suppressor protein of septate junctions. Proc Natl Acad Sci USA 1993; 90:7834-7838.
10. Green CR, Bergquist PR. Phylogenetic relationships within the invertebrata in relation to the structure of septate junctions and the development of 'occluding' junctional types. J Cell Sci 1982; 53:279-305.
11. Loewenstein WR, Kanno Y, Socolar SJ. The cell-to-cell channel. Fed Proc 1978; 37:2645-2650.
12. Eernisse DJ, Albert JS, Anderson FE. Annelida and Arthropoda are not sister taxa: A phylogenetic analysis of spiralian metazoan morphology. Syst Biol 1992; 41:305-330.
13. Dallai R, Xue' L. Cell junctions in the gut of Protura. Tissue Cell 1992; 24:51-59.
14. Noirot-Timothee C, Noirot C. Septate and scalariform junctions of the arthropods. Int Rev Cytol 1980; 63:97-140.
15. Lane NJ, Flores V. Actin insertion into septate junctions. Tissue Cell 1988; 20:211-217.
16. Lane NJ, Flores V, Harrison JB, Lee WM. Septate junction-associated actin is anchored onto microtubules. Cell Biol Int Reports 1987; 11:250.
17. Woods DF, Bryant PJ. The discs-large tumor suppressor gene of *Drosophila* encodes a guanylate kinase homolog localized at septate junctions. Cell 1991; 66:451-464.
18. Duvert M, Gros D, Salat C. The junctional complex in the intestine of *Sagitta setosa* (Chaetognatha): The paired septate junction. J Cell Sci 1980; 42:227-246.
19. Green CR. Septate junctions of the phylum Hemichordata. J Ultrastruct Res 1981; 75:1-10.

20. Dallai R. Septate and continuous junctions associated in the same epithelium. J Submicr Cytol 1976; 8:163-174.
21. Skaer H leB, Harrison JB, Lee WM. Topographical variations in the structure of the smooth septate junction. J Cell Sci 1979; 37:373-389.
22. Dallai R, Trastullo E, Lupetti P, Mencarelli C. Unusual cytoskeletal association with the intercellular septate junction in the midgut of Collembola (Insecta: Apterygota). Int J Insect Morphol Embryol 1993; 22:473-486.
23. Colombo A, Bonfanti P, Camatini M. Actin, α-actinin and vinculin are associated with septate junctions in Insecta. Cell Motil Cytoskeleton 1993; 26:205-213.
24. Lane NJ, Dilworth S. Isolation and biochemical characterization of septate junctions. Differences between the proteins in smooth and pleated varieties. J Cell Sci 1989; 93:123-131.
25. Fain-Mauriel MA, Cassier P. Une nouveau type de jonctions: les jonctions scalariformes. etude ultrastructurale et cytochimique. J Ultrastruct Res 1972; 39:222-238.
26. Lane NJ. Freeze-fracture and tracer studies of the intercellular junctions of insect rectal tissues. Tissue Cell 1979; 11:481-506.
27. Burighel P, Dallai R Martinucci GB. Scalariform junctions in the gut of Tunicates. Biol Cell 1985; 54:171-176.
28. Lane NJ. Cytoskeletal associations with intercellular junctions in arthropods. In: Lanzavecchia G, Valvassori R, eds. Form and function in Zoology. Modena (I): Mucchi, 1991:87-102.
29. Martinucci GB, Dallai R, Burighel P, Lane NJ. Different functions of tight junctions in the ascidian branchial basket. Tissue Cell 1988; 20:119-132.
30. Claude P, Goodenough DA. Fracture faces of *zonulae occludentes* from 'tight' and 'leaky' epithelia. J Cell Biol 1973; 58:390-400.
31. Schneeberger EE, Lynch RD. Structure, function and regulation of cellular tight junctions. Am J Physiol 1992; 262:L647-661.
32. Gonzalez-Mariscal L. The relationship between structure and function of tight junctions. In: Cereijido M, ed. Tight Junctions. Boca Raton:CRC Press, 1991:67-76.
33. Mollgard K, Malinowska DH, Saunders NR. Lack of correlation between tight junction morphology and permeability properties in developing choroid plexus. Nature (Lond.) 1976; 264:293-294.
34. Lane NJ, Skaer H leB, Swales LS. Intercellular junctions in the central nervous system of insects. J Cell Sci 1977; 26:175-199.
35. Lane NJ. Anatomy of the tight junction: invertebrates. In: Cereijido M, ed. Tight Junctions. Boca Raton: CRC Press, 1991:23-48.
36. Lorber V, Rayns DG. Cellular junctions in the tunicate heart. J Cell Sci 1972; 10:211-227.
37. Lane NJ, Dallai R, Burighel P, Martinucci GB. Tight and gap junctions in invertebrates. J Cell Sci 1986; 84:1-18.
38. Weiss J, Goldman Y, Morad M. Electromechanical properties of the single cell-layered heart of the tunicate *Boltenia ovifera* (sea potato). J Gen Physiol 1976; 68:503-518.
39. Lane NJ, Reese TS, Kachar B. Structural domains of the tight junctional intramembrane fibrils. Tissue Cell 1992; 24:291-300.
40. Burighel P, Lane NJ, Martinucci GB, Fenaux R, Dallai R. Junctional diversity in two regions of the epidermis of *Oikopleura dioica* (Tunicata, Larvacea). Cell Tiss Res 1989; 257:529-535.
41. Burighel P, Martinucci GB, Lane NJ, Dallai R. Junctional complexes of the branchia and gut of the tunicate *Pyrosoma atlanticum* (Pyrosomatida, Thaliacea). Cell Tissue Res 1992; 267:357-364.
42. Bentzel CJ, Palant CE, Fromm M. Physiological and pathological factors affecting the tight junction. In: Cereijido M ed. Tight junctions. Boca Raton:CRC Press, 1991:151-174.
43. Anderson JM, Balda MS, Fanning AS. The structure and regulation of tight junctions. Curr Opin Cell Biol 1993; 5:772-778.
44. Nagano T, Suzuki F. The postnatal development of the junctional complexes of the mouse Sertoli cells as revealed by freeze-fracture. Anat Rec 1976; 185:403-415.
45. Simionescu N, Simionescu M, Palade GE. Open junctions in the endothelium of the postcapillary vesicles of the diaphragm. J Cell Biol 1978; 79:27-44.

46. Dallai R, Burighel P, Martinucci GB, Maci R, Camatini M. Actin localization at the tight junction level in invertebrate ciliated epithelia. Tissue Cell 1989; 21:37-46.
47. van Meer G, Gumbiner B, Simons K. The tight junction does not allow lipid molecules to diffuse from one epithelial cell to the next. Nature (Lond.) 1986; 322:639-641.
48. Mandel LJ, Bacallao R, Zampighi G. Uncoupling of the molecular 'fence' and paracellular 'gate' functions in epithelial tight junctions. Nature 1993; 361:552-555.
49. Staehelin LA. Structure and function of intercellular junctions. Int Rev Cytol 1974; 39:191-183.
50. Cowin P, Franke WW, Grund C, Kaprell P, Kartenbeck J. The desmosome-intermediate filament complex. In: Edelman GM, Thiery JP, eds. The Cell in contact. New York: Wiley and Sons, 1985:427-460.
51. Fawcett DW. Junctional specializations. In: Fawcett DW, ed. The Cell 2nd ed. Philadelphia: WB Saunders Comp, 1981:124-194.
52. Ashhurst DE. An insect desmosome. J Cell Biol 1970; 46: 421-425.
53. Lane NJ, Skaer H leB. Intercellular junctions in insect tissues. Adv Insect Physiol 1980; 15:35-213.
54. Wacker IU, Rickard JE, DeMay JR, Thomas EK. Accumulation of a microtubule-binding protein, pp170, at desmosomal plaques. J Cell Biol 1992; 117:813-824.
55. Geiger B, Avnur Z, Volberg T, Volk T. Molecular domains of adherens junctions. In: Edelman GM, Thiery JP, eds. The Cell in contact. New York: Wiley and Sons, 1985:461-489.
56. Lane NJ, Dallai R, Martinucci GB, Burighel P. Cell junctions in amphioxus (Cephalochordata): a thin section and freeze-fracture study. Tissue Cell 1987; 19:399-411.
57. Tsukita SH, Tsukita SA, Nagafuchi A, Yonemura S. Molecular linkage between cadherins and actin filaments in cell-cell adherens junctions. Curr Opin Cell Biol 1992; 4:834-839.
58. Garrod DR. Desmosomes and hemidesmosomes. Curr Opin Cell Biol 1993; 5:30-40.
59. Bartnik E, Weber K. Widespread occurrence of intermediate filaments in invertebrates; common principles and aspect of diversion. Eur J Cell Biol 1989; 50:17-33.
60. Peifer M. The product of the *Drosophila* segment polarity gene *armadillo* is part of a multi-protein complex resembling the vertebrate adherens junction. J Cell Sci 1993; 105:993-1000.
61. Dallai R, Giusti F. The epithelial cell junctions in Onychophora. Camatini M ed. Myriapod biology. Accademic Press, 1979: 433-443.
62. Lane NJ, Campiglia SS, Lee WM. Junctional types in the tissues of an Onychophoran: the apparent lack of gap and tight junctions in *Peripatus*. Tissue Cell 1994: 26, 143-154.
63. Peracchia C. Low resistance junctions in crayfish. I. Two arrays of globules in junctional membranes. J Cell Biol 1973; 57:54-65
64. Peracchia C. Low resistance junctions in crayfish. II. Structural details and further evidence for intercellular channels by freeze-fracture and negative staining. J Cell Biol 1973; 57:66-76.
65. Stevenson BR, Paul DL. The molecular constituents of intercellular junctions. Curr Opin Cell Biol 1989; 1:884-891.
66. Nicholson B., Dermietzel R, Teplow D, Traub O, Willecke K, Revel JP. Two homologous protein components of hepatic gap junctions. Nature 1987; 329:732-734.
67. Finbow MF, Shuttleworth J, Hamilton AE, Pitts JD. Analysis of vertebrate gap junction protein. EMBO J 1983; 2:1479-1486.
68. Ryerse JS. Isolation and characterization of gap junctions from *Drosophila melanogaster*. Cell Tissue Res 1989; 256:7-16.
69. Finbow ME, Buultjens TEJ, Lane NJ, Shuttleworth J, Pitts JD. Isolation and characterization of arthropod gap junctions. EMBO J 1984; 3:2271-2278.
70. Finbow ME, Eldridge T, Buultjens J, Kam E, Shuttleworth J, Pitts JD. Comparison of the protein components present in vertebrate and arthropod gap junction preparations. In: Gap junctions. Cold Spring Harbour Laboratory 1985: 77-105.

71. Finbow ME, Goodwin S, Meagher L, Lane NJ, Keen J, Findlay JBC, Kaiser K. Genetic and functional analysis of Ductin in insects—a putative component of gap junctions and the vacuolar H^+-ATPase. J Cell Sci 1994; in press.
72. Nelson H, Nelson N. The progenitor of ATP synthases was closely related to the current vacuolar H^+-ATPase. FEBS Lett 1989; 277:147-153.
73. Mandel M, Moriyama Y, Hulmes JD, Pan YE, Nelson H, Nelson N. cDNA sequence encoding the 16 kD proteolipid of chromaffin granules implies gene duplication in the evolution of H^+-ATPases. Proc Nat Acad Sci 1988; 85:5521-5524.
74. Lane NJ. Novel arthropod cell junctions with restricting intercellular 'linkers'. J Neurocytol 1989; 18:661-669.
75. Lane NJ, Abbott NJ. Freeze-fracture evidence for a novel restrictive junction at the blood-brain barrier of the cuttlefish *Sepia officinalis*. J Neurocytol 1992; 21:295-303.
76. Firth JA, Bauman KF, Sibley CP. The intercellular junctions of guinea-pig placental capillaries; a possible structural basis for endothelial solute permeability. J Ultrastruct Res 1983; 85: 45-57.
77. Leach L, Firth JA. Fine structure of the paracellular junctions of terminal villous capillaries in the perfused human placenta. Cell Tissue Res 1992; 268:447-452.
78. Schultze C, Firth JA. Immunohistochemical localization of adherens junction components in blood-brain barrier microvessels of the rat. J Cell Sci 1993; 104:773-782.
79. Abbott NJ, Lane NJ, Bundgaard M. A fibre matrix model for the restrictive junction of the blood-brain barrier in a cephalopod mollusc: implications for capillary and epithelial permeability. J Neurocytol 1992; 21:309-311.
80. Abbott NJ, Bundgaard M, Lane NJ, Hollgard K. Parallels between junctions in invertebrate brain and embryonic mammalian brain. Proc Physiol Soc 1988; 400:72 P.
81. Bryant PS, Watson KL, Justice RW, Woods DF. Tumor suppressor genes encoding proteins required for cell interactions and signal transduction in *Drosophila*. Development 1993; in press.
82. Musil LS, Goodenough DA. Multisubunit assembly of an integral plasma membrane channel protein, gap junction connexin43, occurs after exit from the ER. Cell 1993; 74:1065-1077.
83. Gumbiner B. Breaking through the tight junction barrier. J Cell Biol 1993; 123:1631-1633.
84. Tsukita Sh, Itoh M., Nagafuchi A, Yonemura S, Tsukita Sa. Submembranous junctional plaque proteins include potential tumor suppressor molecules. J Cell Biol 1993; 123:1049-1053.
85. Citi S. The molecular organization of tight junctions. J Cell Biol 1993; 121:485-90.
86. Woods DF, Bryant PJ. Apical junctions and cell signaling in epithelia. J Cell Sci 1993; Suppl. 17: 171-181.
87. Furuse M, Hirare T, Itoh M, Nagafuchi A, Yonemura S, Tsukita Sa, Tsukita Sh. Occludin: a novel integral membrane protein localization at tight junctions. J Cell Biol 1993; 123:1777-1788.
88. Fehon RG, Dawson A, Tsakonas A. The *Drosophila* homolog of membrane skeleton protein 4.1 is a component of septate junctions and is encoded by the *coracle* gene. 1993, in press.

CHAPTER 3

The Roles of Gap Junctions in Early Development

Catherine Leclerc

1. INTRODUCTION

Gap junctions are ubiquitous membrane structures, found in nearly all developing and fully differentiated tissues (review by Caveney[1]), forming channels in apposed plasma membrane regions of adjacent cells (see ref. 2-5 for reviews on structure and functional regulation of the gap junctional channels).

During early development, intercellular communication has been reported, either ultrastructurally, showing the presence of gap junctions, or experimentally, showing dye and/or electrical coupling, between blastomeres, without exception, in a number of embryos. In invertebrates, gap junctions were first detected in mollusc embryos. The early observations of Potter et al[6] in 1966 indicate that squid embryonic cells are all electrically coupled. Later, in the germ-layer stage of two squid (*Loligo pealei* and *Loligo vulgaris*), gap junctions have been found to be ubiquitously distributed throughout and between the germ layers.[7,8] In the mollusc gastropod *Patella* and *Lymnaea*, dye coupling through gap junctions was first observed after the fifth cleavage in *Patella* and the second in *Lymnaea*.[9-11] In echinoderm, ionic coupling has been reported in blastula (32-cell stage) of the starfish *Asterias*[12] while the sea urchin *Paracentrotus*, electron microscopy revealed that gap junctions first appear after the fourth cleavage, at the same time as mitosis become asynchronous.[13] In ascidian embryos (*Ciona* and *Phallusia*), all cells are ionically coupled through gap junctions from the first cleavage onward whereas dye coupling only occurs from the 16- or 32-cell stages, corresponding to blastula stages.[14,15] In vertebrate embryos, the presence of gap junctions has been detected by ionic coupling in blastula of the amphibian *Xenopus* embryo[16] as

Molecular Mechanisms of Epithelial Cell Junctions: From Development to Disease
Edited by Sandra Citi, M.D., Ph.D.;

well as in urodeles (*Triturus*).[17,18] In the teleost fish *Fundulus*, reaggregated blastoderm cells are linked via gap junctions[19] and in intact embryos, similarly to the squid embryo, blastoderm and yolk cells communicate through gap junctions. Uncoupling between these two layers occurs when gastrulation begins.[20] Similar results were obtained with the zebrafish, *Brachydanio*.[21,22] In early embryo of the chick[23] electrical coupling occurs in cells within the same tissue and between different tissues. In the mouse, the acquisition of functional coupling through gap junctions between blastomeres was first detected at the time of compaction.[24,25] However, in human embryos gap junctions were only first detected in late blastocyst stage.[26] Despite this large literature showing that in most cases, gap junctions are associated with special developmental events, little is yet known about the significance of their presence in embryonic development.

Gap junctional channels differ from other membrane channels in many ways. The relevant properties for a role of gap junctions in development are the following:

1. Communication through gap junctions admits passage between cells of cytoplasmic molecules, neutral or charged, without secretion into the extracellular space. The pore size is such that small metabolites (up to about 1.2 kDa[27]), including second messengers such as cAMP, amino acids, or small peptides may be transferred from cell to cell through gap junctions whereas proteins and nucleic acids are excluded. This means that the cells can exchange regulatory molecules involved in the control of cell proliferation and determination while they retain their individuality of differentiation and expression.
2. As will be detailed below, gap junctions appear at critical embryonic stages. However, the presence of intercellular communication through gap junctions immediately prior to and/or during developmental events only suggests that gap junctions might be implicated in development. A good example illustrating this statement is the experiment of Warner and Gurdon[28] during mesodermal induction in the blastula of *Xenopus*. Animal pole cells which normally give rise to ectodermal derivatives can be induced to form mesodermal derivatives, such as muscle, by contact with vegetal pole cells of mid-blastula embryos. Although gap junctional communication is present between animal and vegetal cells shortly after conjugation of the tissues, the suppression of intercellular coupling by injection of antibodies to the 27 kD gap junction protein,[29] does not inhibit the expression of a specific marker for muscle differentiation (the production of cardiac mRNAs) in antibodies-injected embryo.
3. In most species, as soon as gap junctions appear, the embryo initially forms a single gap junctional communication unit. With the progressive commitment of the blastomeres to their respective developmental fates, this single communication unit subdivides in smaller units. The appearance of such restriction in cell coupling, associated with cell determination raised the question of who comes first, cell determination or gap junctions?

According to the pattern of intercellular communication and the time of appearance of this coupling, gap junctions probably fulfill different task during development. One possible role for gap junctions is that the determination of a group of cells to a certain fate might arise from cell-to-cell communication within the group. This may restrict the transfer of developmentally important molecules or "morphogen" within a single yet undifferentiated compartment. Consequently, the communication compartment should match the development compartment. The existence of a single communication unit in early embryos may also allow blastomeres to exchange, through gap junctions, positional information with their neighbors. In such situations, gap junctions may be more directly implicated in pattern formation rather than in cell determination or differentiation.

Although numerous models have been studied, very few have unequivocally demonstrated a direct contribution of gap junctions in development. Therefore, in this chapter, I will only consider embryonic models in which cell-to-cell interactions through gap junctions have been shown to play a major role.

2. THE MOLLUSCAN MODEL

2.1 Development of Molluscan Embryo

Mollusc provide a suitable model to study the relation between the so-called communication compartment and the developmental compartment.

The spiral mode of cleavage which characterizes mollusc development is regular in both time and space, blastomeres positions are therefore constant from embryo to embryo. This allows precise cell lineage studies to trace back to the uncleaved egg, the origin of the different larval and adult structures.[30]

In order to understand the importance of the changes in gap junctional pattern in mollusc, the most relevant aspects of two gastropod embryos, the freshwater *Lymnaea* and the marine limpet *Patella*, will be briefly detailed, from early cleavage stages to trochophore larva.

Figure 3.3.1d gives a schematic representation of the trochophore larva. The larva can be subdivided in two main developmental domains. The pre-trochal (anterior) and the post-trochal (posterior) domains, separated from each other by the prototroch, a circular ring of larval cells which are ciliated and form the locomotory organ of the larva. The

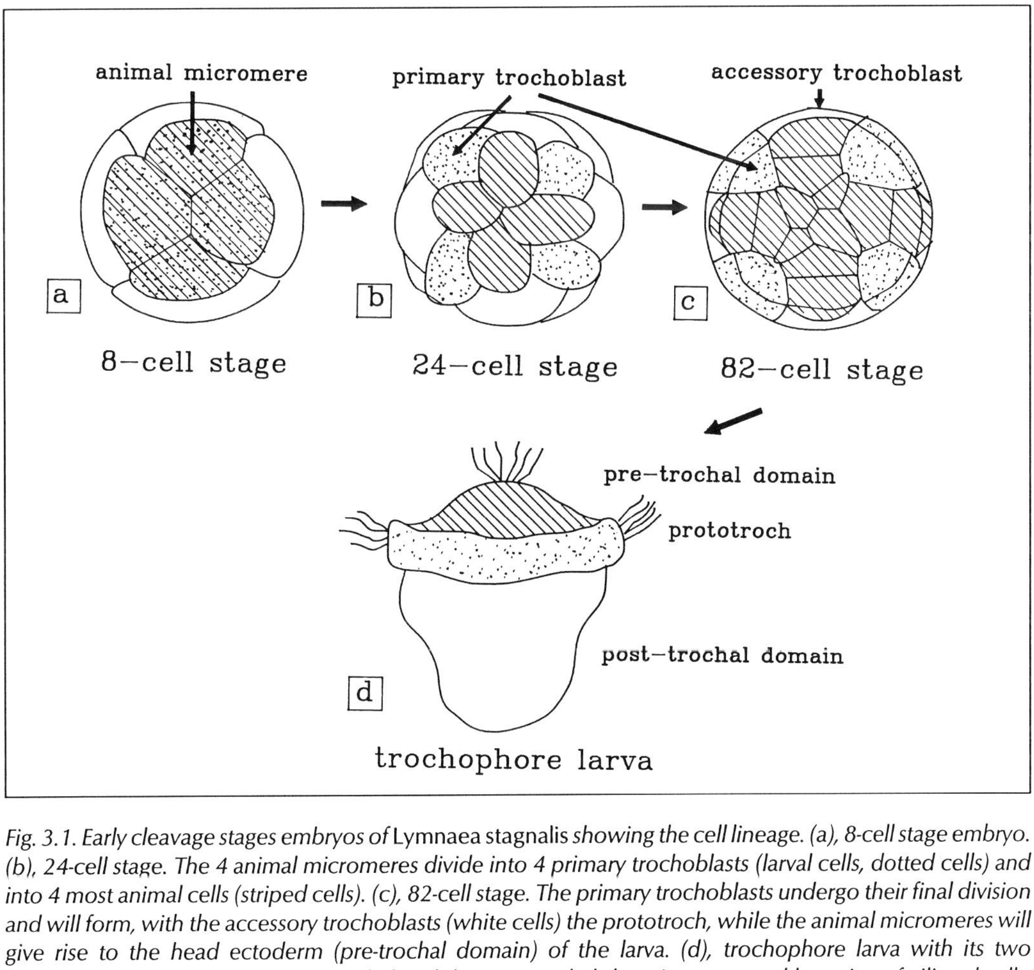

Fig. 3.1. Early cleavage stages embryos of Lymnaea stagnalis *showing the cell lineage. (a), 8-cell stage embryo. (b), 24-cell stage. The 4 animal micromeres divide into 4 primary trochoblasts (larval cells, dotted cells) and into 4 most animal cells (striped cells). (c), 82-cell stage. The primary trochoblasts undergo their final division and will form, with the accessory trochoblasts (white cells) the prototroch, while the animal micromeres will give rise to the head ectoderm (pre-trochal domain) of the larva. (d), trochophore larva with its two developmental domains, the pre-trochal and the post-trochal domains separated by a ring of ciliated cells, the prototroch.*

ectoderm of the trochophore is composed of two types of cells. Small proliferating cells from which will originate the adult ectoderm and large cell-cycle arrest cells, which constitute the larval structures, such as the prototroch and the apical organ cells, the larval sensory organ.

From cell lineage studies, it has been shown that the four animal micromeres of the 8-cell stage embryo will give rise to the pre-trochal ectoderm (Fig. 3.1a). At the next cleavage, these micromeres will divide into four animal micromeres, arranged in a cross pattern and into four primary trochoblasts displaying an interradial position (Fig. 3.1b, c). These trochoblasts will later participate in the formation of the prototroch.

2.2 Pattern of Dye Coupling in the Ectoderm of Lymnaea and Patella Embryos

In the embryos of *Lymnaea*, electron microscopy observations and iontophoretic injections of Lucifer Yellow give evidence that from the 4-cell up to the 24-cell stages, blastomeres are coupled through gap junctions.[31,32] In contrast, in *Patella* embryos, although electron microcopy study indicates the presence of gap junctions at the 2-cell stage,[33] dye coupling only appears at the 32-cell stage.[10,11,34] Whatever the stage of gap junctional communication appearance, in both species, the embryo initially forms a single communication compartment.

The first change in dye coupling pattern in both types of embryo appears after the sixth cleavage; i.e. at the 48-cell stage and 72-cell stage for *Lymnaea* and *Patella* respectively.[32,34] This corresponds to the apparition of a group of cells, the primary trochoblasts which are uncoupled whereas the other cells of the embryos are still dye coupled. At the following stage, the uncoupling of an other group of cells, the accessory trochoblasts will constitute, with the primary trochoblasts the ring of uncoupled prototroch cells (Fig. 3.1d). In *Lymnaea* and *Patella* this restriction in the dye coupling pattern will therefore result in the formation of two distinct communication compartments which match the two developmental compartments, the pre-trochal and the post-trochal domains.[32,35,36] At the gastrula stage, further restriction within each domains has been observed, again correlated with developmental compartmentalization.[35,36] The pre-trochal ectoderm becomes subdivided into two communication compartments strictly corresponding to the apical organ and to the head ectoderm. Similarly, communication restriction boundaries first subdivide the post-trochal ectoderm into two gap junctional communication units localized with the ventral and dorsal domains of the embryo respectively. Then, a further compartmentalization in dye spreading correlates with the mantle epithelium, the mantle edge domains for the dorsal side and with the foot and the ectoderm between the foot and the mantle domains for the ventral side. Despite the loss of dye coupling, electrical coupling is maintained between dorsal and ventral domains.[35,34] Conversely to the situation at the blastula stages, the restriction in dye coupling between dorsal and ventral cells of the post-trochal ectoderm arises without insertion between the two domains of cells with reduced coupling properties, limits of dye spreading are sharp.[15]

Figure 3.2 summarizes the changes in gap junctional communication pattern associated with the development of the embryos of *Lymnaea* and *Patella* up to the trochophore larva. It clearly shows the overlapping of the communication and the developmental compartments. These data clearly illustrate the following concept: Changes in intercellular communication coincide in most cases with important changes in embryonic developmental orientations and the progressive gap junctional uncoupling occurs between cells as their fates diverge. At this point however the question remains whether developmental domains are a cause or a consequence of the appearance of gap junctional communication units? To answer this question, I will now mentioned some precise developmental aspects of the pond snail *Lymnaea* embryo.

In this species, it is precisely known from cell lineages studies at which moment different sets of larval cells are determined.[30] At the 8-cell stage, none of the cells are committed to a particular fate and the embryo behaves as a single communication unit

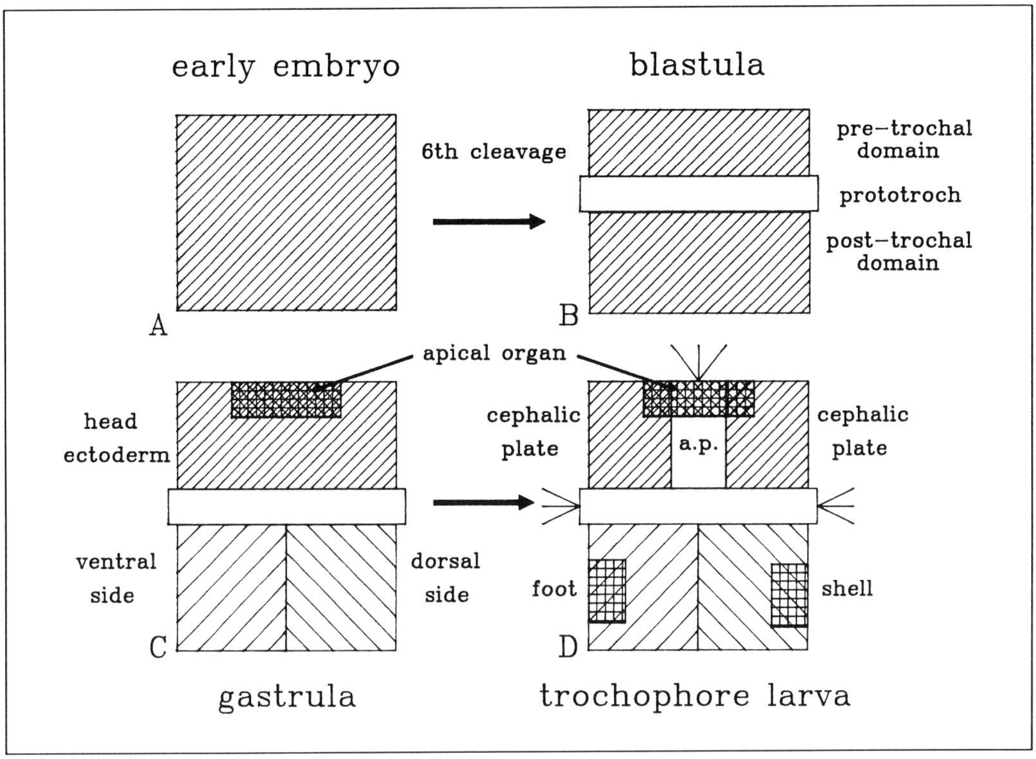

Fig. 3.2. Communication and developmental compartments in mollusc embryo, from early cleavage stages to trochophore larva. As commitment occurs, distinct developmental domains uncouple from each other and become separate communication units (rectangles with different patterns). Uncoupled larval cells have been represented by white rectangles. a.p., apical plate.

(Fig. 3.2A). After the fourth cleavage, the embryo reaches the 16-cell stage and the four primary trochoblasts are committed for larval structures while the other twelve cells still have multiple potentialities. However, the 16-cell stage embryo still constitutes one single compartment of communication. Therefore, determination of the primary trochoblasts precedes cell uncoupling. These cells become dye-uncoupled from each other and from the adjacent cells only at the 48-cell stage, after their last divisions and after having started their differentiation into larval cells. At the 48-cell stage, although the accessory trochoblast cells and the apical rosette cells are already fully committed and are not dividing further, they are still dye- coupled with the surrounding cells (Fig. 3.2B).[9] Again at this stage, cell determination occurs before change in dye coupling.

Despite the substantial work done on the mollusc embryos in order to prove the importance of gap junctions during embryogenesis, most of the evidence provided remains correlative.

With the two following models, the *Xenopus* embryo and the preimplantation mouse embryo, experiments on the morphogenetic effects of inhibition of gap junctional communication give proof for a direct role of gap junctions in development.

3. GAP JUNCTIONS IN AMPHIBIAN EMBRYOS

As in embryos of all other species so far studied, changes in pattern of gap junctions occurs during important embryonic events in amphibian embryos, from anouran and urodele species.

Electrophysiological studies reveal that between early stages (4-cell embryo) to mid-blastula (128-cell embryo), animal pole cells of *Xenopus* are electrically coupled.[16,37] However in *Xenopus*, dye coupling was not observed until the 16-cell stage.[38-40] In this embryo the pattern of cell to cell transfer of Lucifer Yellow was explored extensively from the 16-cell stage through to the 256-cell stage (early blastula).[40] At the 16-cell stage, the frequency of transfer of the dye in animal pole cells is highly asymmetric with respect to dorso-ventral polarity, being low at the future ventral side and high at the future dorsal side. No transfer was observed from animal to vegetal pole cells. This asymmetric pattern became more marked at the 32-cell stage (60% and 24% for dorsal and ventral cells respectively), it is lost when embryo reached 64 cells. At this stage the frequency of dye transfer is about 47% on both sides. At the 128-cell stage this value drops to 20% for dorsal cells and to 10% for ventral cells. In addition, transfer of dye between ventral and dorsal cells is rare. However, the asymmetry is restored at the 256-cell stage, with 65% and 25% of transfer for dorsal and ventral cells, respectively. These results indicates that the asymmetry in the pattern of gap junctional communication is correlated with the dorso-ventral axis of the embryo. However, it arises after specification of dorso-ventral polarity. In *Xenopus*, the dorsal side is marked at the 2- or 4-cell stages by the position of the gray crescent, appearing opposite the sperm entry point (review in Kao and Danilchick[41]). In addition, comparison of dye transfer pattern obtained with different probes (fluorescein, dicyanoargentate and lead EDTA) that permeate gap junctions, suggests that the difference observed between dorsal and ventral regions is due to difference in the selectivity of the channels rather than to more gap junctions in dorsal cell membranes.[40]

It has been known for a long time that cell-to-cell interactions play important roles in embryogenesis. Indeed, upon contact with some neighboring cells, at a precise time during development, a particular group of embryonic cells may be induced towards a specific developmental fate. Such interactions are called embryonic inductions and, probably the best known is the neural induction, most extensively studied in amphibians (for a review see Saxen[42]). During neural induction, a close contact is set between dorsal ectoderm cells and the underlying mesoderm cells, this contact is the trigger for ectoderm cells to differentiate into the nervous system. Dorsal ectoderm cells differentiate into epidermis instead of nervous system when culture isolated from the mesoderm inducing cells. What are the mechanisms involved in the exchange of information between embryonic cells and implicated in the control of cell fate in relation with the surrounding environment? Intercellular exchange of information through gap junctions may constitute one of the pathways involved in cell communication. Therefore, special attention has been devoted to changes in gap junctional pattern in relation with neural induction.

In *Rana pipiens* embryos, freeze fracture and thin sections observations indicate that during neurulation gap junctions progressively appear in the lateral cell membranes of the prospective neural ectoderm.[43] More recently, three mRNAs coding for gap junction proteins have been identified in the oocyte and early embryo of *Xenopus*. In vitro translation of the RNA or deduced amino acid sequences from the cDNAs revealed that these mRNAs code for three different connexins. Connexin30[44] showing 71% homology with rat or human liver connexin32; connexin43[45] with 87% homology with rat heart connexin43 and finally, a third one, connexin38[46] yet expressed only in oocytes and early embryos of *Xenopus*, which shows 32 and 41% homology to rat connexin32 and connexin43 respectively. Moreover, developmental studies indicate that the level of these mRNAs are differentially regulated. Connexin38 mRNA is a maternal transcript; detected in early oocytes through late oogenesis, its level gradually decreased through blastula and gastrula stages and it is absent by early neurula stage. On the other hand, connexin30 mRNA follows a totally inverted profile, it first appears at early neurula.[44-46] The pattern for connexin43 mRNA is more complex, it is present in early and late oocyte and during tailbud stages while it is undetectable from unfertilized egg up to neurula embryo.[45]

However, none of these results give more than strong correlations indicating that gap junctions have crucial implications in neural induction.

The first attempt to elucidate the role of intercellular communication in early development was performed in *Xenopus* embryos with antibodies raised against the 27 kD rat liver gap junctional protein (connexin32). These antibodies were injected into a dorsal well-communicating blastomere of an 8-cell stage embryo.[29] Such injections reduce the transfer of Lucifer Yellow in dorsal cells of 32-cell stage embryos, i.e. after two additional cleavages. In antibodies-injected embryos, failure to transfer to cells other than the sister cell was observed in about 70% to 80% of the tested embryos, while it was at most 30% in control uninjected embryos or injected with pre-immune serum. In addition, failure to transfer Lucifer Yellow was accompanied by inhibition of electrical coupling. Restriction of cell-to-cell communication causes defects in patterning in tadpole stages in relation with the developmental fate of the injected cell. Cells differentiate, but the structures resulting from the progeny of the antibodies-injected cell in tadpole stages (trigeminal ganglion, otic vesicles, eye and part of the brain) were abnormally placed relative to their normal position along the anteroposterior axis. In the most severe cases, these structures were totally absent. These results suggest that gap junctions play a role in pattern formation rather than in cell determination or differentiation. Other experiments performed on amphibian embryos further support this view.

In the urodeles embryos *Ambystoma mexicanum*[47] and *Cynops pyrrhogaster*[48] electrical coupling has been reported between lateral ectoderm and neural plate cells. These cells uncoupled at the time of neural tube closure. In both species, it appears that cells from lateral ectoderm and neural plate are still ionically coupled although they already display distinct membranes properties, strongly suggesting that like in mollusc, cell determination occurs prior to change in gap junctional coupling.

The next question is whether the strong dorso-ventral polarity in gap junctional communication in the 32-cell embryo observed by Guthrie et al[40] reflects the future dorso-ventral organization of the embryo. The following experiments support this relationship.

Mesoderm induction in amphibian embryo has been the subject of extensive studies. Mesoderm results from interactions between the vegetal pole cells and the adjacent equatorial animal pole cells. Furthermore, mesoderm induction is linked to the establishment of the dorso-ventral axis (for reviews see Smith[49] and Dawid[50]). Indeed, dorsal mesoderm will differentiate into notochord and muscle while ventral mesoderm will give rise to blood, mesenchyme and mesothelium. Perturbation in the dorso-ventral pattern of the embryo can be obtained by several means. UV irradiation of fertilized egg is known to enhance ventral structures at the expense of dorsal ones, while lithium treatment of embryos between the 2-cell and the 32-cell-stages produces the opposite effects, enhancing dorsal structures.[51,52] On the other hand, the induction of secondary dorsal structures can be obtained by ectopic expression of the TGF-B related factor, activin B and of some *Wnts* (review in Smith[49] and Moon[53]).

Table 3.1 summarizes the effects of such treatments on gap junctional communication measured in 32-cell-stage *Xenopus* embryos. Treatments able to induce secondary dorsal axis abolish the asymmetry of dye transfer between dorsal and ventral regions and bring up ventral cells at the level of transfer characteristic of dorsal cells.[54-56] Conversely, UV-irradiation leads both to the development of embryos with little or no dorsal axial structures and to a reduction in dye transfer in dorsal regions.[54] Moreover, ectopic expression of bFGF or *XWnt*-5A have no effect on gap junctional communication nor on development of the dorso-ventral axis.[55,56]

Altogether these data strongly support the idea that the existence of regions with different gap junctional communication

Table 3.1

Treatment	Effect on development	Effect on gap junction
Lithium[54] Activin B[56] *Wnt*-1[55,56] *XWnt*-8[55]	enhances dorsal structures in ventral side	abolishes asymmetry increases coupling
UV[54]	enhances ventral structures in dorsal side	abolishes asymmetry decrease coupling
bFGF[56] *XWnt*-5A[55]	no effect on embryonic axis	no effect

properties is needed in order to stabilize in the right position, the regions with dorsal and ventral fate. However, according to the results reported by Guthrie et al,[40] there is no evidence to suppose that the dorsal and ventral sides of the amphibian embryo correspond to communication compartments similar to those found in mollusc embryos[57] and in insects.[58,59]

4. GAP JUNCTIONS IN MOUSE PREIMPLANTATION EMBRYO

4.1 Time course of gap junctional communication appearance in the mouse embryo

In the preimplantation mouse embryo, a temporal correlation between the apparition of functional gap junctional channels and compaction of the blastomeres at the 8-cell stage has been found.[25,60-62]

The 8-cell stage is a crucial stage since it is the starting point of the first differentiation event in mammalian development. Prior to this stage, all the blastomeres seem to retain the same developmental potency.[63] However, after this stage, cellular relations and communications undergo profound modifications. The eight blastomeres reduce their cellular contact to form a compact mass of cells (Fig. 3.3). This phenomenom, called compaction, indeed consists of an extensive reorganization of both the cell surface and cytoplasmic components of the embryo, resulting at the 16-cell stage in the formation of two distinct subpopulations of cell (review in Johnson[64]); a population of large polar cells and a population of small apolar cells arise, the polar ones being located on the outside of the compacted mass, while the apolar are located inside. This situation further leads at the blastocyst stage to the differentiation of the two first cell lineages: The outer layer (the trophectoderm) and the inner layer (the inner cell mass, ICM). The ICM then differentiates into the primitive ectoderm or epiblast from which will develop the fetus and into a peripheral layer, the primitive endoderm or hypoblast. The trophectoderm and the hypoblast form the basis of the placenta and the extra-embryonic yolk sac, required for successful interaction with the mother's uterine membrane.

From the 8-cell stage onwards, blastomeres of the preimplantation mouse embryo communicate through gap junctions. Ultrastructural studies indicate that gap junctions formation in mouse embryos take place in association with compaction and tight junction formation.[24,65,66] Gap junctional plaques have been observed in freeze-fracture replicas of mouse embryos from the time of compaction through to the blastocyst stage. Whereas tight junctions (review in Fleming et al[67]) were only found in outside blastomeres, gap junctions were detected outside as well as inside blastomeres of morula.[66] Electrophysiological studies provide further

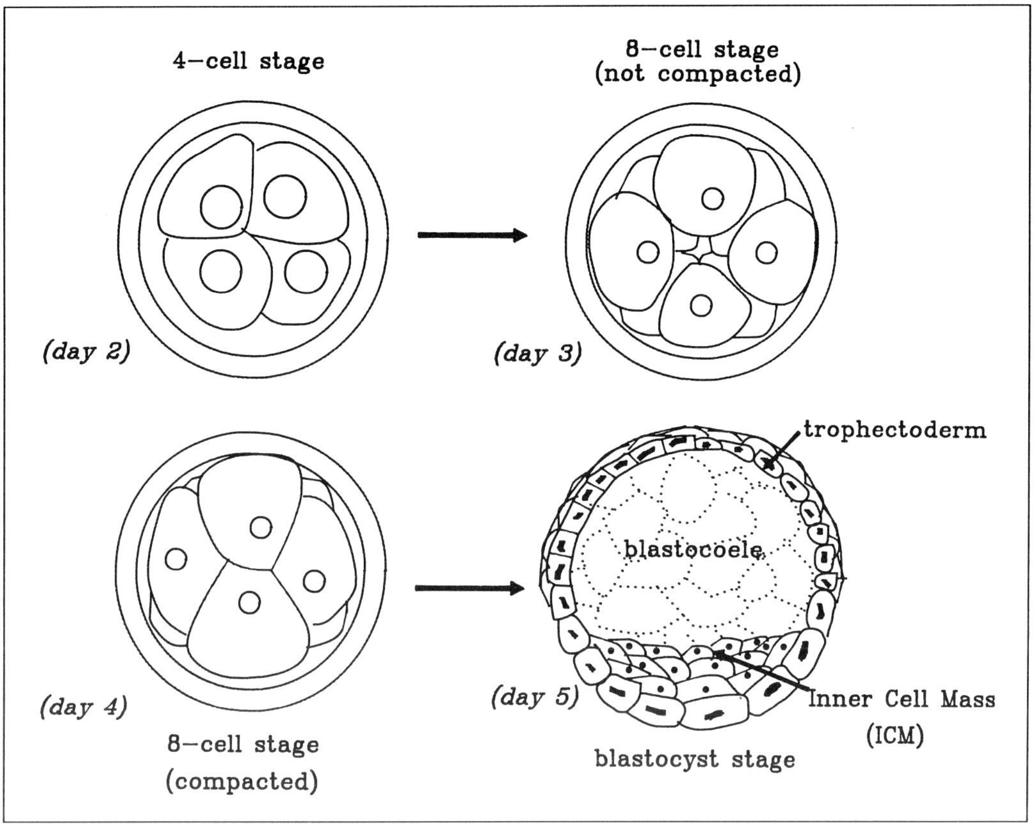

Fig. 3.3. Diagram of the mouse preimplantation developmental stages, from the 4-cell to the blastocyst stages. Within a few hours of the formation of the 8-cell stage, on day 4 after successful insemination, the embryo changes from a loose aggregate of blastomeres into a compact mass, the morula. This process is called compaction. As cleavage proceeds, on day 5, the embryo acquires a fluid-filled cavity, the blastocele. The blastocyst consists of an outer layer, the trophectoderm, surrounding a small cluster of cells at one pole, the inner cell mass (ICM).

evidence that gap junctional communication is acquired prior to the fully compacted stage, about three hours after completion of the third cleavage. The extent of gap junctional communication was measured by monitoring the presence of ionic coupling and low-molecular weight dyes (Lucifer Yellow) transfer between cells of early mouse embryos, from the 2-cell to the blastocyst stages. The early mouse embryo does not possess functional gap junctional channels until early compaction, i.e. at the late 8-cell stage. At early stages (2 cell and 4 cell), cytoplasmic bridges were responsible for ionic and dye coupling of sister blastomeres.[25,60]

With the emergence of antibodies raised against gap junction proteins (for details about gap junction protein structure and regulation, see chapters 10 and 11 in this volume), new tools became available for detailed studies of the pattern of gap junction communication in relation with compaction in the preimplantation mouse embryo.

Gap junctional channels were detected by both western blot analysis and by laser confocal microscopy using affinity purified antibodies prepared against synthetic peptides corresponding to different intra-cytoplasmic domains of connexin43 protein. With such peptide antibodies, plaque-like structures resembling gap junctions could only be detected, between cells at points of contacts, at the late 8-cell stage, after the onset of compaction.[68-71] Figure 3.4 shows

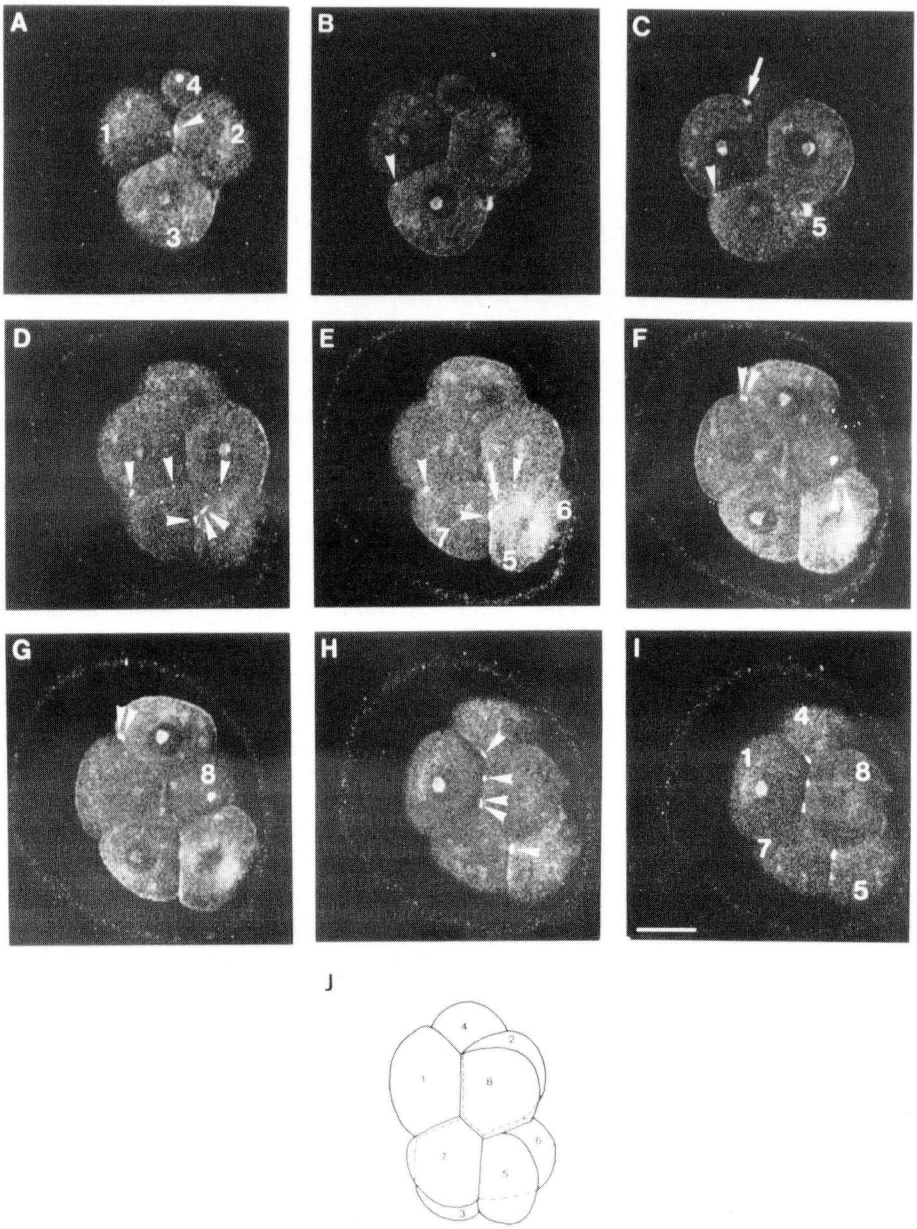

Fig. 3.4. Localization of gap junction protein using antibodies to peptide HJ[98] on methanol-fixed mouse embryos. The successive optical sections give indication of the number and the distribution of gap junctions which can be present in an 8-cell stage embryo. Eighteen gap junctions could be counted. The blastomeres are numbered from 1 to 8 according to appearance on the successive sections (diagram in J). Gap junctions in embryo were viewed with a Bio-Rad MRC 500 laser scanning confocal microscope. Scale bar = 25 μm.

that at the 8-cell stage gap junctions are present at point of contact between membranes of all blastomeres. As compaction progress during the 8-cell stage, adjacent cell membranes flatten against each other and the gap junctional plaques increase in both size and number. Figure 3.5 gives histogram plots of the length of gap junctions in 14 embryos at early and late compacted 8-cell stages with median value of 1.2 and 1.4 µm respectively. In contrast, in 16-cell stage embryo, whereas the number of junctions increases, plaque-like foci become smaller and closer together, therefore creating a pattern lining-up the blastomeres.[70] As development progresses from late morula through to blastocyst stages, connexin43 immunoreactivity becomes differentially distributed among blastomeres with different developmental fates. In late morula, outer blastomeres which will give rise to trophe-ctodermal cells, exhibit a zonular array of connexin43 staining, punctuated by plaque-like foci. Such zonular pattern is not observed in inner blastomeres from which will form the embryo proper, which showed only punctate junctions.[69-71] Such differential staining pattern of connexin43 persists in blastocysts between trophectoderm and ICM cells. However, blastomeres at the blastocyst stage, are ionically and dye coupled and both trophoblasts and cells of the ICM are linked through gap junctions to one another.[25] In addition, with some of the specific synthetic peptide antibodies, connexin43 can be detected in the cytoplasm of both 4-cell embryos and morula.[69,71] Cytoplasmic staining is also observed in 8-cell stage embryos. By northern blot and RT/PCR analysis of gap junction mRNA,

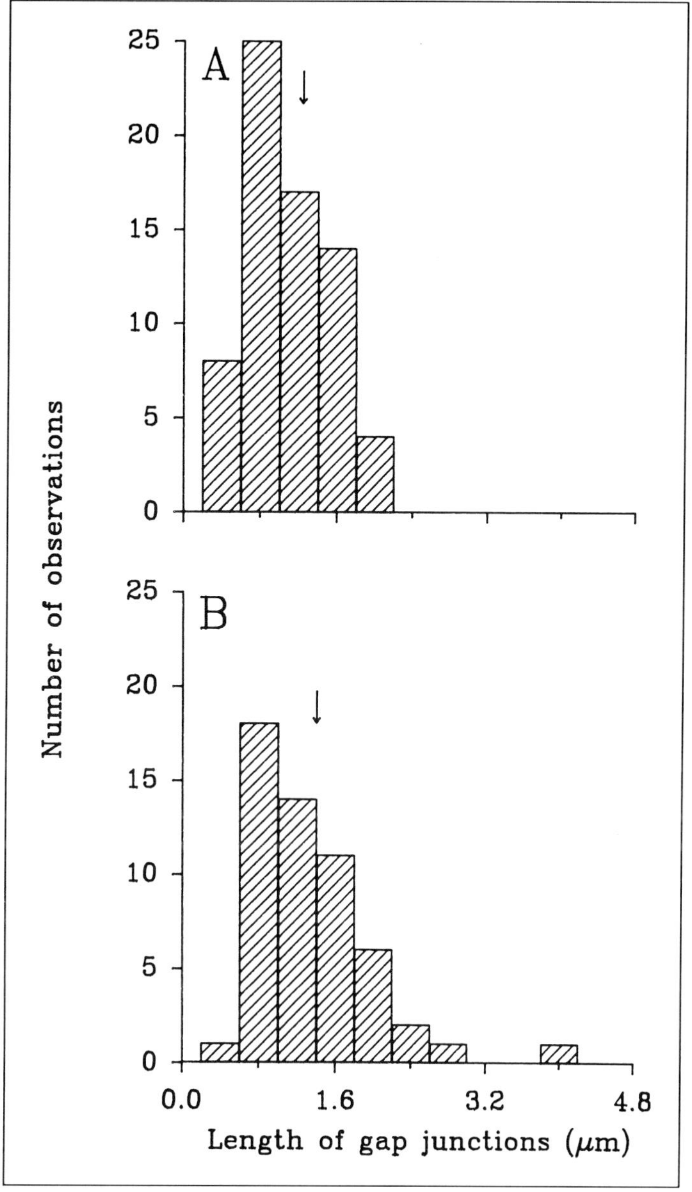

Fig. 3.5. Comparison of the length of gap junctions in 14 normal embryos of early and late 8-cell stages. Ordinate, number of observations; abscissa; size of the fluorescent spot (in µm) corresponding to specific immunolabelling of the gap junction protein with anti-peptide HJ (see legend of Fig. 3.3). A. 8-cell stage embryos at the onset of compaction. B. 8-cell stage, fully compacted embryos. The arrows indicate the median of each distribution, 1.2 and 1.4 µm for early and late 8-cell stages embryos respectively.

connexin43 transcripts were first detected at the 4-cell stage, the signal per embryo then increases through to the blastocyst stage.

It thus appears that connexin43 protein can be detected at early stages, first in cytoplasm of 4-cell embryos then in both the cytoplasm and the plasma membrane at intercellular contacts from the 8-cell stage onwards. In addition, it is the only gap junctional protein present during preimplantation stages. The expression of connexin26, 32 and 46 genes has been examined in preimplantation stages, none has been found to be transcribed or to contribute connexins for the formation of gap junctional plaques.[68,69,72] Detection of connexin43 with antipeptide antibodies along with the finding that connexin43 mRNA is present and translated from the 4-cell stage indeed demonstrate that cytoplasmic pool exists at this early stage. This pool of nascent connexin43 will then be used to form gap junctional plaques in plasma membranes.

What triggers the initiation of gap junction assembly in the 8-cell stage mouse embryo?

Experiments with inhibitors of transcription and protein synthesis have demonstrated that junction assembly does not depend neither upon transcription of connexin43 genes nor upon the translation of its mRNA.[62,73] However, continued gene expression is required for the full extent of junctional coupling to be established from the 8-cell compacted stage onwards. Indeed, McLachlin and Kidder[73] found that unlike ionic coupling, dye coupling was weak or undetectable in the 8-cell compacted embryos treated with either inhibitors from the 4-cell stage. In addition, the onset of gap junctional coupling at the 8-cell stage is not linked to cell flattening itself nor to cytokinesis or mitosis. It does not require either intact networks of microtubules or microfilaments since the assembly of gap junctional channels arises even in presence of microfilaments and microtubule poisons.[74] Recently, De Sousa et al[71] have shown that the appearance, at the late 8-cell stage, of intercellular coupling through gap junction involves intracellular trafficking. Monensin and brefeldin-A that block such traffic interfere with the acquisition of dye coupling by preventing nascent connexin43 to be transferred from intracellular membrane to the plasma membrane to form gap junctional plaques. However, although precursors are present in 4-cell embryos, premature gap junctional channel formation was never obtained when 2-cell or 4-cell embryos were aggregated with communication-competent 8-cell embryos.[62] It seems therefore that the appearance of intercellular communication through gap junctions in mouse embryo follows a precise timing depending on the developmental age of the embryo. This behavior raised the question of the role of cell-to-cell communication during this crucial step of development.

4.2 Role of Gap Junction in Early Mouse Development

While the role of tight junctions and desmosomes in blastocele formation and expansion is well documented,[67,75] the role of gap junctions in preimplantation mouse embryo development is not so clear. Indeed, in the mouse embryo, cell polarization and compaction can occur in the absence of intercellular communication.[76] Anyhow, recent studies provide direct evidence for a major contribution of cell-to-cell communication during preimplantation development.

Inhibition of gap junctional communication by intracellular injection of antibodies raised against the major 27/32 KD rat liver gap junctional protein[77] or by injection of antisense RNA to the same gap junctional protein[78] leads to block of the intercellular communication. As a consequence, the communication-incompetent blastomeres are unable to take part in compaction and the embryos do not form normal blastocysts. From these experiments, it has been proposed by Lee et al[77] that gap junctions are in fact not involved in the initiation of compaction but in its maintenance. At this point, it should be stressed that the maintenance of a compacted state in the mouse embryo is a crucial step for the formation of the blastocyst and therefore for the differentiation of the first two cell lineage: trophectoderm and ICM.

The second body of evidence arises from a study of DDK-defective embryos. The reproductive physiology of DDK females is characterized by semisterility when mated with males from alien strains. Normal fertility occurs in the intrastrain mating of DDK females with DDK males and in interstrain mating of DDK males with females from alien strains.[79] DDK eggs fertilized by alien sperm (C3H strain) decompact toward the late 16-cell stage and fail to form blastocysts, only 5% to 10% survive to implant

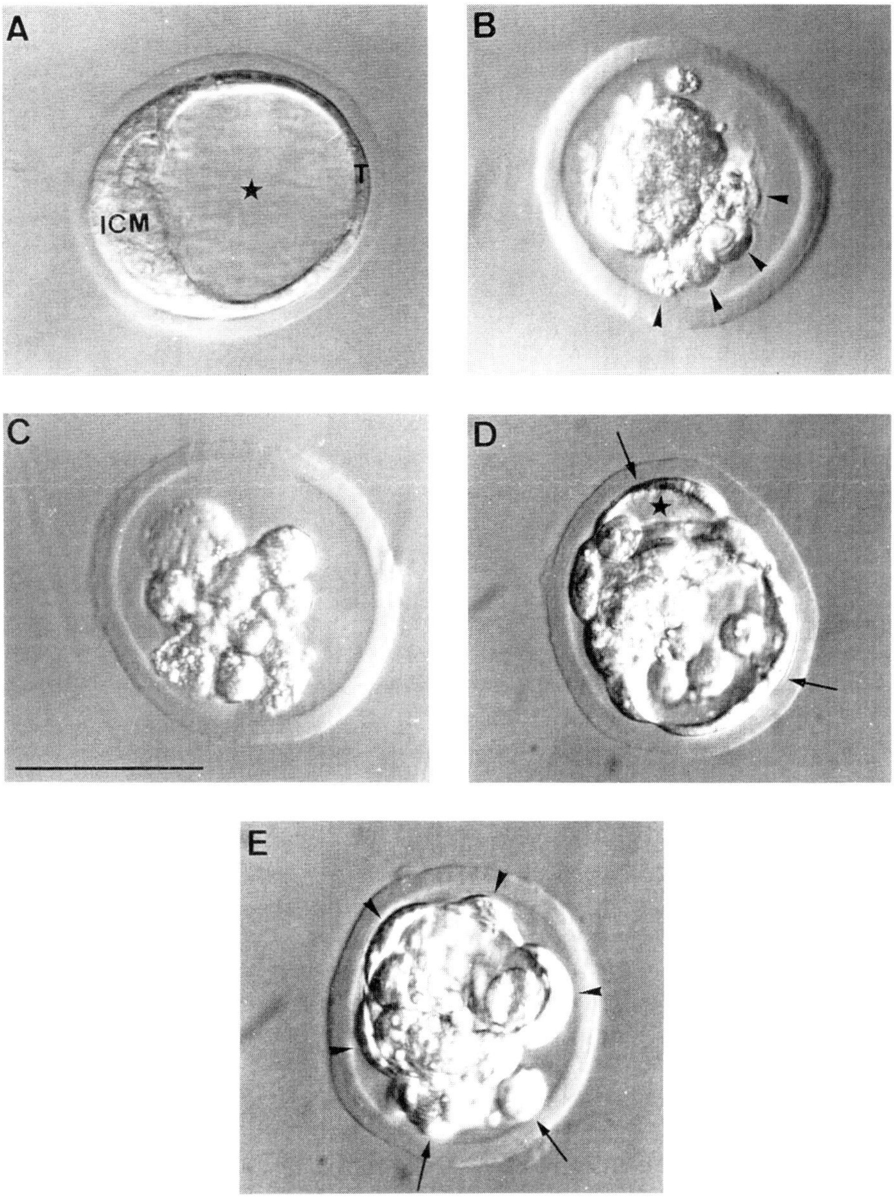

Fig. 3.6. Morphological appearance of in vitro cultured morula derived from the cross of a DDK female with a C3H/Bi male. (A.) normal expanded blastocyst; T, trophectoderm; ICM, inner cell mass; the star indicates the blastocyst cavity. (B-E.) abnormal morula showing diverse degrees of decompaction and abnormal blastocysts. (B.) several blastomeres (arrows) being completely extruded from the resting part of the morula. (C.) degenerated embryo with all blastomeres decompacted. (D.) abnormal blastocyst with two cavities (arrows). One cavity is formed by only one cell (star). (E.) abnormal blastocyst with four cells (arrowheads) which start to accumulate fluid, two decompacted blastomeres (arrows) are clearly visible. Scale bar = 80 μm.

and develop to term (Fig. 3.6). Embryological observations have shown that the defective DDK/C3H embryos are characterized by the absence of trophoblast cells. The underlying cause of this defect in trophoblast formation seems to be due to incompatibility between the egg cytoplasm and the female pronucleus.[80,81]

Concerning intercellular communication through gap junctions, preimplantation mouse embryos, along with the DDK defective embryos provide a unique and simple model in which to observe the developmental consequences of drugs that influence gap junction communication. Indeed, as it will be illustrated below, the modulation of gap junctional conductance has a direct effect on compaction and hence, on the formation of the blastocyst.

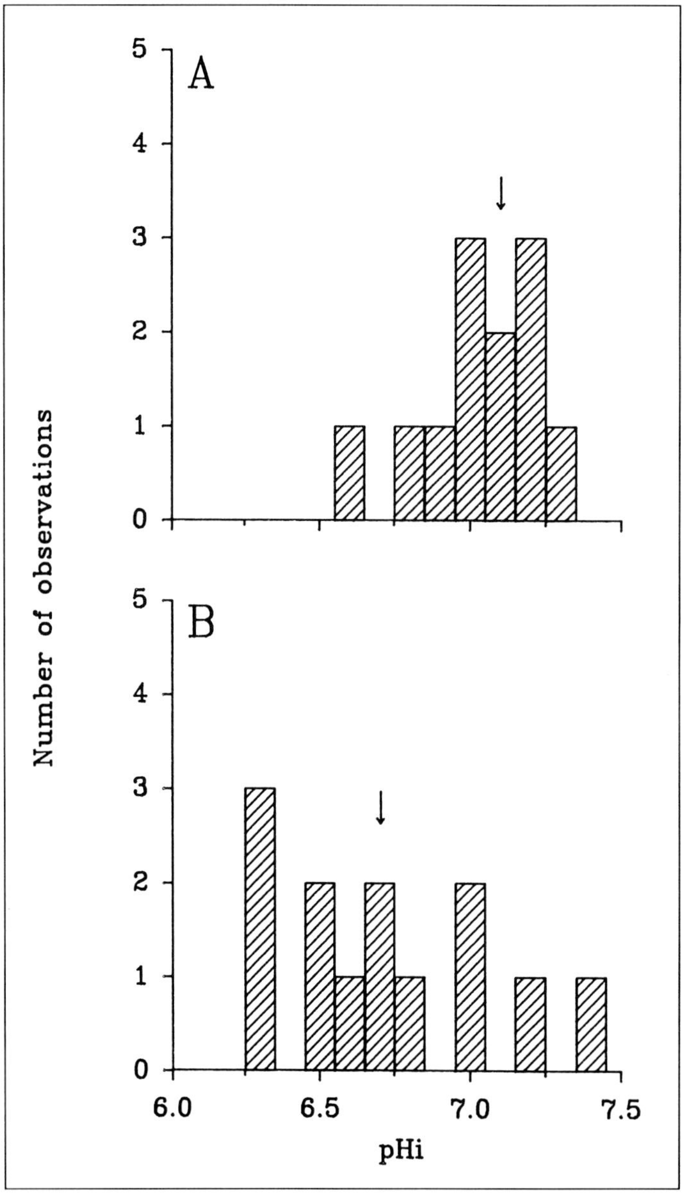

Fig. 3.7. Histogram plots of the intracellular pH_i measure with the pH-sensitive fluorescent probe BCECF, in 8-cell stage embryos. A, pH_i obtained on control DDK/DDK and $C57Bl_{10}/Balb_c$. B. intracellular pH recorded on defective DDK/C3H embryos. The arrows indicate the median of each distribution, 7.1 and 6.7 for control and DDK/C3H embryos respectively.

First, an association has been found between decompaction of the DDK/C3H embryos at the late 16-cell stage and both poor gap junctional communication, as measured by Lucifer Yellow transfer at the 8-cell compacted stage and low intracellular pH at the same stage, calculated with the pH-sensitive dye BCECF[82] (and Leclerc-David et al in preparation). In the DDK/C3H embryos at the 8-cell compacted stage the defect is not yet apparent however, the distribution of dye transfer times is already significantly different from that of the DDK/DDK embryos. The median is 139 and 92 seconds for DDK/C3H and DDK/DDK embryos, respectively. Similarly at this stage, a significant difference in intracellular pH (pHi) has been calculated. Normal DDK/DDK embryos have a pH_i close to pH 7.0 with 8% (n = 12) scored with a pHi of less than 6.8 whereas mean pHi in DDK/C3H embryos is 6.7 with 62% (n = 13) scored with a pHi of less than 6.8 (Fig. 3.7). This low value could account for the reduction of dye transfer through gap junctions observed within the DDK/C3H population.

Second, an increase in gap junction conductance can be achieved either by raising intracellular pH with a weak base (methylamine) or by increasing intracellular levels of cyclic-AMP. This increase in cyclic-AMP was achieved by using either the membrane permeant derivative of cyclic-AMP or by stimulating the adenylate cyclase with forskolin. In such conditions, incubation of the DDK/C3H embryos during the crucial period (i.e. between the 8-cell and the 16-cell stages) in solution containing methylamine (10 to 20 mM) or cyclic-AMP (1 mM) or even forskolin (100 µM) produce both a significant increase in gap junction permeation (dye transfer times became comparable to control DDK/DDK) and prevent the expression of the lethal defect in 40% and 50% to 55% of the DDK/C3H embryos treated with methylamine and forskolin respectively. These embryos develop normally up to the blastocyst stage[82] (and Leclerc-David et al in preparation). However, the effect of both treatments do not persist much beyond the expanded blastocyst stage.

Alternatively, in normal developing DDK/DDK embryos, a decrease in gap junction conductance produced by incubation in 10 mM butyric acid was able to induce defects similar to the DDK/C3H syndrome, where partial or total decompaction of the blastomeres and vesicles like blastocel formation occurs. While 91% of the control embryos formed normal blastocysts, 40% to 70% of the treated embryos failed to maintain compaction and did not reach the expanded blastocyst stage[82,83] (and Leclerc-David et al, in preparation). Finally, the pattern of gap junctions distribution in the defective DDK/C3H embryos also differ from that of normal DDK/DDK embryos. Whereas in control embryos at the 16-cell stage, the outermost cells develop belt gap junctions between them, in DDK/C3H embryos it only forms punctuate spots distributed randomly along the intercellular opposing membranes.[70] Furthermore, the treatment of normal DDK/DDK embryos with the weak acid butyrate mimicked the DDK syndrome in every characteristic. In addition to slow dye transfer and decompaction, the distribution of gap junctions matched that observed in the DDK/C3H-defective population.[70]

Besides an insight into the mechanism involved in the DDK syndrome, which will not be discussed below and from a general point of view, the results presented above indicate that the integrity of the gap junctional intercellular communication is an essential requirement in order to maintain a compacted organization of the mouse blastomeres, up to the onset of cavitation. This compacted state is necessary to allow the divergence of two distinct cell lineages: The placental (originating from the outer polar cells) and the fetal (originating from the inner apolar cells) lineage.

What is the precise role played by gap junctions in this divergence?

Cells need to communicate in order to specify their respective positional situation. In that respect, in preimplantation mouse development, it seems unlikely that gap junctions are directly involved in cell determination. The absence of trophoblast differentiation in the DDK-defective embryos is probably an indirect consequence of the impaired cell-to-cell coupling. In the preimplantation mouse embryo, the time at which the positional determination of cell fate occurs has been determined. At the 8-cell stage, individual blastomeres can still give rise to both ICM and trophectoderm derivatives.[63,84] However, at the fully expanded blastocyst, both ICM and trophectoderm cells are committed to their respective fates.[85-87] These findings can be correlated with the observation of the pattern of cell-to-cell coupling. Gap junctional intercellular communication is maintained between cells (ICM and trophectoderm) until the fully expanded blastocyst and progressive uncoupling occurs after hatching, first between trophectoderm and ICM then between ICM cells themselves. In the postimplantation embryo the pattern of gap junctional communication has been studied on in vitro implanted embryos for up to 72 hours, corresponding to late day 8 in vivo embryos. Although both trophoblasts and ICM cells remains ionically coupled throughout day 8, a progressive restriction of the extent of dye coupling between both cell layers has been found. In the most developed embryos (late day 7 or day 8), the giant trophoblast cells are neither dye coupled to ICM cells nor to each other. In addition, ICM cells at this late stage reveals also a further compartmentalization which may correspond to the cell lineages arising from the ICM.[88] Therefore gap junctional uncoupling arises long after cell determination and even differentiation.

5. CONCLUSION

In molluscan embryos, characterized by determinative development, as well as in embryos with regulative development such as amphibians and mammals, it thus appears that direct cell-to-cell communication through gap junctions is involved in global patterning rather than in cell determination or differentiation.

How can gap junctions participate in this process of patterning? It may be assumed that gap junctions are involved in the transmission of "signal" molecule(s) to all blastomeres, a signal that can specify the positional information required for cell commitment. To come back to the mouse embryo, failure in the transmission of this signal to all blastomeres will lead to decompaction and consequently absence of cells determination and later to differentiation. This is the sequence of events observed in the DDK-defective embryos. Such a developmentally important signal molecule has been isolated in the fresh water coelenterate *Hydra.*[89] In this animal, the body plan is very simple. It is composed of a two-layered epithelial cylinder whose cells communicate through gap junctions,[90] with a head and a foot at opposite ends. The signal or morphogen is a small peptide molecule of 11 amino acids, small enough to diffuse from cell to cell through gap junctions. This molecule is produced by the existing head to prevent the formation of a secondary one in the anterior region. In addition, it is distributed as a monotic gradient, termed the head inhibition gradient, down the body axis. Selective disruption of gap junctional communication with antibodies raised against the rat liver gap junction protein (connexin32) perturbs the head inhibition gradient and, as a result an additional head is formed close to the first one.

In the preimplantation mouse embryo, gap junctions may fulfill a similar role in signal transmission. Indeed, in spite of ionic coupling, the differential gap junctional staining pattern observed between outside and inside blastomeres in late morula and between ICM and trophectoderm in blastocyst stages may also serve to restrict the diffusion of the signal molecule(s). This therefore would create not a gradient as in *Hydra*, but a differential concentration of such molecules between polar and apolar cells, setting up by this means the positional information required for the divergence of the placental and the fetal lineages. However, in the mouse embryo and in other embryos, signal molecules of this kind are still to be found.

In insects, the imaginal wing disc of *Drosophila* illustrates an other aspect of the role of gap junctions in development. Wild-type imaginal discs have numerous gap junctions, their distribution is non-random and changes during development. As shown by Ryerse and Nagel,[91] the number of gap junctions increases in association with cell proliferation throughout the third larval stage. This increase continues in early pupal stage while morphogenesis and cell differentiation occur. Data from *Drosophila* wing disc mutants yield further information. Particularly, in the $C43^{hs1}$ mutant, a significant reduction in the number of gap junctions associated with a functional defect of the gap junctional channels (dye coupling) has been correlated with disc overgrowth at developmental stages where cell proliferation slows down in wild-type *Drosophila.*[92,93] Furthermore, morphological observations indicate that the onset of tissue overgrowth occurs after reductions in gap junctions.[94] These data raise the possibility that during embryonic development, in addition to a control of pattern formation, gap junctions may be involved in the control of cell proliferation. Indeed, recent experiments demonstrate that an increase in intercellular communication through gap junctions, by transfection of tumor cells with cDNA encoding connexins, leads to inhibition of the growth rate of tumor cells in vitro[95,96] and to suppression of the tumor in vivo.[96,97]

ACKNOWLEDGMENTS

The work on the mouse embryos was performed in Professor A.E. Warner laboratory (UCL, London) and funded by the Wsllcome Trust. I am grateful to C. Babinet (Institut Pasteur, Paris) for providing DDK embryos and the hospitality of his laboratory.

REFERENCES

1. Caveney S. The role of gap junctions in development. Ann Rev Physiol 1985; 47:319-335.
2. Beyer EC, Goodenough DA, Paul DL. The connexins, a family of related gap junctions proteins. In: Hertzberg EL, Johnson RG, ed. Gap junctions. New York: Alan R Liss, 1988:167-175.
3. Beyer EC, Paul DL, Goodenough DA. Connexin family of gap junctions. J Membrane Biol 1990; 116:187-194.
4. Bennett MVL, Barrio LC, Bargiello TA, Spray DC, Hertzberg E, Saez JC. Gap junctions: new tools, new answers, new questions. Neuron 1991; 6:305-320.
5. Saez JC, Berthoud VM, Moreno AP, Spray DC. Multiplicity of controls in differentiated and undifferentiated cells and possible functional implications. In: Shenolikar S, Nairn AC, eds. Advances in Second Messenger and Phosphoprotein Research. Vol 27. New York: Raven Press, 1993:163-198.
6. Potter DD, Furshpan EJ, Lennox ES. Connections between cells of the developing squid as revealed by electrophysiological methods. Proc Natl Acad Sci (USA) 1966; 55:328-335.
7. Ginzberg RD, Morales EA, Spray DC, Bennett MV. Cell junctions in early embryos of squid (*Loligo pealei*). Cell Tissue Res 1985; 239:477-484.
8. Marthy HJ, Dale B. Dye-coupling in the early squid embryo. Roux's Arch Dev Biol 1989; 198:211-218.
9. Van den Biggelaar JAM, Serras F. Determinative decisions and dye coupling changes in the molluscan embryo. In: Modern Cell Biology Vol 7. Gap Junctions. New York: Alan R. Liss, 1988:483-493.
10. De Laat SW, Tertoolen LGJ, Dorresteijn AWC, Van den Biggelaar JA. Intercellular communication patterns are involved in cell determination in early molluscan development. Nature 1980; 287:546-548.
11. Dorresteijn AWC, Wagemaker HA, De Laat SW et al. Dye-coupling between blastomeres in early embryos of *Patella vulgata* (Mollusca, Gastropoda): Its relevance for cell determination. Roux's Arch Dev Biol 1983; 192:262-269.
12. Tuppers JT, Saunders JW, Edwards C. Intercellular permeability in the early *Asterias* embryo. Dev Biol 1972 ; 27:546-554.
13. Andreuccetti P, Barone Lumaga MR, Cafiero G, Filosa S, Parisi E. Cell junctions during the early development of the sea urchin embryo (*Paracentrotus lividus*). Cell Diff 1987; 20:137-146.
14. Serras F, Baud C, Moreau M, Guerrier P, Van den Biggelaar JA. Intercellular communication in the early embryo of the ascidian *Ciona intestinalis*. Development 1988; 102:55-63.
15. Serras F, Van den Biggelaar JAM. Progressive restrictions in gap junctional communication during development. In: Robards AW, Jongsma H, Lucas WJ, et al eds. Parallels in cell to cell junctions in animals and plants London: NATO ASI Series, Springer-Verlag, 1989.
16. Palmer JF, Slack C. Some bio-electric parameters of early *Xenopus* embryos. J Embryol exp Morph 1970; 24:535-553.
17. Ito S, Hori N. Electrical characteristics of *Triturus* egg cells during cleavage. J Gen Physiol 1966; 49:1019-1027.
18. Ito S, Loewenstein WR. Ionic communication between early embryonic cells. Dev Biol 1969; 19:228-243.
19. Bennett MVL, Spira ME, Spray DC. Permeability of gap junctions between embryonic cells of *Fundulus*: a reevaluation. Dev Biol 1978; 65:114-125.

20. Kimmel CB, Spray DC, Bennett MVL. Developmental uncoupling between blastoderm and yolk cell in the embryo of the teleost *Fundulus*. Dev Biol 1984; 102:483-487.
21. Kimmel CB, Law RD. Cell lineage of zebrafish blastomeres II. formation of the yolk syncytial layer. Dev Biol 1985; 108:86-93.
22. Kimmel CB, Law RD.Cell lineage of zebrafish blastomeres I. cleavage pattern and cytoplasmic bridges between cells. Dev Biol 1985; 108:78-85.
23. Sheridan JD. Electrophysiological evidence for low-resistance intercellular junctions in the early chick embryo. J Cell Biol 1968; 37:650-659.
24. Ducibella T, Albertini DF, Anderson E, Biggers JD. Intercellular junctions and their appearance during development. Dev Biol 1975; 45:231-250.
25. Lo CW, Gilula NB. Gap junctional communication in the preimplantation mouse embryo. Cell 1979; 18:399-409.
26. Dale B, Gualtieri R, Talevi E Tosti E, Santella L, Elder K. Intercellular communication in the early human embryo. Mol Reprod Dev 1991; 29:22-28.
27. Simpson I, Rose B, Loewenstein WR. Size limit of molecules permeating the junctional membrane channels. Science 1977; 195:294-296.
28. Warner AE, Gurdon JB. Functional gap junctions are not required for muscle gene activation by induction in *Xenopus* embryos. J Cell Biol 1987; 104:557-564.
29. Warner AE, Guthrie SC, Gilula NB. Antibodies to gap junctional protein selectively disrupt junctional communication in the early amphibian embryos. Nature 1984; 311:127-131.
30. Verdonk NH, Van den Biggelaar JAM. Early development and the formation of the germ layers. In: Verdonk NH, van den Biggelaar JAM, Tompa AS, eds.The Mollusca Vol 3. New York: Academic Press, 1983:91-122.
31. Dorresteijn AWC, Van den Biggelaar JAM, Bluemink JG et al. Electron microscopical investigations of the intercellular contacts during the early cleavage stages of *Lymnaea stagnalis* (Mollusca, Gastropoda). Roux's Arch Dev Biol 1981; 190:215-220.
32. Serras F, Van den Biggelaar JAM. Is a mosaic embryo also a mosaic of communication compartments? Dev Biol 1987; 120:132-138.
33. Dorresteijn AWC, Bilinski SM, Van den Biggelaar JAM, Bluemink JG. The presence of gap junctions during early *Patella* embryogenesis: an electron microscopical study. Dev Biol 1982; 91:397-401.
34. Serras F, Dictus WJAG, Van den Biggelaar JAM. Changes in junctional communication associated with cell cycle arrest and differentiation of trochoblasts in embryos of *Patella vulgata*. Dev Biol 1990; 137:207-216.
35. Serras F, Damen P, Dictus WJAG et al. Communication compartments in the ectoderm of embryos of *Patella vulgata*. Roux's Arch Dev Biol 1989; 198:191-200.
36. Serras F, Notenboom RGE, Van den Biggelaar JAM. Communication compartments in the post-trochal ectoderm of the mollusc *Lymnaea stagnalis*. Exp Cell Res 1990; 187:284-291.
37. Turin L, Warner AE. Intracellular pH in early *Xenopus* embryos: its effect on current flow between blastomeres. J Physiol (London) 1980; 300:489-504.
38. Slack C, Palmer JF. The permeability of intercellular junctions in the early embryo of *Xenopus laevis*, studied with a fluorescent tracer. Exp Cell Res 1969; 55:416-419.
39. Guthrie SC. Patterns of junctional communication in the early amphibian embryo. Nature 1984; 311:149-151.
40. Guthrie SC, Turin L, Warner AE. Patterns of junctional communication during development of the early amphibian embryo. Development 1988; 103:769-783.
41. Kao K, Danilchik M. Generation of body plan phenotypes in early embryogenesis. In: Methods in cell biology. Vol 36. New York: Academic Press, 1991:271-284.
42. Saxén L. Neural induction. Int J Dev Biol 1989; 33:21-48.
43. Decker RS, Friend DS. Assembly of gap junctions during amphibian neurulation. J Cell Biol 1974; 62:32-47.
44. Gimlich RL, Kumar NM, Gilula NB. Sequence and developmental expression of mRNA coding for a gap junction protein in *Xenopus*. J Cell Biol 1988; 107:1065-1073.

45. Gimlich RL, Kumar NM, Gilula NB. Differential regulation of the levels of three gap junction mRNAs in *Xenopus* embryos. J Cell Biol 1990; 110:597-605.
46. Ebihara L, Beyer EC, Swenson KI, Paul DL, Goodenough DA. Cloning and expression of a *Xenopus* embryonic gap junction protein. Science 1989; 243:1194-1195.
47. Warner AE. The electrical properties of the ectoderm during induction and early development of the nervous system. J Physiol (London) 1973; 235:267-286.
48. Ito S, Ikematsu Y. Inter and intratissue communication during amphibian development. Dev Growth and Differ 1980; 22: 247-256.
49. Smith JC. Mesoderm induction and mesoderm-inducing factors in early amphibian development. Development 1989; 105:665-677.
50. Dawid IB. Mesoderm induction and axis determination in *Xenopus laevis*. Bioessays 1992; 10:687-691.
51. Kao KR, Elinson RP. Dorsalization of mesoderm induction by lithium. Dev Biol 1989; 132:81-90.
52. Youn BW, Malacinski GM. Axial structure development in ultraviolet irradiated (notochord defective) amphibian embryo. Dev Biol 1981; 83:339-352.
53. Moon RT. In pursuit of the functions of the *Wnt* family of developmental regulators: insights from *Xenopus laevis*. Bioessays 1993; 15:91-97.
54. Nagajski DJ, Guthrie SC, Ford CC, Warner AE. The correlation between patterns of dye transfer through gap junctions and future developmental fate in *Xenopus*: the consequences of UV irradiation and lithium treatment. Development 1989; 105:747-752.
55. Olson DJ, Christian JL, Moon RT. Effect of *Wnt*-1 and related proteins on gap junctional communication in *Xenopus* embryos. Science 1991; 252:1173-1176.
56. Olson DJ, Moon RT. Distinct effects of ectopic expression of *Wnt*-1, activin B and bFGF on gap junctional permeability in 32-cell *Xenopus* embryos. Dev Biol 1992; 151:204-212.
57. Serras F, Kühtreiber WM, Krul MRL, Van den Biggelaar JA. Cell communication compartments in molluscan embryos. Cell Biol Int Rep 1985; 9:731-736.
58. Warner AE, Lawrence PA. Permeability of gap junctions at the segmental border in insect epidermis. Cell 1982; 28:243-252.
59. Weir MP, Lo CW. gap junction communication compartments in the *Drosophila* wing imaginal disc. Dev Biol 1984; 102:130-146.
60. Goodall H, Johnson MH. Use of carboxyfluorescein diacetate to study formation of permeable channels between mouse blastomeres. Nature 1982; 295:524-526.
61. Goodall H, Johnson MH. The nature of intercellular coupling within the preimplantation mouse embryo. J Embryol exp Morph 1984; 79:53-76.
62. McLachlin JR, Caveney S, Kidder GM. Control of gap junction formation in early mouse embryos. Dev Biol 1983; 98:155-164.
63. Kelly S. Studies of the developmental potential of 4- and 8-cell stages mouse blastomeres. J Exp Zool 1977; 200:365-376.
64. Johnson MH. Three types of cell interaction regulate the generation of cell diversity in the mouse blastocyst. In: Edelman GM, Thiery JP, Eds. The cell in contact. Adhesions and junctions as morphogenetic determinants New York: Neurosciences Institute Publication, 1985:27-48.
65. Calarco PG, Brown EH. An ultrastructural and cytological study of preimplantation development of the mouse. J Exp Zool 1969; 171:253-284.
66. Magnuson T, Demsey A, Stackpole CW. Characterization of intercellular junctions in the preimplantation mouse embryo by freeze-fracture and thin-section electron microscopy. Dev Biol 1977; 61:252-261.
67. Fleming TP, Javed Q, Hay M. Epithelial differentiation and intercellular junction formation in the mouse early embryo. Development 1992; supplement:105-112.
68. Nishi M, Kumar NM, Gilula NB. Developmental regulation of gap junction gene expression during mouse embryonic development. Dev Biol 1991; 146:117-130.

69. Valdimarsson G, De Sousa PA, Beyer EC, Paul DL, Kidder GM. Zygotic expression of the connexin43 gene supplies subunits for gap junction assembly during mouse preimplantation development. Mol Reprod Dev 1991; 30:18-26.
70. Becker DL, Leclerc-David C, Warner AE. The relationship of gap junctions and compaction in the preimplantation mouse embryo. Development 1992; supplement:113-118.
71. De Sousa PA, Valdimarsson G, Nicholson BJ, Kidder GM. Connexin trafficking and the control of gap junction assembly in mouse preimplantation embryos. Development 1993; 117:1355-1367.
72. Barron DJ, Valdimarsson G, Paul DL, Kidder GM. Connexin32, a gap junction protein, is a persistent oogenetic product through preimplantation development of the mouse. Dev Genet 1989; 10:318-323.
73. McLachlin JR, Kidder GM. Intercellular junctional coupling in preimplantation mouse embryos: Effect of blocking transcription or translation. Dev Biol 1986; 117:146-155.
74. Kidder GM, Rains J, McKeon J. Gap junction assembly in the preimplantation mouse conceptus is independent of microtubules, microfilaments, cell flattening and cytokinesis. Proc Natl Acad Sci (USA) 1987; 84:3718-3722
75. Magnuson T, Jacobson JB, Stackpole CW. Relationship between permeability and junction organization in the preimplantation mouse embryo. Dev Biol 1978 ;67:214-224.
76. Goodall H. Manipulation of gap junctional communication during compaction of the mouse embryo. J Embryol exp Morph 1986; 91:283-296.
77. Lee S, Gilula NB, Warner AE. Gap junctional communication and compaction during preimplantation stages of mouse development. Cell 1987; 51:851-860.
78. Belilacqua A, Loch-Caruso R, Erickson RP. Abnormal development and dye coupling produced by antisense RNA to gap junction protein in mouse preimplantation embryos. Proc Natl Acad Sci (USA) 1989; 86:5444-5448.
79. Wakasugi N. Studies on fertility of DDK mice: reciprocal crosses between DDK and C57BL/6J strains and experimental transplantation of the ovary. J Reprod Fert 1973; 33:283-291.
80. Mann JR. DDK egg-foreign sperm incompatibility in mice is not between the pronuclei. J Reprod Fert 1986; 76:779-781.
81. Renard JP, Babinet C. Identification of a paternal developmental effect on the cytoplasm of one-cell-stage mouse embryos. Proc Natl Acad Sci (USA) 1986; 83:6883-6886.
82. Buehr M, Lee S, McLaren A, Warner A. Reduced gap junctional communication is associated with the lethal condition characteristic of DDK mouse eggs fertilized by foreign sperm. Development 1987; 101:449-459.
83. Leclerc-David C. The role of gap junctions in preimplantation mouse development. A study of the DDK defective strain. Ph.D. Thesis University of London, London.
84. Balakier H, Pedersen RA. Allocation of cells to Inner Cell Mass and trophectoderm lineages in preimplantation mouse embryos. Dev Biol 1982; 90:352-362.
85. Gardner RL. An investigation of Inner Cell Mass and trophoblast tissues following their isolation from the mouse blastocyst. J Embryol exp Morph 1972; 28:279-312.
86. Rossant J. Investigation of the determinative state of the mouse Inner Cell Mass. I. Aggregation of isolated Inner Cell Mass with morulae. J Embryol exp Morph 1975; 33:979-990.
87. Rossant J. Investigation of the determinative state of the mouse Inner Cell Mass. II. The fate of isolated Inner Cell Lasses transferred to the oviduct. J Embryol exp Morph 1975; 33:991-1001.
88. Lo CW, Gilula NB. Gap junctional communication in the post-implantation mouse embryo. Cell 1979; 18:411-422.
89. Fraser SE, Green CR, Bode HR, Gilula NB. Selective disruption of gap junctional communication interferes with a patterning process in *Hydra*. Science 1987; 237:49-55.
90. Fraser SE, Bode HR. Epithelial cells of *Hydra* are dye-coupled. Nature 1981; 294:356-358.

91. Ryerse JS, Nagel BA. Changes in the distribution of gap junctions in *Drosophila melanogaster* wing discs during the third larval ans early pupal stages of development. Roux's Arch Dev Biol 1984; 193:187-196.
92. Ryerse JS, Nagel BA. gap junction distribution in the *Drosophila* wing discs mutants vg, $l_{(2)}gd$, $l_{(3)}c43^{hs1}$ *and* $l_{(2)}gl^4$. Dev Biol 1984; 105:396-403.
93. Jursnich VA, Fraser SE, Held LI, Ryerse J, Bryant PJ. Defective gap-junctional communication associated with imaginal disc overgrowth and degeneration caused by mutations of the *dco* gene in *Drosophila*. Dev Biol 1990; 140:413-429.
94. Ryerse JS, Nagel BA. Gap junction reductions precede tissue hyperplasia following temperature-upshift in the *Drosophila* ts-mutant $l_{(3)}c43^{hs1}$. Roux's Arch Dev Biol 1985; 194:480-486.
95. Mehta PP, Hotz-Wagenblatt A, Rose B, Shalloway D, Loewenstein WR. Incorporation of the gene for the cell-cell channel protein into transformed cells leads to normalization of growth. J Memb Biol 1991; 124:207-225.
96. Eghbali B, Kessler JA, Reid LM, Roy C, Spray DC. Involvement of gap junctions in tumorigenesis: transfection of tumor cells with connexin32 cDNA retards growth in vivo. Proc. Natl Acad. Sci USA 1991; 88:10701-10705.
97. Rose B, Mehta PP, Loewenstein WR. Gap-junction protein gene suppresses tumorigenicity. Carcinogenesis 1993; 14:1073-1075.
98. Gourdie RG, Harfst E, Severs NJ, Green CR. Cardiac gap junctions in rat ventricule: localization using site directed antibodies and laser scanning confocal microscopy. Cardioscience 1990; 1:75-82.

CHAPTER 4

The Role of Cell Adhesion and Intercellular Junctions During Vertebrate Early Development

Tom P. Fleming

1. INTRODUCTION

Vertebrate development, in essence, can be viewed as a remarkable and complex interaction between two regulatory processes. Firstly, a hierarchy of gene expression activity imposes the body plan and then the details of various cell differentiation programs. Secondly and concurrently, a myriad of cellular interactions and longer-range signals provide information of a positional nature that is used to guide tissue formation and segregation and, via intracellular signaling, to control the precise character of gene expression. This chapter concerns the second component of the developmental equation and deals specifically with the expression, formation and function of cell-cell adhesion complexes and intercellular junctions during vertebrate embryogenesis. The contribution of these systems to development has been studied extensively in amphibian, avian and mammalian embryos, where adhesion and junction formation begin during the earliest stages of egg cleavage. Subsequently, the assembly and disassembly of adhesive junctions characterizes numerous examples of mesenchymal-epithelial interconversion and the reshaping of tissues in response to inductive signals.[1] The cadherin-based adhesion system is the first to function in vertebrate embryos and recent studies have demonstrated that it contributes significantly to spatial organization throughout the whole of development. Glycoproteins belonging

Molecular Mechanisms of Epithelial Cell Junctions: From Development to Disease
Edited by Sandra Citi, M.D., Ph.D.;

to the cadherin superfamily are responsible for adhesion at the adherens and desmosome intercellular junctions (see below). However, cadherin cell-cell adhesion may also be non-junctional, such as along the entire lateral membrane contact site in epithelia. Most studies on cadherins in development do not discriminate between junctional and non-junctional adhesion and so, where necessary, this chapter will focus more on the molecular character of adhesive contacts rather than their structural state.

2. CADHERIN-MEDIATED ADHESION

The concept that cell adhesion events and in particular cadherin-mediated adhesion, may act as an important regulator of morphogenesis and the spatial segregation of cells into distinct tissues is now well established (reviewed in refs. 2-4). The cadherins are a multigene family highly conserved throughout the vertebrates and with invertebrate homologues[2,5-8] (see also chapters 1, 8 and 12 in this volume). Cadherin-mediated cell-cell interactions are dependent upon extracellular calcium and are usually homophilic at the molecular level. At least 15 distinct cadherins have been described so far and the identification of more can be anticipated. The best characterized are the so-called 'classical' cadherins, E-cadherin (epithelial; also known as uvomorulin or L-CAM), N-cadherin (neural; also known as A-CAM) and P-cadherin (placental), but many remain to be fully sequenced. However, database comparisons of existing cadherin sequences have permitted the construction of a hypothetical cadherin evolutionary tree.[9] Despite such diversity, most cadherins are the product of a single gene and evidence of alternative splicing is rare (except for desmosomal cadherins, see chapter 8 in this volume).

As mentioned above, cadherin cell-cell adhesion can occur either at junctional or non-junctional membrane sites. At the adherens intercellular junction (zonula adherens in epithelia) are concentrated the 'classical' cadherins, particularly E- or N-cadherin[10,11] while at desmosomes adhesion is mediated by the desmosomal cadherins, desmoglein and desmocollins.[12,13] Although circumstantial evidence suggested that tight junction adhesion might also be mediated by a cadherin-like molecule,[14] the recent discovery of a novel protein, occludin, as an integral membrane component of the tight junction has discounted this possibility.[15] The vast majority of studies on adhesion during development concern the adherens junction cadherins since these appear to have a primary role compared with tight and desmosomal junctions whose formation occurs secondarily in differentiating epithelia. This review inevitably reflects this bias. The term cadherin will therefore refer to the classical adherens intercellular junction type unless specified otherwise.

To understand how cadherin-based adhesion might regulate morphogenetic processes, it is necessary to emphasize cadherin molecular organization and cytoplasmic domain interactions (see also chapters 1, 8 and 12, this volume). Cadherins possess a conserved structure consisting of five extracellular repeat domains (EC1-5), a single transmembrane segment and a cytoplasmic tail (C-terminal) where indirect linkage to cytoskeleton occurs. The distal EC1 domain contains a recognition sequence required for homophilic interaction[16] while calcium binding and glycosylation sites are found on more proximal EC domains.[8] The cadherin cytoplasmic tail interacts with different sets of intermediary proteins to achieve linkage with different forms of cytoskeleton. Thus, E-cadherin and N-cadherin associate with the catenins (α-, β- and γ-) which in turn link with the actin filament system.[17-19] At the adherens junction, other actin-binding proteins are also present including vinculin, α-actinin, radixin and zyxin and collectively may contribute to junction organization.[5] In addition, E-cadherin may also complex with the fodrin membrane cytoskeleton.[20] In contrast, desmosomal cadherins (desmoglein, desmocollins) form linkages with the cytokeratin filament system via other intermediaries, notably desmoplakins[21,22] (see also chapter 9, this volume). Plakoglobin, however, is found at both adherens and desmosome junction types and may, therefore, complex with more than one cadherin.[23]

Transfection studies have shown that for E- and N-cadherin, cytoplasmic tail interactions and linkage to the cytoskeleton are essential for adhesive function.[17-19,24] Cytoskeletal anchorage of junctional cadherins is therefore important for the generation of intercellular and intracellular forces necessary for embryonic morphogenesis.

3. CADHERIN EXPRESSION PATTERNS DURING DEVELOPMENT

There have been many studies of the distribution of cadherin proteins during vertebrate development. Collectively, these studies have shown that each cadherin is expressed in specific tissues at particular times during embryogenesis with changes in distribution often coinciding with morphogenetic transitions.[2] However, although the overall expression pattern of each cadherin is unique, cadherins may still overlap in their distribution. For example, in the mammalian embryo, E-cadherin is expressed during preimplantation development (see below) and, after implantation, is present in various developing epithelial tissues including the ectoplacental cone, epidermis and nephric tubules. These three tissues also express P-cadherin which in turn is co-expressed with N-cadherin in lateral plate mesoderm and corneal endothelium.[25]

The implication from these and other developmental studies is that the dynamic and selective nature of cadherin expression coupled with the characteristic of homophilic binding regulates tissue formation and boundaries and the relative position of both cells within tissues and tissues within organs. This concept is supported by cell mixing experiments where cells expressing different types and/or relative amounts of cadherins can segregate into distinct aggregates in culture.[26,27] Moreover, the boundaries of cadherin expression in embryos are subject to change by inductive interactions. Such events can lead either to the initiation or loss of cadherin expression in the responding tissue/cells or to the switching of cadherin expression from one type to another. As a consequence, changes in the phenotypic behavior of cells during development, such as occur in tissue remodeling and in cellular migrations, can take place. The remainder of this chapter examines the early stages of vertebrate development to evaluate the role of cadherins and related intercellular junctions in such morphogenetic activities.

3.1 Cleavage and Early Development

Cadherins are the first adhesion molecules to be active as embryogenesis begins and are expressed from maternal as well as embryonic transcripts.[28-32] In *Xenopus*, at least two distinct maternal cadherins are functional in the early embryo, namely U-cadherin ('ubiquitous')[27] and EP-cadherin,[28,30,31] the latter homologous with mouse E- and P-cadherin.[30] U-cadherin is localized at the contact sites between blastomeres from the first cleavage division onwards but is absent from the outer, contact-free, membrane surface.[29] This polarized distribution (basolateral rather than apical) coincides with the generation of the first epithelium, responsible for blastocel formation in the blastula. Tight junction and desmosome formation in the outer cell population, necessary for the vectorial transport of blastocel fluid and stabilization of the epithelium, both take place after cadherin adhesion has begun.[33] However, no distinction has yet been drawn between U- and EP-cadherin in terms of distribution, blastomere adhesion and role in blastocel formation.[29,31] Indeed, both form complexes with the catenins in the early embryo,[34] suggesting both are functional concurrently. Recently, it has been shown that inhibition of maternal EP-cadherin expression by antisense injection caused a significant reduction in adhesion between blastomeres and the failure to form a blastocel.[35] It will be interesting to see whether U-cadherin inhibition causes an equivalent phenotype and how both cadherins contribute towards inductive events during amphibian cleavage, namely the segregation of prospective ectoderm, mesoderm and endoderm germ layers and the specification of dorsal-ventral and anterior-posterior axes.

In the chick[36] and mammalian embryo, E-cadherin is the first cadherin to be expressed. In the mouse, in contrast to *Xenopus*, blastomere adhesion initiates at a precise time during cleavage, so-called compaction at the 8-cell stage, when E-cadherin becomes functional.[32,37] Specific antibodies to the E-cadherin ectodomain binding site[37-39] or a competitive peptide incorporating the binding site sequence[16] inhibit or reverse compaction. E-cadherin is present in oocytes and is synthesized throughout cleavage[32,40] as are both α- and β-catenin (X Lei, B Sheth and T Fleming, in preparation), suggesting that gene expression per se does not regulate initiation of blastomere adhesion. Moreover, the compaction of blastomeres is not inhibited by prior treatment with protein synthesis inhibitors.[41] Since protein kinase C stimulation leads to premature compaction[42] and a number of cellular proteins[43] including E-cadherins[40] become phosphorylated around this time, a post-translational control of adhesion initiation at compaction is implicated, involving cellular kinases.

As compaction occurs in the 8-cell mouse embryo, cell phenotype changes from nonpolar to polar as blastomeres begin to differentiate the trophectoderm epithelium.[44,45] This epithelium forms the wall of the blastocyst, enclosing the inner cell mass (ICM, fetal progenitor) and becomes functional in vectorial transport to generate the blastocel from the 32-cell stage onwards, some 24 hours after compaction (reviewed in refs. 46 and 47). Trophectoderm cells are linked by a typical epithelial junctional complex comprising tight, adherens and desmosomal junctions (Fig. 4.4.1A, B). At compaction, E-cadherin and catenins are localized along cell contact sites between blastomeres[32] (X Lei, B Sheth and T Fleming, in preparation; Fig. 4.4.1C, E), corresponding to the basolateral domain of these proto-epithelial cells (a pattern akin to U-cadherin and catenins in *Xenopus* cleavage, see above). Only later in cleavage as the blastocyst forms do trophectodermal E-cadherin and catenins concentrate apicolaterally to form the zonula adherens junction,[32,48] although both proteins are readily detectable at all cell contact sites in the blastocyst (Fig. 4.4.1D, F). Inhibition of E-cadherin adhesion at compaction (using a neutralizing antibody) does not inhibit cell polarization but does delay its occurrence and randomizes its orientation and, after prolonged treatment, prohibits the formation of ICM in the blastocyst.[38,39] E-cadherin adhesion is therefore necessary for normal blastocyst morphogenesis and tissue segregation but does not directly control the inception of epithelial polarization which appears to be programmed independently. However, it does influence the spatial organization of the trophectoderm epithelial phenotype as discussed below with respect to tight junctions and desmosomes.

As in amphibian and chick early embryos (see above;[49]), tight junction formation in the mouse embryo occurs after adhesion has been initiated by cadherin belonging to the adherens junction type. Zonular tight junctions, with a typical ultrastructure and freeze-fracture image, are established by the 32-cell stage between trophectoderm cells just when blastocel formation begins,[50,51] and at the molecular level contain the cytoplasmic domain tight junction-associated proteins ZO-1[52,53] and cingulin[54,55] (Fig. 4.4.1G-I). Immunological and structural studies have demonstrated that tight junction formation is a gradual process, beginning from compaction 24 hours earlier than blastocel formation. During this period the apicolateral junction matures from a punctate assemblage to a discontinuous linear and finally a true zonular, configuration and concurrently increases in molecular complexity with assembly of ZO-1 preceding that of cingulin[44,51,52,54] (reviewed in refs. 56 and 57). E-cadherin adhesion at compaction appears necessary for tight junction formation. First, both ZO-1 and particularly cingulin are detectable in embryos before compaction but do not assemble tight junctions.[52,54] Second, cingulin synthesis is upregulated significantly when adhesion is initiated at compaction.[55] Third, if E-cadherin adhesion at compaction is inhibited using neutralizing antibody, the membrane assembly of tight junction proteins is disturbed, becoming randomized rather that apicolateral. However, under these conditions, the onset of *expression* of tight junction proteins is not affected.[52]

During cleavage in the mouse embryo, desmosomal cadherins (desmoglein and

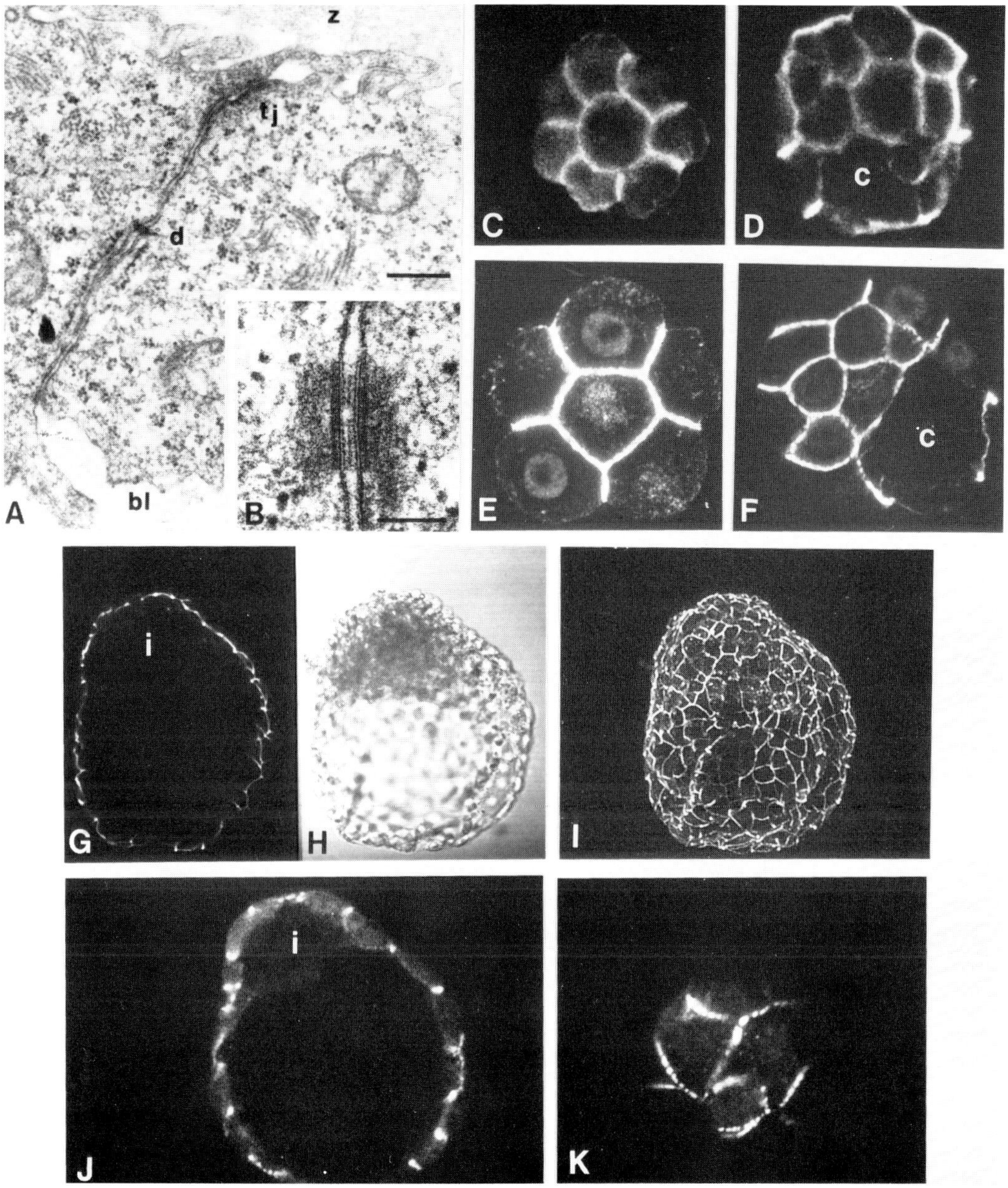

Fig. 4.1. Cell adhesion and junctions in the mouse preimplantation embryo. A,B, electron micrograph of the lateral membrane contact site between trophectoderm cells in a nascent blastocyst. An apicolateral tight and adherens junction site is present (tj) below which are found desmosomes (d and at higher magnification in B); bl, blastocoel; z, zona pellucida. Bar = 1 μm (A) or 0.1 μm (B). C,D, confocal images of E-cadherin localization in a recently compacted 8-cell embryo (C) and a blastocyst (D). E-cadherin is present along all cell contact sites including those of the ICM in the blastocyst (c, blastocoel cavity). E,F, α-catenin localization in compact 8-cell embryo (E) and blastocyst (F) is the same as for E-cadherin. G-I, ZO-1 localization in expanded blastocyst is restricted to the outer trophectoderm. G,H, optical midplane section and corresponding brightfield image; i, ICM. I, ZO-1 in the complete blastocyst (3D reconstruction) showing a zonular distribution around each trophectoderm cell. J,K, desmoplakin distribution in expanded blastocyst is restricted to contact sites between trophectoderm cells; i, ICM. In K, the trophectoderm is shown in tangential plane; punctate desmoplakin sites are visible, probably representing individual desmosomes. Magnification C-F, J, K x 500, G-I x 300.

desmocollins) are first synthesized at the 32-cell early blastocyst stage.[58] This event is coupled with the appearance of the first desmosome junctions in the embryo, between the trophectoderm cells as the blastocyst expands[50,58] (Fig. 4.4.1A, B, J, K). Desmosomal cadherin synthesis occurs after that of other desmosome components (desmoplakin, plakoglobin) and intermediate filaments and so is likely to be permissive for rapid assembly of complete desmosome junctions, presumably necessary to provide structural support for the epithelium as the blastocel expands.[58] Recently, we have shown that desmocollin transcription from the embryonic genome is initiated at the late 16- to 32-cell stage, implicating transcription rather than translation of desmocollin as an important control of desmosome formation.[59] Thus, the embryo appears to utilize distinct mechanisms for tight and desmosome junction formation with the membrane-spanning component being either the first (tight junction) or last (desmosome) to be expressed or assemble, a distinction that influences the rate at which complete junctions are formed (discussed in refs. 56 and 57). However, the formation of both junction types is perturbed in similar ways following inhibition of E-cadherin adhesion from compaction onwards. Thus, the *timing* of desmocollin transcription is unaffected although, at the protein level, the *spatial* organization of desmocollin and desmoplakin membrane assembly is abnormal.[59]

In mammals, implantation of the expanded blastocyst in the uterine wall requires a change in the normal epithelial morphology of the trophectoderm and uterine tissues (discussed in ref. 60, Fig. 4.2). First, the apical membranes of these tissues, normally non-adhesive, must engage in an embryonic-maternal interaction that defines the implantation site (Fig. 4.2A, B). This adhesive reaction appears to be controlled, at least in part, by E-cadherin binding.[61] Next, the trophectoderm cells must lose their epithelial phenotype and junctional integrity and become migratory and invasive trophoblast cells. In mouse, this phase of differentiation coincides with loss of expression of E-cadherin[62] and secretion of proteases (urokinase-type plasminogen activator, metalloproteinases) to effect penetration of the uterine basement membrane and invasion of the stroma[63] (Fig. 4.2C). Once implantation is completed, specific trophoblast lineages re-establish an epithelial organization coinciding with re-expression of cadherin, particularly P-cadherin, as the extra-embryonic placental membranes are formed.[25] Other trophoblast cells become giant cells and adhere to and may engulf maternal decidual cells; here too adhesion is mediated by P-cadherin[61] (Fig. 4.2C). In contrast, human trophoblast continues to express E-cadherin during and after implantation whilst P-cadherin is only minimally expressed in placental tissues.[64] In the human, E-cadherin adhesion regulates the aggregation of cytotrophoblast cells to form the wall of the chorionic villi. Cytotrophoblast cells undergo terminal differentiation by cell fusion to form the syncytiotrophoblast. Syncytia formation coincides with and may be regulated by, the down regulation of E-cadherin expression.[65]

3.2 Gastrulation

Gastrulation and the establishment of segregated germ layers represents the first significant morphogenetic process in development. In all vertebrates, the extracellular matrix has been implicated an important role in the movement of prospective mesoderm into the embryo interior. Thus, peptides containing the cell recognition sequence of fibronectin, or antibodies recognizing the fibronectin receptor, inhibit the migration of mesoderm in urodele amphibians (reviewed in ref. 66). However, cell-cell interactions, in addition to cell-matrix binding, have been shown to influence amphibian gastrulation in different ways. Detailed studies on mesodermal migration in vivo and in vitro have revealed that cells move over the matrix surface in a unidirectional manner as a coherent, multilayered aggregate whilst single cells, generated by disaggregation in calcium-free medium, migrate in a random disorganized fashion[67] (reviewed in ref. 68). E-cadherin is first expressed at the time of gastrulation in amphibians but is localized mainly in ectoderm while mesoderm cells express U-cadherin.[29,69] Antibody to U-cadherin inhibited directed migration of explanted head mesoderm in *Xenopus*.[70] Thus, cadherin-mediated adhesion, presumably associated with adherens type junctions, is essential for coordinated movement of cells along a matrix scaffold during amphibian gastrulation.

Cadherin adhesion and junctions also participate in the activity of bottle cells during amphibian gastrulation. These cells, located at the involuting marginal zone of the blastopore lip, regulate archenteron invagination and mesoderm ingression by undergoing a remarkable change in cell shape.[68] Thus, cell apices on the embryo surface become constricted as the cells elongate along the apico-basal axis and dilate in the basal region, thereby generating the appropriate intercellular forces to enact invagination and ingression. Early morphological studies have shown an intense build up of microfilament-rich apical zonular adherens junctions as bottle cells constrict.[33,71,72] More recently, the apico-lateral junctional zone of constricted bottle cells has been shown to stain intensely with U-cadherin antibody.[29] The implication from these data is that cadherin-mediated adherens junctions provide an anchorage site for actin microfilaments to cause circumferential constriction of bottle cell apices. However, molecular studies are required to confirm this role for junctional organization in amphibian gastrulation. A similar mechanism involving apical constriction of ectoderm epithelial cells occurs during neural plate folding to construct the neural tube; direct molecular evidence for adherens junction cadherins regulating this process has been forthcoming (next section).

In avian and mammalian embryos, mesoderm ingression during gastrulation occurs as individual cells rather than tissue sheets, hence different

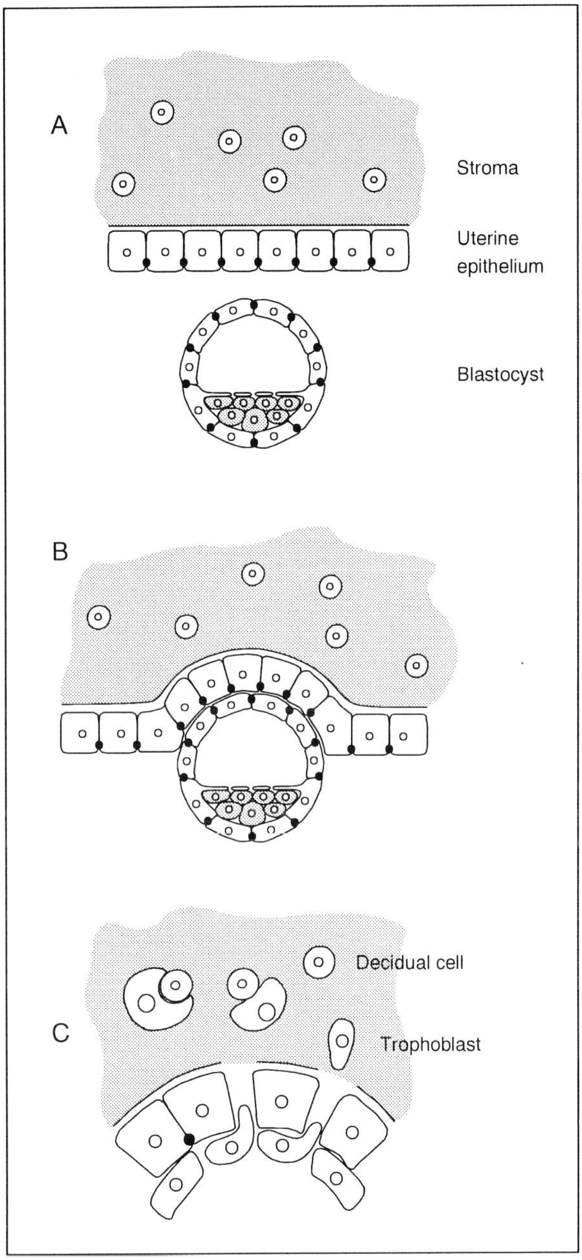

Fig. 4.2. Schematic representation of changes in cell adhesion and junctional integrity during mouse implantation. A, prior to attachment, the trophectoderm and uterine epithelia both express basolateral E-cadherin. B, at attachment, the apical surfaces of both epithelia become adhesive, including E-cadherin-mediated adhesion. C, subsequently, trophectoderm cells lose E-cadherin expression and junctional integrity, become invasive (trophoblast) and penetrate the stroma (shown at higher power). Certain trophoblast cells become giant cells, re-express cadherin (P-cadherin) and adhere to and engulf decidual cells. See text for details and references.

mechanisms of adhesion and junction organization must operate (Fig. 4.3). Ultrastructural studies on chick embryos have shown that as epiblast cells enter Hensen's node and the primitive streak and become primary mesenchyme, they lose their apical junctions.[49] Similarly, in mouse, epithelial-mesenchymal conversion of cells at the primitive streak coincides with loss of junctional adhesiveness and E-cadherin expression.[62] Recently, Burdsal et al[73] have demonstrated using a neutralizing antibody that inhibition of E-cadherin adhesion between explanted mouse epiblast cells caused their premature conversion to migratory cells with attendant loss of expression of epithelial proteins (cytokeratin, E-cadherin itself) and expression of mesenchyme protein (vimentin). Modulation of E-cadherin expression may therefore have a causative role in phenotypic change during gastrulation.

3.3 Neurulation

In the amphibian embryo after gastrulation but prior to neurulation the ectoderm expresses E-cadherin along lateral membranes and at apicolateral adherens junctions.[29,69] Induction of neural plate within the ectoderm by underlying notochord coincides with loss of expression of E-cadherin in presumptive neural plate cells and its replacement by N-cadherin[29,74] (Fig. 4.4). Non-induced ectoderm continues to express E-cadherin and differentiates into epidermis. A similar change in cadherin expression at adherens junctions has also been observed during neurulation in avian embryos.[36,75,76] The switch to N-cadherin expression is a direct and early response to the inductive signal, occurring in isolated *Xenopus* ectoderm only if exposed to a heterologous neural inducer and precedes the morphogenetic response to induction, that of neural plate folding and tube formation and segregation from uninduced ectoderm.[74]

Following N-cadherin expression, the epithelial cells of the neural plate convert from squamous to columnar shape and undergo apical constriction at the level of the zonula adherens as the tissue sheet invaginates (Fig. 4.4). Ultrastructure studies have revealed a marked increase in the thickness of the adherens junction zone and associated circumferential microfilaments precisely as constriction and folding take place, indicating a causal relationship[77,78] as in bottle cells at gastrulation (see above). Indeed, cytochalasin treatment to destabilize microfilaments inhibits both apical constriction and neural plate folding.[78] An actin-mediated constriction of cell apices at the level of the zonula adherens has also been described in avian neural tube formation.[79] Although cytoskeletal activity linked to adherens junctions is thought to generate the forces necessary to reshape neural plate cells, extrinsic forces from surrounding tissues are also likely contributors to the folding process.[80]

The relevance of the change in expression from E- to N-cadherin at the adherens junction prior to amphibian neurulation has been the subject of recent molecular studies. Ectopic expression of N-cadherin in uninduced ectoderm in vivo (by microinjection of synthetic or foreign

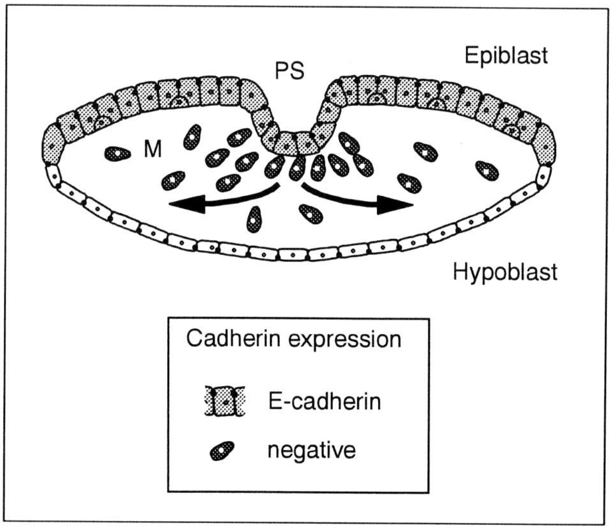

Fig. 4.3. Schematic representation of gastrulation in avian and mammalian embryos. The epithelial cells of the epiblast lose E-cadherin expression and junctional integrity within the primitive streak (PS) and become migratory mesodermal (M) cells. See text for details and references.

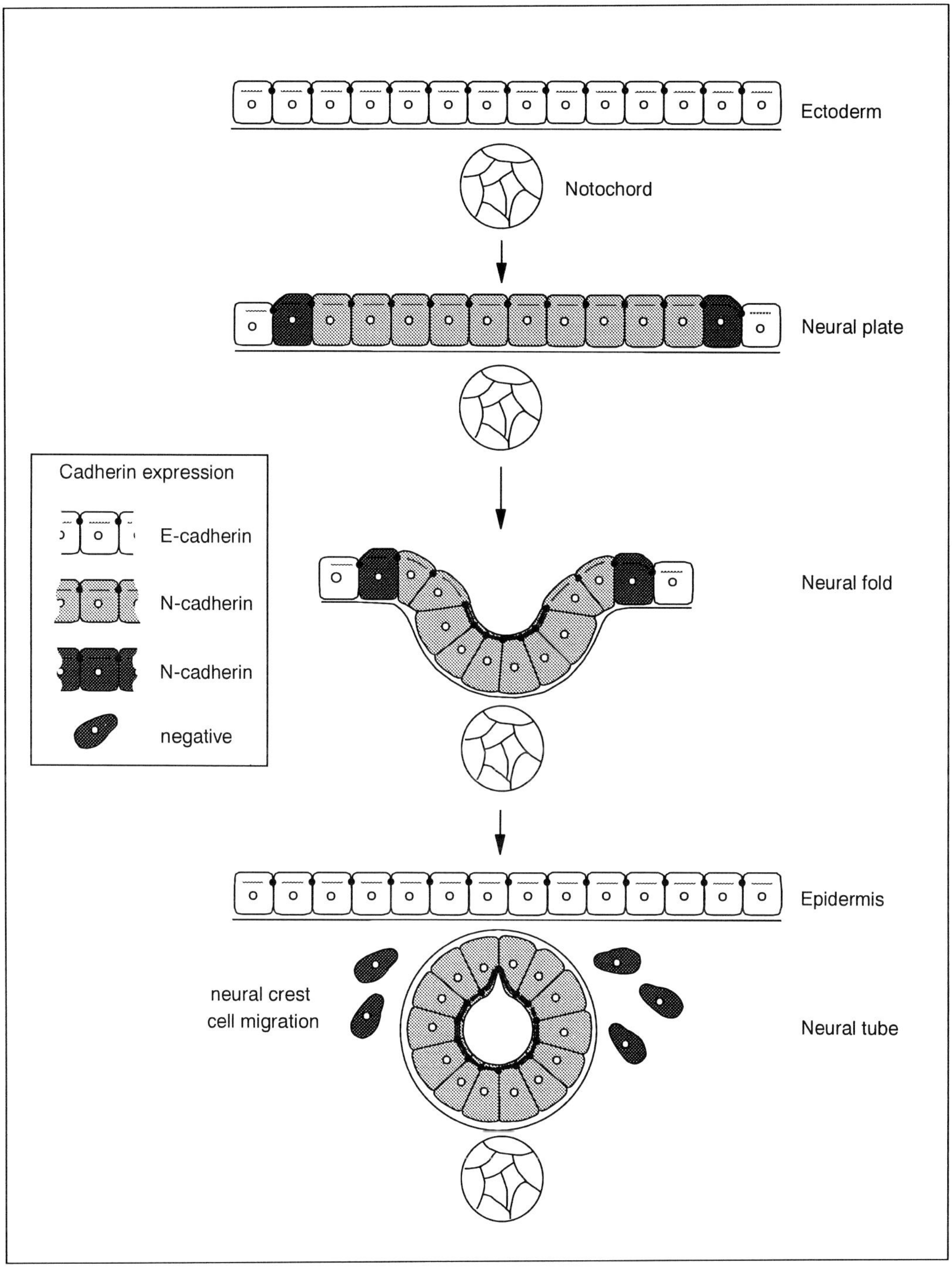

Fig. 4.4. Schematic representation of changes in cell adhesion and junctional organization during amphibian neurulation. Prior to notochordal induction, overlying ectoderm epithelial cells express E-cadherin at adherens junctions and along lateral membrane contact sites. Following induction, the neural plate cells switch from E-cadherin to N-cadherin expression and become more columnar in shape. Subsequently, apical constriction of cells at the level of the zonular adherens contributes to neural plate folding and tube formation. Presumptive neural crest cells cease to express cadherin, lose adherens junctions and epithelial morphology and become migratory. See text for details and references.

N-cadherin RNA transcripts into early blastomeres) caused ectoderm cells either to acquire a columnar neuroepithelial shape or to display more severe structural lesions, hindered ectodermal spreading (epiboly) during gastrulation and further led to neural tube defects including failure to separate from overlying ectoderm.[74,81] These complex pathologies resulting from misexpression of cadherins can best be explained by a corresponding disturbance in the normal behavior of embryonic epithelia. In particular, the switch from E- to N-cadherin expression at neurulation appears necessary for the two epithelia to adopt different morphologies and separate from each other, a natural consequence of homophilic interaction. In support of this model, anti- N-cadherin antibody injection into chick embryos also leads to defects in neural tube formation.[82]

Molecular studies have also shown that the change in epithelial behavior in response to altered cadherin expression during neurulation does not necessarily coincide with a change in the molecular interactions occurring at the cytoplasmic face of the adherens junction. Thus, in embryos in which a mutant truncated form of N-cadherin has been expressed lacking the extracellular domain, not only is N-cadherin adhesion inhibited by a 'dominant-negative' effect but also that of other cadherins.[83] Although this result indicates that the truncated N-cadherin competitively inhibits catenin and microfilament interaction with *all* expressed cadherins, this may be an over-simplification since there is evidence that catenins, like cadherins, are a multigene family.[24] This raises the possibility that cadherin cytoplasmic domain interactions may indeed be highly complex, offering additional mechanisms for regulation of epithelial morphogenesis.

3.4 Neural crest cell migration

Cadherin expression patterns are also implicated in neural crest cell migration from the dorsal surface of the newly-formed neural tube (Fig. 4.4). Following migration, this heterogeneous population of cells differentiates into the peripheral nervous system, skin pigment cells and connective tissue elements in the head (reviewed in ref. 84). In the avian embryo, migration begins as neural tube closure is taking place and is accompanied by the loss of N-cadherin expression and apicolateral adherens junctions as cells switch from neuroepithelial to mesenchymal.[76,82,84,85] Protease digestion of explanted neural tube in calcium-free medium to destroy adhesive contacts has been shown to induce premature migration of presumptive neural crest cells,[86] suggesting that the loss of N-cadherin expression may be permissive for the phenotypic change to occur.

4. MECHANISMS

During vertebrate early development there is a consistent correlation between changes in the distribution or organization of cadherins and cell junctions and changes in cell behavior and phenotype. In most cases, the relationship between adhesive state and morphogenetic activity cannot be defined further, yet in some of the examples of epithelial-mesenchymal interconversion described above there is sufficient evidence to suggest that changes in intercellular adhesiveness may be causal in morphogenetic events. Since the formation of tight and desmosomal junctions tend to follow that of cadherin-based adherens junctions in various examples of epithelial differentiation (see earlier), the mechanisms causing up- or downregulation of cadherins during development are important to define. Three broad mechanisms have been identified.

The first mechanism relies upon direct changes in cadherin expression, either at transcriptional or translational levels. Studies on cultured cells have demonstrated that the synthesis rate of cytoskeletal proteins[87] and the adherens junction component vinculin[88] are upregulated in a reversible manner by the provision of cell contacts. During human cytotrophoblast fusion to form syncytial tissue, downregulation of E-cadherin is achieved by a cessation in its transcription.[65] A similar event occurs as N-cadherin is downregulated during chick retinal development[89] (but see later). However, changing the expression rate is likely to be a slow and unwieldy primary mechanism to modulate cadherin activity in a developmental context. For example, this mechanism does not regulate E-cadherin adhesion at compaction in the mouse embryo[32,40] but the timing of expression of proteins does contribute to the formation of tight junctions and desmosomes during later cleavage (see earlier).

A second and potentially faster, mechanism to alter cadherin activity is by proteinase digestion after membrane insertion. An endogenous enzyme causes separation of a soluble amino-terminal fragment from N-cadherin in cultured chick lens cells and in chick embryo sclerotome cells as they migrate from the compact somite and down-regulate adherens junctions.[90] The remaining truncated membrane fragment of N-cadherin is nonadhesive while the soluble fragment containing the cell-cell recognition sequence is likely to cause de-adhesion of surrounding cells by competitive inhibition, as shown in a related study on the E-cadherin soluble fragment.[91] However, such an autocatalytical mechanism may be counterproductive developmentally if a change in adhesiveness is to be spatially confined.

The most attractive mechanism for modulating adhesion in a highly coordinated developmental situation is by post-translational phosphorylation. The adherens junction is a known site of tyrosine kinases[92] and, in virally-transformed epithelial cells expressing the pp60$^{v\text{-}src}$ oncogene product, it is a target for tyrosine phosphorylation which leads to rapid disassembly and deterioration of the adherens junction (reversible by dephosphorylation) and phenotypic conversion of cells from epithelial to fibroblastic[93](see also chapter 12 in this volume). In related studies, tyrosine-specific phosphorylation of E- and N-cadherin and β-catenin have been reported as the adherens junction is disassembled, indicating these proteins may be substrates for pp60$^{v\text{-}src}$ activity[94,95] or for other cellular kinases in untransformed cells. However, modulation of cadherins and junctional organization by tyrosine phosphorylation has yet to be demonstrated in a genuine developmental context although E-cadherin adhesivity during mouse compaction coincides with PKC-mediated phosphorylation of E-cadherin and other proteins.[40] Perhaps no single mechanism operates in vivo to switch the state of cell adhesivity and phenotype: down-regulation of N-cadherin levels and zonula adherens junctions during chick retinal development appears to be controlled by multiple factors including insulin stimulation, endogenous proteinase activity and a reduction in N-cadherin transcription.[89]

Although diverse intrinsic mechanisms may operate to control cadherin adhesivity and junctional integrity during development, additional 'upstream' mechanisms will control the initiation of such changes as part of an inductive interaction between tissues. Thus, a number of mesenchymally derived factors have been postulated to control epithelial morphogenesis via cell surface receptors. Some of these morphogenetic factors involved in vertebrate development have now been characterized and found to bind tyrosine kinase membrane receptors, e.g., scatter factor/hepatocyte growth factor and the *c-met* receptor (reviewed in refs. 96 and 97) (see also chapter 13 in this volume). It is unclear at present whether these phosphorylation cascades controlling differentiation interact with those identified at the adherens junction (see above) yet the intriguing observation that the *c-ret* proto-oncogene encodes a cadherin-like extracellular domain linked to a tyrosine kinase cytoplasmic tail[98] leaves open such a possibility.

More substantive evidence that the cadherin-mediated adherens junction might be a site for intercellular signal transduction during development has come from parallel studies on *Drosophila* proteins (see also chapter 1 in this volume). The protein encoded by the segment polarity gene *armadillo* is homologous with both vertebrate plakoglobin and β-catenin.[99] Armadillo protein (Arm) is localized at the cytoplasmic face of apicolateral junctions resembling the vertebrate zonula adherens in *Drosophila* embryonic epithelia where it forms a complex with a cadherin-like protein and the homolog of α-catenin.[100] Significantly, Arm is involved in the transduction of the *wingless* signal, a member of the *Wnt* gene family of secreted signaling molecules that contributes to pattern formation in both vertebrates and invertebrates.[101] Recently, injection of β-catenin antibody into *Xenopus* ventral blastomeres was shown to lead to the induction of a second body axis,[102] providing further evidence that the cadherin-catenin junctional site participates in intercellular signaling and pattern formation during development. Lastly, there is evidence that N-cadherin-mediated adhesion activates the second messenger intracellular signaling pathway during neurite outgrowth in differentiating PC12 cells.[103,104]

5. CONCLUSIONS

Taken together, the vertebrate adherens junction system may therefore function as a primary site for inter- and intra-cellular signaling during development as well as an adhesive domain regulating tissue organization and segregation. Although this adhesion system is established and active before that of other junction types during epithelial differentiation, signal transduction properties have recently also been identified at the tight junction (reviewed in ref. 105). Understanding how inductive signals during development lead to changes in cellular adhesivity and junction organization and how the assembly/disassembly state of different junctions is co-ordinated during differentiation will be important goals in future studies.

ACKNOWLEDGMENTS

The support of the Wellcome Trust and the Wessex Medical Trust for research in my laboratory is gratefully acknowledged. Thanks also to many collaborators for sharing precious antibodies.

REFERENCES

1. Fleming TP. ed. Epithelial Organization and Development. London: Chapman and Hall, 1992.
2. Takeichi M. Cadherin cell adhesion receptors as a morphogenetic regulator. Science 1991; 251:1451-55.
3. Edelman GM, Crossin KL. Cell adhesion molecules: implications for a molecular histology. Ann Rev Biochem 1991; 60:155-190.
4. Hynes RO, Lander AD. Contact and adhesive specificities in the associations, migrations and targeting of cells and axons. Cell 1992; 68:303-22.
5. Geiger B, Ayalon O. Cadherins. Ann Rev Cell Biol 1992; 8:307-32.
6. Legan PV, Collins JE, Garrod DR. The molecular biology of desmosomes and hemidesmosomes: 'what's in a name?' BioEssays 1992; 14:385-393.
7. Garrod DR, Collins JE. Intercellular junctions and cell adhesion in epithelial cells. In: Fleming TP, ed. Epithelial Organization and Development. London: Chapman and Hall, 1992:1-52.
8. Grunwald GB. The structural and functional analysis of cadherin calcium- dependent cell adhesion molecules. Curr Opin Cell Biol 1993; 5:797-805.
9. Pouliot Y. Phylogenetic analysis of the cadherin superfamily. BioEssays 1992; 14:743-48.
10. Boller K, Vestweber D, Kemler R. Cell adhesion molecule uvomorulin is localized in the intermediate junctions of adult intestinal epithelial cells. J Cell Biol 1985; 100:327-32.
11. Volk T, Geiger B. A-CAM: a 135kD receptor of intercellular adherens junctions. I. Immunoelectron microscopic localization and biochemical studies. J Cell Biol 1986; 103:1441-50.
12. Koch PJ, Walsh MJ, Schmelz M, Goldschmidt MD, Zimbelmann R, Franke WW. Identification of desmoglein, a constitutive desmosomal glycoprotein, as a member of the cadherin family of cell adhesion molecules. Eur J Cell Biol 1990; 53:1-12.
13. Collins JE, Legan PK, Kenny TP, Macgarvie J, Holton JL, Garrod DR. Cloning and sequencing of desmosomal glycoproteins 2 and 3 (desmocollins): cadherin-like desmosomal adhesion molecules with heterogeneous cytoplasmic domains. J Cell Biol 1991; 113:381-91.
14. Itoh M, Nagafuchi, A. Yonemura S, Kitani-Yasuda T, Tsukita S, Tsukita S. The 220-kD protein colocalizing with cadherins in non-epithelial cells is identical to ZO-1, a tight junction-associated protein in epithelial cells: cDNA cloning and immunoelectron microscopy. J Cell Biol 1993; 121:491-502.

15. Furuse M, Hirase T, Itoh M, Nagafuchi A, Yonemura S, Tsukita S, Tsukita S. Occludin: a novel integral membrane protein localizing at tight junctions. J Cell Biol 1993; 123:1777-88.
16. Blaschuk OW, Sullivan R, David S, Pouliot Y. Identification of a cadherin cell adhesion recognition sequence. Dev Biol 1990; 139:227-29.
17. Nagafuchi A, Takeichi M. Cell binding function of E-cadherin is regulated by the cytoplasmic domain. EMBO J 1988; 7:3679-94.
18. Ozawa M, Ringwald M, Kemler R. Uvomorulin-catenin complex formation is regulated by a specific domain in the cytoplasmic region of the cell adhesion molecule. Proc Natl Acad Sci USA 1990; 87:4246-50.
19. Hirano S, Nose A, Hatta K, Kawakami A, Takeichi M. Calcium-dependent cell-cell adhesion molecules (cadherins): subclass specificities and possible involvement with actin bundles. J Cell Biol 1987; 105:2501-10.
20. Nelson WJ, Shore E, Wang AZ, Hammerton RW. Identification of a membrane-cytoskeletal complex containing the cell adhesion molecule uvomorulin (E-cadherin), ankyrin and fodrin in Madin-Darby canine kidney epithelial cells. J Cell Biol 1990; 110:349-57.
21. Stappenbeck TS, Green K. The desmoplakin carboxyl terminus co-aligns with and specifically disrupts intermediate filament networks when expressed in cultured cells. J Cell Biol 1992; 116:1197-1209.
22. Troyanovsky SM, Eshkind LG, Troyanovsky RB, Leube RE, Franke WW. Contributions of cytoplasmic domains of desmosomal cadherins to desmosome assembly and intermediate filament anchorage. Cell 1993; 72:561-74.
23. Cowin P, Kapprel HP, Franke WW, Tamkun J, Hynes RO. Plakoglobin: a protein common to different types of intercellular adhering junctions. Cell 1986; 46:1063-73.
24. Hirano S, Kimoto N, Shimoyama Y, Hirohashi S, Takeichi M. Identification of a neural α-catenin as a key regulator of cadherin function and multicellular organization. Cell 1992; 70:293-301.
25. Nose A, Takeichi M. A novel cadherin cell adhesion molecule: its expression patterns associated with implantation and organogenesis of mouse embryos. J Cell Biol 1986; 103:2649-58.
26. Friedlander DR, Mege R-M, Cunningham BA, Edelman GM. Cell sorting-out is modulated by both the specificity and amount of different cell adhesion molecules (CAMS) expressed on cell surfaces. Proc Natl Acad Sci USA 1989; 86:7043-47.
27. Nose A, Nagafuchi A, Takeichi M. Expressed recombinant cadherins mediate cell sorting in model systems. Cell 1988; 54:993-1001.
28. Choi Y-S, Sehgal R, McCrea P, Gumbiner B. A cadherin-like protein in eggs and cleaving embryos of *Xenopus laevis* is expressed in oocytes in response to progesterone. J Cell Biol 1990; 110:1575-82.
29. Angres B, Muller AHJ, Kellermann J, Hausen P. Differential expression of two cadherins in *Xenopus laevis*. Development 1991; 111:829-44.
30. Ginsberg D, DeSimone D, Geiger B. Expression of a novel cadherin (EP- cadherin) in unfertilized eggs and early *Xenopus* embryos. Development 1991; 111:315-25.
31. Levi G, Ginsberg D, Girault J-M, Sabanay I, Thiery J-P, Geiger B. EP- cadherin in muscles and epithelia of *Xenopus laevis* embryos. Development 1991; 113:1335-44.
32. Vestweber D, Gossler, A, Boller K, Kemler R. Expression and distribution of cell adhesion molecule uvomorulin in mouse preimplantation embryos. Dev Biol 1987; 124:451-56.
33. Perry MM. Microfilaments in the external surface layer of the early amphibian embryo. J Embryol Exp Morph 1975; 33:127-46.
34. Schneider S, Herrenknecht K, Butz S, Kemler R, Hausen P. Catenins in *Xenopus* embryogenesis and their relation to the cadherin-mediated cell-cell adhesion system. Development 1993; 118:629-40.
35. Heasman J, Ginsberg D, Geiger B, Goldstone K, Pratt T, Yoshida-Noro C, Wylie C. A functional test for maternally inherited cadherin in *Xenopus* shows its importance in cell adhesion at the blastula stage. Development 1994; 120:49-57.

36. Thiery JP, Delouvee A, Gallin WJ, Cunningham BA, Edelman GM. Ontogenetic expression of cell adhesion molecules: L-CAM is found in epithelia derived from the three primary germ layers. Dev Biol 1984; 102:61-78.
37. Vestweber D, Kemler R. Rabbit antiserum against a purified surface glycoprotein decompacts mouse preimplantation embryos and reacts with specific adult tissues. Exp Cell Res 1984; 152:169-78.
38. Shirayoshi Y, Okada TS, Takeichi M. The calcium-dependent cell-cell adhesion system regulates inner cell mass formation and cell surface polarization in early mouse development. Cell 1983; 35:631-38.
39. Johnson MH, Maro B, Takeichi M. The role of cell adhesion in the synchronization and orientation of polarization in 8-cell mouse blastomeres. J Embryol Exp Morph 1986; 93:239-55.
40. Sefton M, Johnson MH, Clayton L. Synthesis and phosphorylation of uvomorulin during mouse early development. Development 1992; 115:313-18.
41. Levy JB, Johnson MH, Goodall H, Maro B. Control of the timing of compaction: a major developmental transition in mouse early development. J Embryol Exp Morph 1986; 95:213-37.
42. Winkel GK, Ferguson JE, Takeichi M, Nuccitelli R. Activation of protein kinase C triggers premature compaction in the 4-cell stage mouse embryo. Dev Biol 1990; 138:1-15.
43. Bloom TL, McConnell J. Changes in protein phosphorylation associated with compaction of the mouse preimplantation embryo. Molec Reprod Devel 1990; 26:199-210.
44. Ducibella T, Anderson E. Cell shape and membrane changes in the 8-cell mouse embryo: prerequisites for morphogenesis of the blastocyst. 1975; 47:45-58.
45. Ziomek CA, Johnson MH. Cell surface interactions induce polarisation of mouse 8-cell blastomeres at compaction. Cell 1980; 21:935-42.
46. Fleming TP, Johnson MH. From egg to epithelium. Ann Rev Cell Biol 1988; 4:459-85.
47. Fleming TP. Trophectoderm biogenesis in the preimplantation mouse embryo. In: Fleming TP, ed. Epithelial Organization and Development. London:Chapman and Hall, 1992:111-36.
48. Reima I. Maintenance of compaction and adherent-type junctions in mouse morula-stage embryos. Cell Differen 1990; 29:143-53.
49. Revel J-P, Yip P, Chang LL. Cell junctions in the early chick embryo—a freeze etch study. Dev Biol 1973; 35:303-17.
50. Ducibella T, Albertini DF, Anderson E, Biggers JD. The preimplantation mammalian embryo: characterization of intercellular junctions and their appearance during development. Dev Biol 1975; 45:231-50.
51. Pratt HPM. Membrane organization in the preimplantation mouse embryo. J Embryol Exp Morph 1985; 90:101-21.
52. Fleming TP, McConnell J, Johnson MH, Stevenson BR. Development of tight junctions *de novo* in the mouse early embryo: control of assembly of the tight junction-specific protein, ZO-1. J Cell Biol 1989; 108:1407-1418.
53. Fleming TP, Hay MJ. Tissue-specific control of expression of the tight junction polypeptide ZO-1 in the mouse early embryo. Development 1991; 113:295-304.
54. Fleming TP, Hay M, Javed Q, Citi S. Localization of tight junction protein cingulin is temporally and spatially regulated during early mouse development. Development 1993; 117:1135-44.
55. Javed Q, Fleming TP, Hay M, Citi S. Tight junction protein cingulin is expressed by maternal and embryonic genomes during early mouse development. Development 1993; 117:1145-51.
56. Fleming TP, Javed Q, Hay M. Epithelial differentiation and intercellular junction formation in the mouse early embryo. Development supplement 1992; 105-112.
57. Fleming TP, Javed Q, Collins J, Hay M. Biogenesis of structural intercellular junctions during cleavage in the mouse embryo. J Cell Science supplement 1993; 17:119-25.
58. Fleming TP, Garrod DR, Elsmore AJ. Desmosome biogenesis in the mouse preimplantation embryo. Development 1991; 112:527-39.

59. Collins JC, Lorimer JE, Garrod DR, Pidsley SC, Buxton RS, Fleming TP. Regulation of desmocollin transcription in mouse preimplantation embryos. 1994 submitted for publication.
60. Denker H-W. Implantation: a cell biological paradox. J Exp Zool 1993; 266:541-558.
61. Kadokawa Y, Fukita I, Nose A, Takeichi M, Nakatsuji N. Expression of E- and P-cadherin in mouse embryos and uteri during the periimplantation period. Dev Growth Differen 1989; 31:23-30.
62. Damjanov I, Damjanov A, Damsky CH. Developmentally regulated expression of the cell-cell adhesion glycoprotein cell-CAM 120/80 in pre- implantation mouse embryos and extraembryonic membranes. Dev Biol 1986; 116:194-202.
63. Strickland S, Richards WG. Invasion of the trophoblasts. Cell 1992; 71:355-57.
64. Shimoyama Y, Yoshida T, Terada M, Shimosato Y, Abe O, Hirohashi S. Molecular cloning of a human Ca^{++}-dependent cell-cell adhesion molecule homologous to mouse placental cadherin: its low expression in human placental tissues. J Cell Biol 1989; 109:1787-94.
65. Coutifaris C, Kao L-C, Sehdev HM, Chin U, Babalola G, Blaschuk OW, Strauss JF. E-cadherin expression during the differentiation of human trophoblast. Development 1991; 113:767-77.
66. Johnson KE, Boucaut J-C, DeSimone DW. The role of the extracellular matrix in amphibian gastrulation. Curr Top Dev Biol 1992; 27:92-127.
67. Winklbauer R. Mesodermal cell migration during *Xenopus* gastrulation. Dev.Biol 1990; 142:155-68.
68. Keller R, Winklbauer R. Cellular basis of amphibian gastrulation. Curr Top Devel Biol 1992; 27:39-89.
69. Choi Y-S, Gumbiner B. Expression of cell adhesion molecule E-cadherin in *Xenopus* embryos begins at gastrulation and predominates in the ectoderm. J Cell Biol 1989; 108:2449-58.
70. Winklbauer R, Selchow A, Nagel M, Angres B. Cell interaction and its role in mesoderm cell migration during *Xenopus* gastrulation. Devel Dynam 1992; 195:290-302.
71. Baker PC. Fine structure and morphogenetic movements in the gastrula of the treefrog, *Hyla regilla*. J Cell Biol 1965; 24:95-116.
72. Perry MM, Waddington CH. Ultrastructure of the blastopore cells in the newt. J Embryol Exp Morph 1966; 15:317-30.
73. Burdsal CA, Damsky CH, Pedersen RA. The role of E-cadherin and integrins in mesoderm differentiation and migration at the mammalian primitive streak. Development 1993; 118:829-44.
74. Detrick RJ, Dickey D, Kintner C. The effects of N-cadherin misexpression on morphogenesis in *Xenopus* embryos. Neuron 1990; 4:493-506.
75. Hatta K, Takeichi M. Expression of N-cadherin adhesion molecules associated with early morphogenetic events in chick development. Nature 1986; 320:447-49.
76. Duband JL, Volberg T, Sabanay I, Thiery J-P, Geiger B. Spatial andtemporal distribution of the adherens junction-associated adhesion molecule A-CAM during avian embryogenesis. Development 1988; 103:325-44.
77. Baker P, Schroeder TE. Cytoplasmic filaments and morphogenetic movement in the amphibian neural tube. Dev Biol 1967; 15:432-50.
78. Burnside B. Microtubules and microfilaments in amphibian neurulation. Am Zool 1973; 13:989-1006.
79. Lee H-Y, Nagele RG. Studies on the mechanism of neurulation in the chick: interrelationship of contractile proteins, microfilaments and the shape of neuroepithelial cells. J Exp Zool 1985; 235:205-15.
80. Schoenwolf GC, Smith JL. Mechanisms of neurulation: traditional viewpoint and recent advances. Development 1990; 109:243-70.
81. Fujimori T, Miyatani S, Takeichi M. Ectopic expression of N-cadherin perturbs histogenesis in *Xenopus* embryos. Development 1990; 110:97-104.
82. Bronner-Fraser M, Wolf JJ, Murray BA. Effects of antibodies against N- cadherin and N-CAM on the cranial neural crest and neural tube. Dev Biol 1992; 153:291-301.
83. Kintner C. Regulation of embryonic cell adhesion by the cadherin cytoplasmic domain. Cell 1992; 69:225-36.

84. Erickson CA. Morphogenesis of the avian trunk neural crest: use of morphological techniques in elucidating the process. Microscopy Research Technique 1993; 26:329-51.
85. Hatta K, Takagi S, Fijisawa H, Takeichi M. Spatial and temporal expression pattern of N-cadherin cell adhesion molecules correlated with morphogenetic processes of chicken embryos. Dev Biol 1987; 120:215-27.
86. Newgreen DF, Gooday D. Control of the onset of migration of neural crest cells in avian embryos: role of Ca^{++}-dependent cell adhesion. Cell Tissue Res 1985; 239:329-36.
87. Ben Ze'ev A. The relationship between cytoplasmic organization, gene expression and morphogenesis. Trends Biochem Sci. 1986; 11:478-81.
88. Bendori R, Salomon D, Geiger B. Contact-dependent regulation of vinculin expression in cultured fibroblasts: a study with vinculin-specific cDNA probes. EMBO J 1987; 6:2897-2905.
89. Roark EF, Paradies NE, Lagunowich LA, Grunwald GB. Evidence for endogenous proteases, mRNA level and insulin as multiple mechanisms of N-cadherin down-regulation during retinal development. Development 1992; 114:973- 84.
90. Volk T, Volberg T, Sabanay I, Geiger B. Cleavage of A-CAM by endogenous proteinases in cultured lens cells and in developing chick embryos. Dev Biol 1990; 139:314-26.
91. Wheelock MJ, Buck CA, Bechtol KB, Damsky CH. The soluble 80-kD fragment of cell-CAM 120/80 disrupts cell adhesion. J Cell Biochem 1987; 34:187- 202.
92. Tsukita S, Oishi K, Akiyama T, Yuji Y, Tadashi Y, Tsukita S. Specific proto-oncogene tyrosine kinases of the *src* family are enriched in cell- to-cell junctions where the level of tyrosine phosphorylation is elevated. J Cell Biol 1991; 113:867-79.
93. Volberg T, Zick Y, Dror R, Sabanay I, Gilon C, Levitzki A, Geiger B. The effect of tyrosine-specific protein phosphorylation on the assembly of adherens-type junctions. EMBO J 1992; 11:1733-42.
94. Hamaguchi M, Matsuyoshi N, Ohnishi Y, Gotoh B, Takeichi M, Nagai Y. $p60^{v\text{-}src}$ causes tyrosine phosphorylation and inactivation of the N-cadherin- catenin cell adhesion system. EMBO J 1993; 12:307-14.
95. Behrens J, Vakaet L, Friis R, Winterhager E, Van Roy F, Mareel MM, Birchmeier W. Loss of epithelial differentiation and gain of invasiveness correlates with tyrosine phosphorylation of the E-cadherin/β-catenin complex in cells transformed with a temperature- sensitive *v-src* gene. J Cell Biol 1993; 120:757-66.
96. Gumbiner B. Epithelial morphogenesis. Cell 1992; 69:385-87.
97. Birchmeier C, Sonnenberg E, Weidner KM, Walter B. Tyrosine kinase receptors in the control of epithelial growth and morphogenesis during development. BioEssays 1993; 15:185-90.
98. Schneider R. The human protooncogene *ret*: a communicative cadherin? Trends Biochem Sci 1992; 17:468-69.
99. Peifer M, McCrea PD, Green K, Wieschaus E, Gumbiner B. The vertebrate adhesive junction proteins β-catenin and plakoglobin and the Drosophila segment polarity gene *armadillo* form a multigene family with similar properties. J Cell Biol 1990; 118:681-91.
100. Peifer M. The product of the Drosophila segment polarity gene *armadillo* is part of a multiprotein complex resembling the vertebrate adherens junction. J Cell Sci 1993; 105:993-1000.
101. Peifer M, Rauskolb M, Williams B, Riggleman B, Wieschaus E. The segment polarity gene *armadillo* affects the *wingless* signaling pathway in both embryonic and adult pattern formation. Development 1991; 111:1028-43.
102. McCrea PD, Brieher WM, Gumbiner B. Induction of a secondary body axis in *Xenopus* by antibodies to β-catenin. J Cell Biol 1993; 123:477-84.
103. Docherty P, Ashton SV, Skaper SD, Leon S, Walsh FS. Ganglioside modulation of neural cell adhesion molecule and N-cadherin-dependent neurite outgrowth. J Cell Biol 1991; 117:1093-99.
104. Saffell JL, Walsh FS, Docherty P. Direct activation of second messenger pathways mimics cell adhesion molecule-dependent neurite outgrowth. J Cell Biol 1992; 118:663-70.
105. Anderson JM, Balda MS, Fanning AS. The structure and regulation of tight junctions. Curr Opin Cell Biol 1993; 5:772-78.

CHAPTER 5

The Molecular Composition of Tight Junctions

Sandra Citi

1. INTRODUCTION

In Vertebrate polarized epithelial cells the junctional "complex" anchoring each cell to the next includes tight junctions (TJ), *zonulae adhaerentes* (ZA) and desmosomes (DS).[1] When cultured epithelial cells are immunofluorescently labeled with antibodies against proteins localized at each of these junctions, the labeling is characteristically confined to the sites of cell-cell contact (Fig. 5.1).

TJ are the most apical element of the junctional complex and form a sealing belt separating different compartments of the extracellular space, thus controlling the permeability of epithelial and endothelial monolayers to ions and molecules that are asymmetrically distributed in the apical versus basolateral spaces (Fig. 5.2) (see also chapters 6 and 7 in this volume and refs. 2-7 for reviews). TJ are therefore fundamentally important in tissues that carry out polarized secretion and/or absorption functions (i.e. glandular and lining epithelia) and maintain a tight separation between blood and interstitial fluids (e.g. endothelia of brain capillaries, "blood-brain barrier").

TJ show two important properties: heterogeneity and modulation. For example, the morphology, permeability and ion selectivity of TJ depends on the type of tissue and species examined (see also ref. 8 for a review). Secondly, the permeability of TJ can be physiologically and pharmacologically modulated (see also chapters 6 and 7 in this volume). These properties are consistent with the existence of a proteinaceous "channel" that bridges the membranes of adjacent cells and is responsible for the "barrier" function of TJ. In addition to membrane protein(s), other protein components located in the cytoplasm may be responsible for many of the functional properties

---

Molecular Mechanisms of Epithelial Cell Junctions: From Development to Disease
Edited by Sandra Citi, M.D., Ph.D.;

of TJ, for example their involvement in maintaining apico-basolateral polarity[9,10] and their regulation by various signaling mechanisms.

The idea that proteins (and not mainly lipids)[11,12] contribute to the formation of TJ was proposed long before specific TJ proteins were identified. For example, it was observed that the morphology of TJ in the electron microscope could be influenced by protein fixatives,[13] and that inhibition of

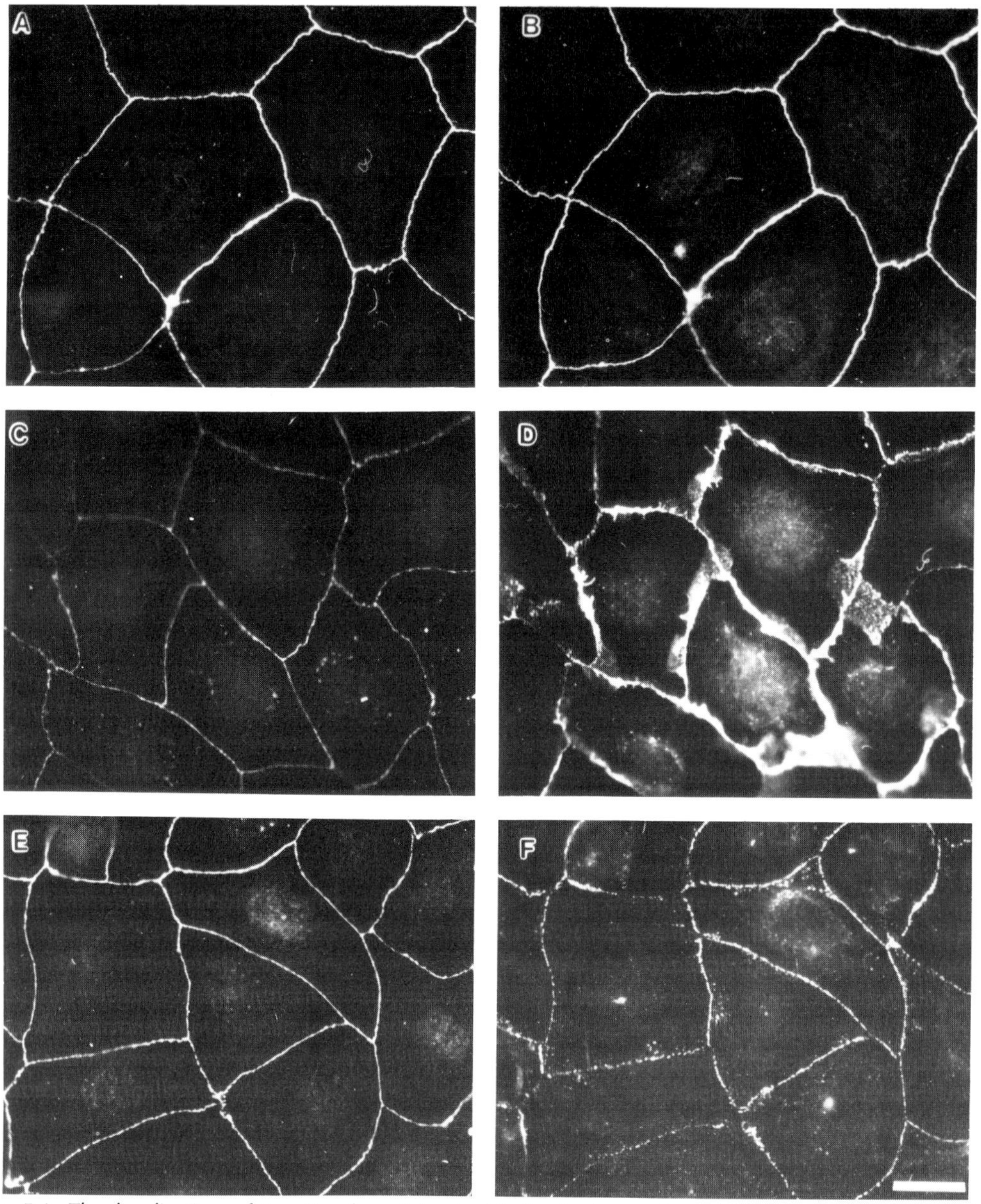

Fig. 5.1. The localization of junctional proteins in Madin Darby canine kidney (MDCK) cells. Double immunofluorescence of cultured MDCK cells stained with antibodies against cingulin and ZO-1 (A and B), cingulin and E-cadherin (C and D), cingulin and desmoplakin (E and F). Fixation, permeabilization and immunolabeling was as described[60]. Cingulin and ZO-1 are localized at TJ, E-cadherin is concentrated in the ZA (although it is also distributed along all the lateral border of these cells) and desmoplakin is localized at DS. Note that proteins belonging to different junctions appear co-distributed at the immunofluorescent level of resolution, underlining the need for immunoelectron microscopy for the definitive localization at one or the other junction. Scale bar = 5 μm.

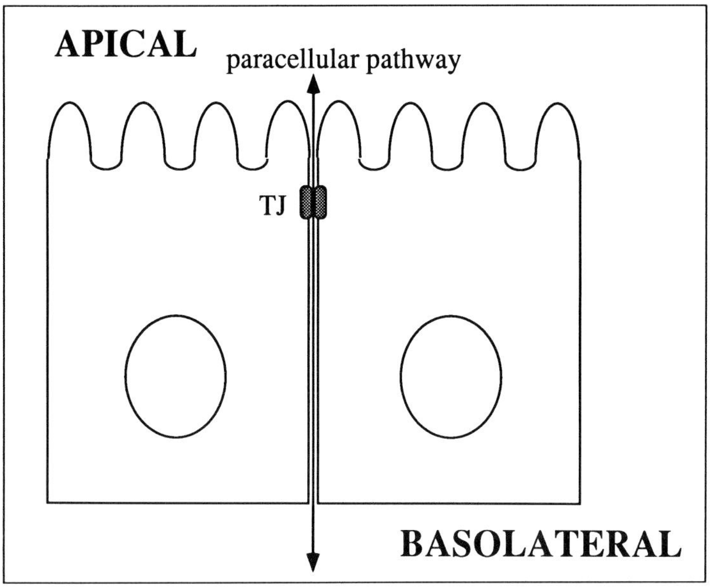

Fig. 5.2. The barrier function of TJ. Schematic diagram representing a TJ between two epithelial cells, as the structure regulating the passage of molecules through the paracellular pathway

Table 5.1. Summary of the properties of known and putative TJ proteins

Name	Apparent MR	Non-epithelial tissues	Im-EM sections	Im-EM membrane protein	Distance from membrane	Phosphorylation
Occludin	65 kD	No	Yes	Yes	integral	n.d.
Cingulin	140 kD	No	Yes	n.d.	40–65 nm	Yes (Ser)
ZO–1	210–225 kD	Yes	Yes	Yes	14-23 nm	Yes (Ser)
ZO–2	160 kD	No	No	Yes	n.d.	Yes
7H6	155–175 kD	n.d.	No	Yes	41 nm	n.d.
BG9.1	192 kD	n.d.	No	Yes	n.d.	n.d.
rab13	41 kD	Yes	n.d.	n.d.	n.d.	n.d.

Apparent Mr = apparent molecular masses based on SDS-PAGE and immunoblotting of different tissues and species
Non-epithelial tissues = occurrence of each protein in non-epithelial tissues and cells, based mainly on immunofluorescence localization in non-epithelial tissue sections, fibroblasts, cardiac myocites and other non-epithelial cells
Im-EM sections = immunoelectron microscopic localization of the protein(s) in the TJ of sectioned epithelial cells
Im-EM membrane proteins = immunoelectron microscopic localization of the protein(s) in the TJ of junctional membrane preparations (typically liver bile canaliculi)
Distance from membrane = distance of the immunoelectron microscopy gold particles from midline of membrane
Phosphorylation = incorporation of ^{32}[P] in protein immunoprecipitated from epithelial cell extracts (Ser = phosphorylation occurring on serine residues)
n.d. = not determined
See text for references and additional information

protein synthesis could prevent the assembly of functional TJ[14] (see also [15]). Furthermore, treatment of TJ preparations with detergents did not dissolve TJ structures, as detected by electron microscopy.[16] However, it is only in the last eight years that several specific TJ proteins have been identified and characterized (see Table 5.1). The identification of TJ proteins has been made possible by the production of specific monoclonal antibodies against relatively impure fractions (for example junctional membrane preparations) and by a coimmunoprecipitation approach, using antibodies to known TJ proteins. The objective of this chapter is to review the recent and rapidly growing information on the structure, function and regulation of specific TJ components.

2. The Membrane Domain of TJ

According to a simplified scheme commonly used for adherens-type junctions,[17] cell-cell junctions can be represented as comprising several "domains": extracellular domain, membrane domain, submembrane domain (often referred to as junctional "plaque" or cytoplasmic "undercoat" domain) and cytoskeletal domain. The extracellular domain contains the extracellular mojeties of membrane proteins involved in the interaction with complementary molecules on adjacent cells. The membrane domain contains the transmembrane portions of such integral membrane proteins. The submembrane domain comprises the intracellular mojeties (cytoplasmic tails) of the transmembrane proteins and other proteins, which may interact with membrane proteins, cytoskeletal proteins, the membrane and/or with one another. Finally, the cytoskeletal domain is formed by actin microfilaments (TJ and *zonulae adhaerentes*) or intermediate filaments (desmosomes) tethered to the submembrane domain. Defining the structural and functional interactions within each domain and between proteins of different domains is a major goal in trying to understand how junctions work.

The membrane domain of TJ has been intensely studied by electron microscopy in the 60s through the early 80s. By transmission EM, the membranes of neighboring cells appear intimately apposed along a series of discrete points.[1] In freeze-fractured TJ membranes these sites of close apposition correspond to a pattern of branching and anastomosing intramembranous fibrils (apparently made of a series of particles aligned together) and complementary grooves[18] (see also chapters 2 and 7 in this volume). For many years the nature of the TJ fibrils and particles seen in freeze-fractured specimens has remained a major unsolved mystery. Several models were proposed, according to which the fibrils were made of proteins, lipids or both.[11,12,14,16,18,19]

The role of membrane lipids in the composition of the membrane domain of TJ has been investigated by a number of approaches, including the perturbation of TJ structure and function by exogeneously added lipids (see chapter 7 in this volume). One interesting series of experiments exploited fluorescently labeled lipid probes and addressed questions concerning the physical continuity between lipids on adjacent cells, or one membrane domains separated by the TJ "fence".[20,21] The results of these experiments indicated that a simple lipid model for TJ structure was unlikely and that lipids in the exoplasmic leaflets of the plasma membrane cannot diffuse across TJ.[20,21]

Nothing was known about specific protein components of the membrane domain of TJ until the recent identification of occludin,[22] a transmembrane protein localized at epithelial TJ. This discovery represents a major breakthrough in the field.[23]

2.1 Occludin

Occludin was identified as a M_r 65 kD antigen recognized by monoclonal antibodies raised against a junctional preparation from chick liver.[22] Immunofluorescence of liver and intestinal epithelia and brain and heart tissue demonstrated that occludin is localized in the junctional regions of epithelial and endothelia, but is absent from nerve cells and cardiac intercalated discs.[22] Immunoelectron microscopy showed the exclusive association of occludin with TJ of epithelial cells. Since the gold particles in the immunolabeled cryosections were mostly distributed on the sides of the TJ membranes, rather than directly over them,[22] it appeared that the epitopes recognized by the antibodies were localized in the submembrane domain of the junction. However, occludin is an integral membrane protein based on biochemical and structural criteria. First, the junctional membrane preparation used as immunogen was "stripped" of peripheral membrane proteins by treatment with 1 M acetic acid. Second, the protein requires ionic detergents (i.e. SDS) to be extracted from membranes. Third, the derived amino acid sequence of occludin shows four predicted membrane-spanning hydrophobic domains.[22]

Although the sequence of occludin does not show significant homology to known proteins, the overall membrane topology of occludin resembles the topology of other integral membrane proteins (proteolipoprotein of myelin, connexins of gap junctions and synaptophysin of synaptic vesicles) that are thought to function by bringing apposed membrane leaflets into close contact.[22] Occludin is predicted to comprise two extracellular loop domains of 43 aminoacids, which have been speculated to be very important for TJ formation.[22] In the model proposed by Furuse et al,[22] the 57-amino acid-long NH2 terminus and the 255-amino acid-long hydrophylic COOH terminus are exposed on the cytoplasmic surface. The high susceptibility of the COOH tail to proteolysis may be responsible for the presence of polypeptides with different electrophoretic mobilities (apparent M_r ranging from 58 kD to 66 kD) in immunoblots stained with anti-occludin monoclonal antibodies.

The immunofluorescent staining of junctions of both epithelial and endothelial cells by anti-occludin antibodies and the detection by Northern blot analysis of a single 2.3-kb message in liver, lung and brain tissue, suggest that the same or highly related occludin molecules are expressed and localized at TJ in epithelial and endothelial cells.[22] With regard to the expression of occludin in species different from chicken, it was reported that three anti-occludin monoclonal antibodies do not recognize TJ in mammalian tissues.[22] Since the epitopes for these antibodies are located in the COOH-terminal portion of occludin, the lack of cross-reactivity may be due to a high interspecies sequence variability in this region. On the other hand, the observation that in mammalian cells transfected with chicken occludin the protein becomes localized at TJ (Furuse et al, manuscript in preparation) suggests significant structural and functional homologies between occludin in different species.

The discovery of occludin has opened a new and exciting area of research, where important questions can now be addressed directly.[23] For example, how do the extracellular loops of occludin assemble to generate a ion-selective pore controlling the paracellular pathway? Does occludin behave as a homophylic adhesion molecule? If occludin is the major constituent of TJ fibrils, what is the mechanism controlling the "polymerization" of occludin molecules into such fibrils? How does occludin interact with other TJ and cytoskeletal proteins and how is occludin regulated? It is likely that many of these questions will be addressed in the near future and the results will be of general interest to researchers in the fields of epithelial and endothelial cell physiology and biology.

3. THE SUBMEMBRANE DOMAIN OF TJ

3.1 ZO-1

3.1.1 Identification and biochemical properties of ZO-1

ZO-1 was the first protein to be localized in the submembrane domain of TJ by immunoelectron microscopy.[24] The identification of ZO-1 was the result of a biochemical approach similar to that used to identify occludin and consists in the production of antibodies against membrane preparations enriched in junctional membranes. The isolation of fractions specifically enriched in TJ is made relatively difficult by the fact that TJ, unlike gap junctions, ZA or DS have a buoyant density similar to that of normal plasma membranes. In addition, TJ proteins are not highly abundant cellular components, as suggested by the observation that little electron-dense material is detectable in the cytoplasmic domain of TJ.[1]

The mouse liver membrane fraction prepared by Stevenson and Goodenough appeared by electron microscopy to contain typical TJ structures, consisting of a network of junctional fibrils.[16] Using this fraction as an immunogen in rats, a monoclonal antibody was produced that recognized a M_r 225 kD antigen, named ZO-1 (for *zonula occludens*-1) and stained by immunoelectron microscopy the cytoplasmic faces of TJ of canaliculus-enriched membrane fraction.[24] Monoclonal antibodies against ZO-1 were also independently generated using a preparation of rat liver ZA-enriched membranes as an immunogen.[25] Initially the antigen (named 220 kD protein) protein was immunolocalized at cadherin-based cell-cell contacts in fibroblasts and cardiac cells and its localization in epithelial cells was not described.[25] Subsequent work has revealed that in epithelial cells the antigen is localized in TJ and is actually identical to ZO-1[26] (see also chapter 12 in this volume).

The biophysical properties of ^{35}S-labeled ZO-1 immunoprecipitated from MDCK cell extracts suggested to Anderson et al an asymmetric monomer shape with a Stokes' radius of 9.0 nm.[27] However purification of ZO-1 from brain tissue and rotary shadowing analysis indicates a spherical shape, with a diameter between 25 and 50 nm.[25]

3.1.2 Sequence of mammalian ZO-1

The complete nucleotide and predicted amino acid sequences of mouse and human ZO-1 have recently been reported.[26,28]

The mouse cDNA sequence (7046 bp, with an open reading frame encoding a 1745 amino acid polypeptide) was obtained from several clones isolated from mouse teratocarcinoma cell libraries.[26] The predicted mouse ZO-1 sequence does not contain long hydrophobic stretches or signal peptides,[26] confirming that ZO-1 is neither a transmembrane nor a secretory protein. It was also noted that a leucine-zipper motif is present at aminoacids 754-775 and that regions with α–helical or β-sheet structure appeared to be absent.[26]

The human cDNA sequence (7888 bp, with an open reading frame encoding a 1736 amino acid polypeptide) was derived from several clones isolated from libraries of human liver and a hepatoma cell line.[28] A homology between the amino-terminal 793 residue of ZO-1 and a family of proteins implicated in signal transduction at specialized cell-cell contacts has been noticed by Anderson's group[28] (see also chapters 1 and 12 in this volume). This family includes the *disc large* (*dlg*) tumor suppressor of Drosophila, PSD-95, a protein from rat brain synapses and the human erythrocyte protein p55. Mutations in the dlg protein, which is localized at septate junctions of insect epithelial cells, result in the loss of epithelial apicobasal polarity and abnormal growth of imaginal discs in fly larvae (see also chapter 1). Members of the dlg family of proteins share a conserved domain of homology with the *src* oncogene homology region 3 (SH3) and a domain homologous to the enzyme guanylate kinase. The identification of these domains in ZO-1 suggests that this protein may be involved in signal transduction events occurring at the TJ. Unlike other members of the dlg family of proteins, ZO-1 contains an additional C-terminal 944 amino acid stretch, which includes an acidic region, an 80 amino acid domain which can be alternatively spliced and a proline-rich region with four tyrosine phosphorylation consensus sequences.[28] The proline-rich motifs represent potential SH3 domain binding sites and the alternatively spliced region generates two ZO-1 isoforms (see below).

Comparison of the sequences of mouse and human ZO-1 reveals that the presumed N-terminal amino acid sequence of mouse ZO-1 contains an additional 12 amino acid-stretch at its beginning. Thus, position 1 of the human sequence corresponds to position 13 of the mouse sequence. Direct chemical determination of the N-terminal sequence of mouse and human ZO-1 would be helpful to determine whether or not this discrepancy is due to alternate start sites. Overall, the sequences of mouse and ZO-1 are very similar (about 90% identity) and the majority of the amino acid substitutions are conservative. About three quarters of the amino acid changes in human versus mouse ZO-1 are located in the C-terminal half of the molecule and more than half of the remaining changes are clustered between residues 300-405. In the N-terminal half, which is the most conserved region of the molecule, residues 406 to 893 show only six substitutions, two of which are nonconservative. The sequence conservation in this region, which comprises the SH3 homology domain, the guanylate kinase and the acidic domains, suggests that these domains play an important functional role in the molecule. The alternatively spliced region that was described in the human sequence is also present in the mouse sequence, but the amino acid identity between human and mouse sequences in this region is lower (67%) than in the rest of the molecule, although most of the substitutions are conservative.

3.1.3 Tissue distribution and isoforms of ZO-1

ZO-1 has been detected by immunofluorescence in junctional complexes of several types of TJ-containing epithelia, such as hepatocytes, enterocytes, kidney and testis epithelial cells,[24,29] and endothelial cells.[90] Immunoelectron microscopy of liver membranes,[24] enterocytes and MDCK cells[30] with anti-ZO-1 antibodies confirms that in these cells ZO-1 is localized in proximity to the TJ membrane.

Recently, the exclusive localization of ZO-1 at TJ has been questioned by several observations. First, ZO-1 was found to be localized along slit diaphragms of the glomerular epithelium, that do not form "classical" TJ, but could be considered as TJ "variants".[31] ZO-1 protein and mRNA could also be detected by immunofluorescence in cells and tissues that are not believed to contain TJ, such as astrocytes, myeloma cells and fibroblasts.[32] Independent immunofluorescence and immunoelectron microscopy studies by the Tsukita laboratory showed that ZO-1 is abundant in brain and is present in fibroblasts and in the intercalated discs of cardiac muscle cells, where it is colocalized with P-cadherin and N-cadherin, respectively.[25,26] Thus ZO-1 in epithelial cells appears to be a bona fide TJ marker, although its presence in non-epithelial cells indicates that its functions are not restricted to TJ (see also[6]).

ZO-1 shows an interspecies difference in its apparent molecular size. ZO-1 from mouse epithelial tissues migrates in SDS gels with an apparent molecular weight of 225 kD and ZO-1 from dog cultured kidney cells (MDCK) with an apparent size of 210 kD.[27] In human ZO-1 a 80-amino acid motif named "α" defines two isoforms of ZO-1 which are differentially expressed in various tissues.[33] In human cells, the two isoforms differ in apparent molecular size (about 220 kD and 215 kD), but when they are coexpressed they appear to be identically localized.[33] In the kidney, the ZO-1 isoform containing motif α (ZO-1α+) is not detected in junctions of glomerular epithelial cells and in the peritubular endothelial cells, whereas it is detected along TJ of renal tubular epithelial cells.[34] Further studies have confirmed that the ZO-1α- isoform is restricted to junctions of endothelial cells and highly specialized epithelial cells, such as Sertoli cells and glomerular podocytes.[91] The functional relevance of motif α is still unclear, since its expression does not appear correlated with the "tightness" of TJ in different epithelia, or their ultrastructural complexity. However, it was proposed that ZO-1α+ is expressed in junctions that show a lesser degree of plasticity, for example less capacity for gross structural remodeling.[91] It is also possible that motif α is involved in the binding of ZO-1 to other TJ proteins and/or its absence may allow ZO-1 to interact with components of junctions present in endothelial cells and glomerular podocytes.

The presence of multiple isoforms of TJ proteins may be one reason for the observed heterogeneity in TJ morphology and permeability properties. However, it cannot be ruled out that TJ heterogeneity and variations across tissues, species and phyla may also be generated by different protein compositions and/or by modulation of the levels of expression, or post-translational modification of specific TJ proteins. For example, ZO-1 homologs have been detected in insect septate junctions, but the expression of other TJ protein homologs in invertebrates has not

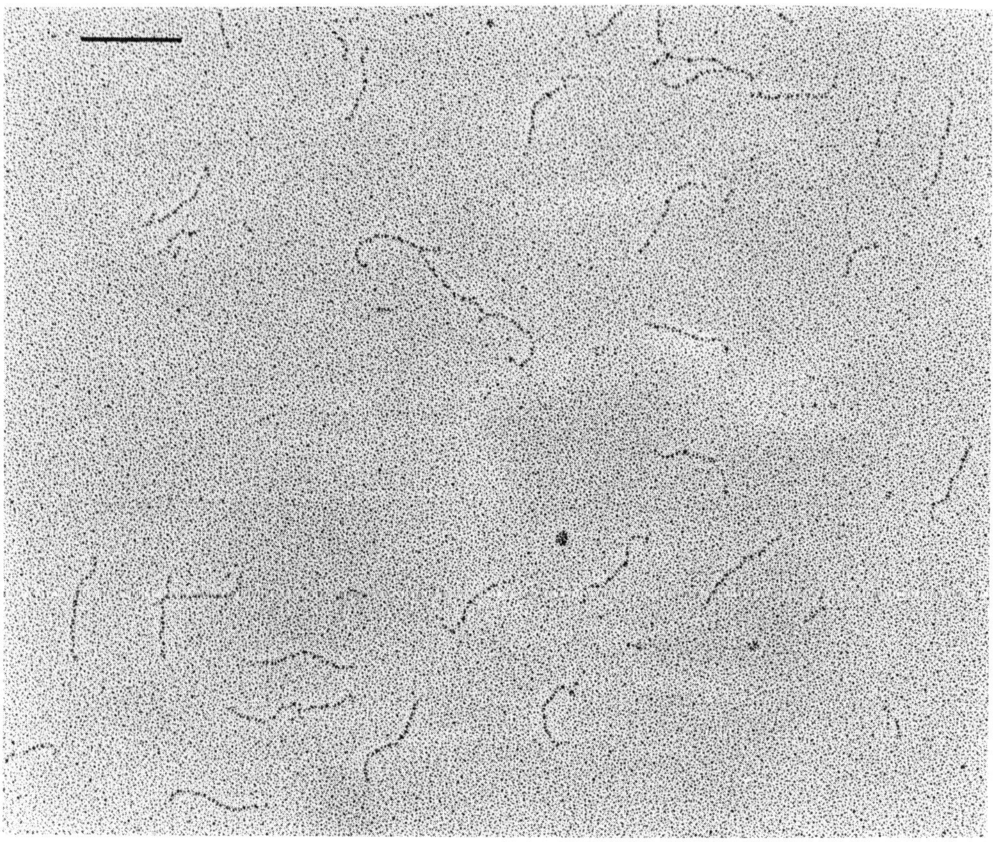

Fig. 5.3. Electron micrograph of rotary shadowed cingulin molecules. Cingulin (M_r 108 kD) purified from chicken intestinal brush border cells was processed for rotary shadowing with platinum as described.[36] Scale bar = 120 nm.

been detected (see also chapters 1 and 2 in this volume). On the other hand, no correlation was found between ZO-1 content and transepithelial resistance in two different strains of MDCK cells,[35] suggesting that it is difficult to make simplified correlations between TJ composition and function. Molecular studies of the expression of specific TJ proteins and their domains and isoforms in a variety of epithelial and non-epithelial cells are likely to clarify these questions.

3.2. Cingulin

3.2.1 Identification, tissue distribution and biochemical properties of cingulin

Monoclonal antibodies against cingulin (M_r 140 kD) were generated while preparing antibodies against nonmuscle myosin from chicken intestinal epithelial cells.[36] Cingulin is a highly asymmetric molecule (Fig. 5.3) and its elution profile from a gel filtration column was found to partially overlap with that of myosin, accounting for the contamination of myosin preparations by cingulin.[37]

The localization of cingulin at epithelial TJ was established on the basis of immunoelectron microscopic localization and immunofluorescent distribution in several chicken epithelial tissues.[36,37] Cingulin was detected in the junctional regions of cultured kidney cells, epithelial cells from intestine, liver, pancreas, kidney, retina and endothelial cells of capillaries of skeletal and cardiac muscle.[36,37] Interestingly, weaker cingulin labeling was also detected in the junctional regions of epithelial cells in non-keratinizing stratified epithelial (esophagus and cornea), where continuous TJ have not been described.[37] Further studies have described the localization of cingulin in junctions of other epithelia and species: dog kidney cells (MDCK),[30,38] human intestine,[39] guinea pig cochlea,[40] rat epididymis,[41] and mouse embryo.[42] Unlike ZO-1, cingulin immunofluorescent labeling was not detected either in cardiac intercalated disks, cadherin-based junctions or kidney glomeruli.[37] Immunoelectron microscopic localization of cingulin in ultrathin frozen sections of chicken intestinal epithelial cells[36,37] and in MDCK epithelial cells[30] demonstrated that cingulin is associated exclusively with the cytoplasmic faces of TJ.

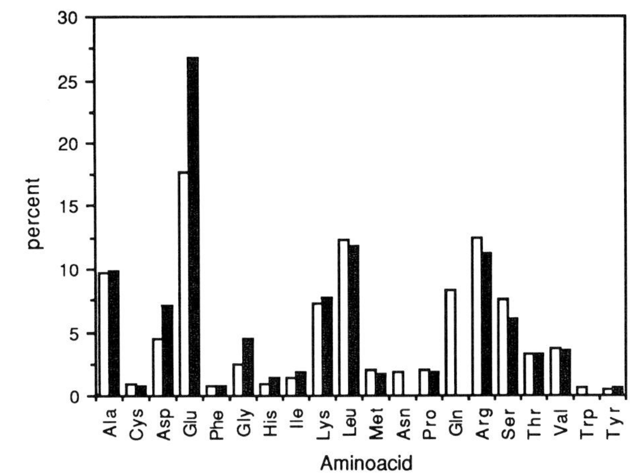

Fig. 5.4. The amino acid composition of cingulin. The histogram shows a comparison of the amino acid composition of purified chicken cingulin, determined as described[36] (black columns) and the amino acid composition of chicken cingulin based on the derived amino acid sequence of a 2.7 kB cDNA clone (white columns). The composition is expressed as percent of total content. Note that determination of amino acid composition by acid hydrolysis results in the transformation of Gln into Glu and Asn into Asp and in the loss of Trp.

The apparent molecular size of cingulin, based on SDS polyacrylamide gel electrophoresis, is 140 kD in all epithelial tissues and species (avian and mammalian) examined so far. However, a M_r 108 kD polypeptide is often labeled in immunoblots of whole extracts of intestinal cells and hepatocytes, suggesting that the molecule is susceptible to proteolysis.

The M_r 108 kD form of cingulin was purified to homogeneity from chicken intestinal epithelial cells and shown by rotary shadowing electron microscopy to have a highly asymmetric, rod-like shape[36] (Fig. 5.3), similar to that of the coiled-coil region (rod, light meromyosin) of myosins. Under nonreducing condition, purified cingulin migrates in SDS gels with an apparent size of about 220 kD, suggesting that it may form dimers by S-S bridges. Furthermore, hydrodynamic measurements confirm that the protein behaves in solution as a dimer, in agreement with its predicted coiled-coil structure.[37] The average contour length of rotary-shadowed cingulin molecules was about 130 nm and their width of about 2-3 nm.[36] There appeared to be no globular "tails" attached to the ends of the molecule (Fig. 5.3), however it can be speculated that the M_r 108 kD form of cingulin may be generated by proteolysis occurring at a "hinge" region between the coiled-coil region and a putative globular head or tail (M_r approximately 30 kD) attached at one end of the rod. Molecular studies suggest that such a domain should be located in the N-terminal portion of the molecule. Determination of the amino acid composition of cingulin (Fig. 5.4) and its isolectric point show that the molecule is rich in hydrophylic aminoacids and overall quite acidic. Furthermore, the protein is resistant to heat and ethanol treatment, consistent with it having the structure of a α-helical, coiled-coil dimer.[36,37]

3.2.2 Sequence of avian cingulin: hypotheses on cingulin function(s)

Several overlapping cDNA clones for cingulin have been isolated from libraries of chicken embryos and chicken intestinal epithelial cells, screened initially with a polyclonal anti-cingulin antiserum (Citi et al, unpublished results).The identity of the clones was established on the basis of sequence matches between the derived amino acid sequence and the sequence obtained from tryptic fragments of purified chicken cingulin. The largest clone (approximately 2.7 kb) contains an open reading frame of approximately 2.4 kb, encoding 811 aminoacids. The 3', polyadenylated end contains an untranslated stretch of 294 bp. The predicted molecular mass of the protein coded by the cingulin cDNA is approximatcly M_r 105 kD. When the cDNA is expressed in bacteria in vitro, using the pET expression system,[43] the recombinant protein migrates with an apparent size of about M_r 105 kD (Citi et al, unpublished results).

The derived amino acid sequence of the cingulin cDNA shows a high content in charged aminoacids and an overall amino acid composition very similar to that of the purified protein (Fig. 5.4). Secondary structure predictions show that the protein has a high content in α-helical structure and hydrophopic regions are absent. The heptad repeat motif (a, b, c, d, e, f, g, where a and d are hydrophobic aminoacids)[44] is detected throughout the sequence, in agreement with the biophysical data and suggesting a coiled-coil structure. Furthermore, comparison of the cingulin amino acid sequence to known protein sequences shows that the highest scores (25-30% overall identity) are obtained with the rod portions of nonmuscle myosins heavy chains (Citi et al, unpublished results).

In summary, the biochemical and molecular characterization of cingulin shows that cingulin contains a large coiled-coil domain, which may allow cingulin molecules to associate together into supramolecular assemblies, like other proteins with such a structure (e.g. myosin tails, tropomyosin, paramyosin, intermediate filament proteins).[44] Preliminary studies in vitro with purified cingulin (M_r 108 kD) indicate that under appropriate conditions cingulin can indeed form aggregates of discrete size, although ordered filaments or paracrystals similar to those obtained with paramyosin were not obtained (Kendrick-Jones and Citi, unpublished observations). If some form of cingulin assembly does occur in vivo, it could be speculated that its role in the organization of the TJ may be to provide a flexible submembrane skeleton which links TJ proteins together or anchors them to the cytoskeleton.

3.3 ZO-2

ZO-2 was identified as a M_r 160 kD polypeptide that coimmunoprecipitated with ZO-1 in MDCK cell extracts.[45] A polyclonal antiserum was recently generated against a ZO-2 fusion protein.[46] This antiserum labeled the cytoplasmic side of TJ in isolated MDCK cell membranes, showing that ZO-2 is localized at TJ.[46] The electron microscopic localization of ZO-2 in sections of epithelial cells and its distance from the TJ membrane have not been established yet. However, since ZO-2 appears to be tightly associated with ZO-1, it is likely that its ultrastructural localization will be very similar to that of ZO-1. The observation that ZO-2 is not accessible to cell surface labeling[45] confirms that ZO-2 is a component of the cytoplasmic domain of TJ, like ZO-1 and cingulin.

A 3.2 kb cDNA clone encoding mammalian ZO-2 was recently isolated using degenerate PCR primers based on a partial tryptic peptide sequence.[46] The partial 3.2 kb clone contains an open reading frame encoding 775 amino acid residues located in the C-terminal region of the protein. ZO-2 sequence shows a strong similarity to the sequence of ZO-1 (overall about 50% identity), which is especially marked (about 70% identity) in the region of homology to the discs-large family of proteins.[46] The similarity also spans the acidic domain, the heptad-repeat region and the proline-rich carboxy-terminal region of ZO-1.[46]

Despite the structural similarity between ZO-1 and ZO-2, their proteolytic maps are distinct,[45] and no immunological cross-reactivity was observed between the two proteins.[45,46] ZO-1 and ZO-2 have similar turnover rates,[45] and display essentially identical immunofluorescent localizations in the junctional regions of several epithelia and in capillary endothelia. However ZO-2 was not detected the adherens junctions of cardiac myocites.[46] Thus ZO-2, cingulin and occludin appear to be more specific TJ markers than ZO-1.

3.4 7H6

The monoclonal antibody 7H6 was generated against a bile canaliculus-rich membrane fraction from rat liver and was shown to react selectively with a M_r155 kD protein in liver and with a M_r 175 kD poplypeptide in MDCK cell extracts.[47] Immunoelectron microscopy demonstrated that 7H6 is localized near TJ in the bile-canaliculus-rich hepatocyte membrane fraction.

The 7H6 antigen was immunofluorescently localized to the junctional complex of several epithelia (liver, intestine, lung and kidney), but its localization in intercalated discs of cardiac myocites has not been reported. Immunoprecipitation and immunoblotting experiments indicate that 7H6 is immunologically distinct from ZO-1, cingulin and ZO-2.[47]

4. THE CYTOSKELETAL DOMAIN OF TJ

4.1 Actin

F-actin has been localized to the cytoplasmic submembrane domain of TJ by immunoelectron microscopy[48] and by visualization of filaments decorated by myosin heads.[49] Although actin is distributed virtually everywhere in the cortical cytoskeleton in epithelial cells, its presence in the TJ suggests a role in the modulation of TJ function. In fact, it has long been established that microfilament-active drugs, such as cytochalasins and phalloidin, have a profound effects on TJ-dependent paracellular permeability (see also ref. 50 for a review and chapter 7 in this volume).

4.2 Spectrin

The spectrin-based cytoskeleton has been implicated in the biogenesis of apico-basal polarity of epithelial cells and in the formation of complexes with ankyrin, Na^+, K^+-ATPase and uvomorulin in MDCK cells.[51,52] The asymmetrical distribution of Na^+, K^+-ATPase in the membrane domains separated by TJ is crucial to maintain the vectorial properties of transporting epithelia. The possible involvement of spectrin as a component of the cytoskeletal domain of TJ has been suggested by the observation that ZO-1 and spectrin co-purify and interact in vitro.[25] Thus, it could be hypothesized that ZO-1 may help to define the polarity of the spectrin cytoskeleton and/or its association with polarized membrane proteins. Further morphological studies and the identification of TJ proteins in spectrin-containing complexes are necessary to clarify the possible roles of spectrin in the organization and function of TJ.

5. OTHER PROTEINS ASSOCIATED WITH TJ

The coimmunoprecipitation approach may help to identify and characterize new TJ proteins. For example, a polypeptide with an apparent size of M_r 130 kD and distinct from cingulin and ZO-2 was recently identified in immunoprecipitates of ZO-1 from MDCK cell extracts.[46,53] Phosphorylated polypeptides with apparent molecular sizes of 180-220 kD and distinct from ZO-1 were detected in cingulin immunoprecipitates from MDCK cell extracts (Citi, unpublished observations).

Other approaches have been used to identify putative additional TJ proteins. Preliminary evidence indicates that a monoclonal antibody (BG9.1) raised against teratocarcinoma cells binds to a M_r 192 kD protein immunolocalized in the cytoplasmic faces of mouse liver TJ in membrane preparations.[54] By confocal microscopy a small rab GTPase protein (rab13) was found to be colocalized with ZO-1 in polarized intestinal epithelial cells (Caco-2), whereas it was distributed in cytoplasmic vesicles in non-epithelial cells.[55] Although the TJ localization of rab13 has not been confirmed by immunoelectron microscopy, these results are extremely interesting in view of recent molecular studies suggesting a relationship of

TJ proteins (ZO-1 and ZO-2) with small GTP-binding proteins involved in membrane traffic and/or signal transduction (see also chapters 1 and 12 in this volume).

6. INTERACTIONS BETWEEN TJ PROTEINS

The relationships between TJ proteins and their interactions with specific cytoskeletal structures and adhesion molecules is crucial to understanding their function(s). Thus far a clear interaction between specific TJ proteins has been demonstrated only for ZO-1 and ZO-2. In fact, ZO-2 was identified by virtue of its coimmunoprecipitation with ZO-1.[45] Such interaction has been confirmed by the complementary experiment, showing that ZO-1 is present in immunoprecipitates prepared with anti-ZO-2 antibodies.[46] Although it is not known how ZO-1 and ZO-2 interact, it has been speculated that the proline-rich carboxyl terminus of ZO-2 may bind to the SH3-like domain of ZO-1, or that the heptad-repeat containing regions in both proteins may mediate their association.[46] It is interesting that aggregates of ZO-1 were occasionally seen crosslinking spectrin tetramers by rotary shadowing,[25] suggesting that ZO-1 may self-associate. The interaction between ZO-1, ZO-2 (and the 130 kD coimmunoprecipitating protein) in a complex was detected even in cells in low extracellular Ca^{2+} (when junctions are disassembled, see section 9), suggesting that preformed TJ subcomplexes are assembled at junctional sites before assembly is triggered by the addition of Ca^{2+}.[53]

ZO-1 has not been detected in 7H6 immunoprecipitates,[47] nor in cingulin immunoprecipitates (Citi, unpublished observations) and cingulin is not found in either ZO-1 or ZO-2 containing immunoprecipitates.[46] These data indicating that cingulin and 7H6 probably do not associate with either ZO-1 or ZO-2.

7. INTERACTIONS OF TJ PROTEINS WITH OTHER JUNCTIONAL OR CYTOSKELETAL PROTEINS

The observation that ZO-1 co-localizes with P- and N-cadherins in non-epithelial cells[25,26] has raised a number of intriguing questions.[6] First, does ZO-1 bind to cadherins or to cadherin-associated proteins? Cotransfection experiments in L-cells showed that cadherins are required to bring ZO-1 into the regions of cell-cell contacts,[26] however there is no evidence that ZO-1 and cadherins interact directly. Second, what is the functional significance of cadherin-ZO-1 association? The function of ZO-1 may be to link cadherins to the cytoskeleton, since ZO-1 has been shown to bind to spectrin, at least in vitro.[25] Conversely, the "organization" of the submembrane cytoskeleton induced by cadherin-mediated cell-cell adhesion may "trap" ZO-1 into the adhesion sites. Third, why is the colocalization of ZO-1 and cadherins restricted to non-epithelial cells, whereas in epithelial cells ZO-1 is only localized at TJ? One explanation may be the existence of cell-specific ZO-1 or cadherin isoforms,[26] or the possibility that ZO-1 displays a higher affinity towards proteins specific of TJ rather than cadherin-based cell-cell contacts.[6]

ZO-1 isolated from rat brain was shown to copurify with spectrin and rotary shadowing electron microscopy showed an interaction between the two proteins in vitro.[25] Although the relevance of this observation to the in vivo situation remains to be established (section 4.2), both ZO-1 and spectrin contain a SH3 homology domain, which may mediate their interaction. Expression of normal and mutated ZO-1 domains in vivo and in vitro will be critical to identify interacting proteins. It is also interesting that ZO-1 may be identical to a 220 kD protein identified as a contaminant in purified clathrin coat preparations deriving from brain coated vesicles,[56] suggesting an association with endocytic or other clathrin-coated vesicles in brain tissue (presumably endothelial cells of brain capillaries and/or astrocytes).

The possibility that cingulin may interact, directly or indirectly, with the actin microfilament system or with other cytoskeletal proteins is suggested by its structure and biochemical properties. For example, cingulin isolated from chicken brush border cells partially copurifies with myosin,[37] suggesting that cingulin

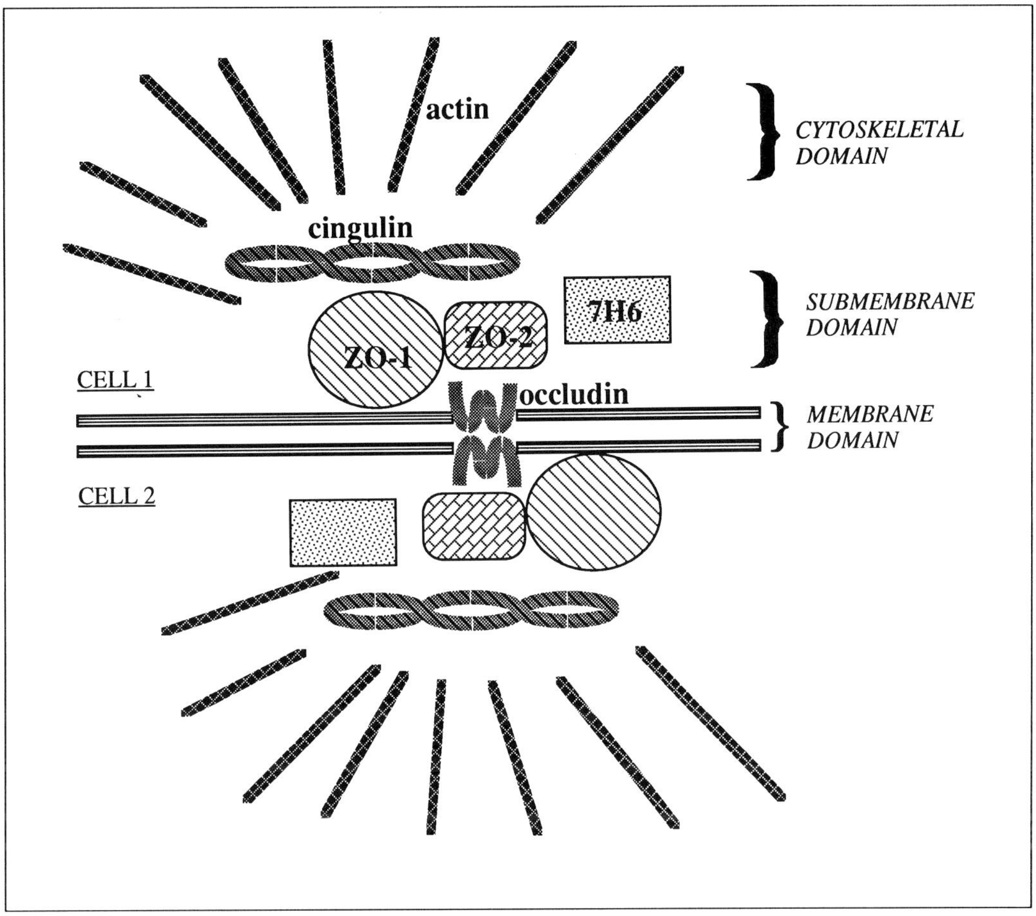

Fig. 5.5. A model for the organization of TJ. This diagram schematically illustrates the organization of TJ into membrane, submembrane and cytoskeletal domains and the relative positions and interactions of occludin, ZO-1, ZO-2, cingulin, 7H6 and actin, based on available data. Note that the scale is approximate and other putative TJ components, such as BG9.1, rab13 and spectrin have been omitted for clarity.

may have an affinity for the myosin-rich and actin-containing fraction of brush border cell extracts. The coiled-coil structure of cingulin and the similarity of its partial cDNA sequence to nonmuscle myosin heavy chains raises the possibility that cingulin may form assemblies with the rod portion of the myosin. However, purified chicken cingulin (108 kD) was found not to coprecipitate with myosin filaments assembled at 50-150 mM NaCl (Citi and Kendrick-Jones, unpublished results). Although purified cingulin (108 kD) does not coprecipitate with purified F-actin in vitro (Citi and Kendrick-Jones, unpublished results), further work is necessary to test the possibility that cingulin (140 kD) may interact with F-actin or other cytoskeletal proteins.

8. THE RELATIONSHIP OF TJ PROTEINS WITH THE PLASMA MEMBRANE

The relationship of TJ proteins with the plasma membrane has been investigated biochemically by studying their extractability from cells and tissues under different in vitro conditions and by examining their fine ultrastructural localization by immunoelectron microscopy.

ZO-1 requires harsh denaturing conditions (pH 11 or 6 M urea) to be effectively solubilized from liver and MDCK membranes, whereas it is not solubilized by 2% Triton X-100, suggesting that it is peripherally, albeit strongly associated with TJ membranes.[27] Cingulin, on the other hand, can be effectively solubilized from brush border cells using a solution containing 0.15-0.3 M NaCl and no detergent, suggesting that it is more loosely associated with the TJ membrane than ZO-1.[37] Neither ZO-1, ZO-2 nor cingulin show a predicted amino acid sequence with hydrophobic domains, confirming that they are all peripherally associated to the TJ membrane. Conversely, the extractability properties and predicted amino acid sequence of occludin indicate an integral membrane protein.[22]

The biochemical data are largely supported by the immunoelectron microscopic localizations of each protein (Table 5.1). For example, the reported average distances of ZO-1 from the midline of the TJ membrane were 23.2 nm in chicken intestinal cells, 14.4 nm in rat kidney cells,[30] and about 15 nm in cardiac intercalated discs.[25] Cingulin labeling was reported to be localized at 40 nm or 65 nm distance from the membrane in intestinal cells, depending on fixation and tissue preparation[30,36] and 50.9 nm in rat kidney cells.[30] For 7H6 the reported mean distance of the gold particles from the membrane was 41.11 nm.[47] Even taking into account possible errors introduced into these measurements by the size of the complex between the first antibody and the second, gold-labeled antibody (or protein A), these results indicate that the submembrane domain of TJ comprises a layer in close contact with the membrane, comprising ZO-1 and ZO-2 and a layer more loosely associated with the membrane domain, comprising 7H6 and cingulin (Fig. 5.5).

9. THE MODULATION OF TJ PROTEIN-MEMBRANE INTERACTIONS BY EXTRACELLULAR CA2

Altering the concentration of extracellular Ca^{2+} changes the state of assembly of cell-cell junctions and the spatial relationship of TJ proteins with the peripheral membrane. Ca^{2+} removal by chelators such as EGTA or BAPTA induces splitting of TJ and adherens-type junctions in several types of cultured epithelial cells, presumably as a result of conformational changes in the Ca^{2+}-dependent cell-adhesion molecules (cadherins) localized at the ZA and DS.[57] When extracellular Ca^{++} is removed, the apposing junctional membranes are pulled apart and the cells tend to become rounded and isolated. TJ opening in MDCK monolayers leads to a rapid (15-30 minutes) decrease in TER.[58] In MDCK cells treated with Ca^{2+} chelators ZO-1 and cingulin labeling is dissociated from the cell peripheries and becomes cytoplasmic, in a diffuse or granular pattern.[29,59,60] It is not clear whether cingulin and ZO-1 remain associated with endocytosed membrane vesicles, as in the case of cadherin,[61] or are internalized in the cytoplasm as "soluble" structural units, as observed for microfilament-associated proteins.[62]

The splitting of cell-cell junctions by removal of extracellular Ca^{2+} is a reversible process, since subsequent addition of Ca^{2+} can induce junction reassembly. When MDCK cells are cultured in low-Ca^{2+} medium (< 50 μM free calcium) for several hours, ZO-1 and cingulin are synthesized and localized diffusely in the cytoplasm,[29,59,60,63] are more soluble,[59] turn over at a faster rate,[29,64] and show higher mRNA levels.[29] Rapid TJ formation can be induced in these cells by the addition of medium with normal Ca^{2+} (Ca^{2+} 'switch'),[65] which initiates cadherin-mediated cell-cell contact.[63,66] Since protein synthesis inhibitors can block TJ assembly if added early during incubation in low-Ca^{2+},[15] it would appear that newly synthesized TJ proteins, rather than proteins derived from previously disassembled junctions, are used to form new junctions.

Why do TJ proteins dissociate from the TJ membranes upon extracellular Ca^{2+} removal and why do they reassemble when physiological extracellular Ca^{2+} concentrations are restored? Several observations suggest that TJ are not intrinsically Ca^{2+}-sensitive, like adherens-type junctions (ZA and DS), presumably because they do not contain cadherin-like molecules or membrane proteins whose conformation and adhesive properties depend on extracellular Ca^{2+}. For example, brief exposure to EGTA separates cell-cell contacts but does not dissociate ZO-1 from the plasma membrane.[59] Secondly, cingulin and ZO-1 labeling are detected in a few areas of cell-cell contact even after long incubation in EGTA and these areas lack staining for adherens junction proteins.[63] Thus, one hypothesis is that TJ proteins are dissociated from peripheral membranes because they may be linked to the submembrane cytoskeleton that in turn is disrupted when Ca^{2+}-sensitive junctions are opened. In other words, TJ may respond "passively" to the pulling forces and the cytoskeletal reorganization generated primarily by the dissociation of cadherins at adherens-type junctions. This idea is supported by the observation that pharmacological blockage of junction dissociation appears to affect all types of junctions.[60,67] To further study this issue and clarify the fate of the membrane domain of TJ in the Ca^{2+} removal model, it will be interesting to define the spatial relationships between occludin and peripheral TJ proteins upon Ca^{2+} removal.

The mechanism by which TJ proteins reassemble at the junctional membrane upon addition of Ca^{2+} could be viewed in a simplistic way as a reversal of junction disassembly, although the two processes are clearly different, for example in their kinetics and requirements for protein synthesis (see also chapters 6 and 7 in this volume). When Ca^{2+} is added, cadherins accumulate at cell-cell contacts and mediate homophylic adhesion and the submembrane cytoskeleton reassembles.[66] These events are strictly temporally associated with the assembly other junction-associated molecules, including TJ proteins, into the cell-cell contact areas. In fact, a hierarchy of junction assembly has been proposed, whereby the accumulation of functional E-cadherin at sites of cell-cell contact is necessary for TJ assembly to occur. This idea was based mainly on the observations that extracellular Ca^{2+}, which is necessary for cadherin function, was required for TJ assembly and that antibodies to the extracellular domain of cadherins prevented TJ formation.[66,68] However, recent studies with protein kinase activators and inhibitors suggest that distinct mechanisms regulate the assembly of proteins of adherens-type junctions and TJ into junctional sites, since they can occur in an apparently independent manner[53,63] (see section 10). In summary,

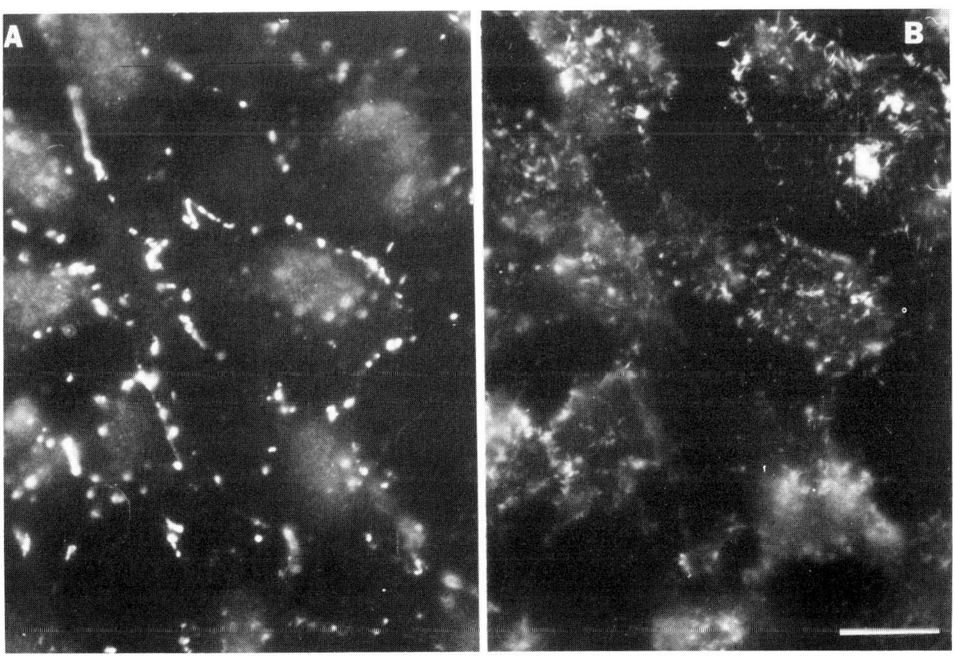

Fig. 5.6. The effect of cytochalasin on the localization of cingulin and actin in MDCK epithelial cells. Double immunofluorescence localization of cingulin (A) and actin (B) in cultured monolayers of MDCK cells treated with 10 μg/ml cytochalasin for 3 hours. Scale bar= 10 μm.

junction disassembly by removal of extracellular Ca^{2+} appears to involve all types of junctions, whereas junction assembly can be experimentally manipulated to favor the assembly of one or the other type of junction (see below).

10. THE ROLE OF THE CYTOSKELETON AND PHOSPHORYLATION IN MODULATING TJ ASSEMBLY AND DISASSEMBLY

A pharmacological approach has proven useful in dissecting the roles of the cytoskeleton and putative signaling pathways in the disassembly and assembly of TJ in a number of experimental models. It has long been known that microfilament-active drugs such as cytochalasins and phalloidin alter the structure and function of TJ in several model systems.[69-72] Cytochalasins are believed to act by fragmenting actin microfilaments and inducing their "collapse", whereas phalloidin is believed to stabilize actin microfilaments.[73] The addition of cytochalasin to monolayers of MDCK cells results in a rapid fall in the TJ-dependent transepithelial resistance.[67,69,70] In cytochalasin-treated MDCK cells the intracellular localization of actin, as determined by phalloidin fluorescence, is altered, with the loss of stress fibers and the appearance of clumps, fragmented "foci" and aggregates (Fig. 5.6). Cingulin labeling along the cell periphery is fragmented and is usually colocalized with actin[67]. The redistribution of cingulin[67] and ZO-1[3] following cytochalasin treatment indicates that actin filaments in the submembrane cytoskeleton of the *zonula adhaerens* and/or TJ are somehow linked to TJ proteins. The molecular basis for such linkage remains to be established. Interestingly, the protein kinase inhibitor H-7 was shown to prevent partially the fall in transepithelial resistance induced by cytochalasin D and to modify the effects of this drug on the intracellular distribution of actin and cingulin.[67] These observations suggest that phosphorylation is involved in mediating the effects of cytochalasin in MDCK cells (see also below).

The notion that an increase in the activity of intracellular kinases is associated with the opening of previously "tight" TJ was introduced in the 80s, when it was shown that protein kinase C activators (phorbol esters) disrupt TJ and increase paracellular permeability in MDCK and other epithelial cell lines.[74,75]

The relationship between Ca^{2+}-dependence of TJ integrity and intracellular protein kinase activity was recently addressed using protein kinase inhibitors. These molecules were found to prevent disassembly of TJ and adherens-type junctions induced by extracellular Ca^{2+} removal.[60] Specifically, several inhibitors (for example H-7, H-8 and staurosporine), when added to confluent MDCK cells prior to extracellular Ca^{2+} chelation, prevented the rapid fall in TJ-dependent transepithelial resistance, the redistribution of junctional proteins (ZO-1, cingulin, E-cadherin and desmoplakin) and the changes in cell shape that were otherwise observed. These observations indicated that phosphorylation is important in mediating the effects of extracellular Ca^{2+} removal and it was suggested that changes in cadherin conformation triggered by Ca^{2+} removal may activate kinase(s) that in turn are responsible for subsequent reorganization of the cytoskeleton and splitting of adherens junctions and TJ.[60] Another possibility is that the kinase inhibitors affect kinase(s) that must be active before Ca^{2+} removal to allow cells to react rapidly to Ca^{2+} chelation. In other words, H-7 may function by lowering the level of phosphorylation of specific cellular protein(s) below the threshold required to obtain effects after Ca^{2+} removal. To identify cellular targets of H-7, its effect on several epithelial cell types, including lens cells, which lack TJ and desmosomes but contain microfilament-associated adherens junctions, were studied.[67] In these cells, H-7 did not influence the trypsin-sensitivity of N-cadherin, suggesting that H-7 does not act by stabilizing the "active" cadherin conformation.[67] This and the observation that H-7 dissolves stress fibers,[63,67,76] and modifies the effects of cytochalasin D on transepithelial resistance and the distributions of actin and cingulin[67] suggests that one major target of H-7 is the adherens junction-associated microfilament cytoskeleton. The effects of H-7 on TJ assembly in the Ca^{2+} removal model may thus be mediated by its effects on the ZA and TJ-associated microfilament network,[67] although it cannot be excluded that kinase inhibitors also act on signaling pathways involving specific TJ proteins. The idea that H-7 acts on the microfilament cytoskeleton is supported by the recent observation that H-7 has multiple effects on cell behavior, including a general reduction in cellular contractility, destruction of stress fibers and an increase in lamellipodial activity (Volberg, Geiger, Citi and Bershadsky, submitted).

Protein kinase inhibitors and other drugs have also been used to investigate the regulation of the assembly and sealing of TJ. Using the Ca^{2+} switch assay and measurement of transepithelial resistance, it was suggested that G proteins, phospholipase C, calmodulin and protein kinase C may participate in TJ assembly.[77] In agreement with these observations, the protein kinase C inhibitor H-7 was shown to reduce the accumulation of cingulin and ZO-1 at sites of cell-cell contact in the Ca^{2+} switch assay.[38] The observation that 1, 2-dioctanoylglycerol (diC8) promotes the assembly of TJ in low extracellular Ca^{2+} further points to the possible roles of protein kinase C in promoting TJ assembly.[53] Several other drugs and agonists have been found to block or enhance TJ assembly (see chapters 6 and 7 in this volume).

The inhibition of TJ assembly by protein kinase inhibitors could in principle depend on their effect on the assembly of adherens-type junctions. As already mentioned, it has been shown that preventing cadherin-mediated cell-cell adhesion prevents cell polarization and the development of TJ.[66,68] To investigate the involvement of adherens-type junctions in TJ assembly, we determined the localization of cingulin and ZO-1 (TJ), E-cadherin (concentrated at the ZA) and desmoplakin (desmosomes) in MDCK cells during junction assembly by Ca^{2+} switch or at normal extracellular Ca^{2+}.[63] H-7 was found to inhibit the accumulation of cingulin and ZO-1 at junctional sites most effectively when added at the time of plating at normal extracellular Ca^{2+}, rather than when added 1 hour prior to the Ca^{2+} switch. However in both assembly protocols H-7 did not prevent the formation of cell-cell contacts nor did it block significantly the accumulation of E-cadherin and desmoplakin at such sites. Double immunofluorescence analysis confirmed that most junctional sites contained adherens-junction proteins, but only a few contained TJ proteins.[63] Electron microscopy supported the notion that H-7 does not significantly prevent adherens junction formation, whereas it reduces TJ. Since treatment with H-7 resulted in a significant decrease in cytoplasmic microfilaments and an overall disorganization of the actin cytoskeleton, it was suggested that one mechanism by which H-7 blocks TJ assembly is by interfering with the phosphorylation-dependent activity of the acto-myosin cytoskeleton.[63] Another important implication of the above results is that cell-cell contact and the accumulation of proteins of adherens-type junctions at these sites may be necessary but is not sufficient to induce the localization of TJ proteins at the cell periphery. The observation that diC8 induces the redistribution of ZO-1, but not E-cadherin, at low extracellular calcium[53] supports the idea that protein kinase C is most critically involved in the assembly of TJ and may be less important in the assembly of adherens-type junction proteins.

11. REGULATION OF TJ PROTEINS BY PHOSPHORYLATION?

In MDCK epithelial cells, both ZO-1[27] and cingulin[78] are phosphorylated on Ser residues. ZO-2 and a coimmunoprecipitating M_r 130 kD protein in MDCK cells are also phosphorylated[53] (Table 5.1). The stoichiometry of phosphorylation of these proteins and the identity of the kinases involved in their phosphorylation in vivo is not known. Recent results show that purified native and recombinant cingulin is phosphorylated in vitro by protein kinase C on two serine residues located in the C-terminal half of the molecule, whereas cingulin was not found to be a substrate for cAMP-dependent protein kinase.[78] ZO-1 displays four potential tyrosine phosphorylation sites,[28] however the phosphorylation of these residues in vitro or in vivo has not been reported.

It is tempting to speculate that phosphorylation may contribute to regulating the structure and/or functions of these two proteins and possibly other TJ proteins. For example, cingulin is structurally similar to nonmuscle myosin heavy chains and in invertebrate nonmuscle myosins phosphorylation occurring in the C-terminus of the rod can significantly influence their ability to assemble into filaments and interact with actin.[79,80] By analogy, it could be proposed that cingulin phosphorylation may regulate its ability to form supramolucular assemblies or interact with other proteins.

One approach to investigating the role of phosphorylation in the function of TJ proteins in vivo is to study the correlation between protein phosphorylation and TJ permeability and assembly in cultured cells. For example, a different phosphate content of ZO-1 in two strains of MDCK cells with different transepithelial resistance has been reported, although a direct correlation between junctional permeability and ZO-1 phosphorylation was not demonstrated.[81] The level of phosphorylation of ZO-1, ZO-2 and the 130 kD coimmunoprecipitating protein did not change after adding Ca^{2+} to MDCK cells grown in low calcium, or after their treatment with the protein kinase C agonist diC8.[53] Furthermore, treatment of MDCK cells at normal Ca^{2+} with phorbol esters does not change the level of cingulin phosphorylation (Citi and Denisenko, unpublished observations). Taken together, these observations suggest that modulation of TJ proteins by phosphorylation is not directly correlated with the assembly and disassembly of TJ induced by extracellular Ca^{2+} and protein kinase C activation. However, it is possible that phosphorylation could be important in the early steps of TJ assembly. For example phosphorylation may be required for the correct folding of TJ proteins and their interaction with other TJ proteins to form complexes which are then translocated as stable units to the junctional membrane.

12. ASSEMBLY OF TJ PROTEINS DURING DEVELOPMENT

Developmental studies provide a useful experimental model to investigate the formation of junctions *de novo*, prior to any previous protein synthesis or junction remodeling. Other chapters in this volume review in detail recent studies on the roles of different junctions and junctional proteins in tissue and organ biogenesis (chapters 1, 3, 4).

The development of TJ in vertebrate embryos can be considered as one aspect of the complex set of changes in cell-cell interactions which determine the fate and spatial organization of developing tissues. The expression of ZO-1 and cingulin in development has been investigated using the mouse preimplantation embryo as a model system.[42,64,82,83] Unlike ZO-1 but similarly to E-cadherin, maternally inherited cingulin expression was detected in unfertilized eggs and in the first stages after fertilization (2- and 4-cell embryos).[42,64] In the eggs, maternal cingulin was found to be localized in the oocyte cortex and corona radiata layer. The significance of such cortical staining is unknown and cortical cingulin labeling was observed in cleavage stages up to 16-cell morulae, after which maternal cingulin appeared to be degraded by endocytic turnover, without contributing to TJ assembly. On the other hand, the synthesis of embryonic cingulin was found to increase at the beginning of compaction (8-cell stage) and its assembly into TJ occurred approximately 10 hours after cell-cell adhesion was initiated and ZO-1 was assembled.[42] The localization of ZO-1 in the regions of cell-cell contact appeared to temporally coincide with, or barely follow, the basolateral adhesion and the development of apical-basal polarity in the compacting 8-cell stage embryo and was inhibited by disruption of the actin cytoskeleton or by inhibition of adhesion.[83] Conversely, preventing protein synthesis at the 8-cell stage or combining single eight-cells with single four-cells perturbed ZO-1 junctional assembly, without preventing adhesion and basolateral polarization. These results suggest that adherens-type junction assembly is not necessarily followed by TJ assembly, in agreement with studies with protein kinase inhibitors in MDCK cells (section 10).

Several additional important questions need to be addressed to understand the precise function of TJ and TJ proteins in epithelial biogenesis. For example, what are the short- and long-term developmental consequences of defects in the structure and function of TJ? What is the role of the membrane component(s) of TJ in determining cell polarization and the assembly of the junctional complex? Now that reagents are available to immunolocalize occludin and to manipulate the expression of several TJ proteins, it should be possible to address these questions and to define more precisely the spatial and temporal pattern of organization of TJ proteins during development.

13. TJ PROTEINS AND DISEASE

The importance of TJ in the barrier function of epithelia suggests that alterations in the expression and function of TJ proteins may lead to changes in junctional permeability and may thus contribute to pathogenesis of diseases of epithelial tissues and organs (see also chapter 7 in this volume and ref. 84). Alterations in the TJ-dependent barrier function of epithelia and endothelia may be responsible for leakage of potentially noxious molecules through the paracellular pathway. For example, the opening of TJ may allow antigens, toxins, carcinogens and other pathological agents to permeate into the blood through the barrier of intestinal epithelium. Reflux of the bile into the blood may be caused by disruption of TJ between hepatocytes lining bile canaliculi. On the other hand, the possibility of increasing TJ permeability at sites such as the blood-brain barrier may be used therapeutically to deliver drugs to the central nervous system.[85]

Reversible changes in TJ structure and function induced by a number of physiological and pathogenic agents (nutrients, cAMP, cells and mediators involved in the immune response, toxins) have been observed in experimental model systems (see chapter 7 in this volume and ref. 86). In addition, studies correlating the localization of ZO-1 and the structure and function of TJ in animal models under physiological or pathological states have been carried out. Cholestasis due to common bile duct ligation in the rat was reported to induce changes in ZO-1 localization without affecting the tissue content of ZO-1 protein,[87] although in a subsequent study a marked increase in ZO-1 protein expression and mRNA levels 9 days after ligation, possibly representing a response to injury, was reported.[88] Activation of Na^+-glucose cotransport in intestinal absorptive cells "opens" TJ and reduces the number of junctional sites displaying "kisses" and fibrils by electron microscopy, while ZO-1 remains localized at sites where TJ no longer exist.[89] These observations suggest that TJ permeability may change dramatically even when the spatial relationship of ZO-1 with the TJ membrane appears unchanged.

Little is known about specific changes in the structure, expression and function of TJ proteins in human disease states. The observation that ZO-1 and ZO-2 have domains with structural homology to the product of the *dlg* (disc-large) tumor suppressor gene in Drosophila raises the possibility that vertebrate TJ may be functionally homologous to insect septate junctions (see also chapters 1 and 12 in this volume). Since mutations of the *dlg* gene result in neoplastic overgrowth of imaginal discs of Drosophila larvae, it could be hypothesized that mutations in TJ proteins may also lead to the loss of the polarized epithelial phenotype and the uncontrolled growth observed in malignant epithelial tumors. In fact, TJ separate the apical and basolateral domains of the plasma membrane in polarized cells and have been implicated in the maintenance of epithelial cell polarity.[10] Although no cause-effect relationship has been established between loss of TJ protein function and malignant growth, studies on E-cadherin and associated proteins (see chapters 8, 12 and 13 in this volume), indicate that proteins associated with the submembrane domain of cell-cell junctions may be critically involved in tumorigenesis, possibly by participating in the transduction of signals regulating cell-cell contact growth inhibition.

So far only one report has documented the expression of TJ specific protein(s) in human tumors.[39] Using immunocytochemistry and immunoblot experiments, cingulin was found to be expressed in human benign and malignant neoplasias of the colon, whereas it was absent from non-epithelial tumors.[39] The expression of cingulin in poorly differentiated colon carcinomas suggested that TJ proteins are present even when the polarized epithelial phenotype is lost and therefore cingulin could be used as an additional epithelial marker in the classification of highly dedifferentiated carcinomas and metastases of unknown origin. Increased levels of cingulin were detected in some colon carcinomas, suggesting that increased expression of TJ proteins may be a cellular response to pathological stress or injury.

14. FUTURE DIRECTIONS

The recent exciting progress in identifying and characterizing TJ proteins now provides an opportunity to study TJ function at the molecular level.[6,7] Bacterial and mammalian expression of cDNAs encoding TJ proteins and their subdomains will be useful to define structure-function relationships in these proteins in vitro and in cultured cells in vivo. Molecular genetic studies on mutations of TJ protein homologs in insects and eventually knockout studies in mice will provide additional information on the physiological role of TJ proteins. Ultimately, a better understanding of the molecular mechanisms that underlie TJ function will be critical to design strategies and reagents for diagnostic and therapeutic purposes.

Acknowledgments

I wish to thank all my collaborators for their contributions and the American Cancer Society, the CNR and MURST for supporting work in my laboratory.

References

1. Farquhar MG, Palade GE. Junctional complexes in various epithelia. J Cell Biol 1963; 17:375-412.
2. Gumbiner B. The structure, biochemistry and assembly of epithelial tight junctions. Am J Physiol 1987; 253:C749-C758.
3. Stevenson BR, Anderson JM, Bullivant S. The epithelial tight junction: structure, function and preliminary biochemical characterization. Mol Cell Biochem 1988; 83:129-145.
4. Madara JL, Hecht G. Tight (occluding) junctions in cultured (and native) epithelial cells. New York: Alan R Liss, 1989:131-163.
5. Schneeberger EE, Lynch RD. Structure, function and regulation of cellular tight junctions. Am J Physiol 1992; 262:L647-L661.
6. Citi S. The molecular organization of tight junctions. J Cell Biol 1993; 121:485-489.
7. Anderson JM, Balda MS, Fanning AS. The structure and regulation of tight junctions. Curr Op Cell Biol 1993; 5:772-778.
8. Reuss L. Tight junction permeability to ions and water. In: Cereijido M, ed. Tight Junctions. Boca Raton: CRC Press, 1992:49-66.
9. van Meer G, van't Hof W, van Genderen I. Tight junctions and polarity of lipids. In: Cereijido M, ed. Tight Junctions. Boca Raton: CRC Press, 1992:187-201.
10. Rabito CA. Tight junction and apical/basolateral polarity. In: Cereijido M, ed. Tight Junctions. Boca Raton: CRC Press, 1992:203-214.
11. Kachar B, Reese TS. Evidence for the lipidic nature of tight junction strands. Nature 1982; 296:464-466.
12. Pinto da Silva P, Kachar B. On tight junction structure. Cell 1982; 28:441-450.
13. van Deurs KG, Luft JH. Effects of glutaraldehyde fixation on the structure of tight junctions. A quantitative freeze-fracture analysis. J Ultrastruct Res 1979; 68:160-172.
14. Griepp EB, Dolan WJ, Robbins ES, Sabatini DD. Participation of plasma membrane proteins in the formation of tight junctions by cultured epithelial cells. J Cell Biol 1983; 96:693-702.

15. Gonzalez-Mariscal L. Biosynthesis of the tight junction. In: Cereijido M, ed. Tight Junctions. Boca Raton: CRC Press, 1992:243-255.
16. Stevenson BR, Goodenough DA. Zonulae occludentes in junctional complex-enriched fractions from mouse liver:preliminary morphological and biochemical characterization. J Cell Biol 1984; 98:1209-1221.
17. Geiger B, Avnur Z, Volberg T, Volk T. Molecular domains of adherens junctions. In: Edelman GM, Thiery JP, eds. The Cell in Contact. New York: John Wiley and Sons, 1985:461-489.
18. Goodenough DA, Revel JP. A fine structural analysis of intercellular junctions in the mouse liver. J Cell Biol 1970; 45:272-290.
19. Hoi Sang U, Saier MH, Elliman LH. Tight junction formation in the establishment of intramembranous particle polarity in aggregating MDCK cells. Exp. Cell Res 1980; 128:223-235.
20. van Meer G, Simons K. The function of tight junctions in maintaining differences in lipid composition between the apical and the basolateral cell surface domains of MDCK cells. EMBO J 1986; 5:1455-1464.
21. van Meer G, Gumbiner B, Simons K. The tight junction does not allow lipid molecules to diffuse from one epithelial cell to the next. Nature 1986; 322:639-641.
22. Furuse M, Hirase T, Itoh M, Nagafuchi A, Yonemura S, Tsukita S, Tsukita S. Occludin: a novel integral membrane protein localizing at tight junctions. J Cell Biol 1993; 123:1777-1788.
23. Gumbiner B. Breaking through the tight junction barrier. J Cell Biol 1993; 123:1631-1633.
24. Stevenson BR, Siliciano JD, Mooseker MS, Goodenough DA. Identification of ZO-1: a high molecular weight polypeptide associated with the tight junction (zonula occludens) in a variety of epithelia. J Cell Biol 1986; 103:755-766.
25. Itoh M, Yonemura S, Nagafuchi A, Tsukita S, Tsukita S. A 220-kD undercoat-constitutive protein: its specific localization at cadherin-based cell-cell adhesion sites. J Cell Biol 1991; 115:1449-1462.
26. Itoh M, Nagafuchi A, Yonemura S, Yasuda-Kitani T, Tsukita S, Tsukita S. The 220kD protein colocalizing with cadherins in non-epithelial cells is identical to ZO-1, a tight junction-associated protein in epithelial cells: cDNA cloning and immunoelectron microscopy. J Cell Biol 1993; 121:491-502.
27. Anderson JM, Stevenson BR, Jesaitis LA, Goodenough DA, Mooseker MS. Characterization of ZO-1, a protein component of the tight junction from mouse liver and Madin-Darby Canine Kidney cells. J Cell Biology 1988; 106:1141-1149.
28. Willott E, Balda MS, Fanning AS, Jameson B, Van Itallie C, Anderson JM. The tight junction protein ZO-1 is homologous to the Drosophila discs-large tumor suppressor protein of septate junctions. Proc. Natl. Acad. Sci. USA 1993; 90:7834-7838.
29. Anderson JM, Van Itallie CM, Peterson MD, Stevenson BR, Carew EA, Mooseker MS. ZO-1 mRNA and protein expression during tight junction assembly in Caco-2 cells. J Cell Biol 1989; 109:1047-1056.
30. Stevenson BR, Heintzelman MB, Anderson JM, Citi S, Mooseker MS. ZO-1 and cingulin: tight junction proteins with distinct identities and localizations. Am J Physiol 1989; 257:C621-C628.
31. Schnabel E, Anderson JM, Farquhar MG. The tight junction protein ZO-1 is concentrated along slit diaphragms of the glomerular epithelium. J Cell Biol 1990; 111:1255-1263.
32. Howarth AG, Hughes MR, Stevenson BR. Detection of the tight junction-associated protein ZO-1 in astrocytes and other nonepithelial cell types. Am J Physiol 1992; 262:C461-C469.
33. Willott E, Balda MS, Heintzelman M, Jameson B, Anderson JM. Localization and differential expression of two isoforms of the tight junction protein ZO-1. Am J Physiol 1992; 262:C1119-C1124.
34. Kurihara H, Anderson JM, Farquhar MG. Diversity among tight junctions in rat kidney: glomerular slit diaphragms and endothelial junctions express only one isoform of the tight junction protein ZO-1. Proc. Natl. Acad. Sci. USA 1992; 89:7075-7079.

35. Stevenson BR, Anderson JM, Goodenough DA, Mooseker MS. Tight junction structure and ZO-1 content are identical in two strains of Madin-Darby Canine kidney cells which differ in transepithelial resistance. J Cell Biol 1988; 107:2401-2408.
36. Citi S, Sabanay H, Jakes R, Geiger B, Kendrick-Jones J. Cingulin, a new peripheral component of tight junctions. Nature 1988; 333:272-276.
37. Citi S, Sabanay H, Kendrick-Jones J, Geiger B. Cingulin: characterization and localization. J Cell Sci 1989; 93:107-122.
38. Nigam S, Denisenko N, Rodriguez-Boulan E, Citi S. The role of phosphorylation in development of tight junctions in cultured renal epithelial (MDCK) cells. Biochem Biophys Res Commun 1991; 181:548-553.
39. Citi S, Amorosi A, Franconi F, Giotti A, Zampi G. Cingulin, a specific protein component of tight junctions, is expressed in normal and neoplastic human epithelial tissues. Am J Pathol 1991; 138:781-789.
40. Raphael Y, Altschuler RA. Reorganization of cytoskeletal and junctional proteins during cochlear hair cell degeneration. Cell Motil Cytosk 1991; 18:215-227.
41. Byers SW, Citi S, Anderson JM, Hoxter B. Polarized functions and permeability properties of rat epididymal epithelial cells in vitro. J Reprod Fert 1992; 95:385-396.
42. Fleming TP, Hay M, Javed Q, Citi S. Localisation of tight junction protein cingulin is temporally and spatially regulated during early mouse development. Development 1993; 117:1145-1151.
43. Studier FW, Rosenberg AH, Dunn JJ, Dubenoff JW. Use of T7 polymerase to direct expression of cloned genes. Meth Enzymol 1990; 185:60-89.
44. McLachlan AD. Structural implications of the myosin amino acid sequence. Ann Rev Biophys Bioeng 1984; 13:167-189.
45. Gumbiner B, Lowenkopf T, Apatira D. Identification of a 160 kD polypeptide that binds to the tight junction protein ZO-1. Proc. Natl. Acad. Sci. USA 1991; 88:3460-3464.
46. Jesaitis LA, Goodenough DA. Molecular characterization and tissue distribution of ZO-2, a tight junction protein homologous to ZO-1 and the Drosophila discs-large tumor suppressor protein. J Cell Biol 1994; 124:949-961.
47. Zhong Y, Saitoh T, Minase T, Sawada N, Enomoto K, Mori M. Monoclonal antibody 7H6 reacts with a novel tight junction-associated protein distinct from ZO-1, cingulin and ZO-2. J Cell Biol 1993; 120:477-483.
48. Drenckhahn D, Dermietzel R. Organization of the actin filament cytoskeleton in the intestinal brush border: a quantitative and qualitative immunoelectron microscope study. J Cell Biol 1988; 107:1037-1048.
49. Madara JL. Intestinal absorptive cell tight junctions are linked to cytoskeleton. Am J Physiol 1987; 253:C171-C175.
50. Madara JL. Relationships between the tight junction and the cytoskeleton. In: Cereijido M, ed. Tight Junctions. Boca Raton: CRC Press, 1992:105-120.
51. Nelson WJ, Shore EM, Wang AZ, Hammerton RW. Identification of a membrane-cytoskeleton complex containing the the cell adhesion molecule uvomorulin (E-cadherin), ankyrin and fodrin in Madin-Darby canine kidney epithelial cells. J Cell Biol 1990; 110:349-357.
52. Nelson WJ, Hammerton RW. A membrane cytoskeletal complex containing Na^+, K^+-ATPase, ankyrin, fodrin in Madin Darby canine kidney (MDCK) cells: implications for the biogenesis of epithelial cell polarity. J Cell Biol 1989; 108:893-902.
53. Balda MS, Gonzalez-Mariscal L, Matter K, Cereijdo M, Anderson JM. Assembly of the tight junction: the role of diacylglycerol. J Cell Biol 1993; 123:293-302.
54. Chapman LM, Eddy EM. A protein associated witht the mouse and rat hepatocyte junctional complex. Cell Tissue Res 1989; 257:333-341.
55. Zahraoui A, Joberty G, Arpin M, Fontaine JJ, Hellio R, Tavitian A, Louvard D. A small rab GTPase is distributed in cytoplasmic vesicles in non polarized cells but colocalizes with the tight junction marker ZO-1 in polarized epithelial cells. J Cell Biol 1994; 124:101-115.

56. Enrich C, Jones GR, Bordas J, Evans WH. A 220 kD polypeptide, immunolocalized to epithelial tight junctions, is associated with brain clathrin preparations. Eur. J Cell Biol 1989; 50:304-312.
57. Geiger B, Ayalon O. Cadherins. Ann Rev Cell Biol 1992; 8:307-332.
58. Martinez-Palomo A, Meza I, Beaty G, Cereijido M. Experimental modulation of occluding junctions in a cultured transporting epithelium. J Cell Biol 1980; 87:736-745.
59. Siliciano JD, Goodenough DA. Localization of the tight junction protein, ZO-1, is modulated by extracellular calcium and cell-cell contact in Madin-Darby canine kidney epithelial cells. J Cell Biol 1988; 107:2389-2399.
60. Citi S. Protein kinase inhibitors prevent junction dissociation induced by low extracellular calcium in MDCK epithelial cells. J Cell Biol 1992; 117:169-178.
61. Kartenbeck J, Schmelz M, Franke W, Geiger B. Endocytosis of junctional cadherins in bovine kidney epithelial (MDBK) cells cultured in low Ca^{2+} ion medium. J Cell Biol 1991; 113:881-892.
62. Volberg T, Geiger B, Kartenbeck J, Franke WW. Changes in membrane-microfilament interaction in intercellular adherens junctions upon removal of extracellular $Ca2+$ ions. J Cell Biol 1986; 102:1832-1842.
63. Denisenko N, Burighel P, Citi S. Different effects of protein kinase inhibitors on the localization of junctional proteins at cell-cell contact sites. J Cell Sci 1994; 107:
64. Javed Q, Fleming TP, Hay MJ, Citi S. Tight junction protein cingulin is expressed by maternal and embryonic genomes during early mouse development. Development 1993; 117:
65. Gonzalez-Mariscal L, Chavez de Ramirez B, Cereijido M. Tight junction formation in cultured epithelial cells (MDCK). J Membr Biol 1985; 86:113-125.
66. Gumbiner B, Stevenson B, Grimaldi A. Role of the cell adhesion molecule uvomorulin in formation and maintenance of the epithelial junctional complex. J Cell Biol 1988; 107:1575-1587.
67. Citi S, Volberg T, Bershadsky AD, Denisenko N, Geiger B. Cytoskeletal involvement in the modulation of cell-cell junctions by the protein kinase inhibitor H-7. J Cell Sci 1994; 107:683-692.
68. Gumbiner B, Simons K. A functional assay for proteins involved in establishing an epithelial occluding barrier: identification of a uvomorulin-like polypeptide. J Cell Biol 1986; 102:457-468.
69. Meza I, Ibarra G, Sabanero M, Martinez-Palomo A, Cereijido M. Occluding junctions and cytoskeletal components in a cultured transporting epithelium. J Cell Biol 1980; 87:746-754.
70. Meza I, Sabanero M, Stefoni E, Cereijido M. Occluding junctions in MDCK cells: modulation of transepithelial permeability by the cytoskeleton. J Cell Biochem 1982; 18:407-421.
71. Elias E, Hruban Z, Wade JB, Boyer JL. Phalloidin-induced cholestasis: a microfilament-mediated change in junctional complex permeability. Proc. Natl. Acad. Sci. USA 1980; 77:2229-2233.
72. Madara JL, Barenberg D, Carlson S. Effects of cytochalasin D on occluding junctions of intestinal absorptive cells: further evidence that the cytoskeleton may influence paracellular permeability and junctional charge selectivity. J Cell Biol 1986; 102:2125-2136.
73. Cooper JA. Effects of cytochalasin and phalloidin on actin. J Cell Biol 1987; 105:1473-1478.
74. Ojakian GK. Tumor promoter-induced changes in the permeability of epithelial tight junctions. Cell 1981; 23:95-103.
75. Mullin JM, O'Brien TG. Effects of tumor promoters on LLC-PK1 renal epithelial tight junctions and transepithelial fluxes. Am J Physiol 1986; 251:C597-C602.
76. Birrell GB, Hedberg KK, Habliston DL, Griffith OH. Protein kinase C inhibitor H-7 alters the actin cytoskeleton of cultured cells. J Cell Physiol 1989; 141:74-84.

77. Balda MS, Gonzalez-Mariscal L, Contreras RG, Macias-Silva M, Torres-Marquez ME, Garcia-Sainz JA, Cereijido M. Assembly and sealing of tight junctions: possible participation of G-proteins, phospholipase C, protein kinase C and calmodulin. J Membr Biol 1991; 122:193-202.
78. Rabino M, Denisenko N, Citi S. Phosphorylation of cingulin in vitro by protein kinase C: identification of two phosphorylation sites. Mol Biol Cell 1993; 4:440a.
79. Kuczmarski ER, Spudich JA. Regulation of myosin self-assembly: phosphorylation of Dictyostelium heavy chain inhibits formation of thick filaments. Proc. Natl. Acad. Sci. USA 1980; 77:7292-7296.
80. Ogihara S, Ikebe M, Takahashi K, Tonomura Y. Requirement of phosphorylation of Physarum myosin heavy chain for thick filament formation, actin activation of Mg^{2+}-ATPase activity and $Ca2^{+}$-inhibitory superprecipitation. J Biochem 1983; 93:205-223.
81. Stevenson BR, Anderson JM, Braun ID, Mooseker MS. Phosphorylation of the tight junction protein ZO-1 in two strains of Madin-Darby canine kidney cells which differ in transepithelial resistance. Biochem J 1989; 263:597-599.
82. Fleming TP, Hay MJ. Tissue-specific control of expression of the tight junction polypeptide ZO-1 in the mouse early embryo. Development 1991; 113:295-304.
83. Fleming TP, McConnell J, Johnson MH, Stevenson BR. Development of tight junctions de novo in the mouse early embryo: control of assembly of the tight junction-specific protein, ZO-1. J Cell Biol 1989; 108:1407-1418.
84. Bentzel CJ, Palant CE, Fromm M. Physiological and pathological factors affecting the tight junction. In: Cereijido M, ed. Tight Junctions. Boca Raton: CRC Press, 1992:151-173.
85. Rubin L. Endothelial cells: adhesion and tight junctions. Curr Opin Cell Biol 1992; 4:830-833.
86. Madara J. Loosening tight junctions. Lessons from the intestine. J Clin Invest 1989; 83:1089-1094.
87. Anderson JM, Glade JL, Stevenson BR, Boyer JL, Mooseker MS. Hepatic immunohistochemical localization of the tight junction protein ZO-1 in rat models of cholestasis. Am J Pathol 1989; 134:1055-1062.
88. Fallon MB, Mennone A, Anderson JM. Altered expression and localization of the tight junction protein ZO-1 after common bile duct ligation. Am J Physiol 1993; 264:C1439-C1447.
89. Madara JL, Carlson S, Anderson JM. ZO-1 maintains its spatial distribution but dissociates from junctional fibrils during tight junction regulation. Am J Physiol 1993; 264:C1096-C1101.
90. Li C, Poznansky MJ, Characterization of the ZO-1 protein in endothelial and other cell lines. J Cell Sci 1990; 97:231-237.
91. Balda MS, Anderson JM. Two classes of tight junctious are revealed by ZO-1 isoforms. Am J Physiol 1993; 264: C918-924.

CHAPTER 6

Extracellular and Intracellular Regulation of Junction Assembly in Epithelial Cells

Lorenza Gonzalez-Mariscal
Jesus Valdes
Marcelino Cereijido
Ruben G. Contreras
Maria R. Garcia-Villegas

1. INTRODUCTION

One of the main roles of epithelial membranes is to act as diffusion barriers between two compartments, one of them being the interstitium in free connection with bloodand the other representing the environment, even if this environment is sometimes reduced to the lumen of extremely small tubulesand is hidden deep in the body, as is for example the lumen of kidney nephrons.[1] This role of epithelia depends on the ability of their cells to establish tight junctions (TJs) at the outermost end of the intercellular space, that seal this space (Fig. 6.1).

However, TJs are not restricted to epithelia, but appear in places where layers of cells separate two compartments of the body with important differences in composition, structure and function, as in the case of endothelia (see refs. 2-4).

2. THE STRUCTURE OF THE TJ

The TJ constitutes a group of 1-10 strands occupying a belt-like region that completely circles the perimeter of the cells at the limit between the apical

Molecular Mechanisms of Epithelial Cell Junctions: From Development to Disease
Edited by Sandra Citi, M.D., Ph.D.;

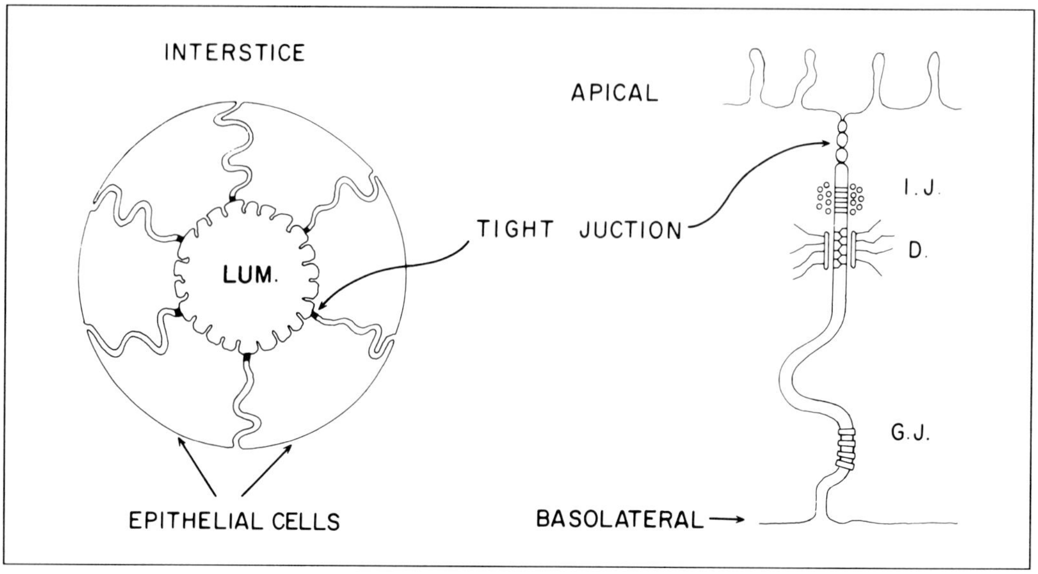

Fig. 6.1. Left, schematic representation of a kidney tubule. Six epithelial cells form tight junctions that severely reduce the escape of substances from the lumen (LUM). Right, two adjacent epithelial cells form tight junctions (TJs), intermediate junctions (IJs), desmosomes (D)and gap junctions (GJ).

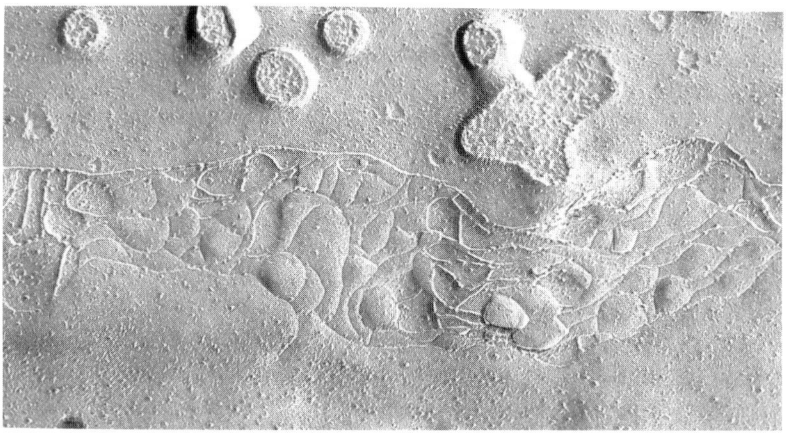

Fig. 6.2. Freeze fracture replica of a monolayer of MDCK cells, showing the strands of TJs, that appear as furrows on E faces and ridges on P ones. (20,000 X 9). The cell has its apical domain at the bottom of the picture, where a number of fractured microvilli can be seen.

and basolateral domains. These strands appear in transmission electron microscopy as a series of punctuated contacts between neighboring cells,[5] and in freeze-fracture replicas as a flat meshwork of anastomosing filaments that resemble ridges in P-faces and furrows in E-faces of glutaraldehyde fixed epithelia (Fig. 6.2). Without this fixative, filaments look like chains of intramembranous particles of some 10 nm in diameter.[6-8]

Although the chemical nature of these strands is still unknown, current models assume that they are composed by rows of proteins.[6,7,9,10] However the possibility exists that each strand is constituted by a cylinder, made of lipid molecules pointing their polar head group towards the aqueous axisand the alkyl chains toward the hydrophobic matrix of the two neighboring cells.[11,12] Studies using diffusive lipid probes[13-15] and resorting to experimental modifications of the lipid composition of the cells[16,17] or to temperature changes tend to discard the lipid model, but fail to provide definitive support for the protein-based one.[18,19]

Although no molecular component of the TJ strand have been found, several proteins associated with this structure have been identified[20-28] (see chapter 5 in this volume).

3. THE TJ AS A DIFFUSION BARRIER

Gonzalez-Mariscal et al[29] have found that, provided a given cell type is able to form TJs among themselves, they are also able to establish TJs when co-cultured with cells from different organs and even different animal species.

In keeping with the role of diffusion barrier of the epithelium, the tightness of the seal offered by TJs is proportional to the difference in composition between the lumen and the blood side. Thus, while the proximal tubule of the kidney, containing plasma-like filtrate, exhibits a transepithelial electric resistance (TER) of 8-10 $\Omega \cdot cm^2$, the collecting ductand the mucosae of the urinary bladder and the colon, whose contents establish sharp electrochemical gradients with interstitial fluid have TERs of several thousand ohms. Furthermore, the value of TER of a given TJ varies in response to physiological conditions and pharmacological challenge.[30-36]

In spite of this wide range of TERs, in all cases the TJ offers a barrier of low ion selectivity, as the ratio between the most and the least preferred cation, or the ratio between sodium and chloride permeabilities (P_{Na}/P_{Cl}, Fig. 6.3) is very low as compared with those of plasma membranes.[37-41] This ion-discriminating ability appears to depend on different molecular components than the ones participating in cell-cell sealing, as Cereijido et al.[39] have found that the value of TER in newly formed monolayers of MDCK cells reaches a maximum with a much faster kinetics than the increase of P_{Na} over P_{Cl} (Fig. 6.3).

The observation that epithelia with a large number of filaments in their TJs tend to have higher electrical resistance across[42] led us to assume that each strand constitutes a resistive element. However, when the electrical resistance of the paracellular route of different epithelia is plotted against the number of strands in their TJs, the increase with each additional strand is not linear, as would be expected from the addition of resistors in series, but exponential. To explain this relationship, Claude[43] has suggested that each strand is traversed by aqueous channels that may be in an *open* or *closed* configuration. As pointed out by Cereijido et al[38,44] this model would not explain the exponential relationship between resistance and number of strands, unless it is assumed that the strand is constituted by segments electrically separated from each other. These authors suggested that this type of electrical compartmentation of resistive (strands) and conductive (aqueous channels) elements in the TJ, may be provided by the frequent anastomosing of strands (for a review of this arrangement see refs. 38, 44 and 45).

4. EXPERIMENTAL MODELS TO STUDY JUNCTION FORMATION

Our understanding of when and how junctions are made is based mainly on work with monolayers of cultured epithelial cells

TIME (h)	0	0.25	4	72
FF				
TER ($\Omega.cm^2$)	0	0	400	250
P_{Na}/P_{Cl}	WATER	≈1	≈1	8

Fig. 6.3. Evolution of the pattern of strands in freeze fracture replicas (FF), TER and P_{Na}/P_{Cl} ratio in monolayers of MDCK cells subject to a Ca-switch at t = 0. Four hours after the switch the sealing capacity, as reflected by TER, has reached its peak, while the ratio between permeabilities still reflects their water mobilities. Three days later TER has decreased to a steady value of 250 $\Omega \cdot cm^2$ and P_{Na}/P_{Cl} reflects a cation selectivity.

that behave in many respects as natural epithelia.[39,44,46-50] In these monolayers cells establish TJs and are polarized into an apical and a basolateral domain.[48,50,51] Harvesting with trypsin-EDTA makes the cells lose TJs and polarity, but these are recovered in a few hours upon plating, in particular if it is made at confluence, so cells do not need to proliferate to cover the whole area.

When these monolayers are prepared on permeable supports, the assembly can be mounted between two chambers containing saline solutions. The passage of an electric current through these monolayers and the voltage deflection it elicits can be used to calculate the value of TER. Current flows through the cytoplasm of the cells (transcellular route) and through the intercellular space (paracellular route). To follow the transcellular pathway current has to traverse twice the plasma membrane, which has an electrical resistance in the order of 10^8 Ω.[52] Simple considerations on the electric circuit thus formed show that the route through a single cell will therefore have at least four time this value (4 x 10^8 Ω). Since a square centimeter of monolayer contains some 10^5 cells constituting resistors arranged in parallel, it follows that the transcellular route presents a resistance of some 4000 · Ω · cm^2. In comparison, the value of the resistance through the whole square centimeter of a monolayer (TER) of MDCK cells is about 200 Ω, i.e. only 1/20 of the transcellular resistance. This indicates that TER reflects primarily the resistance through the paracellular pathway. Resistance through this pathway is afforded by the TJ plus the intercellular space, but this space contributes a mere 10-20 Ω · cm^2.[37,41] Therefore, in practice TER measures the resistance offered by the TJ.[45,46]

The development of TER of a newly plated monolayer is an easy way of gauging the assembly and sealing of the TJ.[39,49,53] However, under these conditions the delay of TER development includes also the time required for attachment of cells to the substrate, recovery from trypsin-EDTA, accommodation of their borders to each other, cell-cell adhesion, etc. To avoid this inconvenience and to study more specifically the assembly and sealing of the TJ, Gonzalez-Mariscal et al.[54,55] have designed a protocol consisting in decreasing the concentration of Ca^{2+} in the bathing medium to 1-5 μM (LC) 20-90 minutes after plating at confluence. Twenty hours later, while monolayers left in media with normal Ca^{2+} (1.8 mM) have achieved their maximum TER, those left in low calcium (LC) media have a negligible resistance. However if Ca^{2+} is added at this time, TJs are formed and TER develops in 2-4 hours. This procedure is referred to as "Ca-switch".

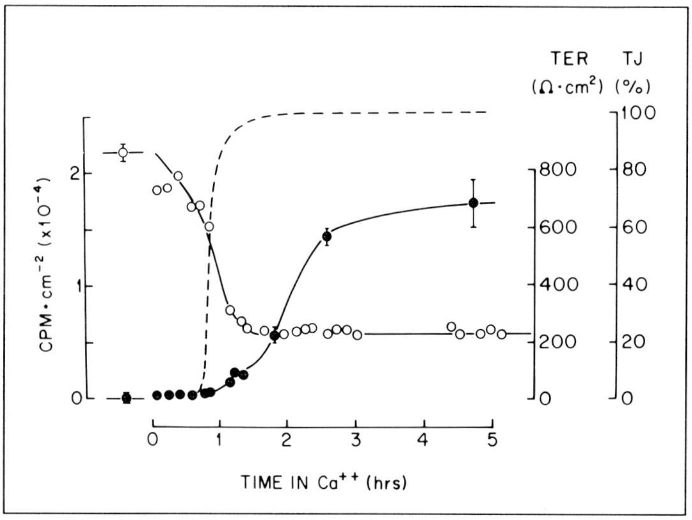

Fig. 6.4. Apical-to-basolateral [³H]-mannitol flux (left ordinate, *empty circles), transepithelial electrical resistance (TER;* middle ordinate, *full circles)and degree of junctional sealing* (right ordinate, *dashed line) as calculated with Eq. 1 at different times of a Ca-switch (time 0). Cells were plated at confluence on Millipore filters in normal calcium (NC) medium for 1.0 hour, then switched to low-calcium medium (LC = 1-5 μM); 20 hours later cells were switched to NC medium (time 0). Taken with kind permission from Gonzalez-Mariscal et al, (1990).*

The decrease of the flux of mannitol across the monolayer during the Ca-switch reflects directly the percentage of junctions sealed at a given time (Fig. 6.4, *filled circles*). The value of TER reflects instead the sealing of the last 1-3% of the strands. Therefore, to transform TER into the fraction of TJs already sealed (f_c) the following equation is used:

$$f_c = [R_\infty (R_o - R_t)] / [R_t (R_o - R_\infty)] \qquad \text{(Eq. 1)}$$

where R_o, $R_t$and R_∞ are the values of TER before, duringand after complete sealing of the TJs.

The use of Ca^{2+} to investigate the intervening factors in junction formation is not surprising, since it was known for several decades that the presence of this ion is required to keep the structure of the epithelium and, in particular, to preserve the TJ in a sealed state.[39,40,47,49,54-61]

These procedures can be used in the presence of inhibitors of transcription, translation, glycosilation, exocytic fusion, polymerization of actin, microtubule assembly, etc. to characterize the cellular mechanisms involved in the making of a TJ.[39,54,55,62-64]

5. THE SITE WHERE CALCIUM TRIGGERS THE ASSEMBLY AND SEALING OF TJS

A switch from LC to NC medium at the 20th hour develops a TER of 676 ± 48 Ω · cm^2 in 5 hours. If instead of the monolayers being switched to 1.8 mM Ca^{2+} they are transferred to a medium with other multivalent cations (Mg^{2+}, Ba^{2+}, Sr^{2+}, Mn^{2+}, Cd^{2+} or La^{3+}), TER remains negligible, suggesting that the ability of Ca^{2+} to trigger junction formation is a highly specific one.

When monolayers are switched to 1.8 mM Ca^{2+}, but in the presence of other ions it is observed that Mg^{2+}, $Ba^{2+}$and Sr^{2+} do not interfere with the effect of Ca^{2+} but that Mn^{2+} and Cd^{2+} inhibit junction formation. Therefore, the selectivity of the site acted upon by Ca^{2+} appears to be $Ca^{2+} > Cd^{2+} \approx Mn^{2+} > Mg^{2+} \approx Ba^{2+} \approx Sr^{2+}$.

If Ca-switch is not made to 1.8 mM Ca^{2+} but to a lower concentration (Fig. 6.7), junction formation can be triggered without modifying its cytosolic chemical activity. This observation is in keeping with results obtained during a Ca-switch in the presence of 1.0 mM La^{3+}, an ion that prevents the penetration of Ca^{2+} into the cytoplasm, without perturbing the assembly and sealing of the TJ.[62] This suggests that Ca^{2+} triggers junction formation by acting on an extracellular site.[65,66]

Using intracellular calcium probes, Gonzalez-Mariscal et al[62] found that MDCK cells incubated for 20 hours in LC have a very low cytosolic concentration of this ion (some 20 nM) and, upon switching them to NC, it increases very sharply, then decreasesand climbs again, albeit at a lower pace (Fig. 6.6). This constitutes a very complex phenomenon, because it involves changes in the influx of $Ca^{2+}$and Ca^{2+} uptake by intracellular compartments.[67,68]

The use of $^{45}Ca^{2+}$ to measure unidirectional fluxes indicated that, during the Ca-switch there is an important penetration of this ion into the cytoplasm.[65,66] Cells incubated in LC have a much higher Ca-permeability than those in NC media, that is quickly reduced during the Ca-switch (Fig. 6.5).[65,66] Randriamampita and Tsien[69] and

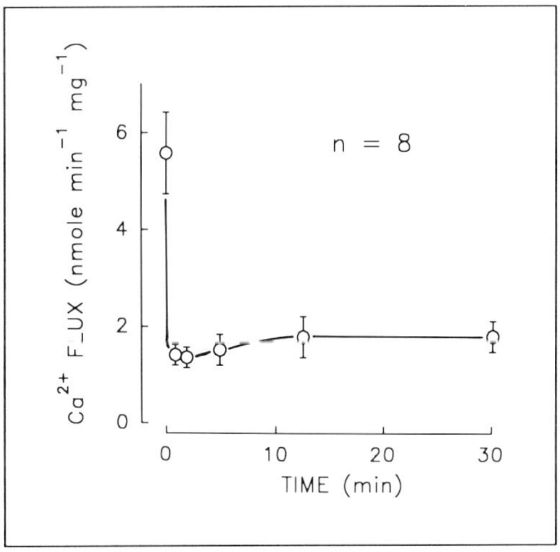

Fig. 6.5. Influx of Ca^{2+} in monolayers preincubated 20 hours in 1-5 μM Ca^{2+}, as calculated through the uptake of ^{45}Ca in MDCK cells. Abscissa represents time after switching from LC to 200 μM Ca^{2+}. Notice the drastic reduction of the influx upon touching the cells with Ca^{2+}. Taken with kind permission from Contreras et al, (1992).

Parekh et al[70] have recently found that cells with depleted intracellular Ca^{2+} compartments, liberate a novel calcium influx factor (CIF) that increases the specific permeability for this ion of the plasma membrane. Therefore, the possibility exists that MDCK cells kept in LC, with the cytoplasmic activity of Ca^{2+} reduced from 130 to 20 nM for several hours, would deplete their intracellular calcium reservoirs, secrete CIF and increase their Ca-permeability. This permeability would be quickly reduced on transferring the cells to NC (Fig. 6.5).

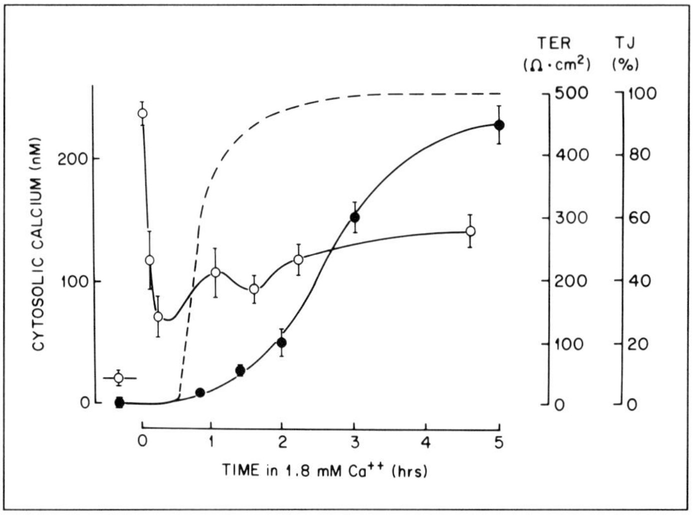

Fig. 6.6. Intracellular Ca^{2+} (empty circles), TER (full circles)and degree of junctional sealing (dashed line) *as a function of time after a Ca-switch. Notice the initial spike of cytosolic Ca^{2+}, followed by a slower and wavy climbing to a new steady state concentration.*

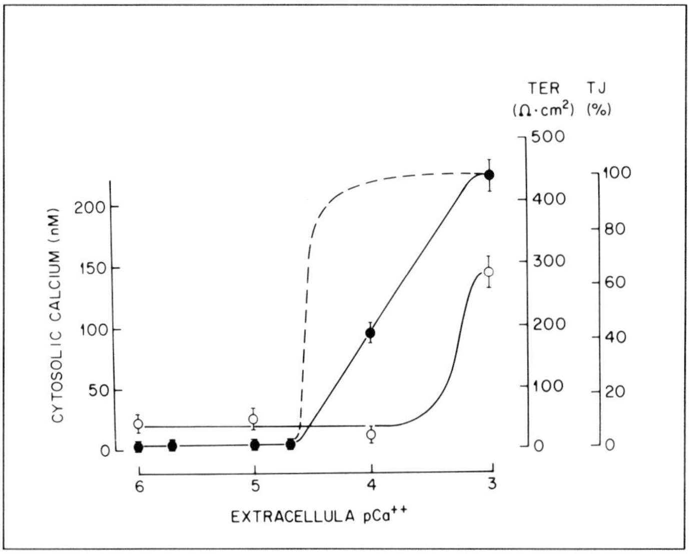

Fig. 6.7. Activity of cytosolic Ca^{2+}, TERand percent of junctional sealing (symbols and definitions as in Fig. 6.6). Monolayers were incubated 20 hours in LC and then switched for an additional period of 5 hours to media with the Ca^{2+} activities specified on the abscissa. Notice that, although 0.4 mM of extracellular Ca^{2+} fails to increase the cytosolic activity of this ion, it nevertheless achieves quasi-complete junctional sealing. Taken with kind permission from Gonzalez-Mariscal et al 1990.

La^{3+} is unable to seal the TJ by itselfand inhibits the influx of Ca^{2+} but produces only a minor reduction of the level of TER achieved by Ca^{2+}, as if to trigger junction formation Ca^{2+} is not required to penetrate through the plasma membrane. In keeping with these observations, verapamil, a powerful inhibitor of Ca^{2+} movement across the cell membrane of MDCK cells, fails to prevent this ion from sealing the TJ.

The fact that Ca^{2+} acts on an extracellular site, but junction assembly and sealing is performed through the series of intracellular events mentioned above, suggests that a transducing mechanisms must be interposed between them. Balda et al[71] have shown that during a Ca-switch pertussis toxin increases several fold the level of TER (Fig. 6.8), while AlF_3 and carbamil choline (carbachol) inhibits it and a nonhydrolyzable GTP analog 5-guanylyl-imido-diphosphate (GTPγs) blocks the development of a cell border pattern of ZO-1, suggesting that G-proteins are involved. Phospholipase C (PLC) and protein kinase C (PKC) probably participate in these process since (i) activation of PLC by thyrotropin-1 releasing hormone increases TERand the inhibition of this enzyme by neomycin blocks the development of this resistance; (ii) 1, 2-dioctanoylglycerol an activator of PKC, stimulates TER development, while polymyxin B and 1-(5-isoquinoline sulfonyl)-2-methyl-piperazine dihydrochloride (H7), which inhibits this enzyme, abolish TER. Addition of 3-isobutyl-1-methyl-xanthine, dB-cAMP or forskolin do not enhance the value of TER, but have just the opposite effect. This was expected, as physiological regulations require that the TER-enhancing cascade involving PLC and PKC must be somehow counterbalanced by another having TER-decreasing ability. Trifluoperazine and calmidazoline inhibit TER development, suggesting that calmodulin (CaM) also plays a role in junction formation. These results indicate that junction formation may be controlled by a network of reactions where G-proteins, phospholipase C, adenylate cyclase, protein kinase C and CaM are involvedand that they may constitute at least part of the mechanisms interposed between the extracellular site where Ca^{2+} acts and the intracellular mechanisms that carry out the assembly and sealing the TJ.[71]

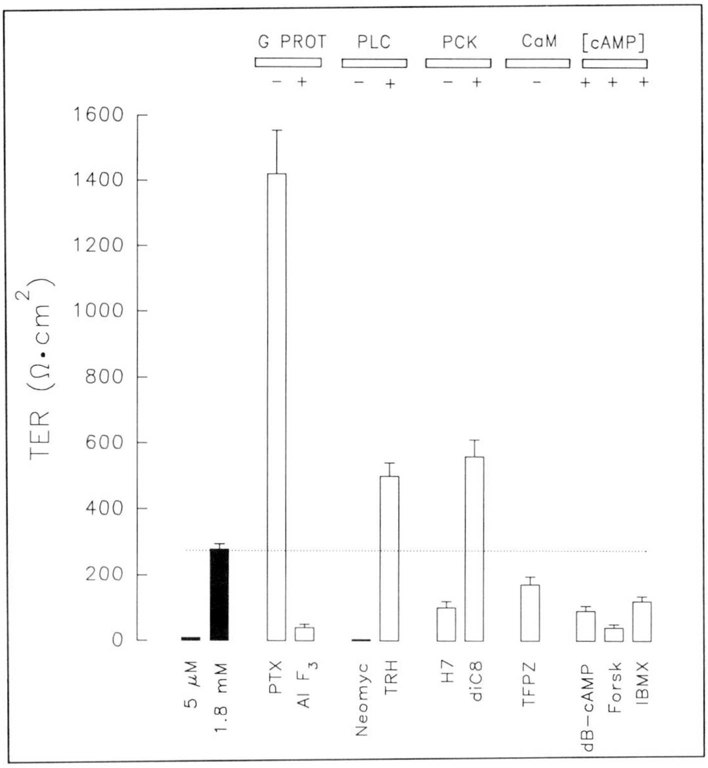

Fig. 6.8. Effect of drugs on TJ formation during Ca-switch, as gauged through the value of transepithelial electrical resistance (TER). Control values before (1st column) and after (2nd column) 4-5 hours of a Ca-switch. To make comparisons easy, the level reached by this control is marked by a dashed line. In the other columns, open ones correspond to a stimulator and hatched ones to an inhibitor. G-prot. (G protein): effect of 5 μM carbachol (open column) and 14 ng/ml pertussis toxin (hatched column). PLC (phospholipase C): effect of 2 μM thyrotropin-releasing hormone (open column) and 110 μM neomycin sulfate (hatched column). PKC (protein kinase C): effect of 100 μg/ml 1,2-dioctanoylglycerol (open column) and 50 μM 1-(5-isoquinolinylsulfonyl)-2-methylpiperazine dihydrochloride (hatched column). CaM (calmodulin): effect of 25 μM trifluoperazine dihydrochloride. Cytosk: effect of 5 μg/ml cytochalasin B.

We have previously suggested[65,66] that this extracellular site may be constituted by E-cadherin, a Ca-dependent cell-cell adhesion protein localized on the lateral membrane of epithelial cells,[72] because the binding of antibodies specific for E-cadherin to MDCK block the formation of TJs during a Ca-switch.[24,72-74] However, recent information, while not discarding the effect on E-cadherin and participation of this molecule, indicate that the scenario may be more complex (see also chapter 5 in this volume).

Other experimental models to study junction formation that do not depend on changes in Ca^{2+} concentration, include immunolocalization studies of ZO-1 and cingulin during blastocyst formation.[75-77] With this approach it was learned for instance that ZO-1 is expressed as early as the 8-cell stage, where it localizes at foci that progressively coalesce into continuous junctions by the 16-cell stage; and that cingulin is presented instead throughout the cortex of the oocyte, but localizes to nascent junctions one or two cell divisions after ZO-1. A novel alternative model to study the biology of the TJ is constituted by couplets of hepatocytes that cooperate to establish a canalicular space limited by TJs.[78,79]

6. CELLULAR EVENTS DURING THE MAKING OF A TJ

Anderson et al[80] have found that Caco-2 cells maintained in a spinner with Ca-free medium for 48 hours have high content of mRNA for ZO-1 and low content of the corresponding protein. Three days after addition of Ca^{2+} and plating the messenger is found to be decreased and the protein ZO-1 increased, as if Ca^{2+} and plating would have promoted translation. Inhibitors of protein synthesis added between 0 to 6 hours after plating at confluence prevent junction formation. When inhibitors are added after this period they fail to block.[39,48] These inhibitors also fail to block if several hours are allowed between harvesting and re-plating,[81] or if they are present during a Ca-switch.[54,55]

The failure of cycloheximide to block TJ formation in a Ca-switch,[55] suggests that during this maneuver the cell contains enough ZO-1 to make the TJ, or else that in spite of its intimate association to TJs, this protein is not essential to assemble and seal it. This ambiguity is compounded by the fact that Howarth et al[82] explored several cell typesand could not find correlation between mRNA for ZO-1, the corresponding protein and the ability of a given cell to make TJs. Furthermore, Itoh et al[83] found the peptide and the mRNA in non-epithelial cells that do not express TJs (see also chapter 5).

Taken together the results of Gonzalez-Mariscal et al would indicate that newly synthesized peptides are required to make TJsand that this synthesis may proceed in the absence of Ca^{2+} and cell-substrate attachment. Whether the peptides needed are those composing or associating with the junctional strands (e.g. ZO-1) is unknown. These peptides seem to be stored in a compartment beyond the Golgi, as 1.0 μM monensin that impairs transit through the cisternae does not prevent junction formation in a Ca-switch.[54,55] The addition of Ca^{2+} causes a transfer of ZO-1 to the vicinity of the plasma membrane, where it forms a continuous ring.[84]

Transfer of junctional components from their trans-Golgi position to the plasma membrane, may also require the participation of the cytoskeleton. Epifluorescence microscopy performed with actin markers indicate that epithelial cells have a framework of microfilaments in close contact with the lateral membrane,[63] and that if this arrangement is perturbed with cytochalasin B junctions are opened and TER vanishes.[64] Furthermore, Madara et al[85,86] have shown that cells from the intestinal mucosa have a delicate meshwork of actin, whose structure varies according to the sealing state of the TJ. We have observed that integrity of the cytoskeleton of actin filaments is also required for the assembly and sealing of TJs, as 5 μg/ml citochalasin B suppresses the development of TER in a Ca-switch.[54,55] Switching to Ca^{2+} also results in an increase of membrane surface, by addition of apical[87] and basolateral membrane.[88] Ca^{2+} may also provoke the incorporation of junctional components, as inhibitors of exocytic fusion like chloroquine prevent the increase in area of the plasma membrane and the development of TER[62] (Fig. 6.9).

7. THE RELATIONSHIP BETWEEN CALCIUM TRIGGERING AND PHOSPHORYLATION

The profound effects of enhancers and inhibitors of PLC and PKC on the level of TER achieved during a Ca-switch, suggests that junctional assembly and sealing depends crucially on the phosphorylation of a junction or junction-related molecular component.

Taking ZO-1, ZO-2 and the 130 kD phosphoprotein as representatives of the TJ-associated peptides, Balda et al[26] investigated whether these are the molecules that change their phosphorylated state during the assembly and sealing of the TJ in MDCK cells.

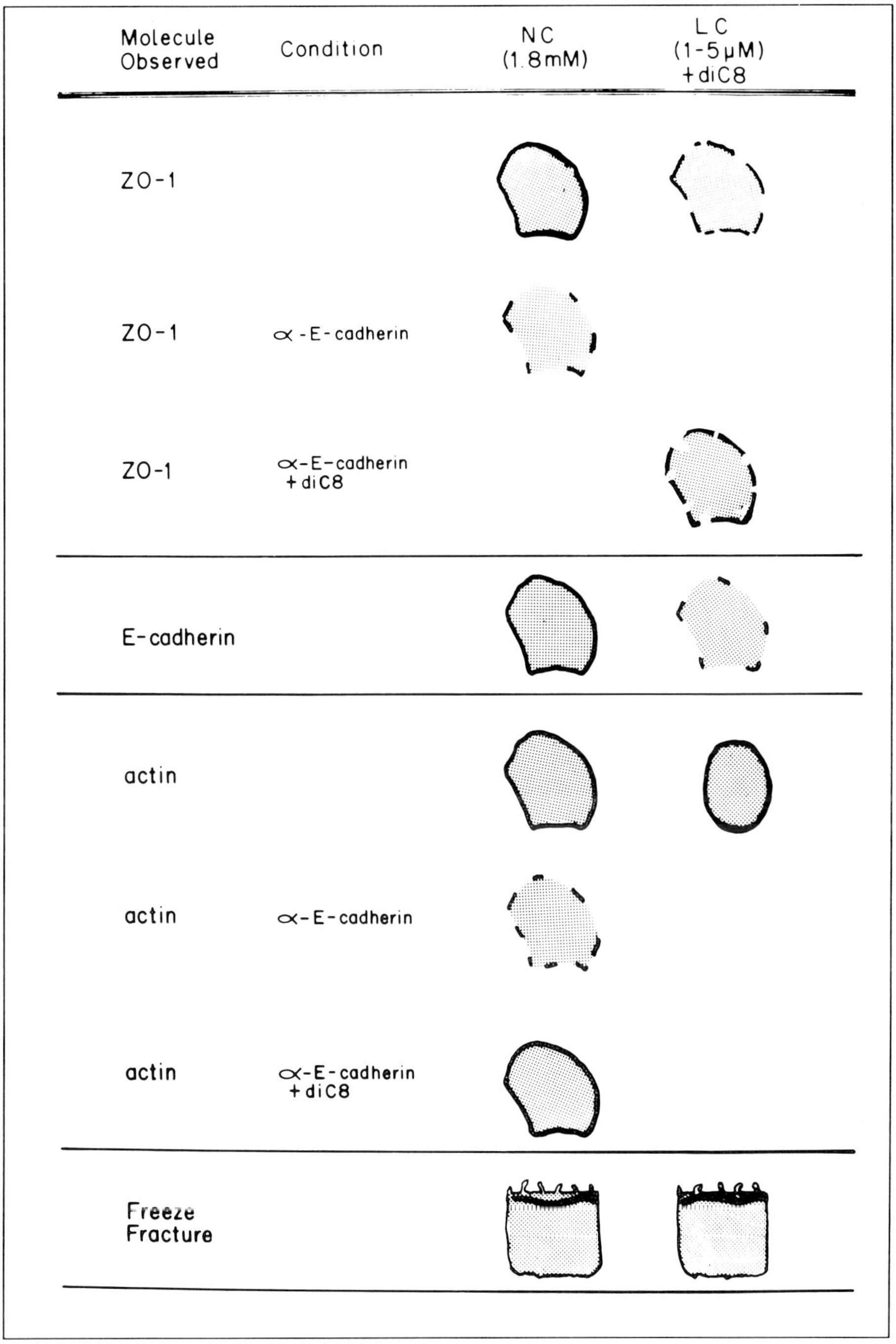

*Fig. 6.9. Schematic representation of the distribution (*bold lines*) of ZO-1 (*rows 1-3*), E-cadherin (*row 4*) and actin (*rows 5-7*), as observed from the apical side in epifluorescence microscopy in MDCK cells (*shadows*) treated with a first antibody against ZO-1, antibody E-cadherin and falloidin respectivelyand distribution of strands as observed in freeze fracture replicas (*bottom row*). Cells in the left column were incubated for 20 hours in LC, switched to NC at this time and incubated for another 2 hours in this medium. Cells in the right column were incubated for 20 hours in LC, switched to fresh LC medium containing 0.5 mM diC8 at this timeand incubated for another 2 hours under this condition. (after Balda et al, 1993).*

They observed that their phosphorylated state is not perturbed by the Ca-switch nor when the effect of this ion is partially mimicked by diC8. These observations cannot be explained as due to a lack of stimulation of some PKC isoforms that depend on the presence of Ca^{2+},[89] since during a Ca-switch the penetration of this ion raises its concentration well above 100 mM. These isoforms might become activated during a Ca-switch, where the penetration of this ion raises its concentration well above of 100 nM.[62]

Besides of participating in de novo making of a TJ, the molecules constituting the cascades mentioned above are likely to be involved in the regulation of TJ permeability.[90] Thus, there are many circumstances in which TJs accommodate the permeability of epithelia and endothelia to physiological requirements, allow migration of leukocytes, or respond to substances derived from tumors or microorganisms.[79,91-98] In this respect, it is interesting that PKC is not only involved in the making of TJs, as mentioned above, but is also required for junction disassembly[99] (see chapter 5)and that PKC agonists can elicit opposite effect on junction resistance in different epithelial cell lines.[100,101]

8. THE RELATIONSHIP BETWEEN CALCIUM TRIGGERING AND CELLULAR EVENTS.

As mentioned above there are several cellular events that occur during the making of a TJ, in particular when this process is triggered by Ca^{2+}: activation of E-cadherin, cell-cell attachment, increase of cytosolic Ca^{2+}, activation of G-proteins, PLC, PKCand calmodulin, exocytic fusion, organization of actin filaments and microtubules. Some of them occur simultaneously with TJ assembly and sealing, but do not seem to be required to perform these tasks. This is the case of the increase in cytosolic Ca^{2+},[62,65,66] and the organization of microtubules.[54,55] Whether there is a causal relationship between the other phenomena and junctional events is still unknown.

Balda et al[26] have recently performed a series of studies with MDCK cells, that casts some light in these relationship, whose results are sketched in Figure 6.9. The upper row (*left*) represents a cell (*shadow*) as observed by immunofluorescence from the apical side, after 2-3 hours of a Ca-switch, showing a continuous ring of ZO-1 (*bold line*) that has been just transferred to the plasma membrane region. On the right side a cell kept in LC, but treated with diC8 develops a reticular pattern resembling the control one, although with discontinuous segments, indicating that the activation of PKC is able to induce translocation of junctional components. This activation does not achieve a transfer of ZO-1 in isolated cells (not shown), indicating that it depends on cell-cell contacts.

The second row of Figure 6.9 illustrates the distribution of ZO-1 in a cell that was treated with anti-E-cadherin and then transferred to NC. The reticular pattern is incomplete, suggesting that E-cadherin is interposed between the Ca-triggered site and the mechanisms distributing ZO-1. In the third, a cell kept in LC but treated with diC8 achieved a ZO-1 segmented pattern, but nevertheless similar to those in NC, indicating that PKC is interposed down stream in the cascade.

The fourth row, shows on the left side the pattern of distribution of E-cadherin in NCand on the right the pattern in LC, demonstrating that the pattern developed by E-cadherin depends on Ca-activation and not on stimulation of PKC.

The fifth row of Figure 6.9 (*left*) illustrates the patterns of microfilaments, as observed with epifluorescence of phaloidin stained cells. Cells in LC have their actin filaments mainly in the perinuclear region (not shown). Treatment with diC8 in LC (*right*) provokes a redistribution of actin filaments toward the periphery of the cell; however, due to the lack of activation of their E-cadherin molecules, cells do not attach each other and remain round. The sixth row represents a cell treated with antibodies anti-E-cadherinand then switched to NC; under this condition the framework of microfilaments is not developed. Finally, in the seventh row, a cell in the same condition as in the 6th row, but treated with diC8, escapes the blockade posed by the antibody and develops a normal pattern of microfilaments. Taken together, these results indicate that PKC activation by diC8 not only induces translocation of ZO-1 but also triggers the reorganization of actin filaments.

Figure 6.9 (*last row*) represents an MDCK cell seen from the lateral side, to illustrate the results obtained with freeze fracture electron microscopy. Confirming observations of Gonzalez-Mariscal et al,[54,55] Balda et al[26] found that as soon as 15 minutes after the Ca-switch, cells already present a very simple TJ, consisting of a single and rather straight strand, almost without branches. Further addition of strands is always made by association of new segments to the initial ones. Balda et al[26] found that essentially the same image can be observed in cells kept in LC, but treated with diC8 (Fig. 6.9, *right*).

Acknowledgments

We would like to thank the pleasant and efficient help of Elizabeth del Oso, Jazmin De Lorenz and Amparo Lazaro. This work was supported by research grants from the National Research Council of Mexico (CONACYT).

References

1. Cereijido M, Ruiz O, Gonzalez-Mariscal L, Contreras RG, Balda MS, Garcia-Villegas MR. The paracellular pathway. A small version of the kidney nephron. In: Kenneth L, Raub TJ, eds. Biological barriers to protein delivery. Plenum 1993:3-21.
2. Rubin L. Endothelial cells: adhesion and tight junctions. Curr Opin Cell Biol 1992; 4:830 833.
3. Schneeberger EE. Heterogeneity of tight junction morphology in extrapulmonary and intrapulmonary airways of the rat. Anat Rec 1980; 198:193-208.
4. Schneeberger EE, Lynch RD. Structure, function and regulation of cellular tight junctions. Am J Physiol 1992; 262 (Lung Cell Mol Physiol 6):L647 L661.
5. Farquhar MG, Palade GE. Junctional complexes in various epithelia. J Cell Biol 1963; 17:375-412.
6. Staehelin LA. Further observations of the fine structure of freeze fracture-cleaved tight junctions. J Cell Sci 1973; 13:763-786.
7. Staehelin LA. Structure and function of intercellular junctions. Int Rev Cytol 1974; 39:191-282.
8. van Deurs B, Luft JH. Effect of glutaraldehyde fixation on the structure of tight junctions. A quantitative freeze-fracture analysis. J Ultrastruct Res 1979; 68:160-172.
9. Chalcroft UP and Bullivant S. An interpretation of liver cell membrane and junction structure based on observation of freeze-fracture replicas of both sides of the fracture. J Cell Biol 1970; 47:49-60.
10. Staehelin LA, Mukherjee TM and Williams AW. Freeze-etch appearance of tight junctions in the epithelium of small and large intestine of mice. Protoplasma 1969; 67:165-184.
11. Kachar B, Reese TS. Evidence for the lipidic nature of tight junctions strands. Nature 1982; 296:464-466.
12. Pinto da Silva P, Kachar B. On tight junction structure. Cell 1982; 28:441-450.
13. Dragsten PR, Blumenthal R, Handler JS. Membrane asymmetry in epithelia: is the tight junction a barrier to diffusion in the plasma membrane?. Nature 1981; 294:718-722.
14. van Meer G, Gumbiner B, Simons K. The tight junction does not allow lipid molecules to diffuse from one epithelial cell to the next. Nature 1986; 322:639-641.
15. van Meer G, Simons K. The function of tight junctions in mantaining differences in lipid composition between the apical and basolateral cell surface domains. EMBO J 1986; 5:1455-1464.
16. Lazaro A, Calderon V, Gonzalez-Mariscal L et al. Tight junctions, lipidsand transepithelial electrical resistance (TER). J Cell Biol 1991; 115:841a.
17. Schneeberger EE, Lynch RD, Kelly CA, Rabito C. Modulation of tight junction formation in clone 4 MDCK cells by fatty acid suplementation. Am J Physiol 1988; 254:C432-440.
18. Gonzalez-Mariscal, L, Chavez de Ramirez B, Cereijido, M. The effect of temperature on the occluding junctions of monolayers of epithelioid cells (MDCK). J Membr Biol 1984, 79:175-184.

19. Ladino C, Schneeberger EE, Rabito CA et al. Inhibition of adenine nucleotide synthesis: effect on tight junction structure and function of clone 4 MDCK cells. Eur J Cell Biol 1991; 55:217-224.
20. Anderson JM, Stevenson BR, Jesaitis LA, Goodenough DA, Mooseker MS. Characterization of ZO-1, a protein component of the Tight Junction from Mouse Liver and Madin-Darby Canine Kidney Cells. J Cell Biol 1988; 106:1141-1149.
21. Stevenson BR, Siliciano JD, Mooseker JD, Goodenough DA. Identification of ZO-1: a high molecular weight polypeptide associated with the tight junction (zonula occludens) in a variety of epithelia. J Cell Biol 1986; 103:755-766.
22. Citi S, Sabanay H, Jakes R, Geiger B, Kendrick-Jones J. Cingulin, a new peripheral component of tight junctions. Nature 1988; 333:272-275.
23. Chapman LM, Eddy EM. A protein associated with the mouse and rat hepatocyte junctional complex. Cell Tissue Res 1989; 257:333-341.
24. Gumbiner B, Lowenkopf T, Apatira D. Identification of a 160 kD polipeptide that binds to the tight junction protein ZO-1. Proc Natl Acad Sci 1991; 88:3460-3464.
25. Zhong Y, Siatho T, Minase T, Sawada N, Enomoto K, Mori M. Monoclonal antibody 7H6 reacts with a novel tight junction-associated protein distinct from ZO-1, cingulin and ZO-2. J Cell Biol 1993; 120:477-483.
26. Balda MS, Gonzalez-Mariscal L, Matter K, Cereijido M. Assembly of the tight junction: the role of diacilglicerol. J Cell Biol 1993; 123:293-302.
27. Willot E, Balda MS, Heintzeltman M, Jameson B, Anderson JM. Localization and differential expression of two isoforms of the tight junction protein ZO-1. Am J Physiol 1992; 262:C1119-1124.
28. Balda MS and Anderson JM. Two classes of tight junctions are revealed by ZO-1 isoforms. Am J Physiol 1993; 264:C918-924.
29. Gonzalez-Mariscal L, Chavez de Ramirez B, Lazaro A, Cereijido M. Establishment of tight junctions between cells from different animal species and different sealing capacities. J Membrane Biol 1989; 107:43-56.
30. Bentzel CJ, Fromm M, Palant CE, Hegel U. Protamine alters structure and conductance of Necturus gallbladder tight junctions without major electrical effects on the apical cell membrane. J Membr Biol 1987; 95:9-20.
31. Bentzel CJ, Palant CE, Fromm M. Physiological and pathological factors affecting the tight junction. In: Cereijido, M, ed. Tight Junctions. Boca Raton: CRC Press, 1992:151-173.
32. Hirsh M, Noske W. The tight junction: structure and function. Micron 1993; 24:325-352.
33. Mazarriegos MR, Tice LW, Hand AR. Alteration of tight junctional permeability in the rat parotid gland after isoproterenol stimulation. J Cell Biol 1984; 98:1865-1877.
34. Mandel, LJ, Bacallao, R, Zampighi, G. Uncoupling of the molecular "fence" and paracellular "gate" functions in epithelial tight junctions. Nature 1993; 361:552-555.
35. Olver RE, Schneeberger EE, Olver RE. Epithelial solute permeability, ion transport and tight junction morphology in the developing lung of the foetal lamb. J Physiol Lond 1981; 315:395-412.
36. Palant CE, Duffey ME, Mookerjee BK, Ho S, Bentzel CJ. Ca^{2+} regulation of tight junction permeability and structure in Necturus gallbladder. Am J Physiol. 1983; 245: C203-C212.
37. Boulpaep EL, Selly JL. Electrophisiology of proximal and distal tubules in the autoperfused dog kidney. Am J Physiol 1971; 221:1084-1096.
38. Cereijido M, Gonzalez-Mariscal L, Contreras RG. Tight junction: Barrier between higher organisms and environment. NIPS 1989; 4:72-75.
39. Cereijido M, Rotunno CA, Sabatini DD. Polarized epithelial cells membranes produced in vitro. In: Hoffman J,ed. Membrane Transport Processes. Raven Press 1978:433-461.
40. Franchi E, Camatini M. Evidence that a Ca^{2+} chelator and calmodulin blocker interfere with the structure of inter-sertoli junctions. Tissue and Cell 1985; 17:13-25.

41. Reuss L. Tight junction permeability to ions and water. In: Cereijido, M, ed. Tight Junctions. Boca Raton: CRC Press, 1992:49-66.
42. Claude P, Goodenough DA. Fracture faces of zonulae occludentes from "tight" and "leaky" epithelia. J Cell Biol 1973; 58:390-400.
43. Claude P. Morphological factors influencing transepithelial permeability: a model for the resistance of the zonula occludens. J membrane Biol 1978; 39:219-232.
44. Cereijido M. Evolution of ideas on the tight junction. In: Cereijido, M, ed. Tight Junctions. Boca Raton: CRC Press, 1992.
45. Gonzalez-Mariscal L. The relationship between structure and function in tight junctions. In: Cereijido, M, ed. Tight Junctions. Boca Raton: CRC Press, 1992:67-76.
46. Cereijido M. Electrical properties of Madin-Darby canine kidney cells. Federation Proc 1984; 43:2230-2235.
47. Cereijido M, Gonzalez-Mariscal L, Avila G, Contreras RG. Tight Junctions. Critical Reviews in Anatomical Sciences 1988; 1:171-192.
48. Cereijido M, Meza I, Martinez-Palomo A. Occluding junctions in cultured epithelial monolayers. Am J Physiol 1981; 240:C96-C102.
49. Cereijido M, Robins ES, Dolan WJ, Rotunno CA, Sabatini DD. Polarized monolayers formed by epithelial cells on a permeable and translucent support. J Cell Biol 1978; 77:853-880.
50. Cereijido M, Stefani E, Martinez-Palomo A. Occluding junctions in a cultured transporting epithelium: Structural and functional heterogeneity. J Membrane Biol 1980; 53:19-32.
51. Cereijido M, Ehrenfeld J, Meza I, Martinez-Palomo A. Structural and functional membrane polarity in cultured monolayers of MDCK cells. J Membrane Biol 1980; 52:147-159.
52. Stefani E, Cereijido M. Electrical properties of cultured epithelioid cells (MDCK). J Membrane Biol 1983; 73:177-184.
53. Misfeldt DS, Hamamoto ST, Pitelka DR. Transepithelial transport in cell culture. Proc Natl Acad Sci USA 1976; 73:1212-1216.
54. Gonzalez-Mariscal L, Borboa L, Lopez-Vancel R., Beaty G, Cereijido M. Electrical properties of MDCK cells. In: Taub M, ed. Tissue culture in epithelial cells. New York: Plenum Press, 1985:25-35.
55. Gonzalez-Mariscal L, Chavez de Ramirez B, Cereijido M. Tight junction formation in cultured epithelial cells (MDCK). J Membr Biol 1985; 86:113-125.
56. Forte JG, Naus AH. Effects of calcium removal on bullfrog gastric mucosa. Am J Physiol 1963, 205:631-637.
57. Galli P, Brenna A, De Camilli P, Meldolesi J. Extracellular calcium and the organization of tight junction in the pancreatic acinar cells. Exp Cell Res 1976; 99:177-183.
58. Martinez-Palomo A, Meza I, Beaty G, Cereijido M. Experimental modulation of occluding junction in a cultured transporting epithelium. J Cell Biol 1980; 87:736-745.
59. Meldolesi J, Castiglioni G, Parma G, Nassivera N, De camilli P. Ca^{2+}-dependent disassembly and reassembly of occluding junctions in guinea pig pancreatic acinar cells: Effect of drugs. J Cell Biol 1978, 79:156-172.
60. Pitelka DR, Taggart BN, Hamamoto ST. Effect of extracellular Ca^{2+} depletion on membrane topography and occluding junctions of mammary epithelial cells in culture. J Cell Biol 1983; 96:613-624.
61. Sedar AW, Forte JG. Effect of calcium depletion on the junctional complex between oxyntic cells of gastric glands. J. Cell Biol 1964; 22:173-188.
62. Gonzalez-Mariscal L, Contreras RG, Bolivar JJ, Ponce A, Chavez de Ramirez B, Cereijido M. Role of Ca^{2+} in tight junction formation between epithelial cells. Am J Physiol 1990; 259:C978-C986.
63. Meza I, Ibarra G, Sabanero M, Martinez-Palomo A, Cereijido M. Occluding junctions and cytoskeletal components in a cultured transporting epithelium J Cell Biol 1980; 87:746-754.
64. Meza I, Sabanero M, Stefani E, Cereijido M. Occluding junctions in MDCK cells: Modulation of transepithelial permeability by the cytoskeleton. J Cell Biochem 1982; 18:407-421.

65. Contreras RG, Gonzalez-Mariscal L, Balda MS, Garcia-Villegas MR, Cereijido M. The role of calcium in the making of a transporting epithelium. News in Physiol Sci 1992; 7:105-108.
66. Contreras RG, Miller JH, Zamora M, et al. Interaction of calcium with the plasma membrane of epithelial (MDCK) cells during junction formation. Am J Physiol 1992; 263:C313-C318.
67. Berridge MJ, Irvine RF. Inositol phosphates and cell signaling. Nature 1989; 341:197-205.
68. Jacob R, Merrit JE, Hallam TJ, Rink TJ. Repetitive spikes in cytoplasmic calcium evoked by histamine in human endothelial cells. Nature 1988; 335:40-45.
69. Randriamampita C, Tsien, R.Y. Emptying of intracellular Ca^{2+} depletion on membrane topography and occluding junctions of mammary epithelial cells in culture. J Cell Biol 1983; 96:613-624.
70. Parekh AB, Terlau H, Stuhmer W. Depletion of $InsP_3$ stores activates a Ca^{2+} and K^+ current by means of a phosphatase and a diffusible messenger. Nature 1993; 364:814-818.
71. Balda MS, Gonzalez-Mariscal L, Contreras RG, Macias-Silva M, Torres-Marquez ME, Garcia-Sainz JA, Cereijido M. Assembly and sealing of tight junctions: possible participation of G-proteins, phospholipase C, protein kinase C and calmodulin. J Membrane Biol 1991; 122:193-202.
72. Boller K, Vestweber D, Kemler R. Cell-adhesion molecule uvomorulin is localized in the intermediate junctions of adult intestinal epithelial cells. J Cell Biol 1985; 100:327-332.
73. Gumbiner B, Stevenson B, Grimaldi A. The role of the cell adhesion molecule uvomorulin in the formation and maintenance of the epithelial junctional complex. J Cell Biol 1988; 107:1575-1587.
74. Vestweber D, Kemler R. Identification of a putative cell adhesion domain of uvomorulin. EMBO J 1985; 139:3393-98.
75. Fleming TP, Hay MJ. Tissue-specific control of expression of the tight junction polypeptide ZO-1 in the mouse early embryo. Development 1991; 113:295-304.
76. Fleming TP, Hay M, Jarved Q, Citi S. Localization of tight junction protein cingulin is temporally and spatially regulated during early mouse development. Development, 1993; 117:1135-1144.
77. Fleming TP, Javed Q, Hay M. Epithelial differentiation and intercellular junction formation in the mouse early embryo. Development 1992(suppl):105-112.
78. Nathanson MH, Gautam A, Bruck R, Boyer JL. Hormonal regulation of paracellular permeability in isolated rat hepatocyte couplets. Am J Physiol 1992; 262:G1079-1086.
79. Tsuneo K, Brauneis U, Gatmaitan Z, Arias I. Extracellular ATP, intracellular calcium and canalicular contraction in rat hepatocyte doublets. Hepatology 1991; 14:640-647.
80. Anderson JM, Van Itallie CM, Peterson MD, Stevenson BR, Carew EA, Mooseker MS. ZO-1 mRNA and protein expression during tight junction assembly in Caco-2 cells. J Cell Biol 1989; 109:1047-1056.
81. Griepp EB, Dolan WJ, Robbins ES, Sabatini DD. Participation of plasma membrane proteins in the formation of tight junctions by cultured epithelial cells. J Cell Biol 1983; 96:693-702.
82. Howarth AG, Hughes MR, Stevenson BR. Detection of the tight junction-associated protein ZO-1 in astrocytes and other nonepithelial cell types. Am J Physiol 1992; 262 (Cell Physiol 31):C461-C469.
83. Itoh M, Nagafuchi A, Yonemura S, Kitani-Yasuda T, Tsukita S, Tsukita S. The 220-kD protein colocalizing with cadherins in non-epithelial cells is identical to ZO-1, a tight junction associated protein in epithelial cells: cDNA cloning and immunoelectron microscopy. J Cell Biol 1993; 121:491-502.
84. Siliciano JD, Goodenough DA. Localization of the tight junction protein, ZO-1, is modulated by extracellular calciumand cell-cell contact in Madin-Darby canine kidney epithelial cells. J Cell Biol 1988; 107:2398-2399.
85. Madara JL. Intestinal absortive cell tight junctions are linked to the cytoskeleton. Am J Physiol 1987; 253:C171-175.

86. Madara JL. Relationships between the tight junction and the cytoskeleton. In: Cereijido, M, ed. Tight Junctions. Boca Raton: CRC Press, 1992:105-119.
87. Vega Salas DE, Salas PJ, Rodriguez-Boulan E. Modulation of the expresion of an apical plasma membrane protein of Madin-Darby canine kidney epithelial cells: cell-cell interactions control the appearance of a novel intracellular storage comparment. J Cell Biol 1987; 104:1249-1259.
88. Contreras RG, Avila C, Gutierrez, B, Bolivar JJ, Gonzalez-Mariscal L, Darzon A, Beaty G, Rodriguez-Boulan E, Cereijido M. Repolarization of Na^{+}-K^{+} pumps during establishment of epithelial monolayers. Am J Physiol 1989; 257(Cell Physiol 26):C896-C905.
89. Kikkawa U, Kishimoto A, Nishizuka Y. The protein kinase C family: heterogeneity and its implications. Ann Rev Biochem 1989; 58:31-44.
90. Duffey ME, Hainau B, Ho S, Bentzel CJ. Regulation of epithelial tight junction permeability by cyclic AMP. Nature 1981; 294:451-453.
91. Huang AJ, Manning JE, Bandak YM, Ratau MC, Hanser KR, Silverstein SC. Endothelial cell cytosolic free calcium regulates neutrophil migration across monolayers of endothelial cells. J Cell Biol 1993; 120:1371-1380.
92. Mamorstein AD, Mortell KH, Ratcliffe DR, Cramer EB. Epithelial permeability factor—a serum protein that condenses actin and opens tight junctions. Am J Physiol 1992; 262:C1403-C1410.
93. Milks LC, Conyers GP, Cramer EB. The effect of neutrophil migration on epithelial permeability. J Cell Biol 1986; 103: 2729-2738.
94. Mullins JM, Laughlin KV, Marano CW, Russo LM, Soler AP. Modulation of tumor necrosis factor-induces increase in renal (LL-PK1) transepithelial permeability. Am J Physiol 1992; 263:F915-F924.
95. Ojakian GK. Tumor promoter-induced changes in the permeability of epithelial cell tight junctions. Cell 1981; 23:95-103.
96. Parsons PE, Sugahara K, Cott GR, Mason RJ, Henson PM. Effects of neutrophil migration and prolonged neutrophil contact on epithelial permeability. Am J Pathol 1987; 129:302-312.
97. Royall JA, Berkow RL, Beckman JS, Cunningham MK, Matalon S. Freeman BA. Tumor necrosis factor and interleukin 1 alpha increase vascular endothelial permeability. Am J Physiol 1989; 257:L399-L410.
98. Wu NZ , Baldwin AL. Transient venular permeability increase and endothelial gap formation induced by histamine. Am J Physiol 1992; 262:H1238-H1247.
99. Citi S. Protein kinase inhibitors prevent junction dissociation induced by low extracellular calcium in MDCK epithelial cells. J Cell Biol 1992; 117:169-178.
100. Ellis B, Schneeberger EE, Rabito CA. Cellular variability in the development of tight junctions after activation of protein kinase C. Am J Physiol 1992; 263:F293-F300.
101. Winter MC, Peterson MW, Shasby DM. Synergistic effects of a Ca^{2+} ionophore and activators of protein kinase C on epithelial paracellular permeability. Am J Respir Cell Mol Biol 1991, 4.470-477.

CHAPTER 7

Tight Junctions: Their Modulation Under Physiological and Pathological States

Eveline E. Schneeberger
Robert D. Lynch

A unique feature of epithelia is their ability to form a continuous, polarized cell monolayer at the interface between two extracellular compartments. This polarized structural organization enables epithelia to regulate both active, vectorial transepithelial transport as well as the passive passage of solutes and water. There are two routes that solute and water may take in traversing an epithelial cell monolayer: the first is across cells or the **transcellular pathway** and the second is between cells or the **paracellular pathway**. As it passively traverses the transcellular route, a small solute encounters the restrictive barriers imposed by the apical and basolateral cell membranes which give rise to a transcellular electrical resistance (TER) ranging from 1000-10,000 $\Omega \cdot cm^2$.[1] The TER measured across most epithelia, however, is up to an order of magnitude less, indicating a substantial passive ion flow via the paracellular pathway.[2] Depending upon the physiological requirements placed on the epithelium, the permeability properties of the paracellular pathway may vary considerably. Since the only continuous, circumferential physical barrier present between cells is the tight junction (TJ), this structure is thought to constitute the major barrier to solute and water flow through the paracellular space. In addition to its *gate or barrier* activity, the TJ may also function as a *fence*. Owing to its unique location in the outer lipid leaflet of the plasma membrane, the TJ separates

Molecular Mechanisms of Epithelial Cell Junctions: From Development to Disease
Edited by Sandra Citi, M.D., Ph.D.;

one subset of mobile proteins and lipids, located in the apical membrane, from another found on the basolateral surface (reviewed in[3]). The aim of this chapter is to review what is known about the regulation of epithelial tight junction structure and activity under a variety of physiological and pathological conditions.

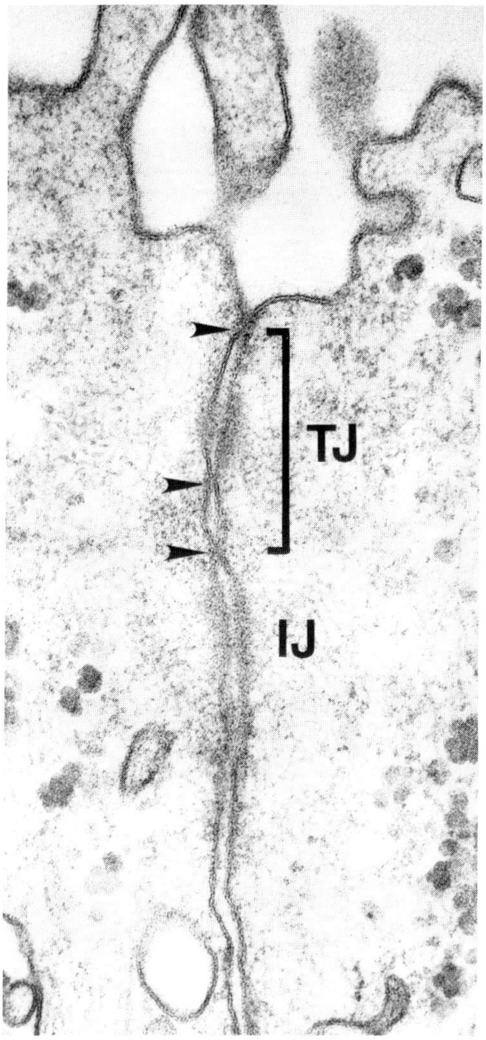

Fig. 7.1. Epithelial cells from the lungs of a fetal lamb. In the region of the tight junction (TJ) arrow heads indicate discrete sites of apparent fusion of the outer lipid leaflet of the adjacent plasma membranes. In the area of the intermediate junction (IJ), the two adjacent plasma membranes are more closely approximated than in the distal portions of the intercellular space. Clusters of cytoplasmic microfilaments border this junction. x 95,200.

1. STRUCTURE-FUNCTION CORRELATES OF TIGHT JUNCTION ACTIVITY

In most epithelia, the TJ or zonula occludens is located at the luminal end of the paracellular space, where it forms a circumferential seal near the apical perimeter of adjacent cells.[4] As seen by transmission electron microscopy, the TJ forms discrete, punctate sites of apparent fusion of the outer lipid leaflet of the plasma membrane (Fig. 7.1). Freeze fracture techniques reveal that these apparent fusion sites are represented by a continuous network of interconnected rows of particles that appear as strands on the protoplasmic fracture face and complementary grooves on the exoplasmic fracture face of the cell membrane (Fig. 7.2);[3] their composition is beginning to be defined with the discovery of the occludin protein.[4a] Although the mechanism of interaction of the extracellular domains of these anatomically defined TJ subunits is at present unclear, it is assumed that a homotypic interaction between TJ particles of adjacent cells is involved. Data obtained from structure-function correlations,[5-7] ion selectivity sequences and charge selectivity of TJs (reviewed in [1,3]) indicate that, while TJ strands themselves may be impermeant, they must contain as part of their structure, highly hydrated pores or channels which, in most epithelia, are lined by negative charges. On theoretical grounds, it has been suggested that these pores, like their counterpart in the cell membrane, may be either in an open or closed configuration (Fig. 7.3).[6] The TJ barrier function may, therefore, be altered in two ways: (1) By increasing or decreasing the number of parallel TJ strands, a process that has been observed under a variety of experimental conditions,[8-10] and/or (2) by altering the net surface charge within TJ channels, a change which affects the permselectivity of TJs.[11] There are, however, instances, for example in two strains of Madin-Darby canine kidney (MDCK) cells, in which large differences in TER (up to 30-fold) cannot be accounted for by differences in the number of TJ strands, in the amount of ZO-1 per cell or in the geometry of the cells.[12] Although the phosphate content of the ZO-1 protein was twice as high in the low as compared to the high resistance strain,[13] the functional significance of this observation remains to be examined.

2. PHYSIOLOGICAL REGULATION OF TIGHT JUNCTION BARRIER FUNCTION

Indications that TJ barrier function might be regulated by intracellular biochemical events came from studies in which intracellular mediators were shown to alter TJ structure and/or permselectivity. Using microelectrode impalement techniques on Necturus gallbladder preparations, Duffey et al[8] showed that when intracellular cAMP was elevated, TJ charge selectivity was altered and TER rose, as did the number of parallel TJ strands. Similar observations were made in goldfish epithelia.[14] More recently, other second messenger-generating systems have also been implicated; application of phorbol esters to two different kidney cell lines, for example, resulted in the activation of protein kinase C and a concommitant decline in TER.[15,16] These early studies indicated that intracellular signals affected TJ barrier function. The precise mechanisms by which they were transmitted from the cytoplasm to the TJ were unclear.

Possible links between the cytoskeleton and the TJ were described using MDCK cells in culture,[17] in intact gallbladder epithelium[10] and in an intestinal epithelial cell line.[18] In all three systems, in which application of cytochalasins (actin filament perturbing agents) resulted in a decline in TER, there was an increase in transepithelial mannitol flux and a decrease in the number of TJ strands. To more directly visualize a physical link between the TJ and the cytoskeleton, Madara detergent extracted intestinal epithelial cells and decorated the retained actin

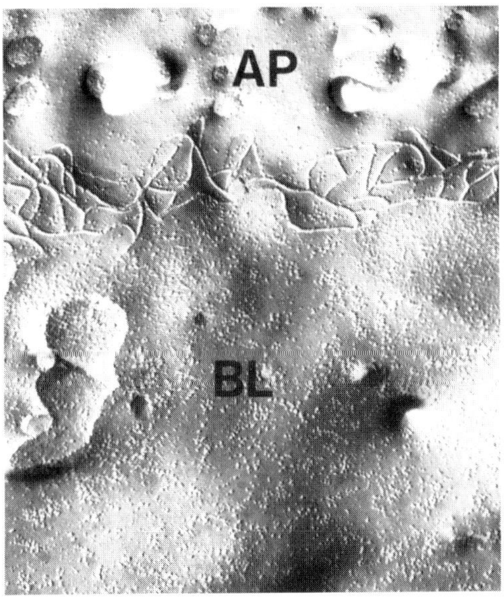

Fig. 7.2. Freeze fracture replica of an MDCK cell from a confluent monolayer. The tight junction forms a network of interconnected strands separating the apical (AP) from the basolateral (BL) plasma membrane domains. x 47,500.

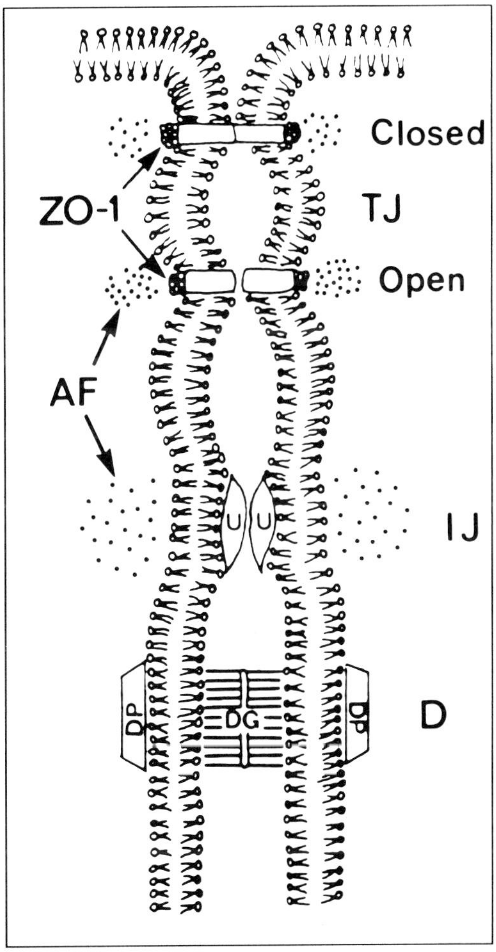

Fig. 7.3. Diagram indicating a possible mechanism whereby tight junction strands from adjacent cells might adopt either an open or closed configuration, depending on the activity of cytoplasmic actin filaments tethered either directly or indirectly via ZO-1 to the tight junction strands.

filaments with heavy meromyosin.[19] In such preparations actin microfilaments appeared to be linked to TJs by means of plaque-like densities that were precisely in register with the focal sites of apparent membrane fusion. The precise relationship of hitherto described TJ associated proteins (ZO-1,[20] cingulin,[21] ZO-2,[22] 7H6[23] to these electron dense plaques and to actin filaments, awaits further immunocytochemical localization studies (see also chapter 5 in this volume). Linking the cell's cytoskeleton to the TJ would provide a mechanism by which TJ barrier function could be finely controlled as part of the cell's response to a perturbing stimulus. It is noteworthy in this respect that both the assembly and maintenance of TJs are energy dependent processes requiring the presence of ATP.[24,25] A distinction between TJ barrier and fence function can be made during early stages of ATP depletion. At that time, with only a modest decline in ATP levels, there is a parallel fall in TER. Interestingly, TJ strands remain intact, as determined by freeze fracture techniques and the ability to maintain a polarized distribution of fluorescent labeled phospholipids, originally incorporated into the apical membrane domain, is retained.[25]

The subtle, cellular control of TJ barrier function has been elegantly documented by Pappenheimer and his associates during nutrient absorption by the small intestine (reviewed in ref. 26). These studies on isolated, perfused intestinal segments became feasible for the first time through the use of fluorocarbon emulsions that have a high oxygen carrying capacity.[27] In these preparations, activation of Na^+-nutrient cotransporters by perfused hexoses or amino acids resulted in dilatations within TJs, condensation of microfilaments in the area of the perijunctional actomyosin ring and enhanced clearance of passively transported solutes.[28] This led to the suggestion that during the absorption of nutrients by the small intestine, high concentrations of hexoses and amino acids are generated in the vicinity of membrane transporters and cell junctions, by the action of membrane bound saccharases and peptidases. As active, concentrative transcellular transport of the enzymatic endproducts occurs, there may be a nonspecific increase of intracellular osmotic pressure that may be the first step in initiating cytoskeletal contraction. Perijunctional actomyosin contraction would then lead to an increase in transjunctional coefficients of osmotic flow and solvent drag.[29] In this view of nutrient absorption, transcellular control mechanisms become coupled to those controlling paracellular permeability.

3. PHYSIOLOGICAL REGULATION OF TIGHT JUNCTION ASSEMBLY

Disruption and reformation of epithelia occurs regularly during such diverse processes as embryogenesis, inflammation and wound healing; these events must be accompanied by the breakdown and reassembly of TJs. To examine events surrounding the assembly of TJs, Gonzalez-Mariscal and colleagues developed the 'Ca^{2+} switch' protocol,[30] a simple assay that affords a means to study agents and/or factors that modulate TJ assembly as well as signal transducing mechanisms (see chapter 6 in this volume). Epithelial cells are plated on a permeable support in the absence of Ca^{2+} for 24 hours, during which time they synthesize the precursors required for TJ formation.[31] Tight junction assembly is triggered by the addition of Ca^{2+} and the course of TJ sealing is then monitored by the sequential measurement of TER. This assay has been useful not only for examining the role of proteins in the mechanisms involved in TJ assembly,[32-33] but also for determining the effect of altered membrane lipid composition on TJ assembly[34-35] (see below).

While the composition of the TJ particles is not known, experiments utilizing Madin-Darby canine kidney (MDCK) epithelial cells, that have been removed from their substratum by trypsin treatment, indicate that protein synthesis is required for the de novo formation of TJs. Addition of cycloheximide (a protein synthesis inhibitor) during the first 6-10 hours after plating, prevents the formation of TJs.[31] It is unclear whether cycloheximide is inhibiting the synthesis of tight junction components themselves or that of adhesion molecules, critical for the initial steps in TJ formation. In either case, when freshly trypsinized epithelial cells are maintained in suspension culture for 6-12 hours without Ca^{2+} and are then plated with Ca^{2+} in the presence of cycloheximide, TJ formation proceeds unimpaired. Epithelial cells, therefore, require 6-10 hours to synthesize the precursors that participate in the assembly of TJs upon addition of Ca^{2+}.

One of the earliest cellular responses triggered by the addition of Ca^{2+} to epithelial cells plated on a substratum, appears to be an exocytic event that results in a 20% increase in the surface area of the cell, as determined by impedance measurements.[36] This is followed by the initiation of cell-cell adhesion, mediated by E cadherin. Evidence for the pivotal role of E cadherin adhesion activity in initiating a cascade of protein interactions leading to the induction of epithelial cell polarity, comes from recent observations made during polarization of Na^+, K^+-ATPase to the basolateral surface of MDCK cells[37] or cadherin transfected fibroblasts.[38] These and other studies have shown a close temporal and spatial relationship between cell-cell contact and the initiation of cell polarity.[39] Once an E cadherin mediated cell-cell contact has been established, there is a gradual assembly of ankyrin-fodrin complexes which become attached to E cadherin. This process also appears to involve the recruitment of ankyrin-fodrin complexes that are linked to other integral membrane proteins, including Na^+, K^+-ATPase.[40] The specificity of E cadherin is highlighted in recent transfection studies, in which Na^+, K^+-ATPase was found to concentrate only at cell-cell contacts in fibroblasts expressing E cadherin, but not in those expressing B cadherin.[38] It is reasonable to draw a parallel between these observations and the events occurring in TJ assembly. The critical role that Ca^{2+} dependent adhesion molecules play in TJ assembly was demonstrated in experiments in which anti-E cadherin antibodies were shown to block TJ assembly in MDCK cells, when they were added simultaneously to cells plated initially in Ca^{2+}-free medium.[41] Furthermore, ZO-1 a TJ-associated cytoplasmic phosphoprotein,[20] requires the presence of both Ca^{2+} and cell-cell adhesion before it will translocate from a juxta-nuclear location to its position on the cytoplasmic side of the TJ.[42] These events parallel those that occur when integral membrane proteins form a complex with ankyrin and fodrin. Clearly, further studies will be required to determine the precise relationship between the TJ and its associated cytoplasmic proteins (ZO-1, ZO-2, cingulin and 7H6 protein) and the actinomyosin cytoskeleton.

The pharmacological studies of Balda et al have provided preliminary data on the possible signal transduction events that may occur in an epithelial cell monolayer stimulated by the addition of Ca^{2+} to assemble TJs.[32] They showed that one of the earliest events to occur upon addition of Ca^{2+} to epithelial cells in a 'calcium switch' assay, is the activation of calcium dependent adhesion molecules, including E cadherin. It was suggested that these cell-cell contacts generate signals that are transduced by GTP binding proteins to phospholipase C which then catalyzes the production of diacylglycerol (DAG) and inositol 1, 4, 5-trisphosphate (IP3). IP3 in turn releases Ca^{2+} from intracellular stores which, together with DAG, activates protein kinase C (PKC). PKC may phosphorylate, among other proteins, the TJ associated phosphoproteins ZO-1[13] and cingulin[43] as well as the cytoplasmic domain of E cadherin, a reaction which promotes its association with actin via catenins.[44] Elevation of intracellular Ca^{2+} also activates calmodulin and facilitates the attachment of the ZO-1 protein to the cell membrane in the region of the TJ.[42] The reader is referred to chapters 5 and 6 in this volume for a more detailed discussion of mechanisms of TJ assembly.

4. PERTURBATION OF MEMBRANE LIPIDS MODULATES TIGHT JUNCTION ASSEMBLY

The biochemical composition of TJ particles has remained remarkably elusive despite attempts, by a number of groups, to generate specific antibodies that would enable the isolation and identification of TJ structural proteins[20,21,23] (see also chapter 5). While no evidence has been obtained to support the original notion that tight junction strands are lipidic in nature,[45] the TJ particles are embedded in the lipid bilayer of the plasma membrane and it is reasonable to suggest that their function might be perturbed by alterations in plasma membrane lipid composition. Membrane lipid perturbations, both in vivo and in vitro, are readily produced by dietary manipulation and supplementation of culture media, respectively.[46] To address the question of the effect of membrane lipid modulation on TJ assembly and barrier function[34,35] the following series of studies was undertaken.

In the first, phospholipid acyl groups were modified by supplementing MDCK cells, grown in serum-free medium, with 18:1(n-9), 18:3(n-3), or 18:3(n-6) complexed to albumin.[34] Although both of these polyunsaturated fatty acids depress the melting point of membrane phospholipids, only 18:3(n-6) contributes significantly to eicosanoid production. The unsaturation indices of the phospholipids of cells receiving either of the 18:3 isomers were more than doubled as compared to cells supplemented with albumin alone or 18:1(n-9) (Table 7.1). The release of PGE_2 into the medium was ten-fold higher in monolayers supplemented with 18:3(n-6) than those receiving 18:3(n-3). Although cells supplemented with 18:3(n-6) required longer periods of time to establish a steady state TER than cells in any of the other groups, no significant difference was observed in the steady state TER or selectivity of the fully formed TJs (Fig. 7.4). The mechanisms

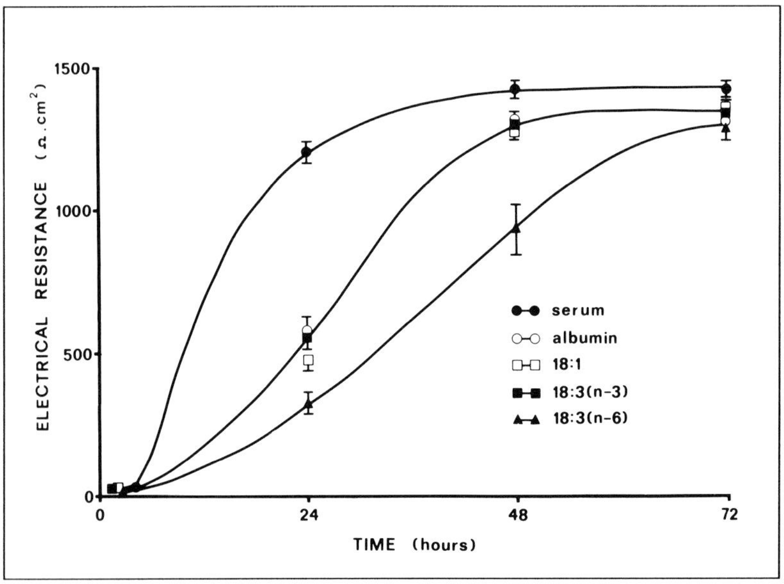

Fig. 7.4. Development of transepithelial electrical resistance (TER) in fatty acid-supplemented and control MDCK cell monolayers after trypsinization. Cells supplemented with 18:3(n-6) achieve a steady state TER more slowly than monolayers supplemented with 18:3(n-3), 18:1(n-9) or albumin added to the serum free medium. The presence of cycloheximide inhibits the development of TER. By permission from Schneeberger et al[34].

Table 7.1. Unsaturation index of phospholipid fatty acids of control, 18:1(n-9), 18:3(n-3) and 18:3(n-6)supplemented clone 4 MDCK cells

Serum	Serum/ lipid free	18:1(n-9)	18:3(n-3)	18:3(n-6)
1.08	0.77	0.79	1.81	1.65

Unsaturation indices were calculated by multiplying% of each fatty acid present in fraction by number of double bonds contained. Products were summed and divided by 100. By permission from Schneeberger et al.[34]

responsible for these alterations in the rate of TJ formation are not certain, but may involve changes in the physical state of membrane lipids (e.g. membrane fluidity) and/or an effect of the synthesized eicosanoids.

Perturbations in membrane phospholipid acyl groups, however, are not confined to the plasma membrane itself and they may have additional effects on the function of cellular organelles. Moreover, homeostatic mechanisms of the cell limit the extent to which membrane phospholipids can be modified. To more specifically restrict the changes in lipid composition to the plasma membrane, a strategy was devised to decrease cell cholesterol, 90% of which resides in the plasma membrane.[47] Cholesterol content was reduced by 25% after incubation of MDCK cell monolayers for 24 hours in a low Ca^{2+} medium supplemented with 1-2 µM Lovastatin, an inhibitor of hydroxymethylglutaryl-coenzyme A (HMG-CoA) reductase. In contrast to the studies with phospholipid modified cells, a reduction in plasma membrane cholesterol content halved the lag time required for Ca^{2+} to induce TJ assembly, as monitored by TER measurements (Fig. 7.5). [^{3}H]Mannitol flux and morphometric measurements of freeze fracture replicas confirmed that the effects on TER reflected changes in the characteristics of the paracellular pathway. Interestingly, peak as well as steady state values of TER were also elevated above control values. The changes in cholesterol content and in the time course of TJ assembly were apparent at levels of Lovastatin which do not affect prenylation of proteins[48] and were abolished if 5 mM mevalonate was present along with Lovastatin. The concentration of Ca^{2+} required to trigger TJ assembly in cholesterol depleted cells was approximately 1/3 of that required to do so in control cells. Paradoxically, $^{45}Ca^{2+}$ uptake by cholesterol depleted cells was actually reduced. These data suggest that cholesterol may modulate the properties of membrane proteins and/or lipids which interact with Ca^{2+}, possibly on the exoplasmic leaflet, during TJ assembly. Studies are currently in progress to address this question further.

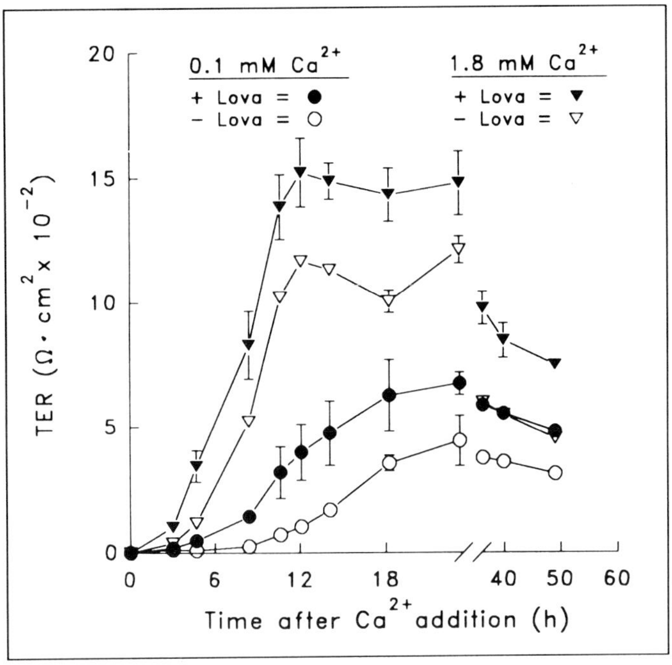

Fig. 7.5. MDCK cell monolayers were cultured in Ca^{++} deficient medium for 6 days. During the last 24 hours they were exposed to 2 µM Lovastatin. Twenty four hours later the Ca^{++} concentration was increased to either 0.1 or 1.8 mM and transepithelial electrical resistance (TER) was measure at frequent intervals. Lovastatin treated cells consistently developed higher steady state levels of TER than normal controls. By permission from Lynch et al[35].

5. HORMONAL AND NEURONAL REGULATION OF TIGHT JUNCTION ACTIVITY

Although the innate barrier properties of TJs vary in different epithelia, TJs must also respond to the varying physiological demands placed upon them. The precise molecular mechanisms involved in the regulation of TJs by hormones and neurotransmitters are beginning to unfold. Based on currently available data, two general patterns have emerged. In the first, hormones and neurotransmitters act via second messenger cascades to directly regulate, by phosphorylation, junctional components including ZO-1 and cingulin, both of which are phosphoproteins.[13,43] In the second, hormone-induced structural alterations in the actin cytoskeleton could impose altered TJ particle contacts, thereby inducing changes in permeability. Several model systems have been employed in these studies: (1) the perfused rat liver stimulated with vasopressin, epinephrine and angiotensin II,[49] and (2) the pancreas and parotid gland undergoing cholinergic stimulation.[50,51] Application of hormones and neurotransmitters results in most cases in a rapid increase in TJ permeability. By contrast, the response to the application of insulin[52] or insulin-like growth factors[53] to the basolateral surface of cultured intestinal epithelial cell monolayers is slow. In such preparations TER declined more than eight-fold over a 3-4 day period and was associated with a subtle condensation of the perijunctional actin ring. The effect was fully reversible once insulin or the insulin-like growth factors were removed from the culture medium. It is of interest that the decline in TER was prevented by the addition of cycloheximide. There is presently no information on what protein(s) are critical for the insulin dependent decrease in TER.

While most neurotransmitters and hormones that have been examined, induced a decline in TJ barrier function, addition of glucocorticoids to cultured mouse mammary epithelial cells, resulted in a more than 10-fold increase in TER and a concomittant reduction in mannitol flux.[54] No alteration in the organization of filamentous actin, ZO-1 or cadherin E was detected. The mechanism of action of glucocorticoids on TJs is unclear and may involve a direct effect on TJ subunits or on factors that regulate their activity. The barrier function of TJs also appears to be augmented following the addition of thyrotropin releasing hormone to MDCK cell monolayers, which is thought to be mediated by its stimulatory effect on phospholipase C.[32]

Indirect evidence indicates that epinephrine, vasopressin and angiotensin II act by inducing changes in Ca^{2+}, inositol polyphosphates and diacylglcerol. Interestingly, the effects of these hormones on TJ permeability in isolated, perfused rat liver preparations can be mimicked by the Ca^{2+} ionophore A23187[55] and an inhibitor of microsomal Ca^{2+} sequestration, 2,5-di-(tert-butyl)-1,4-benzohydroquinone (tBuBHQ).[56] Results from recent studies suggest two possible intracellular pathways that may be involved in the observed vasopressin-induced increase in TJ permeability. In isolated rat hepatocytes, Yamaguchi and colleagues showed that vasopressin induces, within 40 sec, an elevation in intracellular Ca^{2+} concentration and maximal myosin light chain phosphorylation. The same effect was also achieved with the Ca^{2+} ionophore A23187.[57] These results suggest that myosin light chain phosphorylation is at least coincident with and possibly responsible, via microfilament contraction, for the increase in TJ permeability caused by elevations in intracellular Ca^{2+}. In a second study utilizing hepatocyte couplets, Nathanson and colleagues obtained evidence that the vasopressin-induced elevation of intracellular Ca^{2+}, induces increases in paracellular permeability by activation of protein kinase C (PKC), since addition of the PKC inhibitor H-7 abrogated vasopressin's effect on TJ barrier function.[58] Presumably the activation of PKC involves the phosphorylation of proteins other than myosin light chain and its kinase. The pivotal role of PKC activity in TJ barrier function is further borne out by results from two recent studies. In the first, following the removal of extracellular calcium, TJ disassembly was blocked by the addition of PKC inhibitors[43] and in the second, activation of PKC by exogenous diacylglycerol resulted in the association of ZO-1 with the junctional area even in the absence of extracellular Ca^{2+}.[59] While there remain many gaps in our knowledge of tight junction regulation, there is strong evidence that the cytoskeleton is involved in the control of TJ permeability.

Table 7.2. Perturbants of tight junction structure and function

Category	Factor	Reference
Physico-chemical agents	Osmotic reversal	64
	Hypoxia	63
	Hypertonic salt solutions	65
Serum derived	C3a, C5a	72
	Epithelial permeability factor	78
Cell derived	Histamine	75
	Leukotrienes	73
	Interleukin-1	77
	Tumor necrosis factor	77
	Interferon-gamma	80
Bacterial toxins	Pertussis toxin	32
	Cholera toxin	32
	Zonula occludens toxin	87
	Clostridium difficile toxin A	85
	Clostridium difficile toxin B	86
Proteases	Endoproteases	35,66,71
Oxidants	Hydrogen peroxide	89,90

6. PERTURBATION OF TIGHT JUNCTIONS IN PATHOLOGICAL STATES

6.1 AGENTS THAT INTERACT WITH THE CYTOSKELETON

There are a diverse group of factors that are produced in a variety of disease states that affect TJ barrier function via alterations in the cytoskeleton (Table 7.2). Direct action of pharmacological agents on the cytoskeleton has long been known to adversely affect TJ activity. Cereijido and his colleagues showed that the addition of cytochalasin B (a perturbant of microfilaments), but not colchicine (a perturbant of microtubules) to monolayers of MDCK cells with established TJ, resulted in the abolition of TER.[17] These observations were confirmed and extended by Madara et al who showed that exposure of guinea pig ileum to cytochalasin D, induced not only a loss in TER, but also a change in ion selectivity of the TJ.[60] Phalloidin (a toxin, isolated from the green fungus *Amanita phalloides*) when administered to hepatocytes, stabilized filamentous actin and resulted in disorganization of the tight junction as well as a reduction in the number of TJ strands, an increase in junctional permeability to electron dense tracers[61] and an accumulation of microfilaments around bile canaliculi.[62]

6.2 PHYSICO-CHEMICAL AGENTS

There have been a number of interesting observations made, in seemingly disparate experimental models, on the rapid induction of TJ strands. One of these was the observation that epithelial cells in excised rat prostatic glands incubated at 37°C in cacodylate buffer (an arsenic derivative), developed within 5 minutes large numbers of aberrant TJ strands on the basolateral surface.[63] These strands formed in the presence of cycloheximide and dinitrophenol (an uncoupler of oxidative phosphorylation). Subsequent studies indicated that this was evoked in response to a variety of abnormal stimuli. For example, osmotic reversal, induced by filling the toad bladder with Ringer's solution (220 mOsm) and immersing the bladder in distilled water, resulted within 2 mins in the assembly of TJ strands that bridged the space between the basal surface of the epithelial cells and the apical pole of basal cells.[64] These changes occurred in the presence of both cycloheximide and cytoskeleton perturbers (colchicine and cytochalasin B).

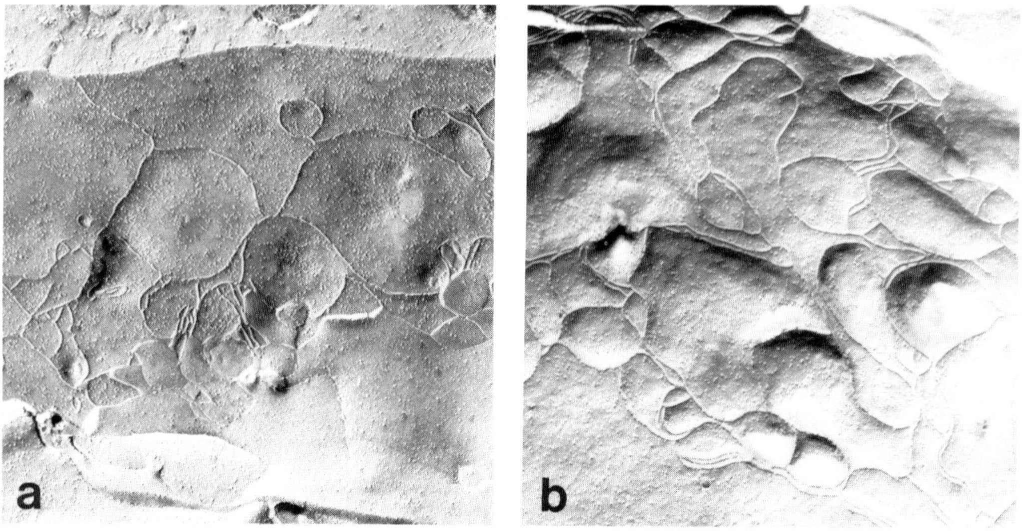

Fig. 7.6. a. Freeze fracture replica of HT 29 cells treated with 0.1% trypsin for two mins shows the rapid assembly of disorganized strands on the surface of these cells. b. Freeze fracture replica of HT 29 cells treated with 160 mM ammonium sulfate. Similar to the effect of trypsin, disorganized arrays of tight junction strands rapidly appear on the lateral surface. a, b x 37,500

The adenocarcinoma cell line, HT 29, has provided an interesting model with which to examine the rapid induction of TJ strands. Under standard culture conditions, these cells do not form TJs. However, within minutes of adding hypertonic salt solutions, numerous, disordered arrays of TJ strands appear on the plasma membrane (Fig. 7.6b).[65] The fact that only certain salts were effective suggests that this is not simply a general response to alterations in osmolarity and ionic strength. All alkali sulfates to varying degrees, as well as calcium and magnesium chloride were potent inducers. Although the relative potencies of different anions were similar in ammonium and sodium salts, the absolute numbers of strands were consistently lower in the sodium series. The results of these studies, suggest that to achieve the maximum induction of TJ strands a specific combination of cations and anions is required. It should be noted that because of their stabilizing effects, anti-chaotropic salts appear to be most effective in triggering TJ strand formation.[65]

6.3 Endoproteases

In acute inflammation, a number of endoproteases are released from infiltrating polymorphonuclear leukocytes (PMN). It is, therefore, of interest to note that endoproteases, but not ectoproteases, induce the de novo assembly of TJ strands in HT 29 cells (Fig. 7.6a),[66] and the formation of aberrant TJ strands in epithelial cells with preformed TJs.[35,67] While this process does not involve protein synthesis, the cells require a period of time during which they synthesize new proteins, before a second round of protease treatment produces additional strands. In studies with HT 29 cells, the protease was administered in the absence of Ca^{2+}, a condition in which cadherins become susceptible to proteolytic cleavage.[69] This suggests that proteases not only induce TJ strand formation, but also digest Ca^{2+}-dependent cell-cell adhesion molecules which then require resynthesis.[70] However, the presence of calcium does not inhibit the induction of new TJ strands. When trypsin is added in the presence of physiological concentrations of Ca^{2+} to MDCK cell monolayers with intact TJs, additional TJ strands are rapidly assembled.[68] That cellular proteases may be involved in the physiological induction of TJ strands has recently been suggested by observations made in Caco-2 cells, an adenocarcinoma cell line which normally forms TJs. Addition of the protease inhibitors, leupeptin or antipain, suppressed the formation of TJs as determined by freeze fracture techniques and TER measurements.[71] In MDCK cells, addition of leupeptin to the medium at the beginning of a calcium switch protocol results in a three- to four-fold reduction in the peak TER attained (unpublished observations). Clearly, further studies are required to identify the endogenous proteases that may be responsible for processing of molecules needed for the assembly of TJs.

In recent studies, Lynch and colleagues reported that the application of 1% trypsin, in the presence of 1.8 mM Ca^{2+} to the basolateral, but not the apical surface of MDCK cell monolayers with intact tight junctions, resulted in the formation of large numbers of aberrant strands and a significant increase in TER.[68] Some of these appeared to originate from the preformed tight junction and extended downward along the basolateral cell surface, while others were present as single strands in the basolateral domain. Within two minutes of exposure to trypsin the TER began to increase and attained peak values 40-60% above baseline (200 $\Omega \cdot cm^2$) by six minutes. Continued incubation resulted in a decline in TER to zero as proteolysis continued. Interestingly, application of the same concentration of trypsin to the apical surface for up to an hour, produced little change in TER and no aberrant TJ strands were observed in freeze fracture replicas. These observations suggest either that there is an intrinsic polarity in the TJ strands, those at the luminal border being refractory to proteolysis or, alternatively, that access of trypsin to the TJ is blocked at the luminal cell surface. In a teleological sense, this lack of susceptibility of the TJ to luminal protease digestion may represent an important protective feature that maintains the integrity of that side of the epithelium directly to environmental insults.

6.4 Serum and Cell Derived Factors and Cytokines

During inflammatory and immune responses, epithelial and endothelial cells are exposed to a number of cell and serum derived factors. The effect of inflammatory and immune mediators on paracellular permeability has been explored more extensively in endothelial than in epithelial cell systems. The mediators examined have included C3a and C5a,[72] leukotrienes,[73] platelet activating factor,[74] histamine,[75] tumor necrosis factor-α and interferon-γ,[76] interleukin 1α,[77] all of which are associated with either an increase in endothelial cell monolayer permeability or with striking cytoskeletal rearrangements.

a. Epithelial permeability factor (EPF). Few studies have systematically examined the ligand/receptor interactions or signal transduction and effector mechanisms that are triggered by the action of cell and serum derived factors acting on the barrier function of TJs. It appears, however, that many of these factors act as cytoskeletal perturbants which in turn produce an altered paracellular permeability. Cramer and her colleagues have identified a circulating serum protein that, when added to MDCK cell monolayers, results in a dose-dependent focal condensation of the perijunctional actin ring at sites where three or more cells are in contact, which is accompanied by a fall in TER.[78] This EPF is heat stable, digestible by pepsin and appears to be a multimer that interacts with the basolateral surface of cells to induce increased paracellular permeability.
b. Interferon-γ (IFN-γ). In a number of chronic disease states, accumulations of mononuclear cells are observed in close proximity to epithelial cells of many organs. T-cells and monocytes/macrophages are the source of a number of potent mediators, some of which have striking effects on the physiological barrier function of both endothelia and epithelia.[76,77,79,80] Interferon-γ, a T-cell derived cytokine, when applied to the basolateral, but not the apical surface, of T84 intestinal epithelial cells is associated with a dose dependent decline in TER which,[79,80] by dual Na^+ and mannitol flux analysis, indicates an effect on TJ permeability.[79] In contrast to its effect on endothelial cells, IFN-γ elicited only subtle condensations of f-actin. After only

a 24 hours exposure, the TER remains depressed over 5 days, but the effect is reversible. The interaction of the cytokine with T84 cells is partially inhibited by the addition of IFN-γ receptor-specific antibodies, indicating that these cells express IFN-γ receptors on their basolateral surface.[80] These studies show that the presence of an immune mediated reaction in proximity to a mucosal surface may have profound effects on the barrier function of the adjacent epithelial cells.

c. Interleukin-1 (IL-1), interleukin-2 (IL-2), tumor necrosis factor-α (TNF-α). These three cytokines increase the paracellular permeability to protein tracers of endothelial cells in culture, by causing major cell shape changes.[77,82] Similar effects in epithelial cells in culture have not yet been reported. Administration of IL-1, IL-2 or TNF-α to T84 cell monolayers did not alter the TER measured across the monolayer.[79] While this suggests that these cells lack the specific cytokine receptors, it does not preclude the possibility that other types of epithelial cells may express such receptors and would respond by altering paracellular permeability. While the precise signal transduction pathway of TNF-α and IL-1 is not completely understood, it is of interest that both of these cytokines appear to stimulate, via a specific phospholipase, the release of diacylglycerol from phosphatidylcholine that activates a Ca^{2+} independent protein kinase C,[83,84] a pathway that is not known to alter TJ activity.[32]

6.5 Bacterial toxins

Depending on their origin, bacterial toxins produced by a variety of microorganisms may enhance or disrupt TJ barrier function. Pertussis toxin added to MDCK cells at the time Ca^{2+} is added to MDCK cells in a calcium switch assay, resulted in a marked increase in TER. It was suggested that pertussis toxin may exert its effect on TER development by relieving a GTP binding protein (G_{pi}) induced inhibition of phospholipase C.[32] By contrast, administration of cholera toxin to a similar MDCK cell preparation produced only a modest reduction in TER, an event thought to be mediated by the ADP-ribosylation of G_s, a reaction which locks this GTP binding protein in its active state, thereby activating adenylate cyclase.[32]

Exotoxin A, produced by *Clostridium difficile*, induces a fall in TER, an increase in passive Na^+, mannitol and inulin flux in mucosal sheets of guinea pig ileum. This was associated with condensation of microfilaments in the zone of the perijunctional actomyosin ring.[85] *Clostridium difficile* toxin B produces similar morphological and functional changes when applied to monolayers of T84 intestinal epithelial cells.[86]

The alpha subunit of *Vibrio cholerae* toxin stimulates adenylate cyclase which subsequently induces the net secretion of copious amounts of fluid into the intestinal lumen. Administration of a recombinant *V. cholerae* vaccine in which the ctx gene was deleted, however, still induced diarrhea. This lead to the search for and discovery of a second enterotoxin produced by *V. cholerae* that specifically affects intestinal TJs. This zonula occludens toxin (ZOT) increased the permeability of segments of rabbit ileum mounted in Ussing chambers and induced a marked simplification of the TJ network, as viewed by freeze fracture techniques.[87]

6.6 Oxidants

Increased permeability of epithelial barriers is one of the hallmarks of acute inflammation. Oxidants are released by inflammatory cells and they constitute one of several inflammatory mediators known to affect paracellular barrier function.[88,89] Application of noncytolytic doses of H_2O_2 either to monolayers of MDCK cells[88,90] or type II alveolar epithelial cells[89] resulted in a decline in TER and a 50% decrease in short circuit current.[89] This effect was only observed when the oxidant was applied to the basolateral surface which led to the suggestion that apically applied H_2O_2 was degraded by intracellular catalase, whereas H_2O_2 applied to the basolateral surface interacted directly with membrane components.[89] Interestingly, application of H_2O_2 to MDCK cells resulted in the stimulation of phospholipase C which had as a consequence an increase in the production of inositol polyphosphates, phosphatidic acid and 1,2 diglycerides.[90] The latter, as discussed above, is a potent stimulator of protein kinase C, the activation of which is associated with a fall in TER.[16]

6.7. Polymorphonuclear Leukocyte (PMN) Migration

Migration of PMN across epithelial barriers is a characteristic feature of acute inflammation. The earliest event in this process, is the adhesion of PMN to endothelium and the subsequent emigration of these cells from the vasculature.[91,92] An inflammatory process occurring in an organ lined by epithelial cells, however, requires that the PMN also traverse the epithelial barrier. The events involved in the migration of PMN across an epithelium and the effect this has on epithelial barrier function is beginning to be addressed.[93,94] In chemotaxis assays in which PMN were stimulated to traverse a T84 intestinal epithelial cell monolayer by a gradient of the chemoattractant n-formyl-methionyl-leucyl-phenylalanine (FMLP), there was a consistent decline in TER and an increase in passive Na^+ and mannitol flux, indicating an effect on the paracellular pathway. The decline in TER was a function of the number of PMN migrating across the monolayer and was reversible when the migration of these cells was stopped.[95] Addition of inhibitors of toxic oxygen metabolites or the use of PMN from patients with chronic granulomatous disease did not prevent transmigration nor the increase in paracellular permeability, indicating that oxygen metabolites are not required in order for the PMN to force its way through the TJ. Similarly, the use of protease inhibitors had no effect on PMN migration suggesting that the release of proteases by PMN is not required for traversing of TJs.[94] However, addition of antibodies to the heterodimeric adhesion glycoprotein CD11b/CD18 (Mac-1), largely prevented the migration of PMN across an epithelial cell monolayer in vitro.[96] By contrast, antibodies that recognized the other two α subunits of this integrin, CD11a or CD11c, were ineffective. Furthermore, it was noted that PMN transmigration was 5-20 times more efficient in the physiological serosal to mucosal direction than it was in the nonphysiological opposite direction. In these studies it was noted that during transmigration, PMN released a factor that elicits chloride secretion by T84 intestinal epithelial cells.[97]

In naturally occurring acute inflammatory processes, epithelial-PMN interactions often occur in the presence of lymphocyte derived cytokines. In this context, Madara and his colleagues examined the effect of IFN-γ treatment of T48 cell monolayers on the transmigration of PMN along a chemotactic gradient. Exposure of T84 intestinal epithelial cells to IFN-γ markedly increased the apical to basolateral transepithelial migration of PMN in a time and dose dependent manner. This phenomenon was specifically an effect on the epithelial cells themselves and not due to the known effects of IFN-γ on TJ permeability. Furthermore, protein synthesis was required and the participation of the CD11b/CD18 integrin was necessary. By contrast, IFN-γ downregulated the transepithelial migration of PMN in the physiological direction, but as with the effect on the migration in the nonphysiological direction, it too was dependent on protein synthesis and was a function of the Mac-1 integrin. These results suggest that IFN-γ may lengthen the retention time of recruited PMN in the paracellular space, by downregulating PMN-epithelial interactions required for subsequent migration. This may serve an important role in mucosal defense.

6.8 Tight Junctions in Human Disease

In addition to changes in TJ structure and function occurring during self-limiting inflammatory and immune responses (see above), there are a few diseases in which TJ dysfunction may be of more permanent duration. Patients with Crohn's disease and their relatives appear to have increased TJ permeability,[99,100] suggesting that this primary defect may allow the more ready access of inflammatory agents into the mucosa, thereby perpetuating the inflammatory bowel disease. In addition, there are two inherited cholestatic liver diseases, Byler's disease[101] and North American Indian childhood chirrhosis[102] which may represent inborn defects in cytoskeletal organization which result in TJ dysfunction. Clearly, the clinical manifestation of TJ dysfunction is an area requiring further study.

7. CONCLUSIONS

The integrity of intercellular tight junctions is pivotal to the maintenance of epithelial barriers between exterior and interior compartments of most organisms. Far from being static structures, TJs are controlled by cellular mechanisms which regulate TJ permeability to meet both physiological requirements and pathological challenges. This regulation is largely mediated by the activity of the cytoskeleton which appears to be physically linked to these intercellular junctions. While considerable progress has been made in the identification of TJ-related proteins, the composition of the TJ particles themselves remains unknown. Identification and characterization of the components of the TJ will greatly facilitate our understanding of the pathophysiological regulation of this important structure.

REFERENCES

1. Powell D. Barrier function of epithelia. Am J Physiol 1981; 241:G275-G288.
2. Frizzell RA, Schultz SG. Ionic conductance of extracellular shunt pathway in rabbit ileum. J Gen Physiol 1972;59:318-346.
3. Schneeberger EE, Lynch RD. Structure, function and regulation of cellular tight junctions. Am J Physiol 1992; 262: L647-L661.

4a. Farquhar MG, Palade GE. Junctional complexes in various epithelia. J Cell Biol 1963; 17:375-412.

4b. Furuse M, Hirase T, Itoh M et al. Occludin: A novel integral membrane protein localizing at tight junctions. J Cell Biol 1993; 123:177701788.

5. Claude P, Goodenough DA. Fracture faces of zonulae occludentes from "tight" and "leaky" epithelia. 1973; 58:390-400.
6. Claude P. Morphologic factors infuencing transepithelial permeability: a model for the resistance of the zonula occludens. J Membr Biol 1978; 39:219-232.
7. Schneeberger EE, McCormack JM. Intercellular junctions in upper airway submucosal glands of the rat: a tracer and freeze fracture study. Anat Rec 1984; 210:421-433.
8. Duffey ME, Hainau B, Ho S, Bentzel CJ. Regulation of epithelial tight junction permeability by cyclic AMP. Nature 1981; 294:451-453.
9. Palant CE, Duffey ME, Mookerjee BK, Ho S, Bentzel CJ. Ca^{++} regulation of tight junction permeability and structure in Necturus gallbladder. Am J Physiol 1983; 245: C203-C212.
10. Bentzel CJ, Hainau B, Ho S, Hui W, Edelman A, Agnastopoulos T, Benedetti EL. Cytoplasmic regulation of tight junction permeability: Effects of plant cytokinins. Am J Physiol 1980; 239:C75-C89.
11. Krasney E, Madara JL, DeBona D, Frizzell R. Cyclic AMP regulates tight junction permselectivity in flounder intestine. Fed Proc 1983; 42:1100 (Abstr)
12. Stevenson BR, Anderson JM, Goodenough DA, Mooseker MS. Tight junction structure and ZO-1 content are indentical in two strains of Madin-Darby canine kidney cells which differ in transepithelial resistance. J Cell Biol 2988; 107:2401-2408.
13. Stevenson BR, Anderson JM, Braun ID, Mooseker MS. Phosphorylation of the tight junction protein ZO-1 in two strains of Madin-Darby canine kidney cells which differ in transepithelial resistance. Biochem J 1989; 263:587-599.
14. Bakker R, Groot JA. cAMP-mediated effects of ouabain and theophylline on paracellular selectivity. Am J Physiol 1984; 246:G213-G217.
15. Ojakian G. Tumor promoter-induced changes in the permeability of epithelial cell tight junctions. Cell 1981; 23:95-103.
16. Mullin JE, O'Brien TG. Effects of tumor promoters on LLC-PK renal epithelial tight junctions and transepithelial fluxes. Am J Physiol 1986; 251:C587-C602.
17. Meza I, Obarra G, Sabanero M, Martinez-Palomo A, Cereijido M. Occluding junctions and cytoskeletal components in a cultured transporting epithelium. J Cell Biol 1980; 87:746-754.

18. Madara JL, Stafford J, Barenberg D, Carlson S. Functional coupling of tight junctions and microfilaments in T84 monolayers. Am J Physiol 1988;254:G416-G423.
19. Madara JL. Intestinal absorptive cell tight junctions are linked to cytoskeleton. Am J Physiol 1987; 253:C171-C175.
20. Stevenson BR, Goodenough DA. Zonulae occludentes in junctional complex-enriched fractions from mouse liver: Preliminary morphological and biochemical characterization. J Cell Biol 1984; 98:1209-1221
21. Citi S, Sabanay H, Jakes R, Geiger B, Kendrick-Jones J. Cingulin, a new periperal component of tight junctions. Nature 1988; 333:272-276
22. Gumbiner B, Lowenkopf T, Apatira D. Identification of a 160-kD polypeptide that binds to the tight junction protein ZO-1. Proc Natl Acad Sci USA 1991; 88:3460-3464.
23. Zhong Y, Saigoh T, Minase T, Sawada N, Enomoto K, Mori M. Monoclonal antibody 7H6 reacts with a novel tight junction-associated protein distinct from ZO-1, cingulin, ZO-2. J Cell Biol 1993; 120:477-483.
24. Muckter H, Ben-Shaul Y, Bacher A. ATP requirement for induced tight junction formation in HT 29 adenocarcinoma cells. Europ J Cell Biol 1987; 44:258-264.
25. Mandel LJ, Bacallao R, Zampighi G. Uncoupling of the molecular 'fence' and paracellular 'gate' functions in epithelial tight junctions. Nature 1993 361:552-555.
26. Pappenheimer JR. On the coupling of membrane digestion with intestinal absorption of sugars and amino acids. Am J Physiol 1993; 265:G409-G417.
27. Geyer RP. "Bloodless" rats through the use of artificial blood substitutes. Fed Proc 1975, 34:1499-1505.
28. Madara JL, Pappenheimer JR. Structural basis for physiological regulation of paracellular pathways in intestinal epithelia. J Membr Biol 1987; 100:149-164.
29. Pappenheimer JR, Volpp K. Transmucosal impedance of small intestine: correlation with transport of sugars and amino acids. Am J Physiol 1992; 263:C480-C493.
30. Gonzalez-Mariscal L, Chavez de Ramirez B, Cereijido M. Tight junction formation in cultured epithelial cells (MDCK). J Membr Biol 1985; 86:113-125.
31 Griepp E, Dolan WJ, Robbins ES, Sabatini DD. Participation of plasma membrane proteins in the formation of tight junctions by cultured epithelial cells. J Cell Biol 1983; 96:693.
32. Balda MS, Gonzalez-Mariscal L, Contreras RG, Macias-Silva M, Torres-Marquez ME, Garcia-Sainz JA, Cereijido M. Assembly and sealing of tight junctions: possible participation of G-proteins, phospholipase C, protein kinase C and calmodulin. J Membr Biol 1991; 122:193-202.
33. Contreras RG, Miller JH, Zamora M, Gonzalez-Mariscal L, Cereijido M. Interaction of calcium with plasma membrane of epithelial (MDCK) cells during junction formation. Am J Physiol 1992; 263:C313-C318.
34. Schneeberger EE, Lynch RD, Kelly CA, Rabito CA. Modulation of tight junction formation in clone 4 MDCK cells by fatty acid supplementation. Am J Physiol 1988; 254:C432-C440.
35. Lynch RD, Tkachuk LJ, Ji X, Rabito CA, Schneeberger EE. Depleting cell cholesterol alters calcium-induced assembly of tight junctions by monolayers of MDCK cells. Eur. J Cell Biol 1993; 60:21-30.
36. Gonzalez-Mariscal L, Contreras RG, Bolivar JJ, Ponce A, Chavez de Ramirez B, Cereijido M. Role of calcium in tight junction formation between epithelial cells. Am J Physiol 1990; 259:C978-C986.
37. Nelson WJ, Shore EM, Wang AZ, Hammerton RW. Identification of a membrane-cytoskeletal complex containing the cell adhesion molecule uvomorulin (E-cadherin), ankyrin and fodrin in Madin-Darby canine kidney epithelial cells. J Cell Biol 1990; 110:349-357.
38. Marrs JA, Napolitano EW, Murphy-Erdosh C, Mays RW, Reichardt LF, Nelson WJ. Distinguishing roles of the membrane-cytoskeleton and cadherin mediated cell-cell adhesion in generating different Na^+, K^+-ATPase distributions in polarized epithelial. J Cell Biol 1993; 123:149-164.
39. Rodriguez-Boulan E, Nelson WJ. Morphogensis of the polarized epithelial cell phenotype. Science 1989; 245:718-725.

40. Nelson WJ, Hammerton RW. A membrane-cytoskeletal complex containing Na^+, K^+-ATPase, ankyrin and fodrin in Madin-Darby canine kidney (MDCK) cells: Implications for the biogensis of epithelial polarity. J Cell Biol 1989; 108:893-902.
41. Gumbiner B, Simons D. A functional assay for proteins involved in establishing an epithelial occluding barrier: identification of a uvomorulin-like polypeptide. J Cell Biol 1986; 102:457-468.
42. Siliciano JA, Goodenough DA. Localization of the tight junction protein, ZO-1, is modulated by extracellular calcium and cell-cell contact in Madin-Darby canine kidney epithelial cells. J Cell Biol 1988; 107:2389-2399.
43. Citi S. Protein kinase inhibitors prevent junction dissociation induced by low extracellular calcium in MDCK epithelial cells. J Cell Biol 1992; 117:169-178.
44. Ozawa M, Ringwald M, Kemler R. Uvomorulin-catenin complex formation is regulated by a specific domain in the cytoplasmic region of the cell adhesion molecule. Proc Natl Acad Sci USA 1990; 87:4246-4250.
45. Kachar B, Reese TS. Evidence for the lipidic nature of tight junction strands. Nature 1982; 296:464-466.
46. Lynch RD, Locicero J, Schneeberger EE. Metabolism and incorporation into glycerolipids of exogenous 18:3(n-3) and 18:3(n-6) by MDCK cells. Lipids 1986 ;21:447-453.
47. Lange Y, Swaisgoood MH, Ramos BV, Stech TL. Plasma membranes contain half the phospholipid and 90% of the cholesterol and sphingomyelin in cultured human fibroblasts. J Biol Chem 1989; 264:3786-3793.
48. Repko EM, Maltese WA. Post-translational isoprenylation of cellular proteins is altered in response to mevalonate availability. J Biol Chem 1989; 264:9945-9952.
49. Lowe PJ, Miyai K, Steinbach JH, Hardison GM. Hormonal regulation of hepatocyte tight junctional permeability. Am J Physiol 1988; 255:G454-G461.
50. Kuijpers GAJ, Von Nooy IGP, Vossen MEM, Stadhouders AM, Van Uyen A, De Pont JJ, Bonting SL. Tight junctional permeability of the resting and carbachol stimulated exocrine pancreas. Histochemistry 1985; 83:257-264.
51. Mazariegos MR, Tice LW, Hand AR. Alteration of tight junctional permeability in rat parotid gland after isopropanol stimulation. J Cell Biol 1984; 98:1865-1877.
52. McRoberts JA, Aranda R, Riley N, Kang H. Insulin regulates the paracellular permeability of cultured intestinal epithelial cell monolayers. J Clin Invest 1990; 85:1127-1134.
53. McRoberts JA, Riley NE. Regulation of T84 cell monolayer permeability by insulin-like growth factors. Am J Physiol 1992; 262:C207-C213.
54. Zettl KS, Sjaastad MD, Riskin PM, Parry G, Machen TE, Firestone GL. Glucocorticoid-induced formation of tight junctions in mouse mammary epithelial cells in vitro. Proc Natl Acad Sci USA 1992; 89:9069-9073.
55. Kan KS, Coleman R. The calcium ionophore A23187 increases the tight junctional permeability in rat liver. Biochem J 1988; 256:1039-1041.
56. Llopis J, Kass GEN, Duddy SK, Farrell GC, Gahm A, Orrenius S. Mobilization of the hormone-sensitive calcium pool increase hepatocyte tight juncitonal permeability in the perfused rat liver. FEBS Lett 1991; 280:84-86.
57. Yamaguchi Y, Dalle-Molle E, Hardison WGM. Vasopressin and A23187 stimulate phosphorylation of myosin light chain-1 in isolated rat hepatocytes. Am J Physiol 1991; 261:G312-G319.
58. Nathanson MH, Gautam A, Ng OC, Bruck R, Boyer GL. Hormonal regulation of paracellular permeability in isolated rat hepatocyte couplets. Am J Physiol 1992; 262:G1079-G1086.
59. Balda MS, Gonzalez-Mariscal L, Matter K, Cereijido M, Anderson JM. Assembly of the tight junction: The role of diacylglycerol. J Cell Biol 1993; 123:293-302.
60. Madara JL, Barenberg D, Carlson S. Effects of cytochalasin D on occluding junctions of intestinal absorptive cells: Further evidence that the cytoskeleton may influence paracellular pemeability and junctional charge selectivity. J Cell Biol 1986; 102:2125-2136.

61. Elias E, Hruban Z, Wade JB, Boyer JL. Phalloidin-induced cholestasis: A microfilament-mediated change in junctional complex permeability. Proc Natl Acad Sci USA 1980; 77:2229-2233.
62. Gabbiani G, Montesano R, Tuchweber B. Phalloidin-induced hyperplasia of actin filaments in rat heptocytes. Lab Invest 1975; 33:562-569.
63. Kachar B, Pinto da Silva P. Rapid massive assembly of tight junction strands. Science 1981;213:541-544.
64. Chevalier J, Pinto da Silva P. Osmotic reversal induces assembly of tight junction strands at the basal pole of toad bladder epithelial cells but does not reverse cell polarity. J Membr Biol 1987; 95:199-208.
65. Faff O, Mitreiter R, Muckter H, Ben-Shaul Y, Bacher A. Rapid formation of tight junctions in HT 29 human adenocarcinoma cells by hypertonic salt solutions. Exp Cell Res 1988; 177:60-72.
66. Cohen E, Talmon A, Faff O, Bacher A, Ben-Shaul Y. Formation of tight junctions in epithelial cells. 1. Induction by proteases in a human colon carcinoma cell line. Exp Cell Res 1985; 156:103-116.
67. Orci L, Amberdt M, Henquin JC, Lambert AE, Unger RH, Renold AE. Pronase effect on pancreatic beta cell secretion and morphology. Science 1973; 180:647-649.
68. Lynch RD, Vu Q, McCormack JM, Schneeberger EE. Basolateral but not apical application of protease results in a rapid rise in transepithelial electrical resistance (TER) and formation of aberrant tight junction (TJ) strands in MDCK cells. Molec Biol Cell 1993; 4:440A.
69. Takeichi M. Cadherins: a molecular family important in selective cell-cell adhesion. Annu Rev Biochem 1990; 59:237-252.
70. Ophir I, Cohen E, Bacher A, Ben Shaul Y. Effect of protein synthesis inhibitors on formation and degradation of tight junctions in HT 29 adenocarcinoma cells. Eur. J Cell Biol. 1989; 49:116-122.
71. Bacher A, Griebl K, Mackamul S, Mitreiter R, Muckter H, Ben-Shaul Y.. Protease inhibitors suppress the formation of tight junctions in gastrointestinal cell lines. Exp Cell Res 1992; 200:97-104.
72. Bjork J, Hugli TE, Smedegard G. Microvascular effects of anaphyatoxins C3a and C5a. J Immunol 1985; 134:1115-1119.
73. Dahlen S, Bjork J, Hedqvist P, Arfors K, Hammarstrom S, Lindgren JA, Samuelson B. Leukotrienes promote plasma leakage and leukocyte adhesion in postcapillary venules: in vivo effects with relevance to the acute inflammatory response. Proc Natl Acad Sci USA 1981; 78:3887-3891.
74. Handley, DA, Arbeeny CM, Lee ML, Van Valen RG, Saunders RN. Effect of platelet-activating factor on endothelial permeability to plasma molecules. Immunopharmacology 1985; 8:137-142.
75. Majno G, Shea S, Leventhal M. Endothelial contraction induced by histamine-type mediators: an electron microscopic study. J Cell Biol 1969; 42:647-672.
76. Stolpen AH, Guinan EC, Fiers W, Pober JS. Recombinant tumor necrosis factor and immune interferon act singly and in combination to reorganize human vascular endothelial cell monolayers. Am J Pathol 1986; 123:16-24.
77. Royall JA, Berkow RL, Beckman JS, Cunningham MK, Matalon S, Freeman BA. Tumor necrosis factor and interleukin 1-alpha increase vascular endothelial permeability. Am J Physiol 1989; 257:L399-L410.
78. Marmorstein AD, Mortell KH, Ratcliffe DR, Cramer EB. Epithelial permeability factor: a serum protein that condenses actin and opens tight junctions. Am J Physiol 1992; 262:C1403-C1410.
79. Madara JL, Stafford J. Interferon-gamma directly affects barrier function of cultured intestinal epithelial monolayers. J Clin Invest 1989; 83:724-727.
80. Adams RB, Planchon SM, Roche JK. IFN-gamma modulation of epithelial barrier function. J Immunol 1993; 150:2356-2363.
81. Royall JA, Berkow RL, Beckman JS, Cunningham MK, Matalon S, Freeman BA. Tumor necrosis factor and interleukin 1a increase vascular endothelial pemeability. Am J Physiol 1989; 257:L399-L410.
82. Downie GH, Ryan US, Hayes BA, Friedman M. Interleukin-2 directly increases albumin permeability of bovine and human vascular endothelium. Am J Resp Cell Molec Biol. 1992; 7:58-65.

83. Dinarello CA. Interleukin-1 and interleukin-1 antagonism. Blood 1991; 77:1627-1652.
84. Vilcek J, Lee TH. Tumor necrosis factor. J Biol Chem 1991; 266:7313-7316.
85. Moore R, Pothoulakis C, LaMont JT, Carlson S, Madara JL. C. difficile toxin A increases intestinal permeability and induces Cl^- secretion. Am J Physiol 1990; 259:G165-G172.
86. Hecht G, Koutsouris A, Pothoulakis C, LaMont JT, Madara JL. Clostridium difficile toxin B disrupts the barrier function of T84 monolayers. Gastroenterol. 1992; 102:416-423.
87. Fasano A, Baudry B, Pumplin DW, Wasserman SS, Tall BD, Ketley JM, Kaper JB. Vibrio cholerae produces a second enterotoxin, which affects intestinal tight junctions. Proc Natl Acad Sci USA. 1991; 88:5242-5246
88. Winter M, Wilson JS, Bedell K, Shasby DM. The conductance of cultured epithelial cell monolayers: oxidants adenosine triphosphate and phorbol dibutyrate. Am J Respir Cell Mol Biol 1990; 2:355-363.
89. Kim KJ, Suh DJ. Asymmetric effects of H_2O_2 on alveolar epithelial barrier properties. Am J Physiol 1993; 264:L308-L315.
90. Shasby DM, Winters DM, Shasby SS. Oxidants and conductance of cultured epithelial cell monolayers: inositol phospholipid hydrolysis. Am J Physiol 1988; 255:C781-C788.
91. Osborn L. Leukocyte adhesion to endothelium during inflammation. Cell 1990; 62:3-6.
92. Springer TA 1990. Adhesion receptors of the immune system. Nature 1990; 346:196-197.
93. Milks LC, Conyers GP, Cramer EB. The effect of neutrophil migration on epithelial permeability. J Cell Biol 1986; 103:2729-2738.
94. Nash S, Stafford J, Madara JL. The selective and superoxide-independent disruption of intestinal epithelial tight junctions during leukocyte transmigration. Lab Invest 1988; 59:531-537.
95. Nash S, Stafford J, Madara JL. Effects of polymorphonuclear leukocyte transmigration on the barrier function of cultured intestinal epithelial monolayers. J Clin Invest 1987; 80:1104-1113.
96. Parkos CA, Delp C, Arnaout MA, Madara JL. Neutrophil migration across a cultured intestinal epithelium. Dependence on a CD11b/CD18-mediated event and enhanced efficiency in physiological direction. J Clin Invest 1991; 88:1605-1612.
97. Parkos CA, Colgan SP, Delp C, Arnaout MA, Madara JL. Neutrophil migration across a cultured epithelial monolayer elicits a biphasic resistance response representing sequential effects on transcellular and paracellular pathways. J Cell Biol 1992; 117:757-764.
98. Colgan SP, Parkos CA, Delp C, Arnaout MA, Madara JL. Neutrophil migration across cultured intestinual epithelial monolayers is modulated by epithelial exposure to IFN-gamma in a highly polarized fashion. J Cell Biol 1993; 120:785-798.
99. Marin ML, Geller SA, Greenstein AJ, Marin RH, Gordon RE, Aufses AH. Ultrastructural pathology of Crohn's disease: Correlated transmission elelctron microscopy, scanning electron microscopy and freeze fracture studies. Am J Gastrol 1983; 78:355-364.
100. Hollander D. Crohn's disease—permeability disorder of the tight junction. Gut 1988; 29:1621-1624.
101. DeVos R, DeWolf-Peters C, Desmet V. Progressive intrahepatic cholestasis (Byler's disease): Case report. Gut 1975; 16:943-950.
102. Weber AM, Tuchweber B, Yousef I, Brochu P, Turgeon C, Gabbiani G, Morin CL, Roy CC. Severe familial cholestasis in North American Indian children—a microfilament defect. Gastroenterology 1981; 81:653-662.

CHAPTER 8

Desmosomal Cadherins and Their Cytoplasmic Interactions

Pamela Cowin
Susan Mechanic

1. INTRODUCTION

Desmosomes are symmetrical structures that connect cells to each other. They are found within the intercalated discs of heart, between the dendritic cells of lymph nodes and in all epithelia. In simple epithelia they are sparse and have a polarized distribution, being concentrated at the border of the apical and basolateral cell surface, where they form the third member of the junctional complex known as the terminal bar.[1] In stratified epithelia desmosomes are numerous, occupying up to 70% of the plasma membrane of suprabasal cells. In addition to their adhesive function, desmosomes anchor intermediate filaments to the plasma membrane. Their positioning and number determines the arrangement of this cytoskeletal mesh thereby critically influencing cell shape and tissue resilience.

Desmosomes were first isolated over twenty years ago as the pelletable membrane residue of citric acid solubilization of epidermal keratin.[2] Three major SDS-PAGE bands contained in these subcellular fractions obeyed criteria expected of adhesive components; they were found to be glycosylated to bind calcium ions and were substantially increased in a modified isolation procedure that enriched for the intercellular region of desmosomes.[3-5] It is now known that these SDS-PAGE bands comprise a mixture of structurally related polypeptides. Those in the upper band are referred to as desmogleins (Mr~135,000-165,000) while those in the two lower bands are called

Molecular Mechanisms of Epithelial Cell Junctions: From Development to Disease
Edited by Sandra Citi, M.D., Ph.D.;

desmocollins (Mr~115,000 and 130,000).[6] The contribution of desmocollins and desmogleins to the electron dense midline and cross-bridging structures of the desmosomal adhesive core was confirmed by immunoelectron microscopy.[5] Moreover, monovalent fragments of desmocollin antibodies were shown to inhibit desmosome formation by cultured cells, indicating the crucial role played by these glycoproteins in the initiation of junction formation.[7] A similar adhesive function was suggested for the desmogleins from studies of a range of autoimmune disorders known as *pemphigus*. The sera of patients with pemphigus contain antibodies against desmogleins and in some cases against desmocollins.[8,9] These sera produce blistering of the epidermis and mucous membranes in vivo and disrupt desmosomes between keratinocytes in vitro. (see also chapters 14 and 15 in this volume).

2. COMPLEXITY OF DESMOCOLLINS AND DESMOGLEINS

Early studies using antibodies to examine the expression of a range of desmosomal epitopes indicated that proteins present in the submembranous plaques of desmosomes were immunochemically identical in all tissues and highly conserved across species.[10] In contrast, many desmoglein and desmocollin epitopes displayed significant variation in their pattern of tissue expression.[10-12] This observation provided the first clue that desmosomal glycoproteins might be products of alternative splicing or of a gene superfamily. Cloning and sequencing have since confirmed both of these hypotheses.

Three desmogleins (Dsg1, 2, 3) have been described.[9,13-16] Bovine and human Dsg1 transcripts which are thought to encode the *pemphigus foliaceus* antigen have been localized predominantly to suprabasal layers of the epidermis and are also found in tongue and esophagus.[17,18] Human Dsg2 was cloned from a colon carcinoma cell line and therefore is presumably expressed in intestine.[16] Its transcripts are far less abundant in epidermis than those for Dsg1 and have been described as restricted to basal cells.[17] Human Dsg3 transcripts, which encode the *pemphigus vulgaris* antigen are found in tongue, esophagus and the lower spinous layers of epidermis.[9,17]

Three pairs of desmocollin isoforms (Dsc 1, 2, 3) are produced by an identical alternative splicing pattern in three different genes (Fig. 8.1).[19-23] Bovine and human Dsc1 transcripts have been detected in suprabasal layers of epidermis and tongue.[24,25] Bovine Dsc2 transcripts have been detected in the lower spinous layers of epidermis, throughout the epithelial layers of tongue and also in a number of cultured cell lines derived from simple epithelia.[22,24] Human Dsc3 transcripts, which are so closely related to both bovine Dsc2 and Dsc3 (see Fig. 8.1) that the true homolog is uncertain, are found in the basal layers of epidermis, exocervix and esophagus.[24]

Studies of epitope and mRNA expression patterns indicate that many more desmogleins and desmocollins exist than those described above.[10,18] It is likely that as sequence similarities between new members are clarified, expression patterns of transcripts and proteins may have to be revised.

3. THE CADHERIN GENE SUPERFAMILY

Cloning and sequencing of cDNAs encoding desmosomal glycoproteins revealed that both desmogleins and desmocollins are similar in their ectodomains to members of the classical cadherin family of calcium-dependent cell-cell adhesion proteins (Table 8.1). In fact, when the protein sequences of their ectodomains are compared, desmocollins show a greater resemblance (~40% identity) to classical cadherins than to desmogleins (28% identity). The desmosomal glycoproteins

Fig. 8.1. (Opposite) Multiple sequence alignment of desmocollins. Bovine and human Dsc 1, 2 and 3 protein sequences are aligned. Asterisks indicate identical amino acid residues which are conserved in all subtypes. Open arrows represent conserved cysteines. Closed arrows indicate cysteine residues in EC 5 that are also strictly conserved in the classical cadherins. Vertical boxes enclose known and putative calcium binding pockets. h= human, b= bovine. EC= extracellular domain, TM= transmembrane region, cyto= cytoplasmic domain. Alternate splice forms a and b are outlined in horizontal boxes.

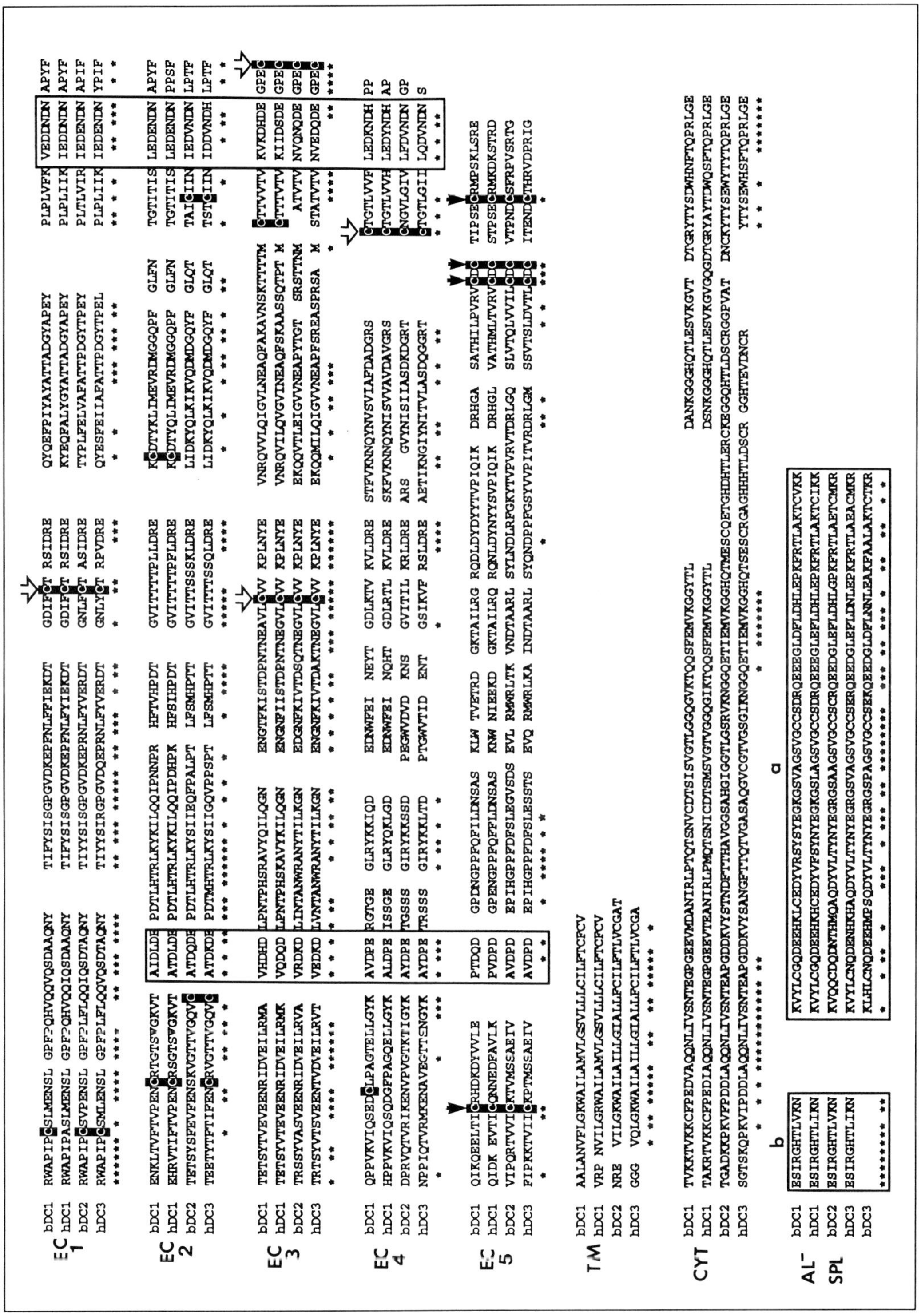
EC 1
EC 2
EC 3
EC 4
EC 5
TM
CYT
AL⁻
SPL
a
b

are sufficiently different in their cytoplasmic sequences from each other as well as from the classical cadherins to form distinct subsets within this superfamily (Fig. 8.2).[9,13-16,19-22,26]

Classical cadherins are a large family of type 1 transmembrane glycoproteins that are differentially expressed during embryogenesis.[27] The complexity of this family combined with the strictly homophilic interaction of its members creates a large ensemble of highly selective cell-cell binding mechanisms that form the molecular basis of cellular recognition.[27] Several classical cadherins are found concentrated in adherens junctions, which bear a superficial resemblance to desmosomes but are biochemically distinct and associate with the actin cytoskeleton.[28] These classical cadherins have been shown to influence the differentiated phenotype of cells by inducing reorganization and polarization of the subcortical and filamentous cytoskeleton; an event which significantly alters cell shape and adhesive behavior and in some cases involves second messenger signaling pathways.[29-32] Together with their associated regulatory proteins, cadherins have been implicated in the control of growth and the invasive properties of transformed and tumor cell lines and in the implementation of patterning cues.[27,33,34] (see also chapters 4, 12 and 13 in this volume) The sequence similarity between the ectodomain of desmosomal

Table 8.1. Cadherin gene superfamily

	Reference
CLASSICAL CADHERINS	27
DESMOGLEINS	13-16
DESMOCOLLINS	19-22
PROTOCADHERINS	45
CRET	43
PTPμ	88
FAT	44
DACHSOUS	44
LETHAL GIANT	87

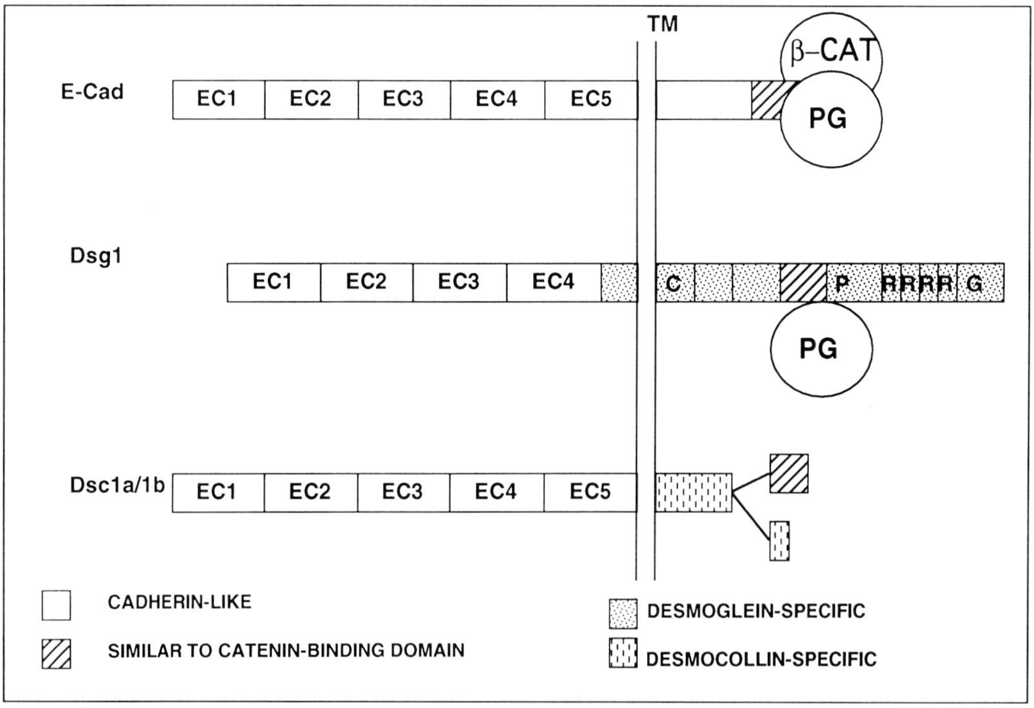

Fig. 8.2. Schematic representation of the structure of the desmosomal and classical cadherins showing the binding site for plakoglobin (PG) on Dsg1 and the putative binding sites for plakoglobin and beta-catenin (β-CAT) on E-cadherin (E-Cad). TM= transmembrane, EC1-5 =extracellular regions, C=cysteine rich, P=proline rich, R= cytoplasmic repeat units, G=glycine rich sequences of Dsg1 cytoplasmic sequence. Dsc1a and 1b alternatively spliced forms are indicated.

glycoproteins and classical cadherins raises the possibility that they mediate a similar adhesive mechanism; their complexity and differential expression suggests they contribute to cell differentiation and possibly to cell recognition.

In addition to the sequence similarity among classical and desmosomal cadherins, a comparison of their gene structure further supports the notion that they have evolved from a common ancestor. The genes for bovine Dsg1 and those of several classical cadherins shows an identical pattern of intron/exon boundaries.[35] This pattern is distinctive—the exon boundaries migrate with respect to the repeat unit of the protein and bear no relation to functional domains. The positioning of the exon boundaries appears to have evolved to constrain unequal intragenic recombination events between sister chromatids to which genes encoding repetitive sequences are highly susceptible.[35] The most probable order of events was that the ancestral gene, comprising a single repeat unit, duplicated twice to give rise to the four repeat structure. The modern form of its exon boundaries were established before further duplication and divergence produced the three major vertebrate cadherin gene families.

Several human desmoglein genes (DSG1, 2 and 3), a desmocollin gene (DSC1) and the gene encoding their closest relative N-cadherin, have been mapped to the same chromosome (chr 18).[6] Although the distance between these human desmosomal and cadherin genes has not been defined, mouse DSC3 and DSG1 are reported to be closely linked.[36] The genes for chick classical cadherins L-CAM and K-CAM are separated by less than 700 bp and may share the same regulatory machinery.[37] So far, no linkage has been found between desmosomal genes and loci for genetic diseases such as Darier's disease (chr 12) or Hailey-Hailey disease, the pathology of which suggest possible defects in keratinocyte adhesion.

4. STRUCTURE AND BIOSYNTHESIS OF DESMOSOMAL CADHERINS

Like most classical cadherins, desmogleins and desmocollins are synthesized as precursor proteins which are cleaved to the mature form just prior to leaving the Golgi compartment.[38-41] Desmosomal glycoproteins, unlike the classical cadherins, enter a Triton X-100 insoluble pool at this time, suggesting they may oligomerize or form a complex with a cytoskeletal component.[42] All three types of cadherin reach the cell surface rapidly (~1-2 hours). If junction formation is inhibited by low calcium culture conditions, desmosomal cadherins are turned over rapidly, with a half life of about 1hr, but when permitted to assemble into junctions they are stable (half life > 24 hours).[38-41] There is little similarity between the desmogleins, desmocollins and classical cadherins in the sequences of either their signal sequence or precursor segment. The function of the prosegment is not certain but it is likely that it serves to prevent adhesive interaction within the biosynthetic compartments of the cell. The protease cleavage site LxRxKR at the amino terminus of the mature proteins is highly conserved in members of all three families; thus the processing of these proteins to the mature form may be carried out by the same protease.

Both desmosomal and classical cadherins show the same division of the extracellular region into four repeats of 110-120 amino acids and a fifth more variable domain adjacent to the membrane. The greatest sequence similarity is found among stretches of acidic residues that in classical cadherins form both known and putative binding sites for calcium ions (see Fig. 8.1). The calcium binding sequences together with several prominent repetitive motifs of unknown function such as LDRE are shared not only with classical cadherins but also with the more distantly related proto-cadherins as well as the Drosophila gene FAT (see chapter 1 in this volume) and the human proto-oncogene *ret* (Table 8.1).[43-45]

When sequences from the extracellular repeats of desmosomal glycoproteins are compared to a consensus derived from seven different classical cadherins, two major differences can be seen.[13,19] Firstly, prolines, which are highly conserved in classical cadherins and desmocollins, are conspicuously absent from both Dsg1 and Dsg3. Secondly, desmocollins and desmogleins contain many more cysteine residues than classical cadherins. Four of the nine extra cysteines first noted in the bovine Dsc1 sequence are conserved in all three types of desmocollins (Fig. 8.1). Likewise, four of the five extra cysteines in bovine Dsg1 are shared by human Dsg3.[13] As prolines and cysteines can produce significant changes in the folding and association between the extracellular domains of proteins, it will be interesting to determine if they contribute to the characteristic structural features seen in the electron microscope such as crossbridges and midline density that distinguish desmosomal adhesive cores from the amorphous intercellular region of adherens junctions.

5. THE ADHESIVE MECHANISM OF CADHERINS

Classical cadherins mediate homophilic binding when transfected into otherwise nonadherent fibroblasts.[29] How desmosomal proteins interact is far from clear. Desmogleins and deomocollins do not produce convincing aggregation or change in the adhesive behavior of these cultured cells (Mechanic and Cowin, unpublished results). Chimeric proteins consisting of the Dsg3 ectodomain and classical cadherin cytoplasmic region have been reported to support weak homophilic adhesion.[89] The psysiological relevance of this finding is, however, not clear, at it is well established that desmosomes mediate very strong adhesion.

Studies in which various extracellular repeats have been exchanged between classical cadherins to create chimeric proteins have shown that sequences in the first extracellular repeat are important for the specificity of the homophilic adhesive mechanism.[46] Site directed mutagenesis identified amino acids flanking residues 78-83 as responsible for these recognition properties.[46] The desmosomal cadherins shows considerable variation in sequence of their corresponding regions.[13,19]

The epitopes for several antibodies which disrupt adhesion mediated by classical cadherins have been mapped to the highly conserved amino-terminal 35 residues and the more diverse fifth extracellular region which flanks the plasma membrane.[46-48] Some progress has been made mapping the epitopes recognized by the adhesion-disrupting pemphigus sera. *Pemphigus vulgaris* sera have been shown to recognize epitopes present in EC1 of Dsg3.[49] Antibodies affinity purified on recombinantly expressed forms of this domain have mild acantholytic activity when assayed in newborn mice. However, extensive absorption of the sera with EC1 did not deplete its acantholytic activity, suggesting that other epitopes important for the adhesive mechanism are expressed elsewhere on Dsg3. Similar studies with *pemphigus foliaceus* sera have indicated that PF sera recognize epitopes in EC5 of Dsg1, although their pathogenicity has not been tested.[50]

On the basis of the studies described above two models of cadherin-mediated adhesion have been suggested. In one, molecules on neighboring cells are proposed to engage head to head through binding of EC1 domains.[46] In the second model EC1 on one molecule is thought to bind to EC5 on molecules from the neighboring cell in a head to tail fashion.[47] While it is possible that the adhesion-disrupting antibodies are identifying epitopes directly involved in the binding site, it is equally plausible that they may interfere with adhesion indirectly. For example antibodies against EC5 may sterically hinder the lateral packing of cadherins within the plane of the membrane. At the present time the chimera studies strongly favor homophilic EC1 interaction.[46] Furthermore in bovine desmogleins EC5 is highly polymorphic, again suggesting a less important role for this region.[35]

6. INTERACTIONS OF THE CYTOPLASMIC DOMAIN OF DESMOSOMAL CADHERINS

It is well known that the highly conserved cytoplasmic domain of classical cadherins exerts a regulatory role on the capacity of the ectodomain to manifest its adhesive properties. Wild-type classical cadherins induce striking changes in transfected fibroblasts causing them to adopt epithelial-like morphology and adhesive behavior.[29,30,51] When mutant cadherins, deleted in their cytoplasmic domain, are transfected into cells, they are expressed on the plasma membrane with normal conformation and binding properties but fail to bring about cell adhesion.[52,53] Further evidence for a regulatory role for the cytoplasmic region comes from studies of the expression of mutants deleted in their ectodomain. Such mutants cause a dominant negative effect on cell adhesion in *Xenopus* embryos principally through their ability to bind and deplete three cytoplasmic regulatory proteins called catenins.[54]

Several observations suggest it is likely that desmoglein and desmocollin cytoplasmic sequences play a similar regulatory role. For example, expression of the desmoglein cytoplasmic region as a chimera attached to the transmembrane region of the gap junction protein connexin produces a dominant negative effect on desmosome assembly by cultured cells.[23] The cytoplasmic region of desmogleins are much larger and complex than those of other cadherins (Fig. 8.2). The 469 amino acid sequence of the Dsg1 cytoplasmic domain has been divided into a number of subdomains based upon prominent sequence features and exonic boundaries. The first exon comprises a cysteine rich sequence (c) and is followed by two exons that show no significant features or homology to other proteins. The fourth and last, exon is very large.[35] It begins with a region of 68 residues that show significant sequence similarity to the catenin binding domain of classical cadherins. This is followed by an equally long proline rich sequence (P) then four 30 residue repeats (R) and ends with a glycine rich (G) repetitive sequence.

The sequences of the first two exons of the cytoplasmic domain of desmocollins are again different from desmogleins and classical cadherins. The desmocollins are produced as two isoforms by alternative splicing of their last exon (Fig. 8.1 and 2). The smaller "b" form comprises 11 residues which are highly conserved in all three types of desmocollin. The larger "a" form of all three types of desmocollin show significant similarity to the catenin-binding domain of classical cadherins. Moreover, this last exon of Dsc1a is able to recruit plaque components and intermediate filaments to associate with the plasma membrane, when expressed as a chimera coupled to the gap junction protein connexin.[23]

These observations have focused attention on the proteins associated with the cytoplasmic domain of desmosomal cadherins. Several components have been localized immunochemically to desmosomal plaques. These include two major components desmoplakins and plakoglobin, as well as a number of minor components such as band 6, desmocalmin, desmoyokin, IFAP-300, a 170kD microtubule associated protein and a 140 kD component.[10,55-60] The interactions among these minor components are poorly understood. Recently, it has been shown that plakoglobin binds to the central region of the cytoplasmic sequence of Dsg1 that shares homology to the catenin-binding domain of classical cadherins (Fig. 8.2).[61] The 19 residue sequence, identified in this study, has considerable similarity to a sequence contained in Dsg3 as well as to the last ten residues of the carboxyl termini of E-cadherin (Fig. 8.3)[61] which have also been reported to interact with plakoglobin.[62,63] The sequence similarity among these proteins in a region identified as the plakoglobin binding site on Dsg1 suggests they employ a similar binding mechanism and provides a logical explanation for the cellular localization of plakoglobin.

Plakoglobin was first described as a component of desmosome preparations (band 5) and was initially immunolocalized to these junctions.[55,64] Subsequently it was found in a variety of cell-cell adherens junctions and in considerable amounts in a cytosolic form.[56,65,66] As the only component shared by the two major types of adhesive cell-cell junction, i.e. desmosomes and adherens junctions, it is likely that plakoglobin plays a pivotal role in regulating intercellular adhesion. A further reason for interest in the associations of plakoglobin stems from what we may infer about its function from studies on related proteins (Table 8.2). Two proteins called beta-catenin and armadillo have been cloned and shown to have strong homology (63-69% identity) to plakoglobin.[34,67] Partial protein sequence of desmosomal band 6, suggests that it too has significant sequence similarity to plakoglobin.[68] Another subset of proteins, GDS, APC, p120 and SRP1 have low sequence similarity to each other and to the other members of the plakoglobin family but conform to a consensus sequence derived from comparisons of plakoglobin, armadillo and beta-catenin.[69-71]

Beta-catenin (63% identity) is the first protein to bind to classical cadherins, complexing to the precursor protein while it is still in the endoplasmic reticulum.[72] Biochemical analysis of the cadherin-catenin complex suggests that beta-catenin is directly associated with the cadherin cytoplasmic sequence.[72,73] Deletion of the catenin binding domain causes classical cadherins to lose their ability to make cells adhere to one another.[52,53] From these experiments it has been suggested that catenins takes part in a chain of proteins that ultimately anchor cadherins to the actin cytoskeleton and thereby concentrate them into sites of cell-cell contact. It has also been proposed that they may take part in signaling cascades that trigger junction assembly.[53,74,75]

Recently beta-catenin has been found in the cytoplasm complexed to the product of the tumor supressor gene APC which also bears homology to plakoglobin.[76,77] APC is the product of a gene linked to an inherited disease of the colon called familial adenopolyposis.[69] Mutations resulting in truncation of this gene product have been implicated in a number of sporadic colon carcinomas.[78] The association between beta-catenin and APC lends support to the possibility that beta-catenin takes part in a pathway that controls cell growth. One possibility is that the dynamic equilibrium between the pool of beta-catenin bound to E-cadherin and the pool that is bound to APC may coordinately regulate cell adhesion and contact inhibition of cell growth. Two observations are pertinent in the context of this hypothesis. Firstly, the putative binding site for plakoglobin on E-cadherin lies in close proximity to that for its close relative beta-catenin so that the two proteins might sterically hinder each other from binding to E-cadherin at the same time [61] Secondly, in a preliminary report, Nelson and colleagues have demonstrated that E-cadherin forms complexes with plakoglobin or beta-catenin but not with both at the same time.[79] Clearly, models can be envisaged in which displacement of beta-catenin by plakoglobin and vice versa may subtly modulate the membrane bound and cytosolic pools of beta-catenin and so provide the cell with a mechanism to link junction formation to contact inhibition of growth. While such a role for

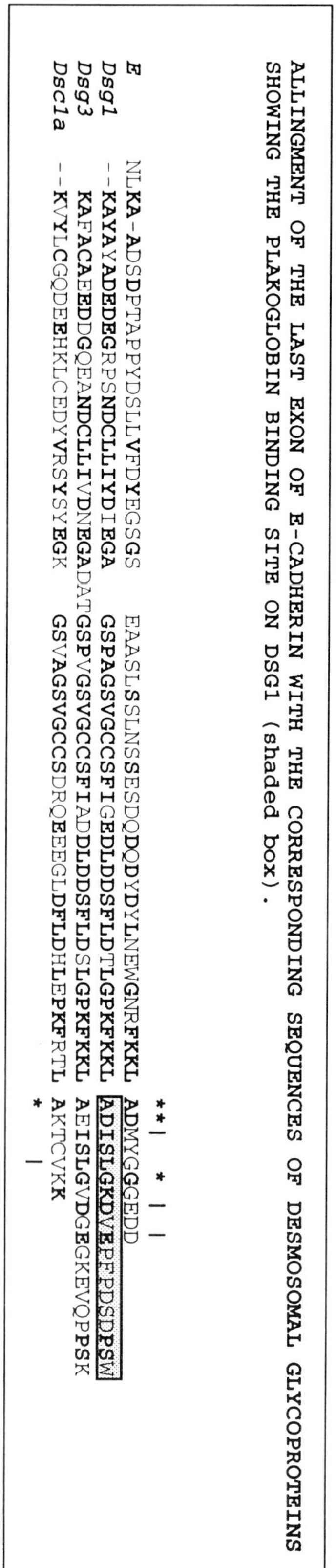

Fig. 8.3. Alignment of the last exon of E-cadherin with the corresponding sequences of desmosomal glycoproteins showing the plakoglobin binding site on DSC1 (shaded box).

plakoglobin and beta-catenin is purely speculative at the present time, it is interesting to note that other proteins with related sequence take part in more conventional signaling pathways. For example, one such protein called GDS is a nucleotide exchange factor.[70] Furthermore a membrane associated phosphoprotein called p120 is thought to play a key role in producing the transforming effects of src (Table 8.2).[71]

Finally, the product of a Drosophila gene called *armadillo* has 63% identity to plakoglobin and 69% identity to beta-catenin.[34] This gene is one of several segment polarity genes which are thought to act in concert to specify patterning of the epidermis across the body segment of Drosophila embryos (Fig. 8.4)[80,81] (see also chapter 1 in this volume). It is currently thought that expression of a transcription factor called Engrailed (*en*) leads to secretion of a protein called Hedgehog (hh) which diffuses establishing a morphogenetic gradient across the segment. Binding of Hedgehog to Patched (ptc), a transmembrane protein on the surface of neighboring cells negates the inhibitory effect of the latter on a protein called Wingless (wg). In the absence of *patched* activity, Wingless is secreted, binds to neighboring cells and negatively regulates the action of zeste-white (zw), a serine threonine kinase.[80,81] The result of this action is to upregulate postranscriptional expression of armadillo (arm) and to cause a critical portion of this protein to associate with the membrane.[82] Given the strong homology among armadillo, plakoglobin and beta-catenin and the finding of armadillo in a complex with alpha-catenin, it is likely that further transmission of patterning information involves engagement of the cadherin based cell adhesion system (Fig. 8.5).[83] Supporting this we have shown that PC12 pheochromocytoma cells expressing Wnt-1, the human homolog of wingless, show a post-transcriptional elevation in plakoglobin.[84] A shift is seen in the relative amounts of plakoglobin from a cytosolic fraction to a membrane bound form. This is accompanied by a significant alteration in the cellular morphology and adhesive behavior giving rise to an epithelial appearance.[84] This demonstrates that the mechanisms of patterning in the Wingless pathway in Drosophila have been highly conserved in vertebrates and cause remarkable changes in cell phenotype. The localized reorganization of the cytoskeleton induced by cadherins activated by the Wingless/wnt-1 route may itself produce the phenotypic change in cellular morphology observed in Drosophila

Table 8.2. Plakoglobin superfamily

Protein	Location/Function	Associations
Plakoglobin	desmosomes, adherens junctions cytosol	Dsg1, Dsg3 classical cadherins homodimer
Band 6	Desmosomes	Dsg1
β-Catenin	basolateral cell surface	classical cadherins
Armadillo	segment polarity gene	cadherin-like complex with α-catenin
APC	tumor suppressor gene for colon cancer and FAP	β-catenin
p120	substrate of Src	
GDS	nucleotide exchange factor	
SRP1	suppressor of RNA Pol 1 mutations	

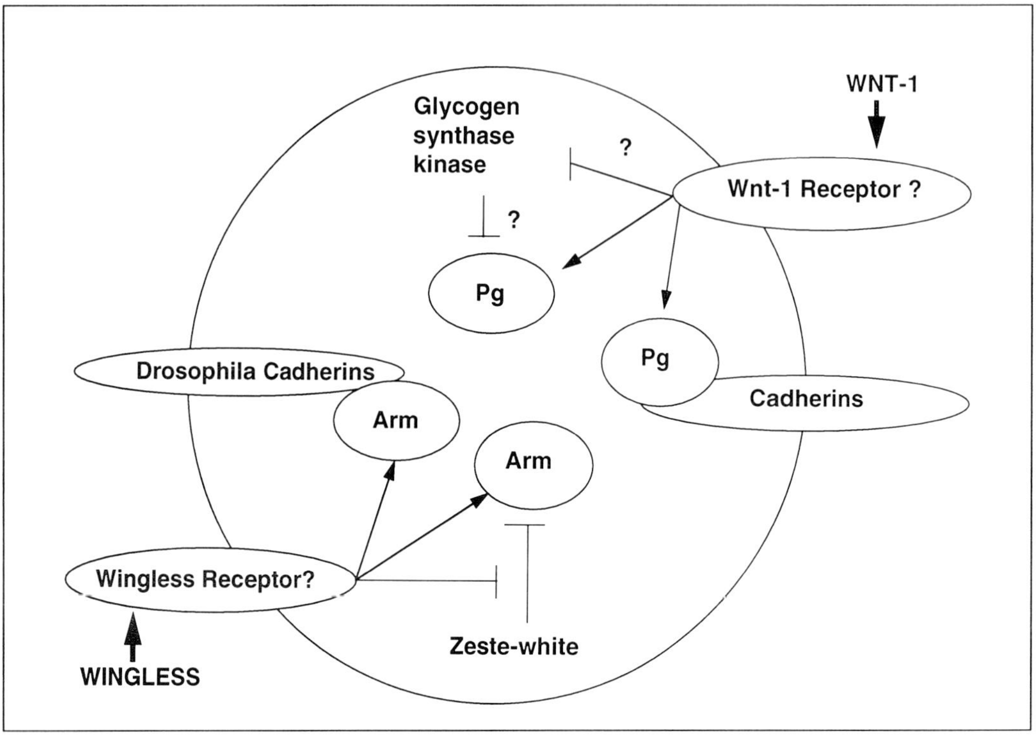

Fig. 8.4. Schematic diagram showing how the wingless, zeste-white and wnt-1 gene products and glycogen synthase may act through plakoglobin (Pg) and the armadillo *(Arm) gene products (see text).*

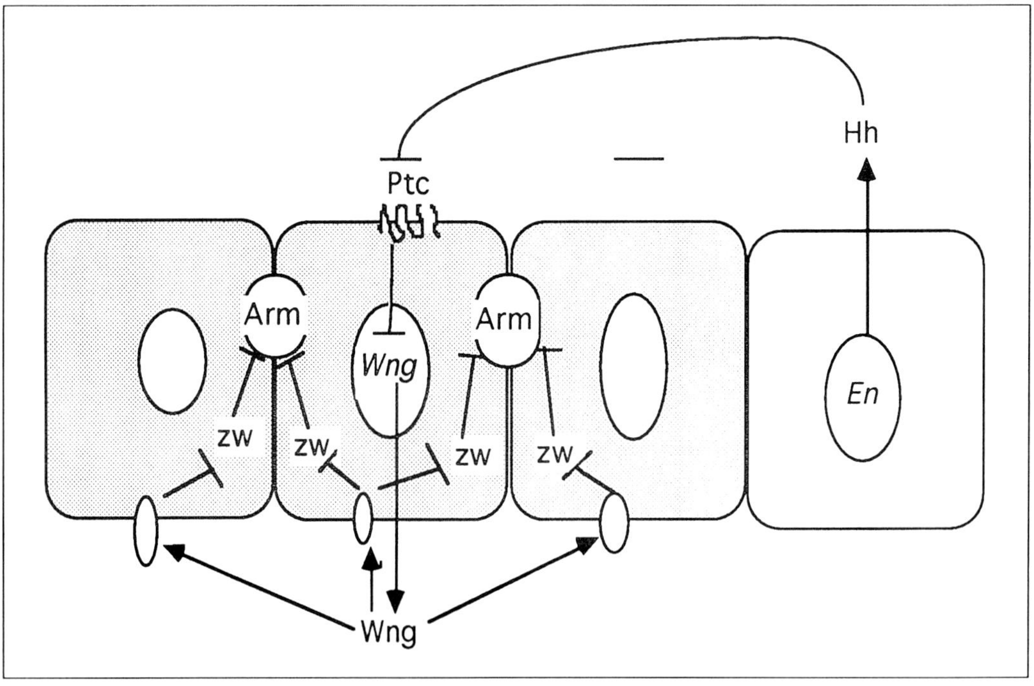

Fig. 8.5. Schematic diagram illustrating the relationships between Wingless (Wng), Armadillo (Arm), zeste-white (zw), patched (Ptc), Hedgehog (Hh) and Engrailed (En) (see text).

epidermis and PC12 cell respectively. Alternatively, cadherin mediated adhesion may be required for other receptors or gap junction to engage and transmit further positional information among the community of adherent cells.[85,86]

7. CONCLUSIONS

A wealth of data has implicated the desmocollins and desmogleins as mediators of strong intercellular adhesion, including: the increase of desmosome number during epiboly; the frequency of these structures in mechanically stressed tissues; the capacity of antibodies directed against desmogleins and desmocollins to disrupt cell adhesion in vitro and to cause blistering diseases in vivo and the sequence similarities of desmosomal glycoproteins to the cadherins. Yet this adhesive function has proven difficult to demonstrate in cultured cells transfected with either desmoglein or desmocollin cDNAs, probably due to a requirement for the formation of a multi-component receptor. However, the interactions of the desmosomal cadherins with cytoplasmic proteins are now beginning to yield to a combination of molecular and genetic techniques. In particular the interactions between the cadherin and plakoglobin superfamilies are proving to be important not only for cell adhesion but also for the processes of patterning, transformation and contact inhibition of growth.

ACKNOWLEDGMENTS

We thank Sailaja Puttagunta for critically reading the manusript and help with the figures. Work in our laboratory is supported by National Institutes of Health RO1-GM47429 and The PEW Charitable Trusts.

REFERENCES

1. Farquhar MG, Palade GE. Junctional complexes in various epithelia. J Cell Biol 1963; 17:375-412.
2. Skerrow CJ, Matoltsky AG. Isolation of epidermal desmosomes. J Cell Biol 1974; 63:515-523.
3. Skerrow CJ, Matolsky AG. Chemical characterization of isolated epidermal desmosomes. J Cell Biol 1974; 63:524-531.
4. Gorbsky G, Steinberg MS. Isolation of the intercellular glycoproteins of desmosomes. J Cell Biol 1981; 90:243-248.
5. Miller K, Mattey D, et al. Localization of the protein and glycoprotein components of bovine nasal epothelial desmosome by immunoelectron-microscopy. EMBO J 1987; 6:885-889.
6. Buxton RS, Cowin P, Franke WW, Garrod DR, Green KJ, King IA, Koch PJ, Magee AI, Dees DA, Stanley JR et al. Nomenclature of the desmosomal cadherins. J Cell Biol 1993; 121:481-483.
7. Cowin P, Mattey D, Garrod D.R. Identification of desmosomal surface components (desmocollins) and inhibition of desmosome formation by specific FAB'. J. Cell Sci. 1984; 70:41-60.
8. Dmochowski M, Hashimoto T, et al. Desmocollins I and II are recognized by certain sera from patients with various types of pemphigus. J Invest Dermat 1993; 100:380-385.
9. Amagai M, Klaus KV, Stanley JR. Autoantibodies against a novel epithelial cadherin in pemphigus vulgaris, a disease of cell adhesion. Cell 1991; 67:869-877.
10. Cowin P, Garrod D. Antibodies to epithelial desmosomes show wide tissue and species cross-reactivity. Nature 1983; 302:148-150.
11. Cohen SM, Gorbsky G, Steinberg MS. Immunochemical characterization of related families of glycoproteins in desmosomes. J. Biol. Chem. 1983; 258:2621-2627.
12. Giudice GJ, Cohen SM, Patel NH, Steiberg MS. Immunological comparison of desmosomal components from several bovine tissues. J Cell Biochem 1984; 26:35-45.
13. Goodwin L, Hill JE, Raynor K, Raszi L, Manabe M, Cowin P. Desmoglein shows extensive homology to the cadherin family of cell adhesion molecules. Biochem Biophys Res Comm 1990; 173:1224-30.
14. Koch PJ, Walsh MJ, Schmelz M, Goldschmidt MD, Zimbelmann R, Franke WW. Identification of desmoglein a constititive desmosomal glycoprotein as a member of the cadherin family of cell adhesion molecules. Eur. J Cell Biol 1990; 53:1-12.

15. Wheeler GN, Parker AE, Thomas CL, Ataliotis P, Poynter D, Arnemann J, Rutman AJ, Pidsley SC, Watt FM, Rees DA et al. Desmosomal glycoprotein DGI, a component of intercellular desmosome junctions, is related to the cadherin family of cell adhesion molecules. Proc Natl Acad Sci. USA 1991; 88:4796-800.
16. Koch P, Goldschmidt MD, Walsh MJ, Zimbelmann R, Franke WW. Complete amino acid sequence of the epidermal desmoglein precursor polypeptide and identification of a second type of desmoglein gene. Eur J Cell Biol 1991; 55:200-208.
17. Arnemann J, Sullivan KH, Magee AI, King IA, Buxton RS. Stratification-related expression of isoforms of the desmosomal cadherins in human epidermis. J Cell Sci 1993; 104:741-750.
18. Razsi L, Raynor K, Aharon Y, Manabe M, Cowin P. Molecular cloning of desmoglein. J Invest Dermatol 1990; 94:570 (Abstr.).
19. Mechanic S, Raynor K, Hill JE, Cowin P. Desmocollins form a novel subset of the cadherin family of cell adhesion molecules. Proc Natl Acad Sci USA 1991; 88:4476-4480.
20. Collins JE, Legan PK, Kenny TP, MacGarvie J, Holton JL, Garrod DR. Cloning and sequence analysis of desmosomal glycoproteins 2 and 3 (desmocollins): cadherin-like desmosomal adhesion molecules with heterogeneous cytoplasmic domains. J Cell Biol 1991; 113:381-391.
21. Parker AE, Wheeler GN, Arnemann J, Pidsley SC, Ataliotis P, Thomas CL, Rees DA, Magee AI, Buxton RS. Desmosomal glycoproteins II and III. Cadherin-like junctional molecules generated by alternative splicing. J Biol Chem 1991; 266:10438-10445.
22. Koch PJ, Goldschmidt MD, Zimbelmann R, Troyanovsky R, Franke WW. Complexity and expression patterns of the desmosomal cadherins. Proc Natl Acad Sci USA 1992; 89:353-357.
23. Troyanovsky SM, Eshkind LG, Troyanovsky RB, Leube RE, Franke WW. Contributions of cytoplasmic domains of desmosomal cadherins to desmosome assembly and intermediate filament anchorage. Cell 1993; 72:561-574.
24. Theis D, Koch PJ, Franke WW. Differential synthesis of type 1 and type 2 desmocollin mRNAs in human stratified epithelia. Int J Developm Biol 1993; 37:101-110.
25. Arnemann J, Sullivan, et all. Stratification o-related expression of isoforms of the desmosomal cadheins in epidermis. J Cell Sci 1993; 104:741-750.
26. Nilles LA, Parry DA, Powers EE, Angst BD, Wagner RM, Green KJ. Structural analysis and expression of human desmoglein: a cadherin-like component of the desmosome. J Cell Sci 1991; 99:809-821.
27. Takeichi M. Cadherin cell adhesion receptors as a morphogenetic regulator. Science 1991; 251:1451-1455.
28. Boller K, Vestweber D, Kemler R. Cell adhesion molecule uvomorulin is localised in the intermediate junctions of adult intestinal epthelial cells. J Cell Biol 1985; 100:327-332.
29. Nose A, Nagafuchi A, Takeichi M. Expressed recombinant cadherins mediate cell sorting in model systems. Cell 1988; 54:993-1001.
30. McNeil H, Ozawa M, Kemler R, Nelson WJ. Novel function of the cell adhesion molecule uvomorulin as an inducer of cell surface polarity. Cell 1990; 62:309-316.
31. Doherty P, Rowett LH, Moore SE, Mann DA, Walsh FS. Neurite outgrowth in response to transfected N-CAM and N-cadherin reveals fundamental differences in neuronal responsiveness to CAMs. Neuron 1991; 6:247-58.
32. Saffell JL, Walsh F, Doherty P. Direct activation of second messenger pathways mimics cell adhesion molecule-dependent neurite outgrowth. J Cell Biol 1992; 118:663-670.
33. Birchmeier W, Behrens J, Weidner KN, Frixen UH, Schipper J. Dominant and recessive genes involved in tumor cell invasion. Curr Op Cell Biol 1991; 3:832-840.
34. Peifer M, Weischaus E. The segment polarity gene armadillo encodes a functionally modular protein that is the drosophila homolog of human plakoglobin. Cell 1990; 63:1167-1178.

35. Puttagunta S, Mathur M, Cowin P. Structure of Dsg1 , the bovine desmosomal cadherin gene encoding the pemphigus foliaceus antigen.J Biol Chem 1994 269:1949-1955
36. Arnemann. in press 1994;.
37. Sorkin BC, Gallin WJ, Edelman GM, Cunningham BA. Genes for two calcium dependent cell adhesion molecules have similar structures and are arranged in tandem in the chicken genome. Proc Natl Acad Sci 1991; 88:11545-11549.
38. Pasdar M, Nelson WJ. Regulation of desmosome assembly in epithelial cells: kinetics of synthesis, transport and stabilization of desmoglein I, a major protein of the membrane core domain. J Cell Biol 1989; 109:163-77.
39. Pasdar M, Krzeminski KA, Nelson WJ. Regulation of desmosome assembly in MDCK cells: coordination of membrane core and cytoplasmic plaque domain assembly at the plasma membrane. J Cell Biol 1991; 113:645-657.
40. Penn EJ, Hobson C, Rees DA, Magee AI. Structure and assembly of desmosome junctions: biosynthesis, processing and transport of the major protein and glycoprotein components in cultured epithelial cells. J Cell Biol 1987; 105:57-68.
41. Penn EJ, Hobson C, Rees DA, Magee AI. The assembly of the major desmosome glycoproteins of Madin-Darby canine kidney cells. Febs Lett. 1989; 247:13-6.
42. Shore EM, Nelson WJ. Biosynthesis of the cell adhesion molecule uvomorulin (E-cadherin) in Madin-Darby canine kidney epithelial cells. J Biol Chem 1991; 266:19672-80.
43. Iwamoto T, Taniguchi M, Asai N, Ohkusu K, Makashima I, Takahashi M. cDNA Cloning of mouse ret proto-oncogene and its sequence similarity to the cadherin family. Oncogene 1992; 8:1087-1091.
44. Mahoney PA, Weber U, Onofrechuk P, Biessmann H, Bryant PJ, Goodman CS. The fat tumor suppressor gene in Drosophila encodes a novel member of the cadherin gene superfamily. Cell 1991; 67:853-68.
45. Suzuki S, Sano K, Tanihara H. Diversity of the cadherin family: evidence for eight new cadherins in nervous tissue. Cell Regul. 1991; 2:261-270.
46. Nose A, Tsuji K, Takeichi M. Localization of specificity determining sites in cadherins cell adhesion molecules. Cell 1990; 61:147-155.
47. Ozawa M, Hoschutzky H, Herrenknecht K, Kemler R. A possible new adhesive site in the cell-adhesion molecule uvomorulin. Mech. Devel. 1990; 33:49-56.
48. Geiger B, Ginsberg D, Salomon D, Volberg T. The molecular basis for the assembly and modulation of adherens-type junctions. Cell Differ Developm 1990; 32:343-53.
49. Amagai M, Karpati S, Prussick R, Klaus KV, Stanley JR. Autoantibodies against the aminoterminal cadherin-like binding domain of pemphigus vulgaris antigen are pathogenic. J Clin Invest 1992; 90:919-926.
50. Allen EM, Guidice G, Diaz L. Production of human DG1 fusion proteins and their reactivity with PF autoantibodies. J Invest Dermatol 1992; 98:590.
51. Marrs JA, Napolitano EW, Murphy-Erdosh C, Mays RW, Reichardt LF, Nelson WJ. Distinguishing roles of the membrane-cytoskeleton and cadherin mediated cell-cell adhesion in generating different Na^+, K^+-ATPase distributions in polarized epithelia. J Cell Biol 1993; 123:149-164.
52. Nagafuchi A, Takeichi M. Cell binding function of E-cadherin is regulated by the cytoplasmic domain. EMBO J 1988; 7:3679-3684.
53. Ozawa M, Baribault H, Kemler R. The cytoplasmic domain of the cell adhesion molecule uvomorulin associates with three independent proteins structurally related in different species. EMBO J 1989; 8:1711-1717.
54. Kintner C. Regulation of embryonic cell adhesion by the cadherin cytoplasmic domain. Cell 1992; 69:225-236.
55. Cowin P, Kapprell HP, Franke WW. The complement of desmosomal plaque proteins in different cell types. J Cell Biol 1985; 101:1442-54.
56. Cowin P, Kapprell H-P, Franke WW, Tamkun J, Hynes RO. Plakoglobin: A protein common to different kinds of intercellular adhering junctions. Cell 1986; 46:1063-1073.

57. Kapprell HP, Owaribe K, Franke WW. Identification of a basic protein of Mr 75,000 as an accessory desmosomal plaque protein in stratified and complex epithelia. J Cell Biol. 1988; 106:1679-91.
58. Cartaud A, Ludosky MA, Courvalin JC, Cartaud J. A protein antigenically related to nuclear lamin B mediates the association of intermediate filaments with desmosomes. J Cell Biol 1990; 111:581-588.
59. Tsukita S, Tsukita S. Desmocalmin: a calmodulin-binding high molecular weight protein isolated from desmosomes. J Cell Biol 1985; 101:2070-80.
60. Wacker IU, Rickard JE, De Mey JR, Kreis TE. Accumulation of a microtubule binding protein, pp170, at desmosomal plaques. J Cell Biol 1992; 117:813-824.
61. Mathur M, Goodwin L, Cowin P. Interactions of the cytoplasmic domain of the desmosomal cadherin, Dsg1, with plakoglobin. J Biochem 1994; 269: 14075-14080.
62. Piepenhagen PA, Nelson WJ. Defining E-cadherin-associated protein complexes in epithelial cells: plakoglobin, Beta and gamma catenin are distinct components. J Cell Sci 1993; 104:751-762.
63. Korman NJ, Eyre RW, Klaus-Kovtun V, Stanley JR. Demonstration of an adhering junction molecule (plakoglobin) in the autoantigens of pemphigus foliaceus and pemphigus vulgaris. N Engl J Med 1989; 321:631-635.
64. Gorbsky G, Cohen SM, Shida H, Giudice GJ, Steinberg MS. Isolation of the non-glycosylated proteins of desmosomes and immunolocalization of a third plaque protein: desmoplakin III. Proc Natl Acad Sci USA 1985; 82:810-4.
65. Franke WW, Goldschmidt MD, Zimbelmann R, Mueller HM, Schiller DL, Cowin P. Molecular cloning and amino acid sequence of plakoglobin, the common junctional plaque protein. Proc. Natl Acad Sci. USA 1989;86:4027-4031.
66. Kapprell HP, Cowin P, Franke WW. Biochemical characterization of the soluble form of the junctional plaque protein plakoglobin from different cell types. Eur J Biochem 1987; 166:505-517.
67. McCrea PD, Turck CW, Gumbiner B. A homolog of the armadillo protein in Drosophila (plakoglobin) associated with E-cadherin. Science 1991; 254:1359-61.
68. Shafer S, Troyanovsky SM, Heid HW,et al. Cytoskeletal architecture and epithelial differentiation: molecular determinants of cell interaction and cytoskeletal anchorage. 1993. 316: 1316-1323.
69. Kinzler KW, Nilbert MC, Su LK, Vogelstein B, Bryan TM, Levy DB, Smith KJ, Preisinger AC, Hedge P, McKechnie D et al. Identification of FAP locus genes from chromosome 5q21. Science 1991; 253:661-669.
70. Peifer M, Berg S, Reynolds A. A repeating amino acid motif defines a new relationship among proteins with roles in signaling tumor suppression cell-cell adhesion and guanine nucleotide exchange. Cell 1994; 76:789-791.
71. Reynolds AB, Herbert L, Cleveland JL, Berg ST, Gaut JR. p120, a novel substrate of protein tyrosine kinase receptors and of p60 vsrc, is related to cadherin-binding factors beta-catenin, plakoglobin and armadillo. Oncogene 1992; 7:2439-2445.
72. Ozawa M, Kemler R. Molecular organization of the uvomorulin-catenin complex. J Cell Biol 1992; 116:989-96.
73. McCrea PD, Gumbiner BM. Purification of a 92-kD cytoplasmic protein tightly associated with the cell-cell adhesion molecule E-cadherin (uvomorulin). Characterization and extractability of the protein complex from the cell cytostructure. J Biol Chem 1991; 266:4514-20.
74. Nagafuchi A, Takeichi M. Transmembrane control of cadherin-mediated cell adhesion: a 94 kD protein functionally associated with a specific region of the cytoplasmic domain of E-cadherin. Cell Regul 1989; 1:37-44.
75. Ozawa M, Ringwald M, Kemler R. Uvomorulin-catenin complex formation is regulated by a specific domain in the cytoplasmic region of the cell adhesion molecule. Proc Natl Acad Sci USA 1990; 87:4246-4250.
76. Su LK, Vogelstein B, Kinzler KW. Association of the APC tumor suppressor protein with catenins. Science 1993; 262:1734-1737.

77. Rubinfeld B, Souza B, Albert I, Muller O, Chamberlain SH, Masirz FR, Munemitsu S, Polakis P. Association of the APC gene product with beta-catenin. Science 1993; 262:1731-1734.
78. Nishisho I, Nakamura Y, Miyoshi Y, Miki Y, Ando H, Horii A, Koyama K, Utsunomiya J, Baba S, Hedge P. Mutations of chromosome 5q21 genes in FAP and colorectal cancer patients. Science 1991; 253:665-669.
79. Nathke I, Hinck L, Swedlow JR, Papkoff J, Nelson WJ. Modulation of cadherin and catenin interaction. Mol Biol Cell 1993; 4:473a (Abstr.).
80. Ingham P. Segment polarity genes and cell patterning within the Drosophial body segment. Curr Biol 1991; 1:261-267.
81. Nusse R, Varmus H. Wnt genes. Cell 1992; 69:1073-1087.
82. Riggleman R, Schedl P, Wieschaus E. Spatial expression of the Drosophila segment polarity gene *armadillo* is post-transcriptionally regulated by *wingless*. Cell 1990; 63:549-560.
83. Oda H, Uemura T, Shiomi K, Nagafuchi A, Tsukita S, Takeichi M. Identification of a Drosophila homolog of alpha-catenin and its association with the armadillo protein. J Cell Biol 1993; 121:1133-1141.
84. Bradley RS, Cowin P, Brown AMC. Expression of Wnt-1 in PC12 cells results in modulation of plakoglobin and E-cadherin and increased cellular adhesion. J Cell Biol 1993; 123:1857-1865.
85. Cowin P, Brown AMC. In: Bernfield M, ed. Molecular Basis of Morphogenesis. New York: Wiley-Liss, Inc., 1993:49-66.
86. Olson DJ, Christian JL, Moon RT. Effect of Wnt-1 and related proteins on gap junctional communication in *Xenopus* embryos. Science 1993; 252:1173-1176.
87. Klambt C. Muller S, Lutzelschwab R, Rossa R, Totzke F, Schmidt O. The *Drosophila melanogaster* l(2)gl gene encodes a protein homologous to the cadherin cell-adhesion molecule family. Dev Biol 1989; 133:425-436.
88. Brady-Kalnay SM, Flint AJ, Tonks NJ. Homophilic binding of PTPμ, a receptor-type protein tyrosine phosphatase, can mediate cell-cell aggregation. J Cell Biol 1993; 122:961-972.
89. Amagai M, Karpati S, et al. Extracellular domain of pemphigus vulgans antigen (desmoglein 3) mediates weak homophilic adhesion. J Invest Dermatol 1994; 102: 402-407.

CHAPTER 9

The Desmosomal Plaque: Role in Attachment of Intermediate Filaments to the Cell Surface

Kathleen J. Green
Thaddeus S. Stappenbeck

1. ABSTRACT

Cytoskeletal-membrane attachments play an important role not only in regulating cell motility, adhesion and maintenance of cell shape, but also in transducing signals generated at the cell surface. These attachments are often mediated by intercellular junctions that comprise macromolecular complexes specialized for interaction with a specific filament type. The most prominent cell surface attachment site for intermediate filaments (IF) is the desmosome, an intercellular junction found in abundance in tissues that experience mechanical stress. A number of molecules in the desmosomal plaque have been suggested as candidate IF linkers, including desmoplakins, plectin, 300 kD IFAP, desmocalmin, band 6 protein, and a 140 kD lamin-like protein. Of these, the most abundant and ubiquitous constituents are the desmoplakins (DPs).

Desmoplakin (DP) I and II are two major related proteins located in the inner portion of the desmosomal plaque. The predicted sequence of DP I suggests it will form a homodimer comprising a central α-helical coiled-coil rod and two globular end domains. The smaller spliced form, DP II, apparently has identical end domains but lacks two thirds of the central rod. The C-terminus contains three regions with significant homology, each of which is made up of a 38-residue motif also found in two other molecules involved in organization of IF, bullous pemphigoid antigen and plectin. When polypeptides including the C-terminus of DP I are ectopically expressed in cultured

Molecular Mechanisms of Epithelial Cell Junctions: From Development to Disease
Edited by Sandra Citi, M.D., Ph.D.;

cells the proteins align with keratin and vimentin IF, whereas those lacking the C-terminus do not. The last 68 amino acids of DP are required for alignment along keratin but not vimentin IF, and residues 48-68 from the C-terminal end are critical for this interaction. These results suggest that the C-terminus of DP plays a role in the attachment of IF to the desmosome and that the presence of a specific site is necessary for interaction with keratin IF. A sequence at the most N-terminal end of DP appears to be required for efficient incorporation into the desmosomal plaque. Interestingly, this region has not been reported in the homologous bullous pemphigoid antigen or plectin molecules and may represent a desmosomal targeting sequence.

2. INTRODUCTION

Cytoskeletal-membrane attachments play a critical role in maintenance of cell shape, the regulation of cell movement and adhesive potential, and in transmission of environmental cues generated by interaction with extracellular matrix molecules, growth factors and neighboring cells. These attachments are usually mediated by macromolecular protein complexes which often take the form of morphologically distinct junctional complexes at cell-cell or cell-substrate borders.[1]

At cell-cell interfaces of mature epithelial tissues the actin-containing microfilament cytoskeleton is linked primarily through the adherens junction,[2,3] whereas the intermediate filament cytoskeleton is linked through the desmosome.[4,5] Although these attachment sites are functionally distinct, the recent dissection of each at the molecular level has revealed several similarities. Both junctions contain transmembrane glycoproteins belonging the cadherin class of calcium-dependent cell adhesion molecules which interact through their cytoplasmic tails with cytoplasmic molecules of the adhesion complex.[6,7] In turn, these cytoplasmic molecules of the junctional plaque are proposed to link and/or stabilize the interaction of the cell surface adhesion molecules with specific cytoskeletal filaments. Other than plakoglobin, which has been reported to be a common component of these junctions,[5,8] the plaque components have thus far been shown to be unique to each junction. These junction specific plaque components are thought to be responsible for the specificity of interaction with a particular cytoskeletal element. This chapter will focus on the desmosomal plaque and its role in specifically linking IF to the cell surface, with particular emphasis on the most abundant plaque proteins of desmosomes, desmoplakins I and II (DP I and II).

3. DESMOSOME STRUCTURE AND MOLECULAR COMPOSITION

Desmosomes are constructed of mirror image plaques at cell-cell borders that sandwich a membrane core of transmembrane glycoproteins called desmogleins and desmocollins. These molecules have recently been shown to make up a new subclass of cadherins (see chapter 8 in this volume) (Fig. 9.1). The plasma membranes of neighboring cells are separated by an ~30 nm space bisected by an electron dense central midline with lateral projections extending to the membrane (for a review of desmosome structure and composition see refs. 9-16.)

The tripartite plaque consists of an extremely electron dense region or "outer plaque" next to the plasma membrane that comprises in large part the C-terminal cytoplasmic tails of the desmogleins and desmocollins, as well as the desmosomal cadherin associated protein plakoglobin. A less dense layer separates this region from a fibrillar layer called the "inner plaque" through which IF bundles appear to loop. Although desmosomes are most prominent in the epidermis, where keratin-containing IF are attached to the plaque, these junctions are also present in cardiac myocytes where they are associated with desmin, and the arachnoidal cells of the meninges and dendritic reticular cells of lymph nodes, where they are associated with vimentin.[15] The most abundant constituents of the inner plaque are the DPs which, as discussed below, may in part comprise a set of fine "traversing filaments" that extend between the IF and the plasma membrane.[17] By linking IF at sites of adhesion between cells,

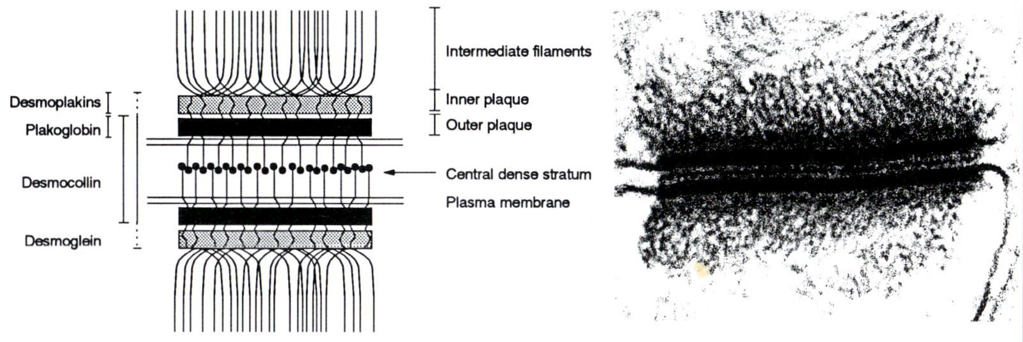

Fig. 9.1. Graphical depiction and ultrastructure of a desmosome from bovine tongue epithelium. The desmosome contains mirror image tripartite plaques consisting of a narrow electronlucent zone sandwiched between outer and inner plaques to which intermediate filaments are attached. The cytoplasmic plaques lie on either side of two opposed lipid bilayers, bisected by the central dense stratum of the intercellular space. The major molecular components found in each region are labeled at the left.

desmosomes effectively create a continuous filamentous network throughout a tissue that is thought to be highly resistant to the forces of mechanical stress. The importance of this network for epithelial tissue integrity has been underscored by the discovery that mutations in genes encoding IF structural proteins result in severe blistering diseases of the skin.[18-20] It follows that the attachment points themselves are just as critical for the integrity of the network; thus, it will be important to understand the molecular mechanism by which this attachment is assembled, maintained and regulated throughout development and differentiation.

4. THE DESMOSOMAL PLAQUE

4.1 Desmoplakins

Desmoplakins (DPs) are the most abundant desmosomal components, comprising ~35% of total desmosomal protein.[21] Furthermore, the DPs are localized at a potentially critical interface between the IF cytoskeleton and the cytoplasmic domains of the desmosomal cadherins,[22] suggesting they might serve as adapters that provide a specific linkage between IF and the transmembrane core. Thus, to gain insights into the mechanism of IF-desmosomal attachment, our laboratory began by investigating the structure and function of these prominent plaque proteins. Indeed, the results outlined in this section suggest that DP is a multifunctional molecule in which the C-terminus is involved in association with the IF cytoskeleton, the rod facilitates formation of the innermost fibrillar plaque, and the N-terminus is required for targeting DP to the desmosomal core at the cell surface.

Analysis of cDNA and genomic clones indicates that DP I and II are likely to be encoded by a single gene that gives rise to predicted polypeptides of 332 and 260 kDa, respectively.[23-27] The smaller DP II molecule appears to be generated from an alternatively spliced message via an internal donor site, and demonstrates a more restricted tissue distribution than the apparently obligate DP I molecule.[28]

Based on analysis of predicted amino acid sequence, DP is likely to form a homodimer consisting of a central heptad repeat-containing α-helical coiled-coil rod domain of ~130nm in length, flanked by two globular end domains[23-25,29] (Fig. 9.2). The polypeptide chains of a homodimer are furthermore predicted to be organized in parallel and in register.[23] This model is consistent with biochemical evidence that purified and cross-linked DP I exists as a dimer in vitro. Furthermore, rotary shadowed images of purified

DP I are strikingly similar to the predicted structure, appearing dumbbell shaped with a central rod of ~130 nm.[30] DNA sequence analysis also reveals that DP II lacks approximately two thirds of the central domain, thus containing a shorter rod of ~43 nm. It is likely that DP II will also form a homodimer in vivo, although in vitro evidence of dimers has not yet been reported.[30] Based on the pattern of acidic and basic residues along the central coiled-coil helix, it is likely that DP I will not only dimerize, but also aggregate with itself or similar molecules into higher order filamentous structures. Such a prediction is consistent with the possibility that DP may at least in part make up the fine 4-5nm "traversing filaments" that appear to link IF with the plaque or the plaque with the plasma membrane.[31]

The C-terminal globular domain of DP can be further divided into three subdomains, termed A,B and C, each containing 176 residues built of a 38 residue repeating motif.[23,29] Each of these regions are predicted to fold into a compact globular structure stabilized by intrachain ionic interactions (Fig. 9.2). The periodicity of 9.4 in acidic and basic residues of the 38 residue motif is the same as that found in the 1B rod domain of IF proteins. This observation suggests the intriguing possibility that DP might interact with IF or IF-like proteins via their common periodicity of charged residues. Like the rod domain, the N-terminus of DP is characterized by heptad repeat substructure over much of its length.[25] However, in contrast to the extensive series of unbroken heptads in the rod, the stretches of heptad in the N-terminus are shorter and, due to the increase in the ratio of apolar to charged residues, are predicted to form a series of compact bundles rather than a coiled rope (Fig. 9.2).

The major structural characteristics displayed by DP are also found in the 230-kD bullous pemphigoid (BP230) antigen[32-34] and the intermediate filament associated protein (IFAP) plectin.[35] BP230 is the target antigen of autoimmune antibodies circulating in patients with the blistering disease for which it is named (see also chapters 14 and 15 in this volume). The antigen has been localized to the hemidesmosomal plaque, in the same relative location as DP in the desmosomal plaque. Both BP230 and plectin have been implicated as playing a role in interacting with IF networks.[36,37] The structural plan of all three members of this emerging gene family is similar to that described for DP, that is, a central coiled-coil rod domain flanked by two globular ends. Comparison of the N- and C-termini among these molecules reveals extensive similarity, suggesting an important evolutionarily conserved function in the organization of IF networks.[29]

The facts that DP I is an obligate constituent of desmosomes, that DPs are the most abundant desmosomal components, and that they are located at the IF-plaque interface, have led investigators to suggest a role for DP in cytoskeletal linkage to the plaque. Consistent with this is the reported colocalization of DP with IF throughout calcium-mediated desmosome assembly.[38-40] In addition, IF polypeptides associated with the desmosomal plaque are apparently incapable of polymerizing in the presence of DP.[41]

However, in vitro biochemical experiments using highly purified components have so far failed to identify a direct interaction between DP and any core glycoprotein or IF polypeptides.[30,42] One possible explanation for this is that the harsh denaturing conditions required to extract individual desmosomal components from desmosomes could affect the in vitro binding capabilities DP. On the other hand, it is also possible that accessory or linking proteins may be necessary to stabilize or mediate the association of DP with IF.

As an alternative approach to in vitro binding experiments, the potential for interactions between DP and IF networks and/or the cell surface desmosomal plaque has been tested by introducing plasmids encoding specific DP domains into cultured cells.[26,43] Constructs encoding full length or individual domains of DP I were transiently transfected into cultured COS-7 or HeLa epithelial cells, and the fate of ectopically expressed protein was investigated by immunocytochemical and biochemical approaches. In certain instances, a peptide tag encoding a 10 amino acid epitope of c-myc[44] was introduced at the

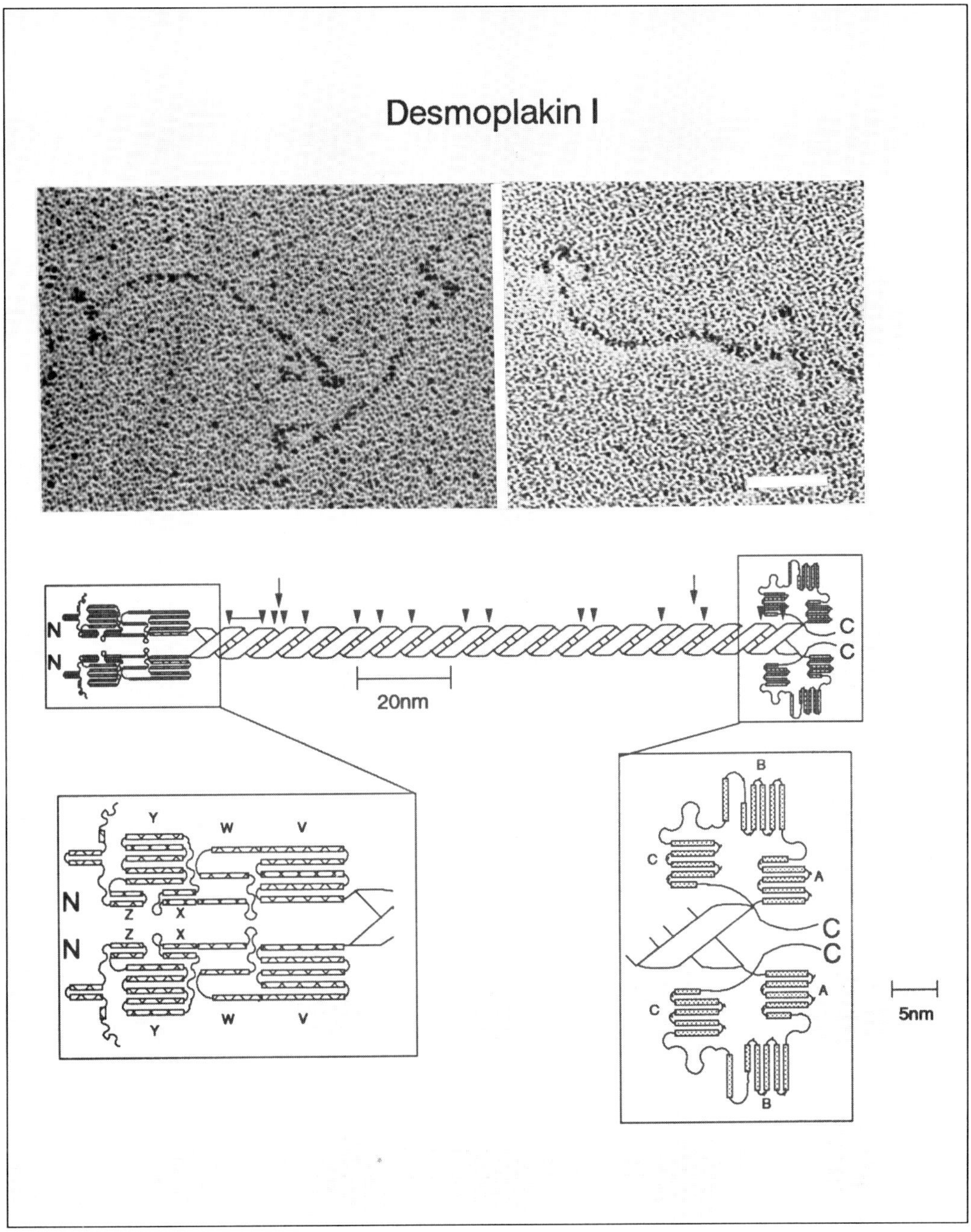

Fig. 9.2. Structure of a desmoplakin homodimer. The predicted model of DP structure is shown in comparison with rotary shadowed images of purified DPI. Note that the rod domain consists of a coiled-coil α-helix ~130 nm in length. Possible break points and stutters predicted by the amino acid sequence are designated by small arrows above the rod domain, and an extended region of predicted flexibility is designated by a horizontal bar. Large arrows define the boundaries of the DPI specific region. Rectangles in the C- terminus contain the 38 residue repeat motif and represent predicted α-helical regions that turn upon themselves and are stabilized by ionic interactions. Rectangles in the amino terminus have a heptad substructure and also represent predicted α-helical regions that turn upon themselves. Rotary shadowed images of DPI were prepared as described.[30] Note the existence of an extended, flexible central rod flanked by two more globular ends. Scale bar = 50nm (Rotary shadowed EM images kindly provided by Dr. Ed O'Keefe).

C-terminus to allow one to distinguish between endogenous DP and ectopically expressed polypeptides using the readily available monoclonal antibody 9E10.2.

In cultured cells, ectopically expressed DP polypeptides containing the C-terminus but lacking the N-terminus (DP.ΔN) are capable of aligning along IF networks (Fig. 9.3).[43] At the highest levels of expression both vimentin and keratin IF networks are disrupted. Other cytoskeletal networks such as microfilaments and microtubules are unaffected, however, indicating the association with IF networks is a specific one. This result suggests that the DP C-terminal domain is involved in attaching IF networks to the desmosomal plaque. The truncated DP molecules also align with vimentin IF in fibroblasts, which are incapable of assembling desmosomes and presumably do not express other desmosomal-specific proteins. This observation increases the likelihood that the observed DP-IF interaction is direct, although the contribution of proteins with a more widespread tissue distribution cannot be ruled out.

In order to further define sequences in the C-terminus required for alignment with IF networks, constructs encoding various deletions of the C-terminus were introduced into several cell lines to test the ability of the truncated proteins to coalign with and disrupt vimentin (3T3 fibroblasts), keratin (HaCaT keratinocytes), or both IF types in the same cell (HeLa cells). Based on these studies, the 68 residues at the extreme C-terminus of the DP molecule contains regulatory sequences that both positively and negatively affect the interaction of DP with keratin, but not vimentin, IF networks.[26,45] The presence of positive regulatory sequences is reflected by the fact that interaction of DP.ΔN (rod plus C-terminus) with IF networks is dependent on the presence of the 68 residues, and that amino acids 48-68 from the C-terminus are critical for the interaction. Constructs missing these residues are unable to align with keratin networks as assessed by immunofluorescence analysis, and this inability to align is correlated with a dramatic increase in Triton solubility of the

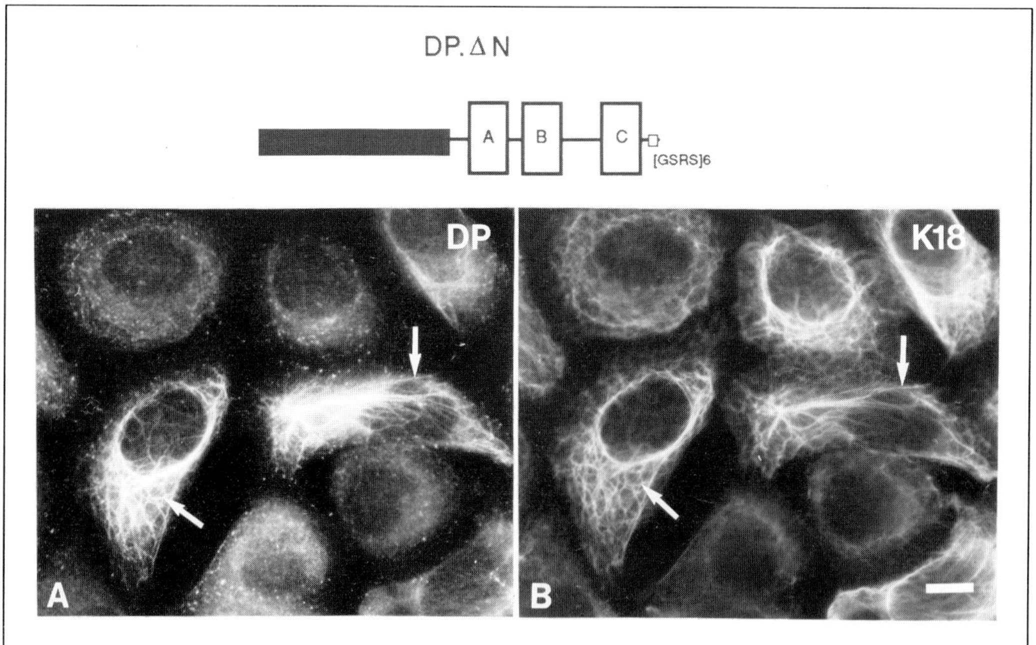

Fig. 9.3. Alignment of DP.ΔN with keratin IF in transfected HeLa cells. Cells were transfected with a construct encoding the rod plus C-terminus of DP (DP.ΔN). 24 hours after glycerol shock, cells were fixed and reacted with a rabbit polyclonal to (A) DP (NW 38) and (B) a mouse monoclonal to keratin 18 (KSB17.2). Note coalignment of the DP polypeptide with keratin IF in transfected cells (arrows). Scale bar = 10μm.

DP polypeptide.[26] The presence of negative regulatory sequences is reflected by the fact that truncations that remove a serine residue located 23 amino acids from the C-terminus of DP, or alteration of this residue to a glycine, results in an increase in the ability of DP.ΔN to align with keratin IF networks. Intriguingly, this serine residue is in a perfect consensus site for phosphorylation by cAMP-dependent kinase. Furthermore, in cultured cells treated with forskolin to activate cAMP-dependent kinase, DP.ΔN.ser is incapable of alignment with keratin IF, whereas DP.ΔN.gly can still associate with the filament system.

Although the interaction of DP polypeptides with vimentin IF networks does not require the terminal 68 residues, it does require the presence of at least two of the C-terminal repeat subdomains. Furthermore, extraction with non-ionic detergents such as Triton indicates that DP associating with vimentin IF netwoks is significantly more extractable than DP associating with keratin IF networks, suggesting that the latter interaction may be of a higher affinity. Although the rod domain alone does not associate with IF networks, addition of the rod to the C-terminus (DP.ΔN) is required for efficient association with vimentin IF, perhaps due to the need for DP dimerization which would increase the number of available C-terminal domains.

The potential for differential interactions between DP and different IF types is consistent with reports that keratins preferentially interact with desmosomes in cells expressing both filament systems.[46] It is possible that the last 68 residues of DP may not only be necessary for specifying an interaction with keratin IF networks, but may promote interaction with keratin at the expense of interactions with vimentin. The differential extractability of DP.ΔN in cells expressing keratin versus cells expressing vimentin only is consistent with this observation. The physiological significance of possible differences in affinity between DP and different filament networks remains unclear, however.

Ultrastructural analysis of aggregates formed as a result of the highest levels of expression in cells transfected with DP.CT (C-terminus only), DP.ROD (rod only) and DP.ΔN (rod plus C-terminus), reveals striking and important differences. Expression of the rod alone does not significantly affect the IF network. However, intact IF are no longer present in cells expressing constructs containing the C-terminus. They are replaced instead with aggregates that contain mutant DP and IF throughout the entire structure (see Stappenbeck and Green, ref. 43, Figs. 7 and 8). Thus, the IF network does not simply collapse around a preformed aggregate of mutant DP.

The ultrastructure of the IF-containing DP.ΔN aggregates is quite different from the DP.CT and DP.ROD aggregates, however. While aggregates of DP.CT and DP.ROD are electron-dense with no apparent filamentous substructure, aggregates of DP.ΔN are composed of a fine filamentous meshwork (Fig. 9.4C). The filaments of this meshwork are shorter and smaller in diameter than the 8-10 nm cytoplasmic IF found in neighboring non-transfected cells, and resemble the meshwork of 4-5nm fine filaments that makeup the innermost desmosomal plaque. It is important to emphasize two critical observations here. First, the meshwork arises only in conjunction with IF polypeptides, as accumulations of DP.ΔN lacking IF are found in certain regions of these cells, and they exhibit an amorphous, non-filamentous ultrastructure (Fig. 9.4D). Second, although the rod domain alone neither forms a meshwork nor associates with IF, its presence on the C-terminus is absolutely required for meshwork formation.

Thus, it is possible that by overexpression of DP.ΔN which can associate with itself and IF polypeptides in some way not yet elucidated, a large plaque-like structure is assembled in the cytoplasm. The presence of the N-terminus on fcull length DP may facilitate the assembly of this meshwork at the cell cortex in normal cells. This model implies that association of IF with the desmosomal plaque occurs not via intact, full caliber IF, but via a mat or meshwork of interwoven IF and DP. Such a model is similar in certain respects to the suggestion of Leloup

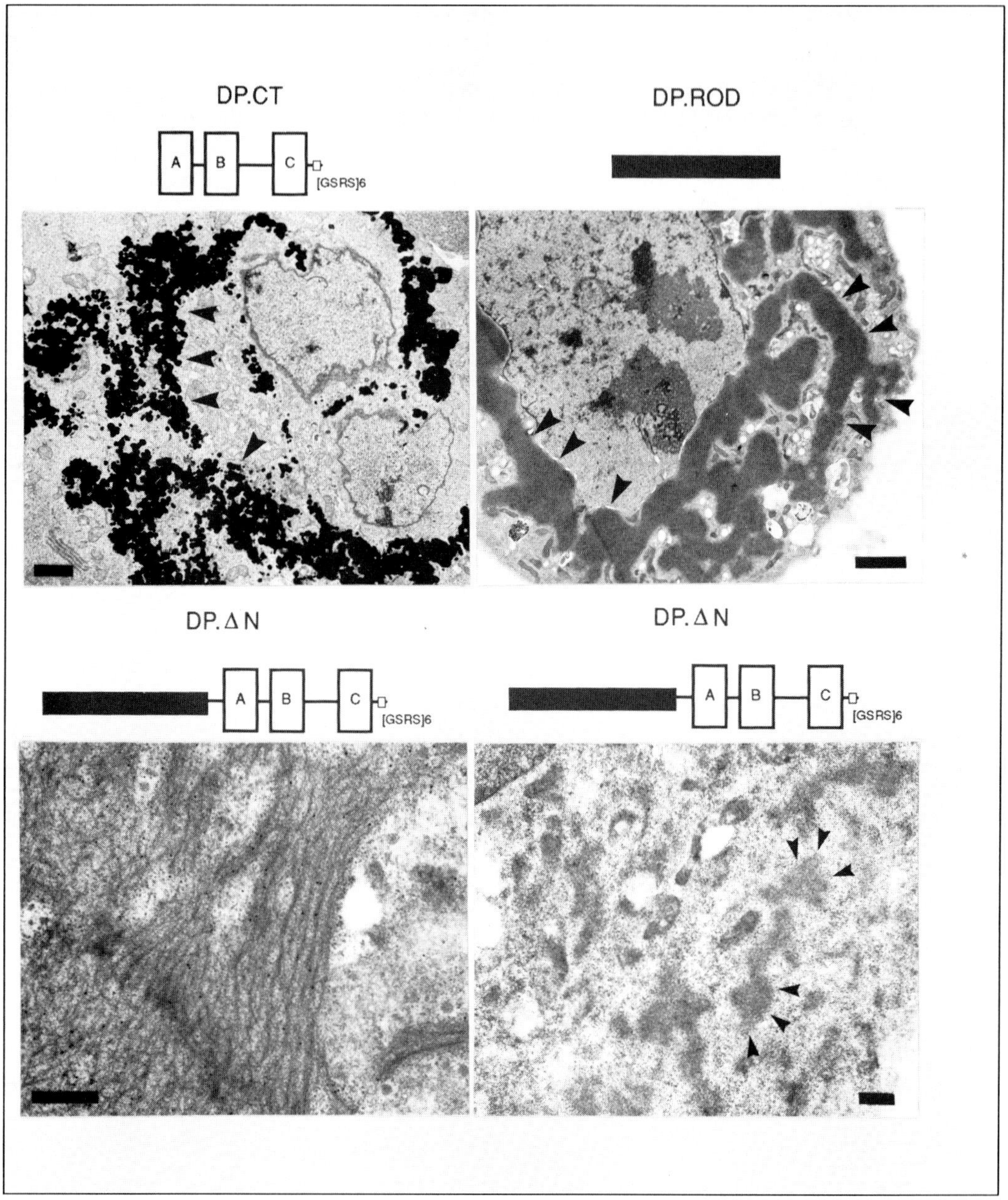

Fig. 9.4. Ultrastructural analysis of transfected cells. EM of COS-7 cells transfected with constructs encoding DP.CT (A), DP.ROD (B) or DP.ΔN (C,D). Cells were fixed either for conventional (A,B) or immunogold EM (C,D). In A, clusters of very electron-dense aggregates (arrowheads), and no intact IF, are present. See Stappenbeck and Green[43] for immunogold analysis of DP.CT transfected cells which demonstrates presence of both DP and IF in these aggregates that do not contain IF. In B, expression of DP.ROD also results in dense (although less so than those in A) aggregates that do not contain IF (arrowheads). In C, DP.ΔN transfected cells were reacted with 9E10.2 which reveals the presence of tagged, truncated DP. Separate labeling experiments reveal the presence of vimentin in these meshworks as well.[43] In D, DP.ΔN transfected cells were reacted with a vimentin antibody. The cytoplasmic electron-dense spots designated by arrowheads are vimentin-negative, but contain tagged DP.ΔN as shown in separate labeling experiments.[43] (Portions of this figure are modified from Stappenbeck and Green J. Cell Biol. 1992; 116: 1197-1209). Scale bars: (A, B) 2.5 μm ; (C, D) 0.5 μm.

who hypothesized that the fine traversing filaments of desmosomes might be unraveled IF protofilaments.[47] Our model adds to Leloup's theory that the "unraveled" IF are also associated in some higher order structure with DP polypeptides.

Consistent with this model is the fact that ectopically expressed polypeptides containing the DP N-terminus do indeed localize to the desmosomal plaque, suggesting that sequences required for incorporation into desmosomes are present in this domain (Figs. 5, 6). Interestingly, c-myc-tagged full length DPI appears in some cells to associate with desmosomes (Fig. 9.5A-D) and in others to interact with IF (Fig. 9.5E). Factors that allow the function of either the C- or the N-terminus to be manifested preferentially in individual cells are not yet understood. However, it is suspected that the incorporation-competent state may be dependent on the stage of cell cycle (e.g. transfected dividing daughter cells often appear to incorporate tagged DP at newly forming desmosomes at cell-cell interfaces) or other physiological states.

Although a partner for DP on the inner surface of the membrane has not been identified, the cytoplasmic domains of one or more of the transmembrane desmosomal cadherins are excellent candidates, as suggested by recent observations reported by Troyanovsky, et al.[48] These investigators developed constructs encoding chimeric fusion proteins in which the transmembrane region of the gap junction protein, connexin 32 (Co32), is fused to the tails of the long ("a") and short ("b") spliced forms of desmocollins (Dsc) and with the tail of desmoglein (Dsg) 1. It was hypothesized that upon expression of these plasmids in cultured cells, oligomerization of the connexin portion of these proteins in the plane of the membrane would mimic clustering thought to occur during the normal process of desmosome assembly. This system would then allow a test of the function of the desmosomal cadherin tails, that is, their role in assembling the desmosomal plaque.

When expressed in A431 cells, the tail of the long ("a" form), but not the short, splice form of Dsc leads to the recruitment of DP, plakoglobin and keratin IF to gap junction-like membrane domains. Perhaps surprisingly, the fusion protein containing the Dsg tail does not form gap junction structures or facilitate IF anchorage to the plasma membrane. Rather, its expression has a dominant negative effect that results in dissolution of endogenous desmosomes and entry of DP into a diffuse, presumably soluble, pool in the cytoplasm. One possible explanation for such a dominant negative effect is that the availability of a limiting binding partner such as plakoglobin or DP might be reduced by association with ectopically expressed protein. However, expression of control constructs in which the Dsg1 tail is fused with synaptophysin, resulting in localization to small cytoplasmic vesicles rather than to the plasma membrane, does not interfere with the formation of endogenous desmosomes. This protein also appears capable of interacting with plakoglobin. The underlying basis for these observations is not yet fully clear.

The results of Stappenbeck, et al indicate that sequences required for interaction with the inner cell surface appear to reside in the N-terminal most 194 amino acids of DP, as molecules lacking this region are unable to associate with the desmosomal plaque.[43] It is interesting to note that this region may be unique to DP, as similar sequences have so far not been identified in BP230 or plectin. It has not yet been determined whether this region is sufficient for targeting to the desmosomal plaque, only that it is necessary.

Deletion of the C-terminus from the full length DPI clone results in a protein still able to associate with desmosomes but un able to interact with IF networks (Fig. 9.6). The inability of DP.ΔC to coalign with and disrupt IF provides additional support for the idea that the C-terminus of DP contains sites necessary and sufficient for interaction with IF. Because DP.ΔC can apparently incorporate into desmosomes, interaction with IF is unlikely to be required for localization of DP to desmosomes. While this result does not rule out the possibility that IF provide one route for assembly of DP into the desmosomal plaque,[38,40,49] it suggests that

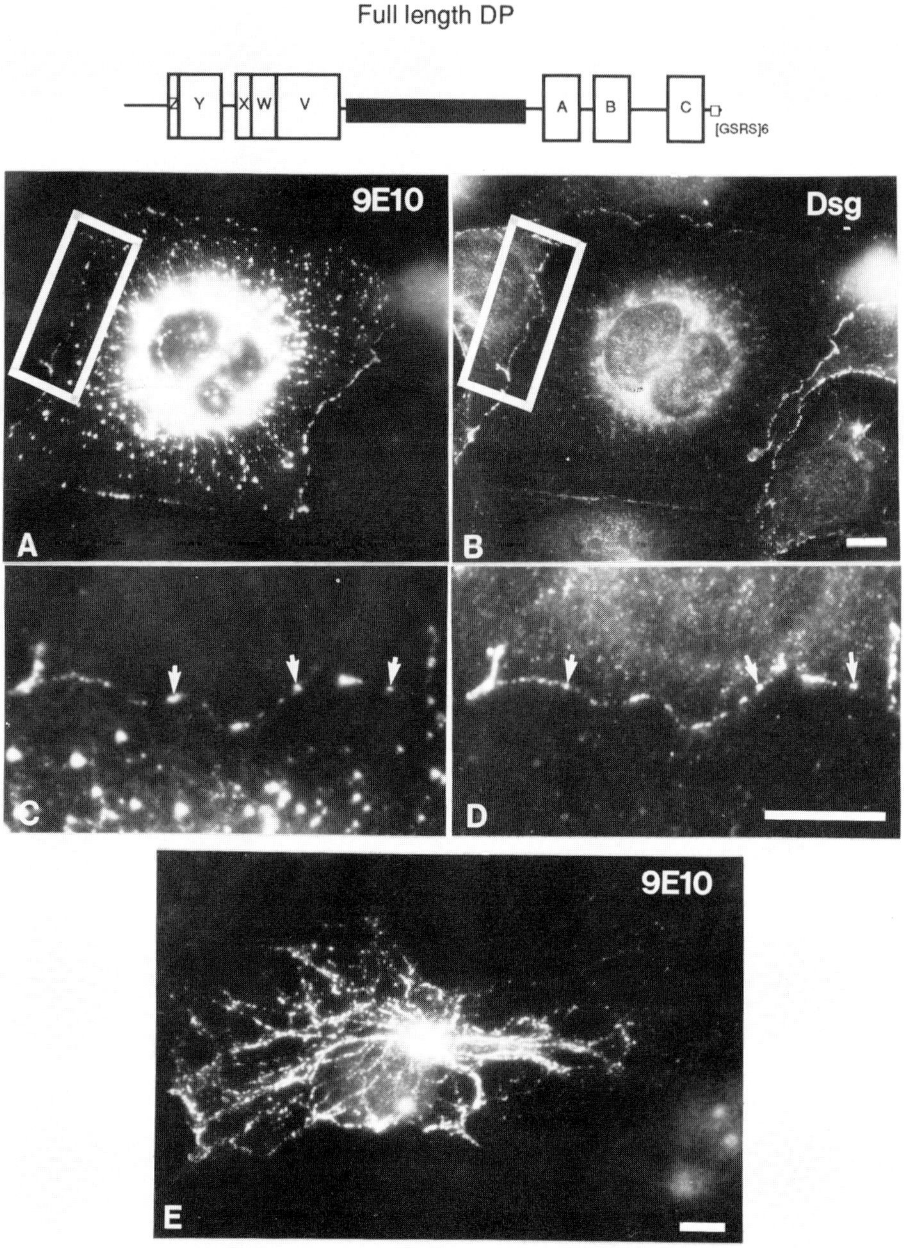

Fig. 9.5. COS-7 cells transfected with a plasmid encoding full length DPI. A-D: 24 hours after glycerol shock cells were fixed and reacted with the mouse monoclonal, 9E10.2, directed against the c-myc tag (A and C), and a rabbit polyclonal antibody directed against bovine desmoglein (Dsg) (B and D). The rectangle in A and B corresponds to B and D, respectively (each rectangle is rotated 70° clockwise). Arrowheads in C and D indicate examples of 9E10.2 and Dsg spots that colocalize at cell-cell borders, suggesting that tagged DP can associate with desmosomes. Scale bars = 10 µm. (As modified from Stappenbeck et al J. Cell Biol. 1993; 123: 691-705). E: Indirect immunofluorescence of COS-7 cells transiently transfected with tagged full length DP, extracted with 0.5% Triton for 15 minutes, fixed and reacted with 9E10.2. Note in this case the punctate, linear staining pattern, reflecting alignment with endogenous IF. Scale bars = 10 µm.

other mechanisms can be utilized, consistent with previously reported observations.[50-52] The existence of an IF independent pathway for delivery of DP to the cell surface could explain the apparently normal assembly of the desmosomal plaque in IF deficient simple epithelia or in epithelia with disrupted IF.[53,54] Together, these results suggest that DP is a multifunctional molecule in which the N-terminus is required for targeting to the inner cell surface, while the C-terminus of DP is required for association with IF networks and together with the rod leads to the assembly of the fibrillar meshwork making up the innermost plaque.

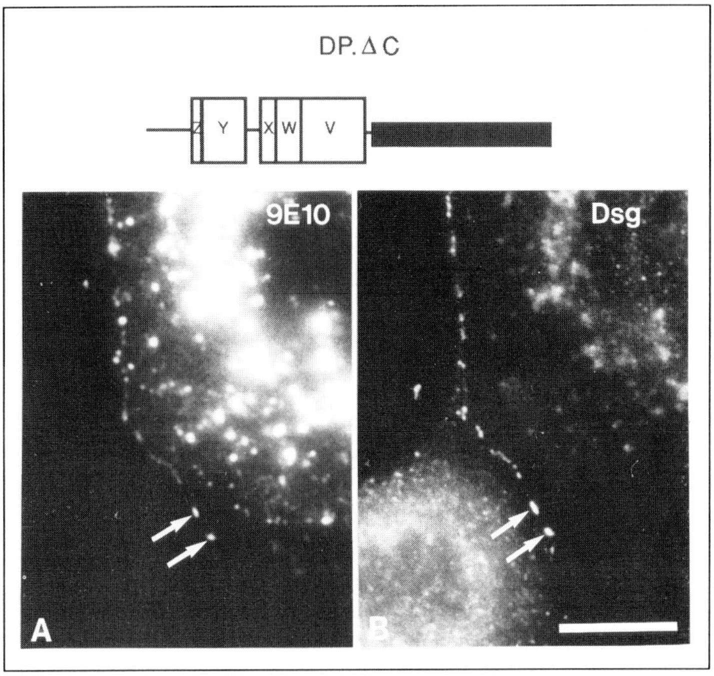

Fig. 9.6. Indirect double label immunofluorescence of COS-7 cells transfected with the DP.ΔC. construct 24 hours after glycerol shock cells were reacted with (A) the mouse monoclonal, 9E10.2, directed against the c-myc tag and (B) a rabbit polyclonal antibody directed against bovine desmoglein (Dsg). Shown here is a high magnification of cell borders. Arrows indicate 9E10.2 and Dsg spots along cell-cell borders that colocalize, suggesting that DP.ΔC can associate with the desmosomal plaque.

4.2 Other Proteins That May Function in Linking/ and or Stabilizing IF at the Desmosomal Plaque

The possible involvement of additional accessory proteins has not been ruled out in the DP-IF interactions reported here. In fact, other molecules localized to desmosomes have been demonstrated to bind IF in vitro. Some of these are expressed in a specific subset of epithelial tissues and thus are not obligate constituents of the desmosome. Examples are keratocalmin and desmocalmin which are thought to be the human and bovine homologues, respectively, of a desmosomal calmodulin-binding protein.[55,56] Desmocalmin is a protein of ~240 kD isolated from bovine muzzle, that cosediments with reconstituted IF. The human form of this molecule appears to be expressed only in stratified epithelial tissues.[56] Another stratification-specific protein is band 6, which binds to radiolabeled type I cytokeratin in overlay assays.[57]

A third candidate, the 140 kD lamin B related protein,[58] is found in very small amounts in desmosome-bearing cells only, and binds to radiolabeled vimentin IF in solid substrate overlay assays. Ouyang and colleagues have also identified a 140 kDa protein in MDCK cells located on the innermost extreme of the desmosomal plaque, situated internal to DP.[59] Based on the delayed time course of redistribution of this component during calcium-induced desmosome assembly, these investigators suggest a role for this protein in stabilizing IF-plaque interactions once they are already established. Whether or not these two 140 kD proteins are identical has not yet been investigated.

In light of the fact that the DP C-terminus aligns along vimentin filaments in fibroblasts, cells which presumably are not expressing other desmosomal components that could act as accessory or linking molecules, it is unlikely that the candidate linkers discussed above are always absolutely required for a DP-IF association. On the other hand, if associated proteins with a widespread tissue distribution such as plectin[60] or 300 kDa IFAP[61] which are present in fibroblasts but have also been reported in desmosomes could be involved in the interaction. Additional work will be necessary to experimentally address the possibility that a macromolecular complex including DP, IF, and plectin or 300 kD IFAP forms the structural link with the cell surface.

5. REGULATION OF DESMOSOME-IF INTERACTIONS AND FUTURE PROSPECTS

The work described in this chapter begins to elucidate structure-function relationships of individual desmosomal plaque constituents. However, less is known regarding regulation of desmosome assembly, adhesion and association with the IF cytoskeleton. Desmosomes are dynamic structures that are modulated not only during differentiation, but also during wound healing, metastasis, and in response to certain growth factors.[62] In these latter processes junctional protein markers disappear from the cell surface, and IF retract into a cap around the nucleus. Preliminary work indicates the likelihood that phosphorylation of certain desmosomal components such as plakoglobin and desmoplakin may play a role in modulating protein-protein interactions and thus the assembly state and function of desmosomes. Thus fertile areas for future investigation are in defining how these molecules are regulated during assembly and disassembly, and how desmosomes function and are regulated in higher levels of tissue organization.

ACKNOWLEDGMENTS

The authors would like to thank all those who have contributed antibodies, cell lines and other reagents for experiments from our laboratory that are reported here including: Pamela Cowin, Orest Blaschuk, Joe Gall, and Peter Steinert. Special thanks go to Dr. Ed O'Keefe who contributed the rotary shadowed images of purified desmoplakin I and to Laura Nilles for expert assistance with the figures. Thanks go also to E. Bornslaeger, M.L.A. Virata, D.A.D. Parry, R. Wagner, C. Corcoran, H. Luu, J. Lamb, and S. Guy for their contribution to various aspects of this chapter. Work from the authors' laboratory is supported by grants from the NIH (AR41836), The American Cancer Society (CB-110A and a Faculty Research Award), The March of Dimes Birth Defects Foundation, The Council for Tobacco Research, USA, Inc. (#2432A) and a Focused Giving Award from Johnson and Johnson Co.

REFERENCES

1. Luna EJ, Hitt AL. Cytoskeleton-plasma membrane interactions. Science 1992; 258:955-964.
2. Geiger B, Ginsberg D. The cytoplasmic domain of adherens-type junctions. Cell Motil Cytoskel 1991; 20:1-6.
3. Tsukita S, Tsukita S, Nagafuchi A, Yonemura S. Molecular linkage between cadherins and actin filaments in cell-cell adherens junctions. Curr Opin Cell Biol 1992; 4:834-839.
4. Jones JCR, Green KJ. Intermediate filament-plasma membrane interactions. Curr Opin Cell Biol 1991; 3:126-131.
5. Kapprell H, Duden R, Owaribe K et al. Subplasmalemmal plaques of intercellular junctions: common and distinguishing proteins. In: Edelman GM, Cunningham BA, Thiery JP, ed. Morphoregulatory Molecules. New York: John Wiley and Sons, 1990:285-314.
6. Buxton RS, Magee AI. Structure and interactions of desmosomal and other cadherins. Sem Cell Biol 1992; 3:157-167.
7. Geiger B, Ayalon O. Cadherins. Ann Rev Cell Biol 1992; 8:307-332.
8. Cowin P, Kapprell H, Franke WW, Tamkun J, Hynes RO. Plakoglobin: a protein common to different kinds of intercellular adhering junctions. Cell 1986; 46:1063-1073.
9. Farquhar MG, Palade GE. Junctional complexes in various epithelia. J Cell Biol 1963; 17:375-412.
10. Staehelin LA. Structure and function of intercellular junctions. Int Rev Cytol 1974; 39:191-283.
11. Hull BE, Staehelin LA. The terminal web: a reevaluation of its structure and function. J Cell Biol 1979; 81:67-82.

12. Arnn J, Staehelin LA. The structure and function of spot desmosomes. Dermatology 1981; 20:330-339.
13. Cowin P, Franke WW, Grund C et al. The desmosome-intermediate filament complex. In: Edelman GM, Thiery J, ed. The Cell in Contact. New York: John Wiley and Sons, 1985:427-460.
14. Steinberg MS, Shida H, Giudice GJ Shida M, Patel NH, Blaschuck OW. On the molecular organization, diversity and functions of desmosomal proteins. In: Bock G, Clark S, ed. Junctional complexes of epithelial cells. Chichester: John Wiley and Sons, 1987:3-19.
15. Schwarz MA, Owaribe K, Kartenbeck J, Franke WW. Desmosomes and hemidesmosomes: constitutive molecular components. In: Palade GE, Alberts BM, Spudich JA, ed. Ann Rev Cell Biol 1990:461-491.
16. Garrod DR. Desmosomes and hemidesmosomes. Curr Opin Cell Biol 1993; 5:30-40.
17. Kelly DE, Shienvold F. The desmosome: fine structural studies with freeze fracture replication and tannic acid staining of sectioned epidermis. Cell Tiss Res 1976; 173:309-323.
18. Fuchs E, Coulombe PA. Of mice and men: genetic skin diseases of keratin. Cell 1992; 69:899-902.
19. Epstein EH. Molecular genetics of epidermolysis bullosa. Science 1992; 256:799-804.
20. Steinert PM. Structure, function and dynamics of keratin intermediate filaments. J Invest Derm 1993; 100:729-734.
21. Skerrow CJ, Matoltsy AG. Isolation of epidermal desmosomes. J Cell Biol 1974; 63:515-523.
22. Miller K, Mattey D, Measures H, Hopkins C, Garrod D. Localisation of the protein and glycoprotein components of bovine nasal epithelial desmosomes by immunoelectron microscopy. EMBO J 1987; 6:885-889.
23. Green KJ, Parry DAD, Steinert PM, Virata ML, Wagner RM, Angst BD, Nilles LA. Structure of the human desmoplakins: implications for function in the desmosomal plaque. J Biol Chem 1990; 265:2603-2612.
24. Green KJ, Stappenbeck TS, Parry DAD, Virata ML. Structure of desmoplakin and its association with intermediate filaments. J Dermatol 1992; 19:765-769.
25. Virata MLA, Wagner RM, Parry DAD, Green KJ. Molecular structure of the human desmoplakin I and II amino terminus. Proc Natl Acad Sci 1992; 89:544-548.
26. Stappenbeck TS, Bornslaeger EA, Corcoran CM, Luu HH, Virata ML, Green KJ. Functional analysis of desmoplakin domains: specification of the interaction with keratin versus vimentin intermediate filament networks. J Cell Biol 1993; 123:691-705.
27. Virata MLA. Identification and characterization of the cDNAs encoding the human desmoplakins I and II, major plaque components of the desmosome involved in intermediate filament organization. Northwestern University, Ph.D. Thesis 1993:1-194.
28. Angst BD, Nilles LA, Green KJ. Desmoplakin II expression is not restricted to stratified epithelia. J Cell Sci 1990; 97:247-257.
29. Green KJ, Virata MLA, Elgart GW, Stanley JR, Parry DAD. Comparative structural analysis of desmoplakin, bullous pemphigoid antigen and plectin: members of a new gene family involved in organization of intermediate filaments. Int J Biol Macromol 1992; 14:145-153.
30. O'Keefe EJ, Erickson HP, Bennett V. Desmoplakin I and desmoplakin II: Purification and characterization. J Biol Chem 1989; 264:8310-8318.
31. Kelly DE, Kuda AM. Traversing filaments in desmosomal and hemidesmosomal attachments: freeze fracture approaches toward their characterization. Anat Rec 1981; 199:1-14.
32. Stanley JR, Tanaka T, Mueller S, Kaus-Kovtun V, Roop D. Isolation of complementary DNA for Bullous pemphigoid antigen by use of patients' autoantibodies. J Clin Invest 1988; 82:1864-1870.
33. Amagai M, Klaus-Kovtun V, Stanley JR. Autoantibodies against a novel epithelial cadherin in pemphigus vulgaris, a disease of cell adhesion. Cell 1991; 67:869-877.

34. Sawamura D, Li K, Chu M, Uitto J. Human bullous pemphigoid antigen (BPAG1). Amino acid sequences deduced from cloned cDNAs predict biologically important peptide segments and protein domains. J Biol Chem 1991; 266:17784-17790.
35. Wiche G, Becker B, Luber K, Weitzer G, Castanon MJ, Hauptmann R, Stratowa C, Stewart M. Cloning and sequencing of rat plectin indicates a 466-kD polypeptide chain with a three-domain structure based on a central alpha-helical coiled coil. J Cell Biol 1991; 114:83-99.
36. Klatte DH, Kurpakus MA, Grelling KA, Jones JC. Immunochemical characterization of three components of the hemidesmosome and their expression in cultured epithelial cells. J Cell Biol 1989; 109:3377-3390.
37. Foisner R, Leichtfried FE, Herrmann H, Small JV, Lawson D, Wiche G. Cytoskeleton-associated plectin: in situ localization, in vitro reconstitution, and binding to immobilized intermediate filament proteins. J Cell Biol 1988; 106:723-733.
38. Jones JCR, Goldman RD. Intermediate filaments and the initiation of desmosome assembly. J Cell Biol 1985; 101:506-517.
39. Pasdar M, Nelson WJ. Kinetics of desmosome assembly in Madin-Darby canine kidney epithelial cells: temporal and spatial regulation of desmoplakin organization and stabilization upon cell-cell contact I. Biochemical analysis. J Cell Biol 1988; 106:677-685.
40. Pasdar M, Nelson WJ. Kinetics of desmosome assembly in Madin-Darby canine kidney cells: temporal and spatial regulation of desmoplakin organization and stabilization upon cell- cell contact. II. Morphological analysis. J Cell Biol 1988; 106:687-695.
41. Jones SM, Jones JCR, Goldman RD. Association of the 240kD and the 210kD desmosomal proteins (DPI and DPII) with keratin filaments. J Cell Biol 1986; 103:420a.
42. Pasdar M, Krzeminski KA, Nelson WJ. Regulation of desmosome assembly in MDCK epithelial cells: coordination of membrane core and cytoplasmic plaque domain assembly at the plasma membrane. J Cell Biol 1991; 113:645-655.
43. Stappenbeck TS, Green KJ. The desmoplakin carboxyl terminus coaligns with and specifically disrupts intermediate filament networks when expressed in cultured cells. J Cell Biol 1992; 116:1197-1209.
44. Munro S, Pelham HRB. A C-terminal signal prevents secretion of luminal ER proteins. Cell 1987; 48:899-907.
45. Stappenbeck TS, Bornslaeger E, Virata MLA, Corcoran CM, Green KJ. Sequences in desmoplakin required for interaction with intermediate filaments and incorporation into desmosomes. J Invest Derm 1993; 100:523.
46. Kartenbeck J, Schwechheimer K, Moll R, Franke WW. Attachment of vimentin filaments to desmosomal plaques in human meningiomal cells and arachnoidal tissue. J Cell Biol 1984; 98:1072-1081.
47. Leloup R, Lawet L, Ronveaux MF et al. Desmosomes and desmogenesis in the epidermis of calf muzzle. Biol. Cell 1979; 34:137-152.
48. Troyanovsky SM, Eshkind LG, Troyanovsky RB, Leube RE, Franke WW. Contributions of cytoplasmic domains of desmosomal cadherins to desmosome assembly and intermediate filament anchorage. Cell 1993; 72:561-574.
49. Trevor KT, Steben LS. Distribution of desmosomal proteins in F9 embryonal carcinoma cells and epithelial cell derivatives. J Cell Sci 1992; 103:69-80.
50. Bologna M, Allen R, Dulbecco R. Organization of cytokeratin bundles by desmosomes in rat mammary cells. J Cell Biol 1986; 102:560-567.
51. Mattey DL, Garrod DR. Calcium-induced desmosome formation in cultured kidney epithelial cells. J Cell Sci 1986; 85:95-111.
52. Duden R, Franke WW. Organization of desmosomal plaque proteins in cells growing at low calcium concentrations. J Cell Biol 1988; 107:1049-1063.
53. Baribault H, Oshima RG. Polarized and functional epithelia can form after the targeted inactivation of both mouse keratin 8 alleles. J Cell Biol 1991; 115:1675-1684.

54. Vassar R, Coulombe PA, Degenstein L, Albers K, Fuchs E. Mutant keratin expression in transgenic mice causes marked abnormalities resembling a human genetic skin disease. Cell 1991; 64:365-380.
55. Tsukita S, Tsukita S. Desmocalmin: a calmodulin-binding high molecular weight protein isolated from desmosomes. J Cell Biol 1985; 101:2070-2080.
56. Fairley JA, Scott GA, Jensen KD, Goldsmith LA, Diaz LA. Characterization of keratocalmin, a calmodulin-binding protein from human epidermis. J Clin Invest 1991; 88:315-322.
57. Kapprell H, Owaribe K, Franke WW. Identification of a basic protein of Mr 75,000 as an accessory desmosomal plaque protein in stratified and complex epithelia. J Cell Biol 1988; 106:1679-1691.
58. Cartaud A, Ludosky MA, Courvalin JC, Cartaud J. A protein antigenically related to nuclear lamin B mediates the association of intermediate filaments with desmosomes. J Cell Biol 1990; 111:581-588.
59. Ouyang P, Sugrue SP. Identification of an epithelial protein related to the desmosome and intermediate filament network. J Cell Biol 1992; 118:1477-1488.
60. Foisner R, Feldman B, Sander L, Wiche G. Monoclonal antibody mapping of structural and functional plectin epitopes. J Cell Biol 1991; 112:397-405.
61. Skalli O, Jones JCR, Gagescu R, Goldman RD. IFAP 300 is common to desmosomes and hemidesmosomes and is a possible linker of intermediate filaments to these junctions. S Cel Biol 1994; 125:159-170.
62. Boyer B, Thiery JP. Epithelium-mesenchyme interconversion as example of epithelial plasticity. APMIS 1993; 101:257-268.

CHAPTER 10

Structure and Assembly of Gap Junctions

Linda S. Musil

1. SUMMARY

Gap junctions are the cell-cell junctions most specialized for intercellular communication within epithelia and other tissues. A combination of biochemical, morphological, and physiological approaches have been used to establish that gap junctional cell-cell channels are composed of members of a family of closely related integral plasma membrane proteins known as connexins. This chapter focuses on recent advances in our knowledge of the multi-step process whereby connexins are assembled into gap junctions and how gap junction function is regulated at a post-translational, pre-gating level.

2. INTRODUCTION

Gap junctions are collections of plasma membrane channels that directly link the cytosols of two adjoining cells (for recent reviews, see refs. 1-3). In epithelial cells, gap junctions are part of the junctional complex located on the lateral membranes. Found in all animals ranging from coelenterates to man, gap junctions are present in virtually all cell types in higher animals except circulating blood cells and fused skeletal muscle.

Although connexin type-specific differences may exist, gap junctional intercellular channels are generally considered to be relatively non-selective. Among the substances that have been proposed or experimentally demonstrated to pass through gap junctions are ions (including H^+, Ca^{2+}, K^+, Na^+, Cl^-), monosaccharides, amino acids, IP_3, and cyclic nucleotides such as cAMP.[4,5] Experimentally, the most commonly used assays for measuring gap junctional communication are electrical coupling (indicative of the passage

Molecular Mechanisms of Epithelial Cell Junctions: From Development to Disease
Edited by Sandra Citi, M.D., Ph.D.; © 1994, R.G. Landes Company.

of current-carrying ions through gap junctions) and intercellular transfer of microinjected membrane-impermeant fluorescent dyes such as Lucifer Yellow (MW = 457 D). Permeation studies have established that the maximum size of molecules able to pass through gap junctions in vertebrate cells is ~1 kD.[6] This limit assures that complex macromolecules are retained in their cell of origin and that each cell maintains its genetic and structural individuality.

A crucial feature of gap junctional intercellular communication is that it can be reversibly regulated by an enormous variety of effectors, at least under experimental conditions. These include intracellular acidification, elevation of free Ca^{2+}, voltage, and cyclic nucleotides, as well as certain exogenous agents such as phorbol esters, retinoic acid, transforming viral oncogene products ($pp60^{v\text{-}src}$), growth factors (EGF, PDGF) and some classes of lipophilic compounds (e.g., n-heptanol, n-octanol, halothane; arachidonic acid) (reviewed in refs. 7 and 8). Many of these agents alter junctional permeability very rapidly and are thought to mediate direct gating of the gap junctional channel (discussed by Spray in chapter 11 of this volume); other effectors work on a longer time scale and could be involved in gap junction assembly or in regulating the transcription and/or translation of gap junction proteins (for reviews, see Musil and Goodenough,[9] and Stagg and Fletcher[10]). These multiple levels of control emphasize that gap junctions are not passive cell-cell tubes but rather dynamic, highly regulatable structures, and are in this respect similar to many other types of cell surface channels.

As might be expected from the nearly ubiquitous distribution of gap junctions and the variety of substances that can be transferred through them, a wide spectrum of potential functions has been ascribed to these structures. Among the least contentious roles proposed for gap junctions are in electrically excitable tissues, where it is well established that gap junctions serve as low-resistance conduits for the propagation of the electrical signal. In heart[11] and smooth muscle (in particular myometrium),[12,13] gap junctions are recognized as being critically involved in the synchronization of muscle contraction. In nerve, gap junctions form electrical synapses and are components of mixed synapses, both of which are fairly widespread in certain regions of the central nervous system (reviewed by Dermietzel and Spray in ref. 14. In nonexcitable tissues, gap junctions are thought to be generally involved in relaying signals between, as well as maintaining metabolic continuity within, connected cells. In this capacity, gap junctions have been implicated in such basic biological processes as growth control, embryonic development, and tissue differentiation. For example, gap junctional communication is first evident in mammals at the eight-cell stage,[15] and several lines of evidence indicate that perturbing gap junctional coupling in early vertebrate embryos has profound effects on subsequent development.[16-18] In certain insect larvae, gap junctions have been shown by dye transfer studies to delineate communication compartments between cells of different developmental lineages in the absence of anatomical boundaries[19,20] (see also chapters 1, 2 and 3 in this volume). In addition, there is an extensive literature documenting an inverse correlation between gap junctional communication and cell transformation in many (though not all) systems (reviewed by Loewenstein and Rose in ref. 21 and by Yamasaki in ref. 22). It is thought that gap junctions serve as conduits for the intercellular transfer of as yet undefined growth-controlling substances, a possibility that has recently been supported by the demonstration that expression of connexins in previously gap junction-null neoplastic cells reduces their ability to form tumors in vivo.[23-25]

In adult differentiated epithelial tissues, gap junctions appear to serve a variety of important functions. There is extensive evidence that gap junctions coordinate secretion in pancreatic acinar cells, in which a tight correlation between the increased secretion of amylase and the inhibition of junctional communication has been elegantly demonstrated. Conversely, in pancreatic B-cells, stimulus-induced secretion of insulin is associated with an increase in cell-cell coupling (reviewed by Bruzzone and Meda in ref. 26). Physiological and pharmacological data also implicate gap junctions in the regulation of secretion in other epithelial organs, most notably in salivary[27] and lacrimal[28] glands, and in the neural regulation of carbohydrate metabolism in liver.[29,30] Gap junctions are differentially expressed in the various layers of the skin, where they may play a role in the establishment or maintenance of the division/differentiation program of the epidermis.[31] In the lens, gap junctions allow the anterior monolayer of epithelial cells to supply nutrients to, and remove waste products from, the underlying metabolically dormant mature lens fiber cells and are thus thought to be essential for maintaining the organ's optical clarity (reviewed by Goodenough in ref. 32). In addition, gap junctions are required for the propagation of Ca^{2+} waves through airway epithelial cells, where mechanical stimulation is transduced into increased ciliary beat frequency via gap junction-mediated intercellular transfer of IP_3.[33] One recent report even raises the possibility that gap junctions may help to restore normal chloride flux in CFTR-deficient immortalized cystic fibrosis airway epithelial cells after transfection of neighboring cells with the CFTR gene.[34] However, it should be stressed that none of the functions attributed to gap junctions in epithelial (or other) tissues has been unequivocally demonstrated, a general problem in the gap junction field. Unlike the case with many other channel proteins, there is no known pharmacological blocker that is completely specific to gap junctions; all of the inhibitors of intercellular communication cited above alter other cellular processes as well and it is often difficult to definitively distinguish between junctional and non-junctional effects. An alternative approach to manipulating gap junction permeability has been the use of anti-connexin specific antibodies, but these reagents are experimentally impractical in many systems and are prone to artifacts (discussed by Warner in ref. 35). The molecular cloning of connexins (see below) has very recently allowed the development of new techniques and reagents for the specific perturbation of gap junction expression (antisense oligonucleotides, transgenic and knock-out mice; mutant forms of connexins with dominant-negative activity) and is rapidly leading to a more detailed understanding of the role of gap junctions in both epithelial and non-epithelial cells.

3. GAP JUNCTION STRUCTURE

By thin-section electron microscopy, gap junctions are characterized as close appositions of a pair of plasma membranes separated by a lanthanum-permeable 2-4 nm space (the "gap") (reviewed by Bennett and Goodenough in ref. 4). Where the plane of section passes tangentially to the plasma membrane, lanthanum staining reveals an array of 8-9 nm subunits, each of which possesses a central density. Freeze-fracture of gap junctions yields a P-face, containing clustered intramembranous particles that correspond to the subunits visualized by thin-section, and a complementary array of pits on the E-face. The number of subunits per gap junction is quite variable and appears to be somewhat tissue-specific, ranging from the smallest clusters that can be reliably identified as gap junctions (as in pancreatic B cells) to plaques containing tens of thousands of such subunits (e.g., liver).

A critical advance in the understanding of gap junctions was the development of methods for the isolation of gap junctions from rodent liver.[36,37] The original procedure was based on the relative resistance of liver gap junctions to solubilization by various detergents, including rather harsh reagents such as sarcosine. Although the purity of these preparations is excellent, the yield is extremely low (perhaps 0.1%; discussed in Musil and Goodenough in ref. 38), raising the possibility that only a specialized subset of gap junctions is recovered. Hertzberg[39] pioneered a detergent-free isolation procedure that utilizes NaOH to extract peripheral proteins, with a ~10-fold increase in yield over sarcosine-based methods but with a somewhat higher level of non-junctional contamination. Variations of both procedures have been developed for other tissues, notably rodent heart.[40] In contrast to most cell types, lens fiber cells have gap junctions that are soluble even in non-ionic detergents such as Triton X-100.[41] The physical basis for these differences in detergent sensitivity is not known, but may involve the greater degree of disorder in the packing of subunits in fiber cell gap junctions relative to those of liver- or heart- derived junctions.

Negative-stain and freeze-fracture images of purified gap junction preparations revealed that their subunits had become tightly packed into highly ordered, hexagonal arrays as a result of the isolation procedure (reviewed by Caspar et al).[42] The crystalline nature of these lattices permitted structural analysis of isolated gap junctions using Fourier techniques. A combination of X-ray diffraction and electron microscopy was used by Makowski et al[43] in 1977 to develop a low resolution (25 Å) model of mouse liver gap junctions that is still the currently accepted view of gap junction structure (Fig. 10.1). According to this model, gap junctions are collections of tightly packed intercellular channels, with lattice constants ranging from ~7.5 nm to 9.5 nm depending (at least in part) on details of the specimen preparation. Each channel is composed of two half-channels (connexons), one embedded in each cell's plasma membrane, which are joined head-to-head to form a intercellular pore. It is these connexons that are visualized by thin-section EM as stain-excluding subunits or by freeze-fracture as intramembranous particles. Each connexon is an annular oligomer

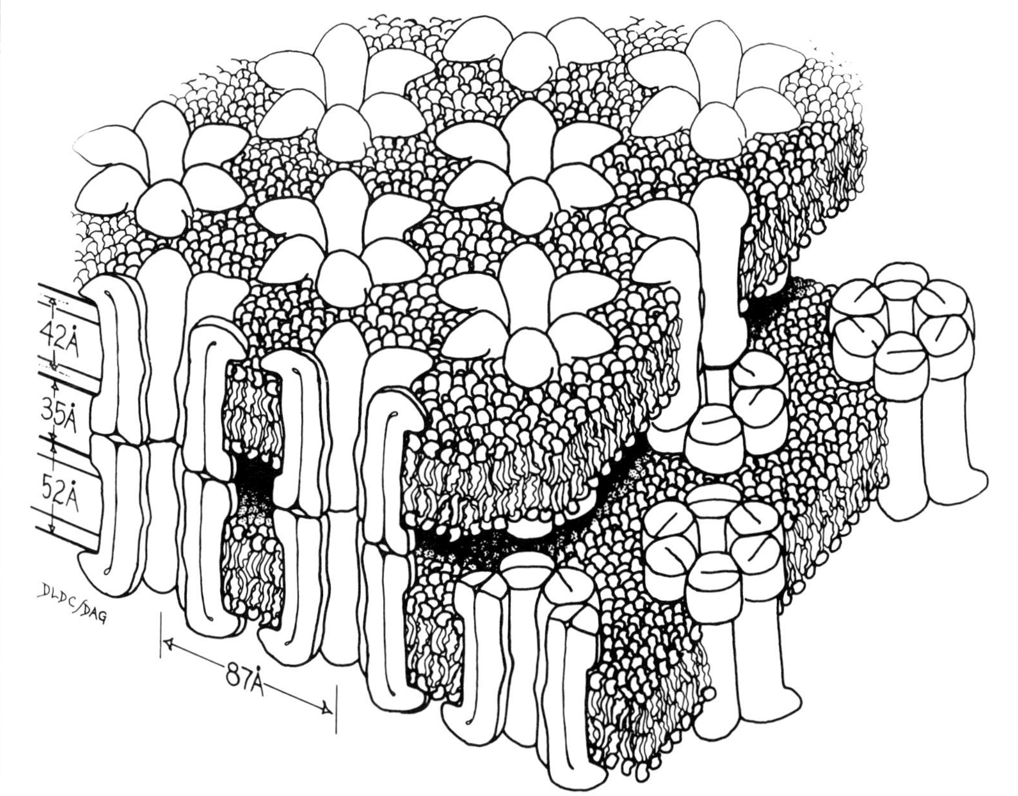

Fig. 10.1. Schematic view of gap junction structure based on correlative EM and X-ray diffraction studies of isolated rodent liver gap junctional plaques. In this model, each connexon is about 8 nm long and has an outer diameter of 6 nm encircling a 2.0 nm (maximum) diameter pore. The mean center-to-center spacing (lattice constant) between connexons ranges from 7.5-9.5 nm in purified gap junctions, but is >10 nm prior to gap junction isolation. The intercellular distance between apposed plasma membranes in the gap junctional plaque is between 2 and 4 nm. From Makowski et al.[43]

of integral membrane proteins that delineate the central aqueous pore. When the molecular weight of a mouse liver gap junction connexon deduced from X-ray diffraction (~140-170 kD) was compared to the then available SDS-PAGE estimate for the mass of the major component of these gap junctions (~27 kD; see below), a model of six integral membrane protein molecules per connexon was established. This stoichiometry was supported by low-dose Fourier microscopy[44,45] and by correlation averaging of frozen-hydrated samples of isolated liver gap junctions.[46]

It should be stressed that in the structural studies cited above, the averaged image for a connexon generated by either Fourier or correlation averaging techniques is unequivocally hexameric only after six-fold (p6) symmetry has been imposed. In the absence of such symmetrization, the stoichiometry of a connexon is often much less clear (see, for example, Sosinsky et al, ref. 46). Alternative methods for connexon visualization such as atomic force microscopy[47] are also unable to resolve the substructure of a connexon to the degree required to determine connexon stoichiometry. The difficulties encountered in using averaging techniques to define connexon structure are due to the fact that the hexagonal array into which connexons in isolated rodent liver gap junctions pack is imperfect even under optimal specimen preparation conditions. Specifically, although gap junctional arrays show well-defined long-range order over hundreds of connexon units, the lattices appear to have substantial short-range disorder.[42] This disorder is likely to be composed of both rotational and translational components, a particularly difficult combination with the potential for creating a false image of a connexon using averaging techniques. Preliminary results obtained using an alternative method for determining the stoichiometry of a large population of connexons, chemical cross linking, raise the possibility that connexons are actually composed of five rather than six molecules (LS Musil, L Makowski, and DA Goodenough, unpublished results).

4. COMPONENTS OF GAP JUNCTIONS: THE CONNEXIN FAMILY OF GAP JUNCTION PROTEINS

When gap junctions isolated from rodent liver were dissolved in SDS and analyzed by PAGE, the major component of these preparations was found to be an ~27 kD protein.[37] Antibodies that specifically recognized this species on immunoblots were generated using isolated liver gap junctions as the immunogen and were shown by immunohistochemistry and immunocytochemistry to selectively stain hepatic gap junctions. These antibodies were then used by Paul[48] to screen a rat liver lambda gt11 expression library, from which a clone encoding the major hepatic gap junction protein was isolated. A highly homologous sequence was independently recovered from a human liver cDNA library using synthetic nucleotide probes based on known hepatic gap junction amino acid sequences.[49] Both the rat and human clones encoded a 32 kD protein whose hydrophobicity plot predicted four membrane-spanning domains. This protein was named connexin32, from the Latin "nexus" (bond) combined with the deduced molecular weight of the gene product. This convention remains the predominant nomenclature used to name gap junction proteins, although an alternative classification based on sequence homologies has been proposed.[50] The discrepancy between the predicted size of connexin32 based on the cDNA clone (32 kD) and its apparent molecular weight on SDS-PAGE (27 kD) has been shown to be due to the anomalous electrophoretic migration of connexin32 in low-percentage polyacrylamide gels.[51]

By screening a rat heart cDNA library under reduced stringency conditions with the connexin32 cDNA clone, the next member of the connexin family, connexin43, was identified by Beyer et al.[52] More recently, both conventional and PCR-based cloning techniques have been used to extend the number of known rodent connexins to 12 (Table 10.1) and it is likely that additional connexins exist. Connexin sequences have also been identified from several other vertebrate species including human, cow, chick, and frog. Some of the encoded proteins are clear homologues of rodent connexins (e.g., human connexin43 differs from rat connexin43 by only 9 out of 382 amino acids;[53] chick from rat connexin43 by 34 residues),[54] but in other cases they appear to be novel connexins (e.g., *Xenopus* connexin38)[55] or their relationship to known rodent connexins is ambiguous. All members of the connexin family described to date have sequences predicted to cross the membrane bilayer four times. Topological mapping using a combination of site-specific antibodies and protease digestion have confirmed this orientation for connexin32[56-58] and connexin43,[59,60] and have established that connexins are arranged in the membrane such that both the amino and carboxyl termini face the cytosol and are separated by two small extracellular domains and an intervening, intracellular loop (see Fig. 10.2). The sequences of all known connexins are highly homologous to each other in their transmembrane and extracellular regions and to a lesser extent at the amino terminus, but diverge from each other markedly in the cytosolic loop and carboxyl tail domains (discussed by Bennett et al in ref. 3 and Kumar and Gilula in ref. 50). It is thought that the sequence variability in these latter regions accounts for the differential sensitivity of connexin family members to various junctional perturbants, including $pp60^{v\text{-}src}$ [61] and intracellular pH.[62] Structural features conserved among the various connexins include six extracellular cysteine residues (three on each loop), some of which appear to participate in intramolecular disulfide bond formation.[63,64] In all known connexins, the third transmembrane region can be modeled as an amphipathic α helix and is therefore thought to line the aqueous gap junctional pore.[3,50] This possibility is supported not only by ample precedent that such domains participate in channel formation in other proteins, but also by circular dichroism[65] and X-ray diffraction[66] studies indicative of a significant amount of helical structure in isolated gap junction preparations. Evidence that connexin transmembrane domains form β pleated sheets has been re-interpreted as consistent with tilted α helices.[66] Despite their ability to be gated by changes in transjunctional voltage, connexin molecules lack the classic S4 segment that acts as a voltage sensor in other classes of cell surface channels. Recent mutational analysis has implicated both extracellular and transmembrane regions in conferring connexin type-specific differences in voltage sensitivity.[67,68] No asparagine-linked glycosylation consensus sequences are present in the extracellular domains of connexins, consistent with the lack of biochemical evidence for the presence of carbohydrate in isolated gap junctions.[37] Connexins also do not possess a classic N-terminal signal sequence and do not appear to normally be substrates for signal peptidase, although such processing may occur when connexins are translated in vitro or over-expressed in tissue culture cells.[69]

Curiously, although clearly recognizable connexins have been found in animals as primitive as fish, no proteins homologous to connexins have been identified in invertebrates by either antibody cross-reactivity or by low-stringency molecular cloning despite intensive effort. Thus the differences in the permeability and ultrastructure of gap junctions from invertebrates and from vertebrates appear to be due to very different protein components. Isolation of gap junctions from several invertebrates has been described (e.g. lobster,[70] *Drosophila*;[71] tobacco budworm[72]) (see also chapters 1 and 2 in this volume), making it likely that definitive identification of the molecular constituents of these junctions will soon be achieved.

5. POST-TRANSLATIONAL MODIFICATIONS OF CONNEXINS

As discussed above, there is no evidence that any of the connexins bear either N- or O- linked oligosaccharides. Although highly unusual for an integral plasma membrane protein, the absence of glycosylation might be a necessity for connexins since repulsion between highly charged carbohydrate chains on apposing or adjacent connexons could interfere with gap junction assembly. One report has described preliminary evidence for fatty acylation of connexin32 and connexin26, but this observation has not been pursued

Table 10.1. Connexin family members identified by cloning from rodents

RODENT CONNEXINS

NAME	TISSUE LOCATION	REFERENCE
Cx26	HEPATOCYTES, MAMMARY EPITHELIA, Leptomeninges, Pinealocytes, Kidney, Intestine	Zhang and Nicholson, 1989 J.Cell Biol. 109:3391
Cx30.3	SKIN	Henneman et al., 1992 Eur. J.Cell Biol. 57:51
Cx31	SKIN, Placenta, Harderian gland	Hoh et al., 1991 J.Biol.Chem. 266:6524
Cx31.1	SKIN (Keratinocytes), Esophagus	Haefliger et al., 1992 J.Biol.Chem. 267:2057
Cx32	HEPATOCYTES, EXOCRINE PANCREAS, MAMMARY EPITHELIA	Paul, 1986 J.Cell Biol. 103:123 Kumar and Gilula, 1986 J.Cell Biol. 103:767
Cx33	TESTIS	Haefliger et al., 1992 J.Biol.Chem.267:2057
Cx37	VASCULAR ENDOTHELIUM	Willecke et al. 1991 J.Cell Biol. 114:1049
Cx40	VASCULAR ENDOTHELIUM	Haefliger et al., 1992 J.Biol.Chem. 267:2057
Cx43	SMOOTH AND CARDIAC MUSCLE, Fibroblasts, Ovarian granulosa, Lens and corneal epithelium, Kidney, Endocrine pancreas, Astrocytes, Testis	Beyer et al., 1987 J.Cell Biol. 105:2621
Cx45	Lung, Brain, Skin, Heart, Intestine	Hennemann et al. 1992 Eur. J.Cell Biol. 57: 51
Cx46	LENS FIBERS, Heart, Kidney	Paul et al., 1991 J.Cell Biol. 115: 1077
Cx50	LENS FIBERS, Atrio-ventricular valves	White et al., 1992 Mol.Biol.Cell 3: 711

Tissues in which the indicated connexin is expressed at high levels are listed in capitals. The tissue distributions given for the various connexins are by no means complete and are provided only to give the reader a general sense of the location of various connexins throughout the body. Note that these distributions are, for the most part, complex and overlapping; many organs, and probably individual cells as well, simultaneously express more than one type of connexin. For example, double-labeling immunocytochemistry has shown co-localization of connexin32 and connexin26 in the same mouse liver gap junctional plaque.[73] Similarly, connexins 43, 40, and 45 have been reported to codistribute in adult dog heart intercalated discs.[74] References contain sequence information for the indicated connexin.

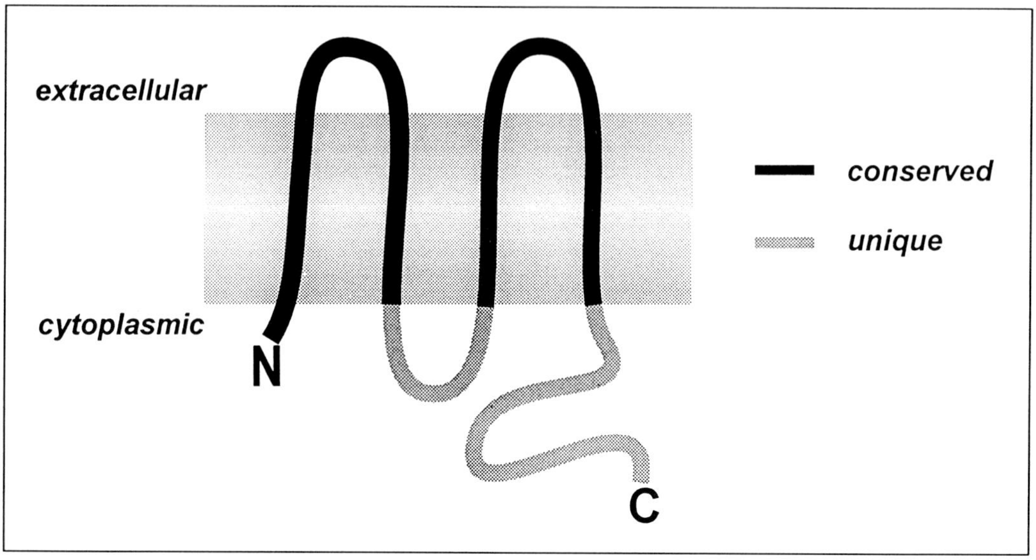

Fig. 10.2. Generalized connexin topology. Regions of greater (solid black line) and lesser (gray stippled line) sequence homology among connexin family members are indicated. The lengths of the intracellular amino terminal region (~23 amino acids) and of the two extracellular domains (~35-40 residues) are very similar for all members of the connexin family. In contrast, the cytosolic loop varies from ~36 (rat connexin32) to ~78 (chick connexin45) amino acids, and the carboxyl tail region from a minimum of 16 residues (connexin26) up to 187 residues (rat connexin46).

further.[73] The only other non-phosphorylation post-translational modification reported for connexins is the proteolytic cleavage of connexin50 (MP70) in mature fiber cells by a calcium-dependent lens protease.[75] Proteolytic processing does not appear to be a general feature of the connexin family but rather to be specific to the lens, where many other classes of proteins are also cleaved within the lens nucleus.

By far the most characterized post-translational modification of connexins is phosphorylation, which can occur (depending on the connexin and the conditions) on either serine or tyrosine residues, with perhaps a minor component of threonine processing in some instances. Serine phosphorylation has repeatedly been demonstrated for both connexin43 and connexin32 (see below) and has also been reported for the lens-type connexins connexin50 (MP70)[76] and connexin46.[77] Preliminary reports suggest that most members of the connexin family are phosphoproteins; in fact, the only connexin for which lack of phosphorylation has been reported is connexin26.[73,78]

Serine phosphorylation was first demonstrated for connexin32 in primary cultures of rodent hepatocytes,[79,80] where the observation that cAMP increased both connexin32 phosphorylation and gap junctional conductance led Saez et al[79] to speculate that protein kinase A-mediated connexin phosphorylation is involved in the gating of junctional channels. More recently, correlative in vivo and in vitro mapping studies have identified ser233 as the major site phosphorylated in intact hepatocytes in response to cAMP.[78] Site-directed mutagenesis of this residue in conjunction with ser240 apparently does not influence the permeability or pH-dependent gating properties of connexin32 when expressed in paired *Xenopus* oocytes,[81] suggesting that the role (if any) of ser233 phosphorylation may be subtle or specific to particular effectors or cell types. TPA, an activator of protein kinase C, also increases the phosphorylation of ser233 as well as of additional serines that are not modified in the presence of cAMP.[78] The fact that TPA, unlike cAMP, does not increase junctional conductance in hepatocytes has led Saez et al[78] to suggest that phosphorylation of these cAMP-insensitive serine residues accounts for the different physiological responses to the two effectors. Although unknown in intact cells, the stoichiometry of phosphorylation of connexin32 in purified gap junctions by either protein kinase A or C is relatively low (~5% for PKA; 10-30% for PKC).[78,82]

Connexin43 is synthesized as a ~42 kD species and is post-translationally converted to two forms with slightly increased apparent molecular weights (connexin43 -P_1 and -P_2, in the nomenclature of Musil and Goodenough)[83] by the addition of phosphate onto serine residues.[54,83,84] Phosphorylation to the P_2 form occurs after transport of connexin43 to the cell surface and is concomitant with incorporation of connexin43 into gap junctional plaques.[38] The phosphorylation sites have not yet been fully mapped, but are likely to be clustered between residues 271-382 of the carboxyl tail domain based on preliminary biochemical analysis[60] and on the presence of consensus sequences for protein kinases in this region.[53] Unlike connexin32, phosphorylation of connexin43 has not been reported to be significantly stimulated by activators of protein kinase A, despite the ability of cAMP to increase junctional communication in some connexin43-expressing cells (e.g., cardiac myocytes).[85] Instead, enhanced serine phosphorylation of connexin43 induced by either TPA[86,87] or EGF[88] is temporally associated with a decrease in junctional permeability. In these instances, the electrophoretic mobility and phosphopeptide map of [^{32}P]-labeled connexin43 from treated cells differs markedly from that of basally phosphorylated controls, suggesting modification of novel serine residues. The relationship between connexin43 serine phosphorylation and gap junctional permeability is therefore not a simple one and involves multiple phosphorylation sites as well as tissue-specific differences in the response of cells to various kinases (discussed by Bennett et al in ref. 3 and Musil and Goodenough in ref. 9). In addition, there is also evidence that serine phosphorylation may influence the permeability of connexin43 channels in a more subtle but potentially as physiologically important manner by altering their unitary conductance.[89,90]

A possible role for basal phosphorylation of connexin43 on serine residues in the establishment and/or maintenance of functional gap junctions was suggested by Musil et al.[38,83] They demonstrated that cell lines deficient in gap junctional communication synthesized connexin43 and transported it to the cell surface, but did not phosphorylate connexin43 to the mature (P_2) form nor accumulate it in morphologically detectable gap junctions. Conversion of such cells to a communication-competent phenotype by transfecting them with a cDNA encoding the intracellular adhesion molecule L-CAM led to both connexin43-P_2 phosphorylation and the formation of junctional plaques. Furthermore, communication-competent cell types treated with known inhibitors of gap junction permeability no longer detectably processed newly synthesized connexin43 to the P_2 form. Together, these results established a close correlation between the ability of cells to phosphorylate connexin43 to the P_2 form and to assemble functional gap junctional plaques. This possibility is supported by a report by Oh et al[91] which describes epithelial-derived rat liver cell mutants with apparent defects in both connexin43 phosphorylation and junctional communication.

In comparison to serine phosphorylation, the role of connexin43 phosphorylation on tyrosine residues is relatively well understood. Undetectable under basal conditions, tyrosine phosphorylation of connexin43 (usually accompanied by an increase in serine modification) is stimulated in the presence of catalytically active forms of the transforming tyrosine protein kinase pp60$^{v\text{-}src}$.[61,84] Concomitant with enhanced phosphorylation of connexin43 is a rapid decrease in junctional permeability, which does not require protein synthesis or involve a change in the number of morphologically recognizable gap junctions and is therefore likely to be due to direct gating of the junctional channel.[92] A cause-and-effect relationship between tyrosine phosphorylation of connexin43 and the decrease in junctional permeability induced by pp60$^{v\text{-}src}$ was indicated by the studies of Swenson et al[61] using a paired *Xenopus* oocyte expression system programmed to synthesize connexin43, connexin32, and/or pp60$^{v\text{-}src}$. These authors demonstrated that, unlike connexin43, neither the phosphorylation nor junctional permeability of connexin32 was affected by pp60$^{v\text{-}src}$. This difference in the behavior of connexin43 and connexin32 was exploited to identify a potential tyrosine phosphorylation site on connexin43, tyr265, that is absent in connexin32. Site-directed mutagenesis of tyr265 eliminated both connexin43 tyrosine phosphorylation and the downregulation of cell-cell communication in the presence of pp60$^{v\text{-}src}$. These results demonstrate that tyrosine phosphorylation is necessary for v-src-induced reduction of cell-cell communication, although whether this modification is in itself sufficient to decrease junctional permeability is not yet known. Interestingly, inhibition of junctional communication by PDGF has also been reported to induce tyrosine phosphorylation of connexin43,[93] raising the possibility that tyrosine protein kinases less potent and promiscuous than pp60$^{v\text{-}src}$ may also regulate cell-cell communication by affecting phosphorylation of connexin43. Also intriguing is indirect evidence that the phosphotyrosine phosphatase inhibitor pervandate increases the phosphorylation of connexin43 in SHE and V79 cells;[94] whether this pervanadate-sensitive phosphorylation is indeed on tyrosine residues remains to be determined.

6. GAP JUNCTION ASSEMBLY

Various physiological assays have indicated that connexins are both necessary and sufficient for the generation of gap junctional channels. Among the most compelling evidence is the induction of gap junction-type intercellular communication by expression of connexin cRNA or cDNA in heterologous systems (paired *Xenopus* oocytes or transfected tissue culture cells) in which endogenous gap junctional activity is either naturally very low or has been minimized. In addition, insertion into artificial membranes of isolated gap junctional plaques,[95,96] detergent-solubilized purified gap junction channels or connexons,[97] or (in one instance) gel-purified renatured connexin protein[95] has yielded channels with at least some of the properties characteristic of gap junctions in intact cells. This is in keeping with the absence of any conclusive biochemical or morphological evidence[98,99] that gap junctions contain proteins other than connexins. Thus, unlike other cell-cell junctions, gap junctions do not appear to be associated with any known cytoskeletal structures and their formation is not detectably inhibited by cytoskeletal-disrupting drugs at concentrations that block cell division or shape changes.[100,101]

Several studies have examined the lipid composition of gap junctions and have generally concluded that gap junctions are enriched in cholesterol, and depleted in phospholipids, relative to the general plasma membrane (reviewed by Malewicz et al in ref. 102). Interpretation of these results must, however, be tempered by the fact that these analyses were conducted on isolated gap junctions prepared using either sarcosine- or NaOH-based procedures which would be expected to severely and perhaps unpredictably alter lipid content and structure.

Given the evidence that connexins are the only proteinaceous components of gap junctions, gap junction assembly can be envisioned as a multistage process involving (1) oligomerization of connexin monomers into connexons; (2) transport of connexons to the cell surface; (3) the association of two connexons in adjoining cell surfaces into intercellular channels; and (4) the clustering of these channels into gap junctional plaques. Our present understanding of each of these steps is reviewed separately below. Interestingly, covalent intermolecular bonds do not appear to play a role at any level of gap junction assembly.

6.1. Oligomerization of Connexins into Connexons

To date, only one report examining this process biochemically has been published. Musil and Goodenough[103] used a combination of sucrose gradient velocity sedimentation and chemical cross-linking to study the assembly of newly synthesized connexin43 monomers into 9S, 200 kD complexes. Five lines of evidence indicated that these oligomers were authentic connexons: (1) their electrophoretic mobility and sedimentation coefficient was consistent with those expected for a connexon, given the limitations of SDS-PAGE and rate sedimentation analysis; (2) their detection in cell lysates was dilution-resistant and did not require previous sucrose gradient fractionation; (3) they are non-covalently bound complexes in which connexins are the only proteins detectable; (4) they could be isolated from gap junctional plaques and (5) they are present in all vertebrate cell types examined, including cells that endogenously express connexin43 (NRK, CHO, S180, L929, F9 cell lines; primary bovine aortic endothelial cells) as well as *Xenopus* oocytes programmed with connexin43 cRNA. The authors then examined the intracellular site in which connexin43 was oligomerized into connexons. The dogma that has emerged from a large number of studies is that newly synthesized multi-subunit integral plasma membrane proteins are oligomerized, at least in part, within the endoplasmic reticulum (ER). For most of these proteins, it has been demonstrated that assembly in the ER is a prerequisite for transport through the secretory pathway, with incompletely assembled subunits undergoing intracellular degradation as part of the cell's "quality control" mechanism (reviewed by Hurtley and Helenius in ref. 104). Oligomerization of newly synthesized connexin43 into connexons occurred at a temperature (20°C) that stopped detectable transport of nascent proteins within the *trans* Golgi network (TGN), the organelle immediately distal to the stacked Golgi cisternae. Surprisingly, connexon formation was reversibly blocked when ER-to-Golgi transport was inhibited by incubating cells at 15°C, or at 37°C in the presence of either brefeldin A (BFA) or the ATP-depleting ionophore CCCP. Connexin43 assembly was also inhibited at the non-permissive temperature in CHO cell mutants with temperature-sensitive defects in ER-to-medial-Golgi transport. These results define a post-ER assembly pathway for connexin43 and distinguish connexin oligomerization from that of all other known multisubunit integral plasma membrane proteins (summarized in Table 10.2). The finding that brefeldin A inhibits connexon assembly argues that connexon formation occurs after transport through the stacked *cis, medial,* and *trans* Golgi cisternae, since BFA generally does not block processes localized to, or occurring before, these compartments. These data are therefore most consistent with connexon assembly taking place in the TGN, which does not fuse with the ER in the presence of BFA and in which newly synthesized proteins accumulate at 20°C.

Why should connexins, unlike other plasma membrane proteins, be assembled within the TGN? One possibility emerges from studies conducted by Kumar and Gilula in which connexin32 was greatly over-expressed in BHK cells following transfection with connexin32-encoding cDNA. Double-bilayer gap junctions were formed in these transfectants not only at cell-cell interfaces but also, remarkably, within intracellular compartments including the ER and the Golgi.[50,103] One explanation for this phenomenon is that the extreme concentration of newly synthesized connexin32 protein somehow prematurely triggered connexon formation, and that these connexons then associated with each other across the ER or Golgi cisternal lumen to form double-connexon channels which later clustered into junctional plaques. Once incorporated into an intracellular plaque, it would be very difficult for a connexon to dissociate from such a structure and travel to the cell surface to participate in the assembly of functional cell-cell channels. In cells that express endogenous levels of connexins, such futile formation of intracellular gap junctions may be prevented by delaying connexon assembly to the last stage of the secretory pathway, that is, to the TGN. Once formed, these connexons would rapidly be delivered to the cell surface via transport vesicles whose diameter (~100 nm)[105] relative to the length of a connexon (~8 nm) would physically prevent two connexons from associating head-to-head and nucleating an intracellular gap junctional plaque.

Table 10.2

Transport Block	Intracellular Site of Transport Block	Connexon Assembly?	Assembly of IMPs in ER?
20°C	TGN	yes	yes; see 15°C
BFA	before TGN; in fused ER/Golgi compartment	no	yes; T cell receptor (Lippincott-Schwartz et al., 1989) Mannose 6-P receptor (Hille et al., 1990) VSV-G (Doms et al., 1989) F protein (Collins and Mottet, 1991)
40°C, ldlF cells	?; before trans Golgi	no	yes; scavenger receptor (Fisher 1992)
15°C	intermediate compartment	no	yes; VSV-G (Doms et al., 1987) HA (Copeland et al., 1988) Mannose 6-P receptor (Hille et al., 1990)
CCCP	ER	no	yes; HA (Copeland et al., 1988) F protein (Collins and Mottet, 1991)

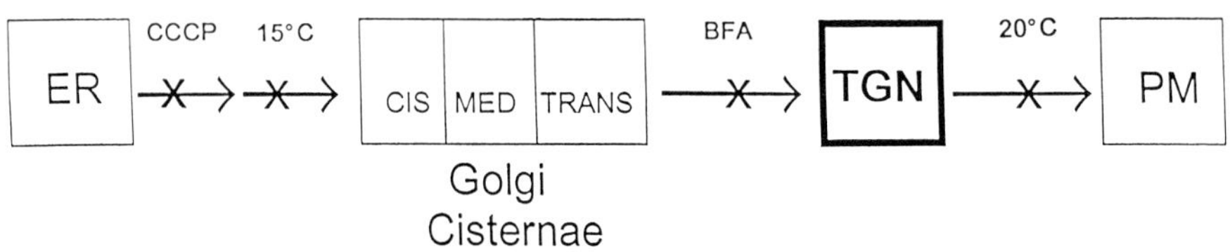

Top, summary of effect of intracellular transport blockers on connexon formation and on oligomerization of other integral plasma membrane proteins (IMPs). Bottom, schematic of the transport of newly synthesized connexin43 protein through the secretory pathway. Full references given in Musil and Goodenough.[103]

6.2. Transport of Connexons to the Plasma Membrane

Using cell-surface biotinylation, Musil and Goodenough[38] demonstrated the presence of a pool of newly synthesized connexin43 on the plasma membrane of NRK cells that was only later incorporated into gap junctional plaques. Since connexon assembly occurs intracellularly (see above), these results were consistent with transport of connexin43 to the plasma membrane in the form of individual connexons. The existence of cell surface extrajunctional connexons (also known as "hemichannels") has long been hypothesized[106] but has been a controversial topic only recently resolved by whole cell physiological studies. De Vries and Schwartz[107] demonstrated that depolarization of single catfish retinal cone horizontal cells in low (≤ 220 μM) calcium buffer induced the opening of large, nonselective cell surface channels with electrophysiological and pharmacological properties indicative of a gap junction hemichannel. Channels with similar properties have been identified in horizontal cells of the skate, although in this species channel opening did not require low extracellular calcium.[108] Perhaps the most dramatic evidence for hemichannels comes from studies conducted by Paul et al,[109] in which expression of high levels of rat connexin46 in single or paired *Xenopus* oocytes resulted in oocyte lysis in low Ca^{2+} medium due to the opening of large numbers of connexon-type cell surface channels.[110] In all three cases, opening of hemichannels was prevented by physiological concentrations of Ca^{2+} and/or maintenance of a normal resting potential; whether additional stimuli could open these channels in vivo remains to be determined. In this vein, there has been one report implicating connexin43 hemichannels as the plasma membrane pores that open in response to extracellular ATP^{-4} in certain cell types.[111] This possibility remains speculative, however, until it can be demonstrated that expression of connexin43 in ATP pore-null cell mutants can confer sensitivity to ATP.

The presence of hemichannels on the cell surface supports a model in which channel precursors are inserted into nonspecialized regions of the plasma membrane from which they can be recruited into areas of cell-cell contact to assemble into gap junctions.[106] Such a cell surface connexon pool is presumably responsible for the formation of gap junctions under conditions (ATP depletion, 4°C) where transport of connexin43 from intracellular compartments is blocked.[112]

6.3. Formation of Intercellular Channels

Once cell-cell contact is established, gap junctional intercellular coupling can reach detectable levels very rapidly (in less than two minutes in some cell types).[106,113] This process requires the specific recognition, and subsequent association, of two connexons on apposing cell surfaces to form an intercellular channel. Antibodies directed against extracellular domains of connexins[114] or synthetic peptides whose sequences correspond to these regions[115] block the establishment of functional gap junctions, presumably by interfering with the assembly of intercellular channels. The residues within the two extracellular loops of connexins that are involved in connexon-connexon association have been studied by expressing cloned wild-type and mutant connexins in the paired oocyte system. Dahl et al[115,116] have reported that site-directed mutagenesis of any one of the six extracellular cysteines that are conserved among all connexins completely blocks the ability of connexin32 to form functional junctions between oocytes. Immunohistochemical analysis indicated that each of these point mutants was transported to the cell surface, suggesting a defect at the level of channel formation and/or permeability. Swenson et al[117] were the first to report that an oocyte programmed to express one connexin could form a functional gap junction when paired with an oocyte expressing a different connexin, as might be expected given the high degree of sequence homology in the extracellular loops of the various members of the connexin family. However, both this as well as later studies[118,119] also demonstrated a surprising selectivity in the ability of different connexins to functionally interact. Although all show some degree of specificity, connexins vary widely in the number of other connexins with which they can form hybrid (heterotypic) channels. For example, connexin46 interacts with itself as well as with connexins 32, 43 and 50 (but not with connexin40), whereas connexin40 functions with itself and connexin37 but not with connexins 32, 43, 46, and 50, at least when expressed in oocytes. Currently it is not possible to predict, based on either sequence similarities or tissue distribution, whether two connexins will be able to form heterotypic channels. In at least some cases, the second extracellular domain appears to be an important determinant: switching the second extracellular loop of connexin50 with that of connexin46 confers upon the chimera the ability to form heterotypic channels with connexin43, a property of wild-type connexin46 but not connexin50.[119] The phenomenon of selective heterotypic gap junction formation is likely, at least in some cases, to play a physiologically important role in dictating whether different types of cells in physical contact can communicate (discussed by Bruzzone et al in ref. 118). Gap junctional communication between heterologous cell types in tissue culture has been described for over 20 years,[120] although it is unclear how widespread this phenomenon is in vivo. Furthermore, the finding that heterotypic gap

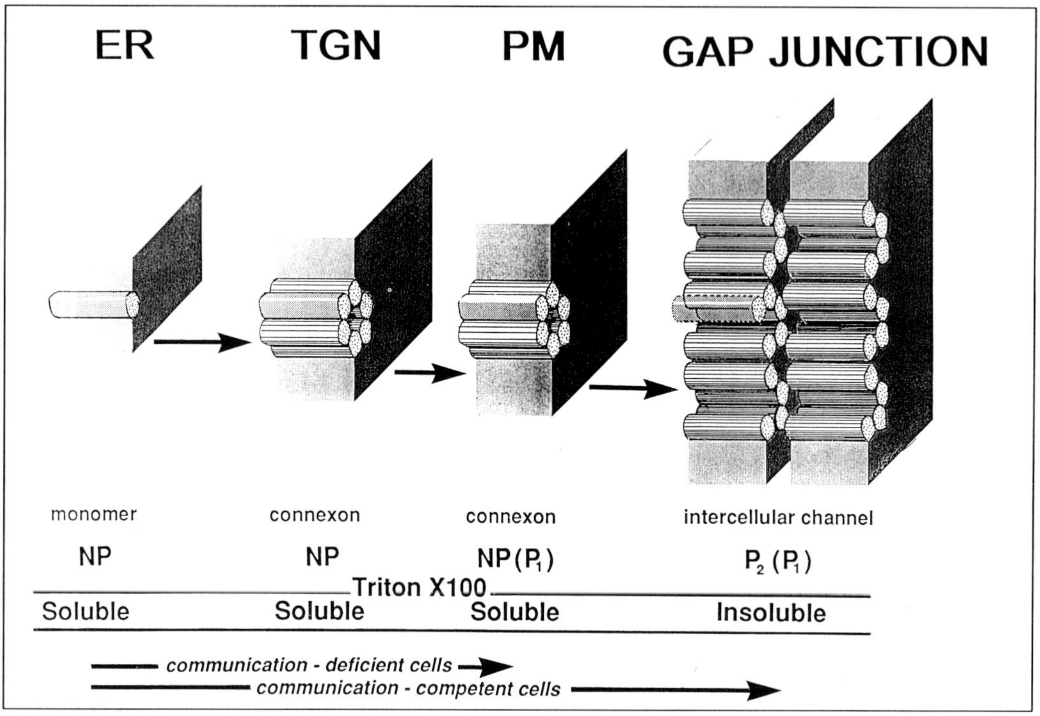

Fig. 10.3. Summary of the biochemically detectable steps in the assembly of connexin43 into gap junctions. Connexin43 is synthesized, presumably in the ER, as a non-phosphorylated (NP) monomer that is soluble in the nonionic detergent Triton X-100 at 4°C. This species is assembled into connexons in a post-ER, pre-plasma membrane compartment that is most likely the trans Golgi network. Connexons are then transported to the cell surface, where at least a fraction of connexin43 is phosphorylated to the intermediate (P_1) form. These assembly and transport steps occur regardless of whether or not the cell is competent to form communicating gap junctions. In communication-competent cells only, cell surface connexin43 is then phosphorylated to the mature P_2 form in a process that is temporally associated with acquisition of Triton insolubility and with accumulation of connexin43 into gap junctional plaques. Data from Musil et al.[38,54,83,103]

junctions often have gating properties (e.g., voltage sensitivity) distinct from those of homotypic junctions composed of either of the parental connexins[121,122] implies that formation of hybrid gap junctions can provide a means of generating novel channel phenotypes. Whether more than one type of connexin can co-assemble to form mixed connexons is not yet known.

6.4. Aggregation of Channels into Gap Junctional Plaques

Once intercellular channels are formed, they become clustered into gap junctional plaques containing up to ~10,000 channels/μm^2, apparently without the benefit of cytoskeletal components or other detectable proteins. How this is accomplished on a molecular level is largely unknown. Abney et al[123] and Braun et al[124] have calculated from statistical mechanical analysis of the positions of intercellular channels in freeze-fracture electron micrographs that interactions between adjacent gap junction channels in a junctional plaque are mutually repulsive at all separations. Plaque assembly therefore cannot be driven by channel-channel attraction. Instead, these authors have proposed that channel clustering is an indirect result of intercellular repulsion between the highly charged cell surface glycoproteins and glycolipids that make up the glycocalyxes of the two participating cells. According to their model, formation of individual cell-cell channels causes the intercellular space between the two cells to be less than the thickness of each cell's glycocalyx. This results in repulsion between the two cell surfaces, which clears the cell-cell interface of most charged plasma membrane components. In fact, such intramembranous particle-poor cell-cell contact areas have been observed by freeze-fracture and are referred to as "formation plaques," although it is not clear if these plaques are an obligatory intermediate in the assembly of all gap junctions.[125] Membrane-membrane repulsion is minimized by reducing the contact area between the two cells, thereby providing the motive force that packs intercellular channels into dense arrays despite the lack of interchannel attractive forces. One consequence of channel clustering is that even if the interactions between the extracellular domains of two apposing connexons are relatively weak, multiplication of these forces by concentration of channels into junctional plaques is likely to result in a high-affinity interaction that stabilizes the gap junction against splitting and other forms of mechanical disruption. Two types of experiments have addressed the role of glycoproteins in the formation of gap junctions. Levine et al[126] have reported that pretreatment of *Xenopus* oocytes with various lectins increases the rate at which junctional conductance increases after microinjection of connexin-encoding cRNAs. In a preliminary report, Wang and Mehta have provided evidence that treatment of various tissue culture cells with the N-linked glycosylation inhibitor tunicamycin also stimulates the ability of cells to form functional gap junctions.[127] In both cases, the enhancement of junctional permeability was attributed to an increased rate of junction formation due to removal of highly charged glycoproteins from the cell surface. Whether these results indicate that normal channel clustering can occur in the absence of glycocalyx repulsion remains to be determined.

Upon incorporation into morphologically recognizable gap junctional plaques, connexin43 undergoes a change in detergent solubility and becomes resistant to Triton X-100 at 4°C.[38] Although a comparable study of detergent sensitivity has not been reported for other connexins, it is clear that at least a fraction of connexin32- and connexin26- containing gap junctional plaques are also insoluble in Triton as well as in most other detergents.[36,37] The physical basis for this detergent resistance is unknown. It is not an inevitable consequence of tight packing of transmembrane proteins in the plane of the plasma membrane since the nicotinic acetylcholine receptor remains Triton soluble even after assembly into extremely high-density cell surface clusters.[128] For several membrane proteins, association with the cytoskeleton is thought to confer resistance to detergent solubilization; however, as already discussed, there is no evidence for interaction between connexins and cytoskeletal elements. One possibility is that junctional plaques are highly enriched in detergent-resistant glycosphingolipids or sphingomyelin, although evaluation of this explanation awaits definitive determination of the lipid composition of gap junctions.[102] It should be mentioned that acquisition of detergent resistance is not a universal feature of gap junction formation, since junctional plaques composed of the lens fiber cell connexin50 (MP70) are dissociated into individual intercellular channels by NP-40 (Triton X-100).[41]

7. FUTURE DIRECTIONS: CROSS-TALK BETWEEN GAP JUNCTIONS AND OTHER CELL-CELL JUNCTIONS

The "precedence hypothesis"[129] states that the establishment of intercellular adhesion via cell-cell adhesion molecules (CAMs) is a necessary prerequisite for the formation of other intercellular junctions. Several lines of evidence indicate that gap junctions follow this rule: when CAM expression or function is inhibited, morphologically and physiologically recognizable gap junctions are reduced.[83,114,130] One role for CAMs in the formation of gap junctions is most likely direct mediation of stable cell-cell contact. An intriguing possibility is that CAMs may also influence gap junction expression by participating in signal transduction cascades, analogous to the signaling function proposed for CAMs in other cellular events[131,132] including tight junction assembly/disassembly (see also chapters 5 and 6 in this volume). Even more provocative is evidence that signaling interactions between CAMs and connexins may be bi-directional. Meyer et al[114] have reported that antibodies to the extracellular domains of connexin43 inhibit not only gap junction assembly but also, unexpectedly, the formation of ultrastructurally recognizable adherens junctions. Paul et al[133] microinjected cRNA encoding a connexin chimera with dominant negative activity into dorsal anterior cells of the eight-cell *Xenopus* embryo and observed (in addition to a reduction in junctional communication) disorganization of the derived ectoderm and eventual delamination and extrusion of the progeny of the injected cell. The extruded cells were viable, indicating that inhibition of gap junction formation and/or function had an effect on cell-cell adhesion rather than on cellular survival. Both types of experiments might suggest a previously uncharacterized mechanical or structural link between gap- and adherens-type junctions; additionally or alternatively, perturbation of gap junction assembly may disrupt as yet unknown signal transduction pathways necessary for the establishment and/or maintenance of adherens junctions. It is anticipated that future research will strengthen the concept that the various classes of cell-cell junctions in epithelia are not autonomous structures but rather part of a interrelated, dynamic network with both mechanical and signaling functions. Given that so many major developmental processes in the body involve direct cell-cell interactions (e.g., compaction, establishment of polarized epithelia; some inductive and pattern-forming events), it is perhaps not surprising that coordination between intercellular junctions occurs on many levels.

References

1. Gilula NB, ed. Gap Junctional Communication. Sem Cell Biol 1992; 3:1-91.
2. Hall JE, Zampighi GA, Davis RM, eds. Gap Junctions. Progress in Cell Research, vol 3. Elsevier, Amsterdam, 1993:1-327.
3. Bennett MVL, Barrio LC, Bargiello TA, Spray DC, Hertzberg E, Saez JC. Gap junctions: new tools, new answers, new questions. Neuron 1991; 6:305-320.
4. Bennett MVL, Goodenough DA. Gap junctions, electrotonic coupling, and intercellular communication. Neurosciences Res Prog Bull 1978; 16:373-486.
5. Saez JC, Connor JA, Spray DC, Bennett MV. Hepatocyte gap junctions are permeable to the second messenger, inositol 1,4,5-triphosphate, and to calcium ions. Proc Natl Acad Sci 1989; 86:2708-2712.
6. Simpson I, Rose B, Loewenstein WR. Size limit of molecules permeating the junctional membrane channels. Science 1977; 195:294-296.
7. Kolb HA, Somogyi R. Biochemical and biophysical analysis of cell-to-cell channels and regulation of gap junctional permeability. Rev Physiol Biochem Pharmacol 1991; 118:1-47.
8. Bennett MLV, Verselis VK. Biophysics of gap junctions. Sem Cell Biol 1992; 3:29-47.
9. Musil LS, Goodenough DA. Gap junctional intercellular communication and the regulation of connexin expression and function. Curr Opin Cell Biol 1990; 2:875-880.
10. Stagg RB, Fletcher WH. The hormone-induced regulation of contact-dependent cell-cell communication by phosphorylation. Endocr Rev 1990; 11:302-325.
11. De Mello WC. Intercellular communication in heart muscle. In: De Mello WC, ed. Intercellular Communication. New York: Plenum Publishing Corp, 1977:87-125.
12. Garfield RE, Blennerhassett MG, Miller SM. Control of myometrial contractibility:role and regulation of gap junctions. Oxford Rev Reprod Biol 1988; 10:436-490.
13. Garfield RE, Sims S, Daniel EE. Gap junctions: their presence and necessity in myometrium during parturition. Science 1997; 198:958-959.
14. Dermietzel R, Spray DC. Gap junctions in the brain: where, what type, how many and why? Trends Neurosci 1993; 16:186-192.
15. Lo CW, Gilula NB. Gap junctional communication in the preimplantation mouse embryo. Cell 1979; 18:339-409.
16. Buehr M, Lee S, McLaren A, Warner A. Reduced gap junctional communication is associated with the lethal condition characteristic of DDK mouse eggs fertilized by foreign sperm. Development 1987; 101:449-459.

17. Warner AE, Guthrie SC, Gilula NB. Antibodies to gap-junctional protein selectively disrupt junctional communication in the early amphibian embryo. Nature 1984; 311:127-131.
18. Bevilacqua A, Loch-Caruso R, Erickson RP. Abnormal development and dye coupling produced by antisense RNA to gap junction protein in mouse preimplantation embryos. Proc Natl Acad Sci USA 1989; 86:5444-5448.
19. Warner AE, Lawrence PA. Permeability of gap junctions at the segmental border in insect epidermis. Cell 1982; 28:243-259.
20. Blennerhassett MG, Caveney S. Separation of developmental compartments by a cell type with reduced junctional permeability. Nature 1984; 309:361-364.
21. Loewenstein WR, Rose B. The cell-cell channel in the control of growth. Seminars in Cell Biol 1992; 3:59-79.
22. Yamasaki, H. Gap junctional intercellular communication and carcinogenesis. Carcinogenesis 1990; 11:1051-1058.
23. Naus CC, Elisevich K, Zhu D, Belliveau DJ, Del Maestro RF. In vivo growth of C6 glioma cells transfected with connexin43 cDNA. Cancer Res 1992; 52:4208-4213.
24. Eghbali B, Kessler JA, Reid LM, Roy C, Spray DC. Involvement of gap junctions in tumorigenesis: transfection of tumor cells with connexin32 cDNA retards growth in vivo. Proc Natl Acad Sci 1991; 88:10701-10705.
25. Rose B, Mehta PP, Loewenstein WR. Gap-junction protein gene suppresses tumorigenicity. Carcinogenesis 1993; 14:1073-1075.
26. Bruzzone R, Meda P. The gap junction: a channel for many functions? Eur J Clin Invest 1988; 18:444-453.
27. Hammer MG, Sheridan JD. Electrical coupling and dye transfer between acinar cells in rat salivary glands. J Physiol 1978; 275:495-505.
28. Randriamampita C, Giaume C, Neyton J, Trautmann A. Acetylcholine-induced closure of gap junctional channels in rat lacrimal glands is probably mediated by protein kinase C. Pflugers Arch 1988; 412:462-468.
29. Seseke FG, Gardemann A, Jungermann K. Signal propagation via gap junctions, a key step in the regulation of liver metabolism by the sympathetic hepatic nerves. FEBS Lett 1992; 301:265-270.
30. Iwai M, Miyashita T, Shimazu T. Inhibition of glucose production during hepatic nerve stimulation in regenerating rat liver perfused in situ. Possible involvement of gap junctions in the action of sympathetic nerves. Eur J Biochem 1991; 200:69-74.
31. Kam E, Pitts JD. Effects of the tumour promotor 12-O-tetradecanoyl phorbol-13-acetate on junctional communication in intact mouse skin: Persistence of homologous communication and increase in dermal-epidermal coupling. Carcinogenesis 1988; 9:1389-1398.
32. Goodenough DA. The crystalline lens. A system networked by gap junctional intercellular communication. Seminars in Cell Biol 1992; 3:49-58.
33. Boitano S, Dirksen ER, Sanderson MJ. Intercellular propagation of calcium waves mediated by inositol triphosphate. Science 1992; 258:292-295.
34. Johnson LG, Olsen JC, Sarkadi B, Moore KL, Swanstrom L, Boucher RC. Efficiency of gene transfer for restoration of normal airway epithelium function in cystic fibrosis. Nature Genetics 1992; 2:21-25.
35. Warner A. The gap junction. J Cell Science 1988; 89:1-7.
36. Goodenough DA, Stoeckenius W. The isolation of mouse hepatocyte gap junctions. Preliminary chemical characterization and x-ray diffraction. J Cell Biol 1972; 54:646-656.
37. Hertzberg EL, Gilula NB. Isolation and characterization of gap junctions from rat liver. J Biol Chem 1979; 254:2138-2147.
38. Musil LS, Goodenough DA. Biochemical analysis of connexin43 intracellular transport, phosphorylation, and assembly into gap junctional plaques. J Cell Biol 1991; 115:1357-1374.
39. Hertzberg E. A detergent-independent procedure for the isolation of gap junctions from rat liver. J Biol Chem 1984; 259:9936-9943.
40. Kensler RW, Goodenough DA. Isolation of mouse myocardial gap junctions. J Cell Biol 1980; 86:755-764.

41. Kistler J, Bullivant S. Dissociation of lens fibre gap junctions releases MP70. J Cell Sci 1988; 91:415-421.
42. Caspar DLD, Sosinsky GE, Tibbitts TT et al. Gap junction structure. In: Hertzberg EL, Johnson RG, eds. Gap Junctions. New York: Alan R. Liss, 1988:117-133.
43. Makowski L, Caspar DLD, Phillips WC, Goodenough DA. Gap junction structures. II. Analysis of the x-ray diffraction data. J Cell Biol 1977; 74:629-645.
44. Unwin PNT, Zampighi G. Structure of the junction between communicating cells. Nature 1980; 283:545-549.
45. Unwin PNT, Ennis PD. Two configurations of a channel-forming membrane protein. Nature 1984; 307:609-613.
46. Sosinsky GE, Baker TS, Caspar DLD, Goodenough DA. Correlation analysis of gap junction lattice images. Biophys J 1990; 58:1213-1226.
47. Hoh JH, Sosinsky GE, Revel J-P, Hansma PK. Structure of the extracellular surface of the gap junction by atomic force microscopy. Biophys J 1993; 65:149-163.
48. Paul DL. Molecular cloning of cDNA for rat liver gap junction protein. J Cell Biol 1986; 103:123-134.
49. Kumar NM, Gilula NB. Cloning and characterization of human and rat liver cDNAs coding for a gap junction protein. J Cell Biol 1986; 103:767-776.
50. Kumar NM, Gilula NB. Molecular biology and genetics of gap junction channels. Seminars in Cell Biol 1992; 3:3-16.
51. Green CR, Harfst E, Gourdie RG, Severs NJ. Analysis of the rat liver gap junction protein: clarification of anomalies in its molecular size. Proc R Soc Lond B 1988; 233:165-174.
52. Beyer EC, Paul DL, Goodenough DA. Connexin43: a protein from rat heart homologous to a gap junction protein from liver. J Cell Biol 1987; 105:2621-2629.
53. Fishman GI, Spray DC, Leinwand LA. Molecular characterization and functional expression of the human cardiac gap junction channel. J Cell Biol 1990; 111:589-598.
54. Musil LS, Beyer EC, Goodenough DA. Expression of the gap junction protein connexin43 in embryonic chick lens: molecular cloning, ultrastructural localization, and post-translational phosphorylation. J Membr Biol 1990; 116:163-175.
55. Ebihara L, Beyer EC, Swenson KI, Paul DL, Goodenough DA. Cloning and expression of a *Xenopus* embryonic gap junction protein. Science 1989; 243:1194-1195.
56. Zimmer DB, Green CR, Evans WH, Gilula NB. Topological analysis of the major protein in isolated intact rat liver gap junctions and gap junction-derived single membrane structures. J Biol Chem 1987; 262:7751-7763.
57. Milks LC, Kumar NM, Houghten R, Unwin N, Gilula NB. Topology of the 32-kD liver gap junction protein determined by site-directed antibody localizations. EMBO J 1988; 10:2967-2975.
58. Hertzberg EL, Disher RM, Tiller AA, Zhou Y, Cook RG. Topology of the M_r 27,000 liver gap junction protein. J Biol Chem 1988; 263:19105-19111.
59. Yancey SB, John SA, Lal R, Austin BJ, Revel JP. The 43-kD polypeptide of heart gap junctions: immunolocalization, topology, and functional domains. J Cell Biol 1989; 108:2241-2254.
60. Laird DW, Revel JP. Biochemical and immunochemical analysis of the arrangement of connexin43 in rat heart gap junction membranes. J Cell Sci 1990; 97:109-117.
61. Swenson KI, Piwnica-Worms H, McNamee H, Paul DL. Tyrosine phosphorylation of the gap junction protein connexin43 is required for the pp60v-src-induced inhibition of communication. Cell Regulation 1990; 1:989-1002.
62. Liu S, Taffet S, Stoner L, Delmar M, Vallano ML, Jalife J. A structural basis for the unequal sensitivity of the major cardiac and liver gap junctions to intracellular acidification: the carboxyl tail length. Biophys J 1993; 64:1422-1433.
63. Rahman S, Evans WH. Topography of connexin32 in rat liver gap junctions. J Cell Sci 1991; 100:567-578.

64. John SA, Revel JP. Connexon integrity is maintained by non-covalent bonds: intramolecular disulfide bonds link the extracellular domains in rat connexin-43. Biochem Biophys Res Comm 1991; 178:1312-1318.
65. Cascio M, Gogol E, Wallace BA. The secondary structure of gap junctions: influence of isolation methods and proteolysis. J Biol Chem 1990; 265:2358-2364.
66. Tibbitts TT, Caspar DLD, Phillips WC, Goodenough DA. Diffraction diagnosis of protein folding in gap junction connexons. Biophys J 1990; 57:1025-1036.
67. Rubin JB, Verselis VK, Bennett MVL, Bargiello TA. A domain substitution procedure and its use to analyze voltage dependence of homotypic gap junctions formed by connexins 26 and 32. Proc Natl Acad Sci 1992; 89:3820-3924.
68. Suchyna TM, Xu LX, Gao F, Fourtner CR, Nicholson BJ. Identification of a proline residue as a transduction element involved in voltage gating of gap junctions. Nature 1993; 365:847-849.
69. Falk MM, Kumar NM, Gilula NB. A novel mechanism for the biosynthetic membrane integration of gap junction proteins (connexins), class III membrane proteins. Molecular Biol Cell 1993; 4:422a.
70. Sikerwar SS, Downing KH, and Glaeser RM. Three-dimensional structure of an invertebrate intercellular communicating junction. J Struct Biol 1991; 106:255-263.
71. Ryerse, JS. Isolation and characterization of gap junctions from *Drosophila* melanogaster. Cell Tissue Res 1989; 256:7-16.
72. Ryerse, JS. Structural, immunocytochemical and initial biochemical characterization of NaOH-extracted gap junctions from an insect, Heliothis virescens. Cell Tissue Res 1993; 274:393-403.
73. Traub O, Look J, Dermietzel R, Brummer F, Hulser D, Willecke K. Comparative characterization of the 21-kD and 26-kD gap junction proteins in murine liver and cultured hepatocytes. J Cell Biol 1989; 108:1039-1051.
74. Kanter HL, Laing JG, Beyer EC, Green KG, Saffitz JE. Multiple connexins colocalize in canine ventricular myocyte gap junctions. Circ Res 1993; 73:344-350.
75. Kistler J, Berriman J, Evans CW, Gruijters WT, Christie D, Corin A, Bullivant S. Molecular portrait of lens gap junction protein MP70. J Struct Biol 1990; 103:204-211.
76. Voorter CEM, Kistler J. cAMP-dependent kinase phosphorylates gap junction protein in lens cortex but not in lens nucleus. Biochim Biophys Acta 1989; 986:8-10.
77. Jiang JX, Paul DL, Goodenough DA. Posttranslational phosphorylation of lens fiber connexin46: a slow occurrence. Invest Ophthalmol Vis Sci 1993; 34:3558-3565.
78. Saez JC, Nairn AC, Czernik AJ, Spray DC, Hertzberg EL, Greengard P, Bennett MV. Phosphorylation of connexin 32, the hepatocyte gap-junction protein, by cAMP-dependent protein kinase, protein kinase C and Ca^{2+}/calmodulin-dependent protein kinase II. Eur J Biochem 1990; 192:263-273.
79. Saez JC, Spray DC, Nairn AC, Hertzberg E, Greengard P, Bennett MV. cAMP increases junctional conductance and stimulates phosphorylation of the 27-kDa principal gap junction polypeptide. Proc Natl Acad Sci 1986; 83:2473-2477.
80. Traub O, Look J, Paul D, Willecke K. Cyclic adenosine monophosphate stimulates biosynthesis and phosphorylation of the 26 kDa gap junction protein in cultured mouse hepatocytes. Eur J Cell Biol 1987; 43:48-54.
81. Werner R, Levine E, Rabadan-Diehl C, Dahl G. Gating properties of connexin32 cell-cell channels and their mutants expressed in *Xenopus* oocytes. Proc R Soc London 1991; 243:5-11.
82. Takeda A, Saheki S, Shimazu T, Takeuchi N. Phosphorylation of the 27-kDa gap junction protein by protein kinase C in vitro and in rat hepatocytes. J Biochem 1989, 106:723-727.
83. Musil LS, Cunningham BA, Edelman GM, Goodenough DA. Differential phosphorylation of the gap junction protein connexin43 in junctional communication-competent and -deficient cell lines. J Cell Biol 1990; 111:2077-2088.
84. Crow DS, Beyer EC, Paul DL, Kobe SS, Lau AF. Phosphorylation of connexin43 gap junction protein in uninfected and Rous sarcoma virus-transformed mammalian fibroblasts. Mol Cell Biol 1990; 10:1754-1763.

85. De Mello WC. Interaction of cyclic AMP and Ca^{2+} in the control of electrical coupling in heart fibers. Biochim Biophys Acta 1986; 888:71-79.
86. Berthoud VM, Ledbetter MLS, Hertzberg EL, Saez JC. Connexin43 in MDCK cells: regulation by a tumor-promoting phorbol ester and Ca^{2+}. Eur J Cell Biol 1992; 57:40-50.
87. Brissette JL, Kumar NM, Gilula NB, Dotto GP. The tumor promoter 12-O-tetradecanoylphorbol-13-acetate and the ras oncogene modulate expression and phosphorylation of gap junction proteins. Mol Cell Biol 1991; 11:5364-5371.
88. Kanemitsu MY, Lau AF. Epidermal growth factor stimulates the disruption of gap junctional communication and connexin43 phosphorylation independent of 12-O-tetradecanoylphorbol 13-acetate-sensitive protein kinase C: the possible involvement of mitogen-activated protein kinase. Mol Biol Cell 1993; 4:837-848.
89. Takens-Kwak BR, Jongsma HJ. Cardiac gap junctions: three distinct single channel conductances and their modulation by phosphorylating treatments. Pflugers Arch 1992; 422:198-200.
90. Moreno AP, Fishman GI, Spray DC. Phosphorylation shifts unitary conductance and modifies voltage dependent kinetics of human connexin43 gap junction channels. Biophys J (Discussions) 1992; 62:51-53.
91. Oh SY, Dupont E, Madhukar BV, Briand JP, Chang CC, Beyer E, Trosko JE. Characterization of gap junctional communication-deficient mutants of a rat liver epithelial cell line Eur J Cell Biol 1993; 60:250-255.
92. Atkinson MM, Menko AS, Johnson RG, Sheppard JR, Sheridan JD. Rapid and reversible reduction of junctional permeability in cells infected with a temperature-sensitive mutant of avian sarcoma virus. J Cell Biol 1981; 91:573-578.
93. Pelletier DB, Boynton AL. Gap junctional communication is inhibited by PDGF-BB. Molecular Biol Cell 1993; 4:126a.
94. Husoy T, Mikalsen S-O, Sanner, T. Phosphatase inhibitors, gap junctional intercellular communication and [^{125}I]-EGF binding in hamster fibroblasts. Carcinogenesis 1993; 14:2257-2265.
95. Young JDE, Cohn ZA, Gilula NB. Functional assembly of gap junction conductance in lipid bilayers: demonstration that the major 27 kD protein forms the junctional channel. Cell 1987; 48:733-743.
96. Spray DC, Saez JC, Brosius D, Bennett MV, Hertzberg EL. Isolated liver gap junctions: gating of transjunctional currents is similar to that in intact pairs of rat hepatocytes. Proc Natl Acad Sci 1986; 83:5494-5497.
97. Harris AL, Walter A, Paul D, Goodenough DA, Zimmerberg J. Ion channels in single bilayers induced by rat connexin32. Mol Brain Res 1992; 15:269-280.
98. Hirokawa N, Heuser J. The inside and outside of gap-junction membranes visualized by deep-etching. Cell 1982; 30:395-406.
99. Peracchia C, Peracchia LL. Gap junction dynamics: reversible effects of divalent cations. J Cell Biol 1980; 87:708-718.
100. Kidder GM, Rains J, McKeon J. Gap junction assembly in the preimplantation mouse conceptus is independent of microtubules, microfilaments, cell flattening, and cytokinesis. Proc Natl Acad Sci 1987; 84:3718-3722.
101. Ito S, Sato E, Loewenstein WR. Studies on the formation of a permeable cell membrane junction. I. Coupling under various conditions of membrane contact. Effects of colchicine, cytochalasin B, dinitrophenol. J Membrane Biol 1974; 19:305-337.
102. Malewicz B, Kumar VV, Johnson RG. Lipids in gap junction assembly and function. Lipids 1990; 25:419-427.
103. Musil LS, Goodenough DA. Multisubunit assembly of an integral plasma membrane channel protein, gap junction connexin43, occurs after exit from the ER. Cell 1993; 74:1065-1077.
104. Hurtley SM, Helenius A. Protein oligomerization in the endoplasmic reticulum. Annu Rev Cell Biol 1989; 5:277-307.
105. Griffiths G, Pfeiffer S, Simons K, Matlin K. Exit of newly synthesized membrane proteins from the trans cisterna of the Golgi complex to the plasma membrane. J Cell Biol 1985; 101:949-964.
106. Loewenstein WR. Junctional intercellular communication: the cell-to-cell membrane channel. Physiol Rev 1981; 61:829-913.

107. DeVries SH, Schwartz EA. Hemi-gap-junction channels in solitary horizontal cells of the catfish retina. J Physiol 1992; 445:201-230.

108. Malchow RP, Qian H, Ripps H. Evidence for hemi-gap junctional channels in isolated horizontal cells of the skate retina. J Neurosci Res 1993; 35:237-245.

109. Paul DL, Ebihara L, Takemoto LJ, Swenson KI, Goodenough DA. Connexin46, a novel lens gap junction protein, induces voltage-gated currents in nonjunctional plasma membranes of *Xenopus* oocytes. J Cell Biol 1991; 115:1077-1089.

110. Ebihara L, Steiner E. Properties of a nonjunctional currant expressed form a rat connexin46 cDNA in *Xenopus* oocytes. J Gen Physiol 1993; 102:59-74.

111. Beyer EC, Steinberg TH. Evidence that the gap junction protein connexin-43 is the ATP-induced pore of mouse macrophages. J Biol Chem 1991; 266:7971-7974.

112. Epstein ML, Sheridan JD, Johnson RG. Formation of low-resistance junctions in vitro in the absence of protein synthesis and ATP production. Exp Cell Res 1977; 104:25-30.

113. Rook MB, de Jonge B, Jongsma HL, Masson-Pevet MA. Gap junction formation and functional interaction between neonatal rat cardiocytes in culture: a correlative physiological and ultrastructural study. J Membrane Biol 1990; 118:179-192.

114. Meyer RA, Laird DW, Revel JP, Johnson RG. Inhibition of gap junction and adherens junction assembly by connexin and A-CAM antibodies. J Cell Biol 1992; 119:179-189.

115. Dahl G, Werner R, Levine E, Rabadan-Diehl C. Mutational analysis of gap junction formation. Biophys J (Discussions) 1992; 62:172-182.

116. Dahl G, Levine E, Rabadan-Diehl C, Werner R. Cell/cell channel formation involves disulfide exchange. Eur J Biochem 1991; 197:141-144.

117. Swenson KI, Jordan JR, Beyer EC, Paul DL. Formation of gap junctions by expression of connexins in *Xenopus* oocyte pairs. Cell 1989; 57:145-155.

118. Bruzzone R, Haefliger JA, Gimlich RL, Paul DL. Connexin40, a component of gap junctions in vascular endothelium, is restricted in its ability to interact with other connexins. Mol Biol Cell 1993; 4:7-20.

119. White TW, Bruzzone R, Wolfram S, Goodenough DA, Paul DL. Selective interactions among the multiple connexin proteins expressed in the vertebrate lens: the second extracellular domain is a determinant of compatability between connexins. J Cell Biol 1994; 125:879-892-.

120. Michalke W, Loewenstein WR. Communication between cells of different type. Nature 1971; 232:121-122.

121. Barrio LC, Suchyna T, Bargiello, T, Xu LX, Roginski RS, Bennett MV, Nicholson BJ. Gap junctions formed by connexins 26 and 32 alone and in combination are differently affected by applied voltage. Proc Natl Acad Sci 1991; 88:8410-8414.

122. Hennemann H, Suchyna T, Lichtenberg-Frate H, Jungbluth S, Dahl E, Schwarz J, Nicholson BJ, Willecke K. Molecular cloning and functional expression of mouse connexin40, a second gap junction gene preferentially expressed in lung. J Cell Biol 1992; 117:1299-1310.

123. Abney JR, Braun J, Owicki JC. Lateral interactions among membrane proteins. Implications for the organization of gap junctions. Biophys J 1987; 52:441-454.

124. Braun J, Abney JR, Owicki JC. How a gap junction maintains its structure. Nature 1984; 310:316-318.

125. Johnson R, Hammer M, Sheridan J, Revel JP. Gap junction formation between reaggregated Novikoff hepatoma cells. Proc Natl Acad Sci 1974; 71:4536-4540.

126. Levine E, Werner R, Dahl G. Cell-cell channel formation and lectins. Am J Physiol 1991; 261:C1025-C1032.

127. Wang Y, Mehta PP. Inhibitors of glycosylation upregulate gap junctional communication in transformed cells. Mol Biol Cell 1992; 3:295a.

128. Miledi R, Molinoff P, Potter L. Isolation of the cholinergic receptor protein of Torpedo electric tissue. Nature 1971; 229:554-556.

129. Edelman GM. Morphoregulatory molecules. Biochemistry 1988; 17:3533-3543.

130. Keane RW, Mehta PP, Rose B, Honig LS, Loewenstein WR, Rutishauser U. Neural differentiation, NCAM-mediated adhesion, and gap junctional communication in neuroectoderm. A study in vitro. J Cell Biol 1988; 106:1307-1319.
131. Doherty P, Ashton SV, Moore SE, Walsh FS. Morphoregulatory activities of NCAM and N-cadherin can be accounted for by G protein-dependent activation of L- and N-type neuronal Ca^{2+} channels. Cell 1991; 67:21-33.
132. Frei T, von Bohlen und Halbach F, Wille W, Schachner M. Different extracellular domains of the neural cell adhesion molecule (N-CAM) are involved in different functions. J Cell Biol 1992; 118:177-194.
133. Paul DL, Yu K, Bruzzone R. Expression of a dominant negative inhibitor of intracellular communication in *Xenopus* embryos causes delamination and extrusion of cells early in gastrulation. Mol Biol Cell 1993; 4:350a.

CHAPTER 11

Physiological and Pharmacological Regulation of Gap Junction Channels

David C. Spray

1. INTRODUCTION

Gap junctions are found in virtually all tissues of multicellular animals and at all stages of development. Even in the few differentiated tissues where gap junctions are expressed not at all (adult skeletal muscles, differentiated spermatozoa, and circulating erythrocytes of most vertebrates) or are not abundant (neurons in the adult brain), the loss of gap junctions is an ontogenetic event which appears to occur as a prelude to establishment of the more differentiated phenotype.[1]

The widespread occurrence of gap junctions suggests that they play an important general role in tissue function, and the discovery that the structures called gap junctions were intercellular channels suggested that this function role was performed by the exchange of ions and small molecules directly between coupled cells. Thus, in embryonic cells, the role of gap junctions has been believed to be in exchange of developmentally relevant molecules, including substances conveying positional and spatial information.[1] And in the vertebrate heart and in giant axons of invertebrates (which are actually separate axonal segments linked by gap junctions), the massive gap junctions comprising intercalated disks and septae have been envisaged to provide the pathway for propagation of active currents (see ref. 2 and 3).

Because the functional importance of gap junction lies in their behavior as channels, our laboratory has concentrated on characterizing the physiological properties of gap junctions formed naturally between various types of vertebrate and invertebrate cells, as well as those channels resulting from

Molecular Mechanisms of Epithelial Cell Junctions: From Development to Disease
Edited by Sandra Citi, M.D., Ph.D.;

exogenous expression of wild-type and mutant gap junction proteins connexins. The intent of this chapter is to summarize what we now know about the properties of gap junctions in various cell types from the perspectives of what goes through the channel (and consequent relevance for tissue function), how the channels open and close, and what physiological and pharmacological stimuli modify whether these channels are in the conducting or nonconducting states.

The flux of a substance through any aggregate of channels is determined by the product of the number of channels, the fraction of time each channel is open, and the flux through each open channel per unit time. Determinants of channel number include regulation of transcription, post-translational processing and membrane insertion which are outside the scope of this review, but are considered in the previous chapter by Musil[4] (see also ref. 5). About a dozen mammalian connexins have now been cloned and sequenced (see ref. 6). Assuming that all of these cDNAs form functional channels (which for most has been demonstrated in the *Xenopus* oocyte expression system; see below), such apparent genomic redundancy might be explained by differences in the properties of channels formed by each connexin type.[7] As is summarized below, there are indeed connexin-specific gating and biophysical properties, and for some of these it is plausible to propose physiological relevance (one of the most straightforward being the weak voltage dependence of channels formed of the major cardiac gap junction protein, connexin43). Although the meaning of other connexin properties is at present obscure, one may hope that these differences may be exploitable in developing therapeutically useful pharmacological reagents or in targeting expression of connexins with novel properties to correct functional deficiencies in selected tissues.

2. SIZES OF GAP JUNCTION CHANNELS

Gap junction channels are big. Even early studies on gap junction channels indicated that they were permeable to fluorescein and lissamine-rhodamine derivatives and therefore had minimal pore sizes exceeding 1 nm. Injection of peptides conjugated to either fluorescein or rhodamine provided evidence that arthropod junction channels were larger than those of vertebrates.[8-11] Although these early studies were qualitative (measuring relative detection of linear or branched dye molecules in adjacent cells at a fixed interval), subsequent studies principally by Weingart, Imanaga, Brink, De Mello and their colleagues have measured rates of intercellular diffusion of labeled spherical or planar molecules of a range of sizes (reviewed in ref. 12). For two preparations (earthworm septate axon and cardiac myocytes), the range of molecular sizes examined is adequate to predict the effective pore radius (see Fig. 11.1). Although it has been customary to express junctional permeability in terms of junctional contact area, the relevant parameter to which intercellular dye flux should be normalized is the conductance of the junctional membrane (see ref. 13). For example, in studies performed under voltage clamp conditions to determine whether the voltage-induced closure of junctional channels in embryos reduced size of the open channel, Verselis et al[14] measured permeability/conductance (P_j/g_j) ratios of about 1.5, 0.7 and 0.2 for TMA, TEA, and TPA (see Fig. 11.2). Using similar methods, Bodmer et al[15] measured slightly lower ratios P_j/g_j for TMA and TEA in cultured *Aplysia* neurons, which are coupled by junctional pathways that are apparently not appreciably permeable to Lucifer Yellow.

Qualitatively, the permeability of junctional membranes of a variety of cell types is illustrated in Figure 11.1, where relative P_j values obtained from cardiac preparations and earthworm septate axons were obtained by normalizing P_j values summarized in ref. 12 to that of K^+. Extrapolating the regression line to intersection of the abcissa at a permeability 10^{-4} that of K^+ predicts a permeability limit at about 1 kD. As has been pointed out previously, although molecules up to this limit are predicted to pass between coupled cells, the latency to detection of the larger molecules can be quite long.[16] Thus, although the lack of detectable fluorescein transfer

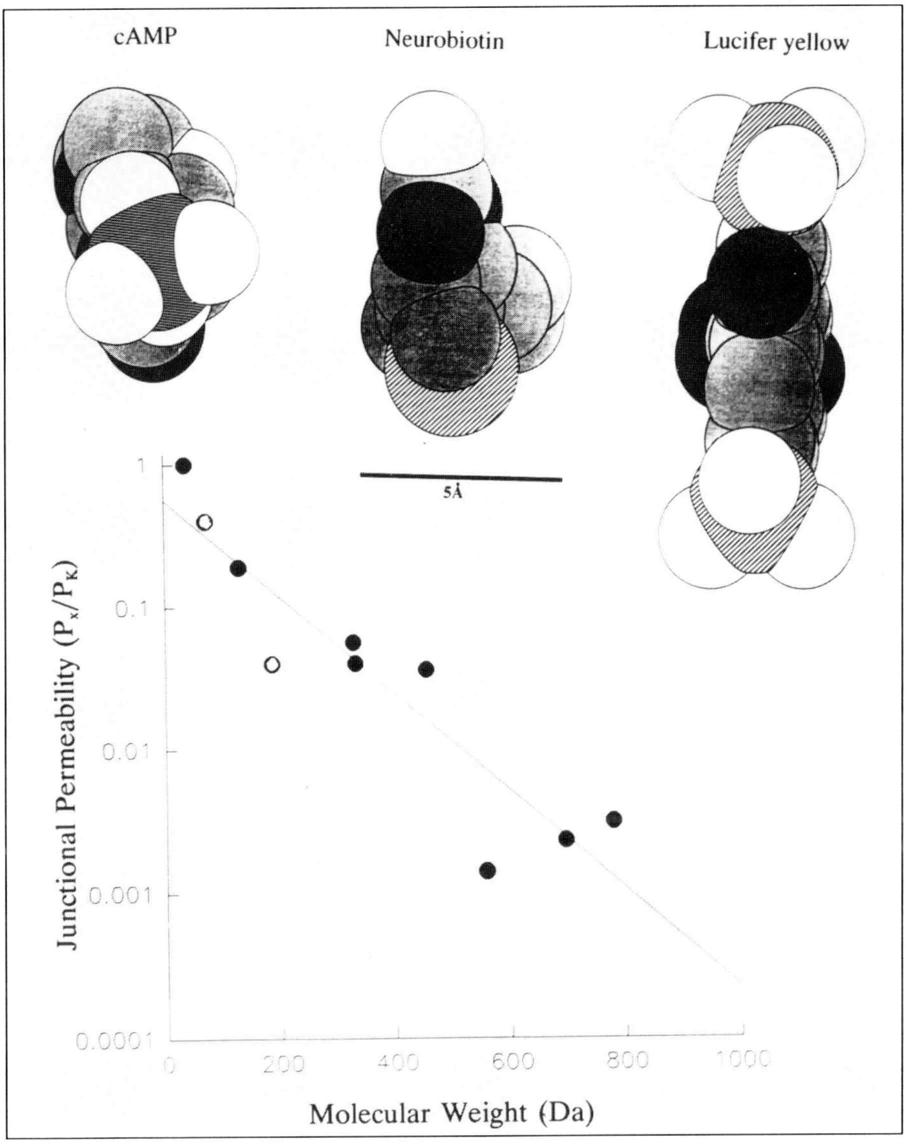

Fig. 11.1. Junctional permeability to molecules of various sizes. Space filling models of two commonly used tracer molecules (neurobiotin and Lucifer Yellow) and a second messenger molecule (cAMP) are illustrated above (kindly provided by Dr. A. Peinado). Below the junctional permeabilities measured for molecules of a wide range of molecular weight are plotted after normalization to a value of 1 for K^+. Solid circles are values summarized by Imanaga[12] on cardiac muscle; open circles are permeabilities of tetramethylammonium and tetrapropylammonium relative to that of tetraethylammonium, measured in pairs of amphibian embryonic cells[14].

between early embryonic cells was initially interpreted as evidence for reduced channel size compared to the adult (see ref. 17), the use of the more fluorescent and less membrane permeant dye Lucifer Yellow indicated that the large size of the recipient cell simply increased the time required for detection.[16] Whether or not volume mismatches (in total cell size or in the asymmetric intervention of processes on which junctions are located) may explain the occasional reports of chemical rectification[18,19,20] is an issue that remains to be explored fully.

The recent applications of biotin and neurobiotin (see Fig. 11.1, top) to studies of

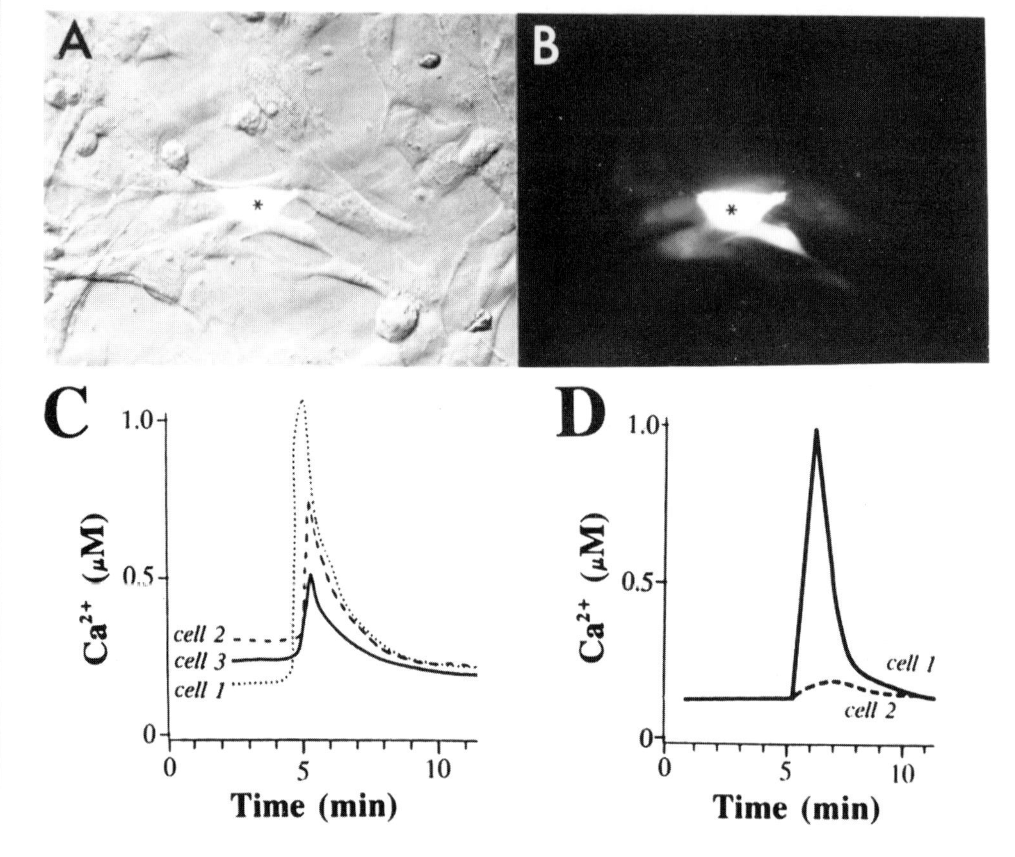

Fig. 11.2. Gap junction mediated intercellular diffusion. A, B. SKHep1 cells transfected with connexin32 cDNA. Lucifer Yellow was injected into the cell marked with an asterisk. One minute later, dye was detected in all contiguous and several second order cells. C, D. Human corpus cavernosum *smooth muscle in which intracellular Ca^{2+} was measured using FURA2 C. Ca^{2+} was injected into cell 1 and increased Ca^{2+} was rapidly detected in two other contiguous cells. D. After treatment of cells with 1 mM heptanol, which reduces junctional conductance, Ca^{2+} injection no longer led to marked elevation in the adjacent cell.*

coupling in the brain and retina have resulted in the striking demonstration of much more extensive intercellular connections than was formerly believed.[21] The greater spread of neurobiotin than Lucifer Yellow (space-filling model in Fig. 11.1; dye spread illustrated in Fig. 11.2A, B) is generally attributed to its smaller size (although other factors such as cytoplasmic inertness may contribute to less binding within the injected cell, such that the diffusional driving force may be greater than for Lucifer Yellow). Note from interpolation in Figure 11.1 that the permeability predicted solely on the basis of size of the neurobiotin molecule is about five times that of Lucifer Yellow, the most commonly used fluorescent tracer.

The largest molecular size to which gap junctions are permeable is thus about 2 nm, and thus the channels would be expected to accommodate most molecules below an M_r of about 1000 Da. This size range includes many physiological second messenger molecules [such as cAMP (top, Fig. 11.1), inositol trisphosphate and Ca^{2+}] as well as current-carrying ions (of which K^+ is most abundant and most mobile; in excitable systems where current carrying capacity of the junctional membrane is its functional role, the gap junction channel is operationally a

K^+ channel). Tissue metabolites and enzyme substrates are also permeant, as demonstrated in recovery of function ("kiss of life") assays.[22,23]

Ca^{2+} was until rather recently believed to play a decisive role in closing gap junction channels (see below) and thus the critical functions that could be fulfilled by intercellular Ca^{2+} diffusion were entirely overlooked. Studies on *Obelia,* a hydroid with eyespot support cells containing an endogenous photoprotein (obelin) provided the first indication for junctional Ca^{2+} permeability, since these cells glowed after Ca^{2+} was elevated by illumination of the photoreceptors to which they were coupled.[24] This observation was soon extended to mammalian gap junctions by Saez et al,[25] working on freshly dissociated rat hepatocytes loaded with the Ca^{2+} indicator Fura2. These studies showed that Ca^{2+} injection into one cell of a pair led to apparently diffusional increase in the adjacent cell. Furthermore, IP_3 injection produced localized Ca^{2+} elevations in both the injected and adjacent cells. These spatially discrete Ca^{2+} elevations suggested that either IP_3 itself or an active metabolite (but not Ca^{2+} in this case, because the Ca^{2+} elevations were spatially remote from each other) was diffusing intercellularly to effect Ca^{2+} release. Gap junctions were strongly implicated in this cell to cell diffusion by the findings that the transfer was unaffected by reducing extracellular Ca^{2+} and was blocked by the uncoupling agent heptanol. Subsequent studies on astrocytes, tracheal epithelial cells, corpus cavernosum smooth muscle (illustrated in Fig. 11.2C, D) and aortic endothelial cells have now reported that evocation of elevated intracellular Ca^{2+} (by pharmacological and mechanical stimuli as well as by direct intracellular Ca^{2+} injection) can lead to propagated "Ca^{2+} waves," which are apparently due to regenerative Ca^{2+} or IP_3 induced Ca release triggered by transfer of these second messenger molecules through junctional channels.[26-31] In brain slices as in these other preparations, these waves cross long distances, with velocities on the order of 10-25 μ/sec. Interestingly, this velocity is similar to that of the spread of refractoriness through brain tissue ("spreading depression") resulting from strong mechanical or electrical stimulation (e.g. ref. 32). Whether the similar time course merely reflects a common mechanism (both types of regenerative responses being due to reaction-diffusion processes) or indicates a causal role of gap junctions in spreading depression remains unexplored. However, the physiological importance of long range signaling by Ca^{2+} waves among glia, and in the maintenance of vascular tone when evoked between smooth muscle, appears to be one of the most important, yet most consistently overlooked, functions of the gap junction channel.

A few studies have been interpreted as providing evidence for either differential permeability of gap junction channels in particular situations (e.g., at developmental boundaries in arthropods: see ref. 19) or as indicating changes in permeability as junctional conductance was reduced (e.g. ref. 33). The rare attempts to quantitatively examine this issue have indicated rather decisively that channel constriction was unnecessary to explain the data, and that all-or-none channel closure seemed a more general principle by which gap junction channels operate.[10,13] Nevertheless, very recent studies have produced what appears to be unequivocal evidence for channel substates (where changes in unitary conductance may or may not be linked to permeability restriction; see below). And studies on transfected mammalian cells (where various connexins can be expressed on the same cellular background) appear to suggest that channels formed of some connexins are less permeant than those formed of others. For an example of the latter, connexin45 channels in SKHep1 cells are impermeant to Lucifer Yellow, whether expressed homotypically or paired with cells expressing connexin43.[34] Clearly, permeability is one property of gap junction channels to which attention of the channel physiologists should be drawn.

Another example of ionic transfer across gap junctional membranes is so-called "energy-transfer", where currents generated by the cell or cell group imposes polarity on the system by inducing current flow from current sources to other regions which are current sinks.[35,36] Finally, osmotic balance may be achieved through intercellular diffusion of ions or solvent; such a function together with buffering K^+ ions (and Ca^{2+} waves mentioned above) are the presumed role of gap junctions between glial cells.

Data on which to evaluate gap junction selectivity are fewer than the abundant literature relevant to permeability estimates, being limited to comparisons of a few charged fluorescein derivatives and to only a few ionic substitutions within patch pipettes. With regard to large molecules, Flagg-Newton and Loewenstein (see ref. 37) initially showed that negatively charged dyes moved more rapidly than positively charged ones and that the permeability was related to both the magnitude and polarity of the charge on the molecule. Brink et al[38] performed similar, though more thoroughly quantified studies on effects of charge on channel permeability to fluorescein derivatives. Relative to the negatively charged compounds, aminofluorescein was found to be markedly restricted, which lead to the hypothesis that a fixed negative charge exists within the gap junction channel. Presumably, location of this site could be approximated by studies of use- and voltage-dependence of the block, and by using site-directed mutagenesis with expression in exogenous systems.

An alternate method for determining channel selectivity is to use electrophysiological methods, either to measure shifts in reversal potential under conditions of asymmetric internal solutions in the two cells or to measure unitary conductances under symmetrical conditions following ionic substitution. Although the range of ionic substitutions have been small, several studies have examined effects on single channel currents. For example, Neyton and Trautman[39] determined ratios of 1:0.8:0.7: for K:Na:Cl in their pioneering study of lacrimal gland cells and Brink et al reported ratios of 1.0:0.7:0.5 for K:Na:Cl in excised patches of presumably junctional membrane in earthworm septate axons.[40] Moderate confusion in the literature has been generated by the use of different pipette solutions by different laboratories measuring connexin43 channels and the initial assumption that permeability to organic (glutamate, aspartate, gluconate) and inorganic ions (Cl^- and F^-) would be similar. In fact, differences in unitary conductances measured by the same or different laboratories with the different solutions indicates that Cl^- is roughly twice as permeable through these junctional membranes as are the organic anions, a finding which is consistent with the predictions on the basis of size shown in Figure 11.1.

Single channel studies in general have revealed the surprising results that unitary conductances of channels formed by the same connexin may vary in different tissues and under different conditions, and that the range of channel sizes that a single connexon type exhibits is connexin-specific. Initial predictions from geometric factors were that unitary conductances should be on the order of 100 pS (a right cylindrical pore with a diameter of 1 nm, a length of about 3 nm and cytoplasmic resistivity: 100 Ohm-cm). The single channel conductances (γ_j) values obtained for gap junctions in various cell types and after transfection with wild-type and mutant connexins are summarized in Table 11.1 (see also Fig. 11.3C, D). These values span the range from 30 pS (connexin45 and substate of connexin43 channels) to more than 300 pS (connexin37 and channels between blastomeres in early sea urchin embryos). Several noteworthy points are apparent from this table. First, the arthropod channels thus far recorded do not exhibit appreciably larger unitary conductances than mammalian channels despite their apparently higher permeability.[9] It should be noted that such a higher ratio of permeability to γ_j might be consistent with longer channels in the anthropod phyla, even if channel diameter were somewhat larger. Second, each connexin type has a characteristic set of γ_j values, although not all channel sizes have been recorded in all cell types in which the same connexin is expressed. Although multiple channel sizes might represent the co-expression of more than one connexin type (the initial and still common interpretation), a few recent reports indicate for connexin43 that γ_j is modifiable by the voltage imposed across the junctional membrane and by the phosphorylation state of the protein. This modulation of single channel properties is unique to gap junction channels and is considered in detail below. Finally, conductance states are distinct for different connexins. Although γ_j values for different connexins overlap, they appear to offer a signature by which the connexins forming the channels can be identified using physiological measurements.

The significance of γ_j for tissue function is not straightforward. However, for a channel

Table 11.1: Unitary conductances of gap junction channels

Cell type	Unitary conductance	Connexin type	Reference
Rat lacrimal gland	~ 120 pS	Cx32, Cx26	39
Rat neonatal heart	20-30, 45-60 pS	Cx43	61,85,86
Rat Schwann cells	40 pS	?	81
Chick embryonic heart	60-80, 165 pS	Cx42,Cx43,Cx45	105
Adult rodent heart	30-40 pS	Cx43?	83,84,129
Pancreatic acinar cells	27, 130 pS	Cx32	56
Human cc smooth muscle	25,40,80,120 pS	Cx43 + ?	87
Human umb. endo. cells	90 pS	Cx43	136
Human airway sm. muscle	90 pS	Cx43	88
Rat astrocytes	50-60 pS	Cx43	126,127
Rat lepto. cells	60, 100 pS	Cx43, Cx26	138
Rabbit art sm	70-80 pS	Cx43	131
Rat Leydig cells	60, 90 pS	Cx43	89
Fish horizontal cells	70 pS	?	135
Skate blastomeres	~100 pS	?	133
Xenopus blastomeres	100 pS	?	136
CHO Cells	22-120 pS	?	56
HeLa cells	20-30 pS	?	90
A7R5 aortic sm. muscle	140,100,70 pS	Cx43,Cx40	139
WB cells	30,100 pS	Cx43 + 34 kD	130
SKHep1 cells	30 pS	Cx45	82
transf hCx43	30,60,90 pS	Cx43	93,120
transf rCx32	120-150 pS	Cx32	128
transf XCx38	25 pS	XCx38	104
N2A cells	40 pS	?	132
transf rCx37	200pS (and less)	Cx37	132
transf cCx42	180 pS	cCx42	95
transf cCx43	70 pS	cCx43	95
transf cCx45	30 pS	cCx45	95
Earthworm septate axon	30-40, 100 pS	?	40
Mosquito cell line	70-125, 175-225 pS	?	106
Tenebrio epidermal cells	288, 50-100 pS	?	110
Crayfish hepatopancreas	40, 80,200 pS	?	111
Squid blastomeres	>250 pS	?	133
Sea urchin blastomeres	220 pS	?	134

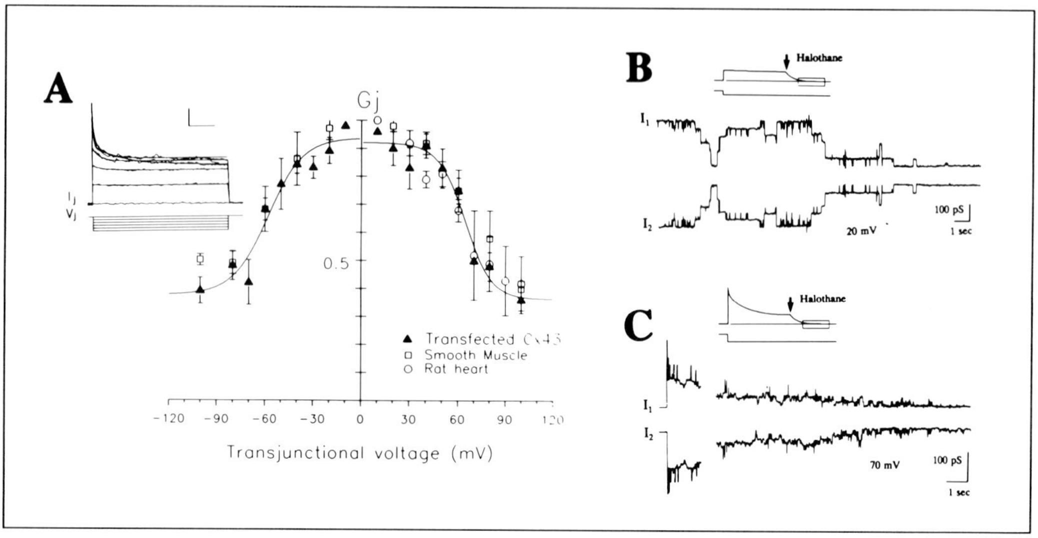

Fig. 11.3. Voltage dependence of junctional conductance in human connexin43 channels. A. Inset shows relaxation of junctional currents in a pair of SKHep1 cells transfected with connexin43. Note that at even large transjunctional voltages (V_j), junctional current (I_j) does not decrease to zero. Plot shows fit of a Boltzmann relation to steady state g_j data obtained in several connexin43-containing preparations at different V_j values. Note that at V_j values > 70 mV in both polarities, a large residual conductance (g_{min}) remains. B,C. The residual conductance is due to a substate of the connexin43 channel. In B, halothane, (an uncoupling agent which does not alter channel size) was added to a cell pair during a sustained V_j of 20 mV (see inset). Channels recorded during halothane uncoupling were 60-70 or 90-100 pS. C. Halothane uncoupling of a cell pair where V_j was large enough to reach g_{min} in the macroscopic g_j recording (see inset) led to detection of only smaller (30 pS) channel activity. Modified from ref. 82.

whose function is mediating current flow, γ_j will govern how many channels are required, when matched with input resistance of the cells. Thus, in cells such as neurons and other small mammalian cells where input resistance may be exceptionally high (e.g. ~10 GOhms), a single open gap junction channel with γ_j = 100 pS provides a coupling coefficient of 0.5,[41] and thus 50% of the steady state current generated in one cell will be transferred across the junctional membrane (transient events are more complicated, due to the time constant of the junctional membrane: see[42]). For large adult cardiac myocytes, by contrast, input resistance is 30 times lower, and the large junctional plaques coupling these cells at intercalated disks provides an enormous safety factor insuring against conduction failure.[2]

3. GATING

The perception of gap junction channels as large pores whose function appeared to be the continuous diffusional exchange of factors largely led to the view that the channels must differ fundamentally from other ionic channels. Although the emphasis by Loewenstein's group on uncoupling cells by Ca^{2+} treatments (see ref. 37) should have changed this perception, it was finally dispelled by recordings between voltage clamped cell pairs, where high resolution measurements demonstrated that gap junction channels continuously open and close (see Fig. 11.3C, D). Despite the fact that the underlying mechanisms remain to be fully elucidated, there are four general categories of agents which operate on junctional channels to change conductance of the junctional membrane. As introduced above, total conductance is the product of number of channels, unitary conductance, and fraction of time that each is open. Although it would be of immense interest to know unambiguously which process is operated on by each type of agent, few studies have succeeded in dissociating the effects. The technical problem to be overcome in such studies is the large number of junctional channels typically connecting cell pairs leading to the situation in which channel behavior can only be observed through the use of uncoupling agents. Ways around this difficulty have been to attempt to seal onto junctional regions in freshly dissociated cells, thereby isolating only a few channels in the patch,[40] or to choose poorly coupled cells, either freshly paired or cell types in which total expression is normally low.

In terms of number of studies, and the generation of controversy, the most studied agents modifying junctional conductance have been the ions H^+ and Ca^{2+}. Numerous publications by Loewenstein's group, primarily in insect salivary gland cells (although also performed in amphibian embryonic and mammalian tissue culture cells) indicated that lowering extracellular or raising intracellular Ca^{2+} led to uncoupling, with estimates of effective intracellular concentrations being in the range above 10 μM. Interestingly from the current perspective regarding voltage sensitivity of junctional membranes, Ca^{2+} uncoupling in salivary gland cells could be reversed by hyperpolarization.[43] Studies on cardiac myocytes and crayfish septate axon have been interpreted as supporting the generality of the Ca^{2+} effect, and an hypothesis of calmodulin involvement has been advanced as an explanation.[44] Despite this abundant literature, however, the ready passage of Ca^{2+} waves throughout coupled networks clearly provides evidence that these levels of Ca^{2+} do not block coupling completely, as does the maintenance of conduction through contracting cardiac and smooth muscle tissue.

What then is the mechanism by which this uncoupling by Ca^{2+} is achieved? One explanation of much of the data obtained in the insect cells is simply that Ca^{2+} injection and manipulations of extracellular Ca^{2+} depolarized the cells, because of the finding that junctional conductance between insect cells is markedly affected by resting potential, as is discussed below. Such an explanation would be consistent with the restoration of coupling by hyperpolarization, the relative insensitivity of mammalian cells to comparable treatments, and might also explain why such high levels of Ca^{2+} are necessary for the reported effects in the salivary gland cells.

Another possible explanation for the data, especially as pertains to vertebrate cells, was raised by the demonstration that acidification of early amphibian embryos led to rapid uncoupling.[45] Soon thereafter, simultaneous measurement of intracellular pH (pH_i) and junctional conductance (g_j) in pairs of amphibian and fish embryonic ells revealed that the pH_i-g_j relation followed a simple titration curve,[46] and subsequent studies have suggested that pH sensitivity may also be a connexin-specific property.[47] There have, however, been challenges to this hypothesis. First, internally perfused crayfish axons lost pH sensitivity (unlike the similar responses to acidification reported for perfused and intact blastomeres)[48] leading to the hypothesized involvement of a soluble second messenger.[49] Second, calmodulin inhibitors have been reported to affect pH sensitive coupling in crayfish and amphibian embryos,[44] although studies by others have indicated a nonjunctional origin of the protective effect.[50] Finally, in cardiac myocytes, it has been proposed that there exists a synergistic action of low pH_i and high Ca^{2+} [51,52] (note, however, than even in the more quantitative of these studies the data analysis required inclusion of pH sensitivity of the Ca^{2+} indicator and that Ca^{2+} measurements were not done simultaneously with conductance measurements). Moreover, Peracchia has recently concluded that low pH_i has no effect on junctional conductance independent of Ca^{2+} in a study on Novikoff hepatoma cells.[53]

It has also been suggested that low pH_i, which is generally induced by exposure to weak acids, exerts a stronger effect on g_j than exposure to strong acids.[54] The hypothesized mechanism involves the lipophylicity of weak acids, such that they act on gap junction channels in the same manner as lipophiles.

Although it can certainly be argued that the pH_i necessary to uncouple cells is unphysiological and that tissues exposed to such profound acidification have little chance of recovery, unraveling the mechanism of action will seem likely to extend our understanding of how these channels are regulated at the level of channel structure. For example, recent site-directed mutagenesis studies suggest that the hinge (or cytoplasmic loop) regions of connexins32 and 43 may interact with the cytoplasmic tail in a ball and chain mechanism.[55] Moreover, the technology is now available to perform simultaneous measurement of intracellular pH and Ca activities while cells are voltage clamped. Such studies should ultimately resolve this still murky issue.

The mechanism of pH_i or Ca^{2+}-induced uncoupling remains to be explored at the level of single channel properties. Although it has been suggested that Ca^{2+} reduces channel conductance or permeability between cells[33] and in vesicles containing isolated gap junction protein,[44] striking changes in γ_j were not observed in experiments where g_j was reduced by exposure to high internal Ca^{2+} [56] and have not been detected in studies on isolated junctions inserted into lipid bilayers.[57] In pairs of cardiac myocytes uncoupled by CO_2 exposure, Burt and Spray[58] noted that open times of channels became very short, while channel size was unchanged. They interpreted this behavior as reflecting a change in channel open probability. Most recently, a small change in unitary conductance has been detected following exposure of N2A cells to CO_2 (M. Waltzmann, M. Chanson, and D.C. Spray, in preparation). Here, endogenous expression is low enough for junctional conductance changes to be followed at the single channel level; it is expected that future studies on this and other systems with low g_j should finally resolve this issue.

Lipophilic molecules provide another set of compounds that affect whether gap junction channels are conducting or nonconducting, and it is conceivable that these molecules may offer the greatest hope for strategies to develop therapeutically useful agents. The generation of oleic acid and arachidonic acids, acetyl carnitines and other lipophilic compounds under ischemic conditions is a possible example of naturally occurring, albeit pathological, action.[2,59] This group of agents also includes the arrhythmogenic anesthetic halothane (see insets Fig. 11.3B, C), certain alcohols (heptanol and octanol are most potent, hexanol is not and can be used to control for some nonspecific alcohol effects) and the doxyl stearate spin resonance probes.[60,66] Although arachidonic acid acts in part through generation of lipoxygenase products, effects also appear to be direct.[67-69]

Several modes of action of the lipophiles have been proposed. Based on studies with heptanol, uncoupling was initially associated with changes in general membrane fluidity,[60] although further experiments indicated closer correlation of uncoupling with fluidity changes in cholesterol-rich membrane domains.[69a] Such studies reinforce the notion that the lipid environment of the gap junction may be unusual, consistent with older studies indicating that junctional membranes were particularly rich in cholesterol (as judged by filipin incorporation into junctional complexes)[70] and by studies quantifying lipid composition in isolated junctional structures.[71,72] Changes in gap junction distribution and function have been produced by changes in membrane lipid induced by diet,[73,74] again suggesting that membrane environment may alter functions of the junctional channel. Alternatively, lipophiles may act by intercalating into hydrophobic domains of connexin molecules, or at the lipid-protein interface surrounding the channel.[61,64] Because connexins are fatty acylated, it is even possible that the lipophilic uncoupling agents destabilize the anchoring function of the myristilate or palmitate linked to the connexin molecule. Presumably, such mechanisms could be illuminated through the use of such techniques as NMR or ESR using lipophilic probes or connexins engineered with modified reporting residues. Although such studies of intrinsic tryptophan fluorescence have been applied to show calmodulin binding to connexins,[44] such protocols could presumably be modified to resolve structural correlates of channel closure at moderate or high resolution.

Burt and colleagues have recently suggested that oleic acid may more effectively block coupling between cells connected by certain connexins than others.[63] Such differential sensitivity was not observed with halothane, and the oleic acid findings may thus indicate that a new subtype designation is required to designate the pharmacological actions of lipophiles. If generally true that different connexins are differentially affected by certain lipophiles, this class of uncoupling agent may hold promise for the "silver bullet" which will selectively reduce junctional conductance without affecting other ionic channels. Furthermore, such a compound would ideally affect channels formed of one or another connexin differentially, and, moreover, related molecules might be expected to be found which increase junctional conductance rather than decreasing it.

A few studies have begun to address the action of lipophiles at the single channel level. Heptanol, octanol, halothane, arachidonic acid, oleic acid and doxyl stearic acid all apparently act on a variety of cell types without reducing γ_j. One reason that these agents are so commonly used to reduced g_j for single channel recordings is that even when g_j is reduced to very low levels the open time of individual events appears to be longer than those obtained with CO_2 uncoupling (see Fig. 11.3B, C). Thus, γ_j can be measured easily at even the modest time resolution of strip chart recordings. Although this channel behavior in the presence of lipophiles has been commented upon only phenomenologically,[60] it suggests the possibility that channel closed time is more effectively reduced than open time.

Voltage dependence of gap junction channels was initially viewed as a curiosity confined to certain "rectifying" gap junctions involved particularly in escape behaviors of vertebrates and invertebrates.[3,75] However, the discovery of voltage dependence in blastomeres from early amphibian embryos,[76] and subsequent analysis of the steady state and kinetic properties[77,78] made it clear that this sensitivity was a true gating property of the channels. Moreover, subsequent studies on a variety of cell types and on connexins exogenously expressed in *Xenopus* oocytes and in mammalian cells have established the point that all mammalian gap junction channels are voltage sensitive, with the degree of voltage sensitivity a distinguishing feature among the connexin channels.

Because each gap junction channel spans two membranes, with hemichannels docking across extracellular space, it presents two distinct regions for voltage drops to be sensed by the channel proteins. One of these is *transjunctional*, such that the potential drops from inside one cell to inside the other cell, along the axis of the channel without influence of the extracellular field. Such a drop will only occur when a potential difference occurs between the two cells (when one cell is depolarized relative to the other). The other possible potential that can be sensed is from the inside to the outside of the cell, and was initially referred to as *inside-outside* voltage dependence ($V_{i\text{-}o}$)[79] to distinguish it from transjunctional (V_j) voltage sensitivity. The field that is presumably sensed is transmural, across the wall of the channel within the extracellular space, which is the region which Loewenstein initially termed the "perijunctional insulation".[37] (It is not generally noted that because this latter field is sensed across the extracellular space, it will be obliterated by high resistance tight junctions encircling the gap junction; although it was once suggested that transjunctional voltage dependence might be attributable to $V_{i\text{-}o}$ gating under such a topology,[37] this is clearly not the general case; see below).

The initial demonstration of gating by transjunctional voltage (V_j) in amphibian embryos described several features which are now known to be generalizable to most mammalian connexins. These studies used the dual voltage clamp technique, where each cell is voltage clamped to the same holding potential. Then, as one cell is stepped to a command potential, the clamp on the other cell injects current (I_2) to hold its potential construct. This current is equivalent to current flowing through the junctional membrane (I_j), but is of opposite sign ($I_j = -I_2$); when I_j is divided by V_j, junctional conductance is measured directly. For small and brief V_j pulses, most gap junctions exhibit linear I_j-V_j relations (the clearest exception appears to be connexin37 expressed in *Xenopus* oocytes.)[80] However, when increasing V_j of either polarity and long duration (> 10 sec, typically) is applied to either cell of the pair, I_j relaxes during the pulse, reaching steady state values at the end of the pulse that depend on the magnitude of the V_j pulse (see inset, Fig. 11.3A). For the amphibian blastomeres (and for most of the connexins expressed either endogenously or exogenously in mammalian cells), the relaxations during the V_j pulses were found to be described well by single exponential declines, indicating that a two state kinetic process was involved in voltage dependent gating.[78] For amphibian blastomeres and most other cell types, when the steady-state g_j is plotted as a function of V_j of both polarities, the decrease is symmetrical about the 0 mV axis (see Fig. 11.3A, graph). Even at high voltages, however, a detectable residual conductance remains, which was originally termed minimum conductance, g_{min}. This residual voltage-insensitive conductance component is apparently a connexin-specific gating characteristic, and the ratio of minimum conductance, g_{min}, to maximal conductance at the start of the pulse, g_{max}, was about 5% in the amphibian blastomeres and varies from a high of about 40% in the case of cardiac myocytes and for connexin43 expressed in numerous other cell types or in transfected cells to a low of 0 in rat Schwann cells, where the connexin awaits identification.[81] After subtraction of the residual conductance, g_{min}, the data for V_j of each polarity can be fit by a form of the Boltzmann equation, where assumptions include a population of channels with a single open and a single closed state, where open probability is governed by the applied voltage.[76,77] Such fits determine V_0, the voltage at which the voltage sensitive component of g_j declines to 50% of its maximal value, and n, the equivalent number of gating charges; moreover, the amount of work on the channel necessary for the open-closed transition can be estimated from the product of these parameters.[77,82]

Quantitative studies of voltage dependence have now been performed on gap junctions between numerous types of mammalian cells (cardiac myocytes;[83-86] Schwann cells;[81] corpus cavernosum and tracheal airway smooth muscle;[87,88] Leydig cells;[89] CHO cells;[90] SKHep1 cells[82] and in poorly coupled mammalian cells in which connexin cDNAs have been introduced.[91-93] In addition, similar studies have been performed on chick myocytes and in mammalian cells transfected with the chick cardiac connexins.[94,95] In these studies, the decays in I_j with increasing V_j of one polarity have generally been found to be described well by single exponential kinetics. V_0 values have ranged from 15 mV or less to more than 60 mV and n has ranged from about 2 to 6 gating charges.

Interestingly, values for the same connexins expressed in *Xenopus* oocytes have differed from those obtained in mammalian cells. For example, in *Xenopus,* neither connexin32 nor connexin43 display simple kinetics of channel closure in response to V_j pulses, and V_0 is much higher than in mammalian cells. Numerous explanations might account for this difference. The most trivial is that in the oocyte gap junctions tend to form on cell processes and to be massive after cRNA injection (see ref. 96). Either of these morphological features would be expected to increase the series or access resistance to the junctional region, thereby reducing the effective V_j imposed across the junctional region under voltage clamp conditions (see refs. 91 and 97). Alternatively, differences could be due to variations in lipid composition of the membrane, to differences in post-translational processing of the connexin proteins in amphibian as compared to mammalian systems (e.g., fatty acid acylation or phosphorylation, see below), or to differences in intracellular pH or other endogenous gap junction channel ligand, or in membrane surface charge. It is highly desirable that the origin of such striking differences be resolved, because the oocyte expression system continues to be widely used without general appreciation of the differences in behavior of the junctional channels as compared to those in the tissues or species of origin. A hint at underlying phenomena which could be pursued is the striking difference in the behavior of *Xenopus* connexin38, which shows complicated kinetics in oocytes but is reliably two state in early embryos of the same species.[77,78,98]

Studies first performed in amphibian blastomeres (and subsequently observed in other systems) indicated that V_j dependent gating involved two gates in series.[77,78] When g_j was first reduced by a strong V_j pulse of one polarity and then V_j was reversed, g_j increased slowly and then decreased to the low conductance state. The kinetics of this process were consistent with those predicted for a three-state model in which a closed channel must first open before it can sense the potential of the other sign and close again. Such fitting satisfies the contingent model in connexin32[91] and connexin45[82] channels as well as the amphibian embryos, but has been challenged with regard to cardiac junctional channels.[85] However, this latter interpretation is complicated by the substantial contribution of the voltage-insensitive component of junctional conductance in these cells.

Recent studies have clarified several issues regarding the polarity and single channel components of voltage sensitivity of g_j. First, acid injection into amphibian embryonic cells led to an asymmetric voltage sensitivity, revealing that the voltage sensor on each hemichannel responded to relative positivity of each cell.[91] Asymmetric pairings of connexin32 and connexin26 expressed in oocytes led to steady stage g_j-V_j relations which were initially reported not to be the geometric sum of each hemichannel's properties in the symmetric case;[100] however, it is now realized that the properties do match those predicted if gates are assumed to possess opposite polarities of voltage sensitivity.[101] Most recently, charge substitutions in the first extracellular loop or in transmembrane domains of the connexin molecule have reversed field sensitivity, suggesting a role of this domain in the voltage sensor.[101,102] However, these extensive domain substitutions have thus far not totally resolved where the sensor lies within the connexin structure.

The other major insight into voltage dependence has come from single channel studies on connexin43 expressed in transfected mammalian cells, in which the nature of g_{min} has been revealed. In its initial description, several possible interpretations of this residual voltage-sensitive component of the g_j were given: that it represented a separate population of channels which were voltage insensitive, that it was due to a nonzero open probability of the channel, which might be caused by reorientation of the channel voltage sensor (or attributable to cycling around a circular reaction scheme)[79] or that it was due to a voltage independent substate of the channel. Spray et al[76] tried to test this latter hypothesis by injecting Lucifer Yellow into one cell while junctional channels were held in the reduced conductance condition. Although dye passage was reduced profoundly at large V_j, the low g_{min} and the poor detectability in these opaque cells precluded an unambiguous conclusion. Verselis et al,[13] using K sensitive electrodes to detect TEA permeability, noted nonzero permeability at low g_j values, seemingly favoring all-or-none closure. However, in connexin43-expressing cells, where g_{min} is large, openings of the main state do not occur at high voltages, and uncoupling by halothane under these conditions reveals that channels are all of smaller size (illustrated in Fig. 11.3B, C).[103] Thus, for mammalian connexin43 (and possibly for *Xenopus* connexin38,[104] chick cardiac channels,[105] insect channels[106] and human connexin45 as well)[82] the residual conductance, g_{min}, corresponds to a voltage insensitive substate of the main channel.

In a few systems, g_j is low under normal conditions and it has now been possible to compare open probability at the single channel level with macroscopic voltage dependence.[81,82,104,107] In all systems, P_o has been found to be between 0.5 and 1 at V_j= 0 mV and to decrease at higher voltages with P_o, following the form of the macroscopic Boltzmann behavior.

By contrast to the V_j sensitivity observed in all vertebrate and most invertebrate preparations, $V_{i\text{-}o}$ sensitivity is largely confined to a few well-studied preparations in arthropods, and in one molluscan and one mammalian system. This property was first characterized in *Chironomus* salivary gland,[108] and subsequently has been detected in squid blastomeres,[79]

Drosophila salivary gland,[109] a cell line derived from mosquito,[106] insect epidermal cells,[110] crayfish hepatopancreatic cells,[111] and in connexin26 expressed in oocytes.[100] In all these invertebrate preparations depolarization of either or both cells decreases g_j such that at 0 mV resting potential the cells are almost completely uncoupled. In all cases except the crayfish hepatopancreas, V_{i-o} and V_j sensitivity coexist, complicating data analysis. As is mentioned above, the field that is presumably sensed by this gating behavior is that across the wall of the channel, most logically at the interface between the docking hemichannels. Although this uncoupling was initially attributed to reduced channel size,[108] Chanson et al[111] did not detect such changes, and it remains possible that the effect on γ_j in the mosquito cells could be due to transjunctional voltage sensitivity.

V_{i-o} sensitivity is substantial in connexin26 expressed in oocytes[100] although strong V_{i-o} dependence was not detected in preliminary experiments performed on pc12 cells transfected with connexin26.[112] Resolution of whether the differences are preparation- or experimental-dependent awaits more detailed examination.

The final gating mechanism to be considered in this brief review is that of phosphorylation of the connexin proteins. Connexin32 was one of the first membrane channel proteins to be demonstrated to be a phosphoprotein,[113] and the major site of phosphorylation in response to either cAMP-dependent kinase or protein kinase c has now been mapped to SER232 using synthetic peptides and matching these peptides to protease fingerprinting of phosphorylated connexin32.[114] The time course of phosphorylation matches that of increased g_j in these cells in response to cAMP-elevating treatments (although comparable studies with kinase c activation remain to be performed). Stoichiometry of ^{32}P incorporation into connexin32 in response to cAMP is rather low, about 10%, although the stoichiometry of incorporation into the corresponding peptides is almost 100% with kinase c.[114,115] Whether the low stoichiometry in vivo represents phosphorylation of only a limited number of connexin molecules within the connexon remains to be determined, as does whether single channel properties of steady state or kinetic parameters of voltage sensitivity are affected.

Connexin26 is apparently not a phosphoprotein in vivo and is not phosphorylated by any of the attempted kinases.[114]

Phosphorylation of connexin43 has received the greatest attention by far, primarily because of the findings that phosphorylation results in altered gel mobility that can be monitored on Western blots without the necessity of autoradiographs.[116] Of the numerous potential phosphorylation sites within the murine connexin43 sequence, there is moderately strong evidence that four are utilized: SER262, TYR286, SER364 and SER368. To begin with the tyrosine residue, Paul's group demonstrated TYR phosphorylation in oocytes injected with src (a tyrosine kinase), which correlated with uncoupling.[117] Evidence that TYR286 was the residue involved was provided by mutation of this site to PHE, after which the protein was no longer phosphorylated and coupling was no longer responsive to src. That such a site might be functionally useful has been indicated by Bruce Nicholson's laboratory, which has shown that transfection of cells with connexin32, which does not contain a src tyrosine phosphorylation site, protects cells from the phenotypic transformation effects of src.[118]

Evidence that phosphorylation of SER364 and SER368 result in physiological effects on connexin43 gap junction channels has been provided. In cardiac cells, g_j increased in response to kinase a or kinase c activation and was inhibited by inhibitors of these kinases.[119,120] In studies using synthetic peptides, these serine residues were found to be superb phosphorylation substrates for kinase c, with stoichiometry of 1 mole phosphate per mole peptide per site.[119] Studies on human connexin43 transfectants indicate that phosphorylation state presumably of these same two SER residues, affects single channel conductance.[120] Treatment with okadaic acid, an inhibitor of phosphatase, results in phosphorylation of the protein and over a similar time course okadaic acid and activators of protein kinases a and c lead to smaller γ_j values (60 vs 90 pS events measured in CsCl internal solution at moderate V_j's). Similarly, intracellularly applied phosphatase or the protein kinase inhibitor staurosporine increased γ_j and led to decreased phosphate incorporation. Interestingly, dephosphorylation by phosphatase treatment resulted in much faster decline in I_j during pulses of all voltages, although the steady state Boltzmann distribution was not found to be noticeably altered.[121] These findings indicate that both opening and closing rate constants were affected to the same extent. Mutational analysis of the human connexin43 sequence supports the assignment of phosphorylation to these SER residues: substitution by ALA or truncation of the molecule to remove the residues renders the channels insensitive to phosphorylating treatments.[93,122]

Evidence that SER262 is phosphorylated in the rat sequence comes from studies in which cGMP stimulation increases phosphate incorporation into connexin43 from myocytes or from rat connexin43 transfectants, but does not affect incorporation into the human sequence, which does not contain this putative cGMP phosphorylation site.[123] Critical studies requiring further experimentation include determining effects on the voltage independent substate, single channel evaluation of junctional conductance of TYR phosphorylation, as well as further mutational analysis of channel responses to these pharmacological agents.

It remains to be determined whether gap junction proteins formed of other connexins will show the remarkable range of plasticity in response to phosphorylation which is exhibited by connexin43. However, comparison of c-terminal sequences does indicate the presence of similar consensual phosphorylation sites, suggesting that such manipulation will be possible (see ref. 6).

In summary, the gating and biophysical properties of gap junction channels endow this family of channel forming proteins with a remarkable range of functional plasticity. During pathological situations such as ischemia, most of these mechanisms are activated: pH_i is reduced, Ca^{2+} is mobilized, lipophiles are generated, transjunctional voltage gradients exist and protein phosphorylation patterns are markedly changed.[2,59] Whether protective pharmacological agents can be developed with knowledge of these properties should be a major goal of physiologists now working on gap junction gating. Moreover, understanding how mutations may affect function[59,93,101,122] is of growing importance, with recent reports that human connexin43 phosphorylation mutants are associated with human teratology,[124] that connexin43 "knockouts" survive until birth but then die (C. Naus and J. Roussant, personal communication) and that loss of connexin32 expression leads to peripheral demyelination.[125]

References

1. Warner A. Gap junctions in development-a perspective. Seminar Cell Biol 1993; 3:81-91.
2. Spray DC, Rook M, Moreno AP, Saez JC, Christ G, Campos de Carvalho AC, and Fishman GI. Cardiovascular gap junctions: Gating properties, function and dysfunction. In: Ion Channels in the Cardiovascular System: Function and Dysfunction. Futura Publishing, Mt. Kisco, NY, 185-217.
3. Furshpan EJ, Potter DD. Transmission at the giant motor synapses of the crayfish. J Physiol 1959; 145:289-325.
4. Musil L. chapter 10. This volume.
5. Spray DC, Saez JC, Burt JM, Watanabe T, Reid LM, Hertzberg EL, Bennett MVL. Gap junctional conductance: multiple sites of regulation In: Hertzberg EL, Johngon RG, eds. Gap Junctions. New York: AR Liss, 1988:227-244.
6. Willecke K, Hennemann M, Dahl E, Jungbluth S. The mouse connexin gene family. Progr Cell Biol 1993; 3:33-40.
7. Dermietzel R, Hwang TK, Spray DC. The gap junction family: structure, function and chemistry. Anat Embryol (Berl) 1990; 182:517-528
8. Simpson I, Rose B, Loewenstein WR. Size limit of molecules permeating the junctional membrane channels. Science 1977; 195:294-296.

9. Schwartzmann G, Wiegandt H, Rose B, Zimmerman A, BenHaim D, Loewenstein W.R. Diameter of the cell-to-cell junctional channels as probed with neutral molecules. Science 1981; 213:551-553.
10. Zimmerman AL, Rose B. Permeability properties of cell-to-cell channels: kinetics of fluorescent tracer diffusion through a cell junction. J Membr Biol 1985; 84:269-283.
11. Flagg-Newton J, Simpson I, Loewenstein WR. Permeability of the cell-to-cell membrane channels in mammalian cell junction. Science 1979; 205:404-407.
12. Imanaga I. Cell-to-cell coupling studies by diffusional methods in myocardial cells. Experientia 1987; 43:1080-1083.
13. Verselis V, White RL, Spray DC, Bennett MVL. Gap junctional conductance and permeability are linearly related. Science 1986; 234:461-464.
14. Verselis V, White RL, Spray DC, Bennett MVL. Relative permeabilities of gap junctions in tetraalklylammonium ions: Comparison of chemical and electrical fluxes. Submitted to J Gen Physiol.
15. Bodmer R, Verselis V, Levitan I, Spray DC. Electrotonic synapses in *Aplysia* neurons in situ and in culture: Characteristics and permeability measurements. J Neurosci 1988; 8:1656-1670.
16. Bennett MVL, Spira ME, Spray DC. Permeability of gap junctions between embryonic cells: A re-evaluation. Devel Biol 1978; 65:114-125.
17. Bennett MVL. Junctional permeability. In: Feldman J, Gilula NB, Potts JD, eds. Intercellular Junctions and Synapses. London: Chapman and Hall, 1978:25-36.
18. Robinson SR, Hampson ECGM, Munro MN, Vaney DI. Undirectional coupling of gap junctions between neuroglia. Science 1993; 262:1072-1074.
19. Caveney S. The role of gap junction in development. Ann Rev Physiol 1985:319-335.
20. Flagg-Newton JL, Loewenstein WR. Asymmetrically permeable membrane channels in cell junction. Science 1980; 207:771-773.
21. Peinado A, Yuste R, Katz LC. Extensive dye coupling between rat neocortical neurons during ther period of circuit formation. Neuron 1993; 10:103-114.
22. Pitts JD, Finbow M. Junctional permeability and its consequences. In: DeMello WC, ed. Intercellular Communication. New York: Plenum, 1977:61-68.
23. Hobbie L, Kingsley DM, Kozarsky KE, Jackman RW, Krieger M. Restoration of LDL receptor activity in mutant cells by intercellular junctional communication. Science 1987; 235:69-73.
24. Dunlap K, Takeda K, Brehm P. Activation of a calcium-dependent phosphoprotein by chemical signaling through gap junctions. Nature 1987; 325:60-62.
25. Saez JC, Connor JA, Spray DC, Bennett MVL. Hepatocyte gap junctions are permeable to the second messenger, inositol 1,4,5-triphosphate, and to calcium ions. Proc Natl Acad Sci U.S.A. 1989; 86:2708-2712.
26. Charles DC, Merrill JE, Dirksen ER, Sanderson MJ. Intercellular signaling in glial cells: Calcium waves and oscillations in response to mechanical stimulation and glutamate. Neuron 1991; 6:983-992.
27. Sanderson MJ, Charles AC, Dirksen ER. Mechanical stimulation and intercellular communication increases intracellular Ca^{2+} in epithelial cells. Cell Regulation 1990; 1:585-596.
28. Nathanson M, Padfield P, O'Sullivan A, Burgstahler A, Jamieson J. Mechanism of Ca^{2+} wave propagation in pancreatic acinar cells. J Biol Chem 1992; 267:18118-18121.
29. Christ GJ, Moreno AP, Melman A, Spray DC. Gap junctions mediate intercellular diffusion of Ca^{2+} in cultured human corporal smooth muscle cells. Amer. J Physiol 1992; 263 (Cell Physiol 32):C373-C383.
30. Cornell-Bell A, Finkbeiner S, Cooper M, Smith S. Glutamate induces calcium waves in cultured astrocytes: Long-range glial signaling. Science 1990; 247:470-473.
31. Demer LL, Wortham CM, Dirksen ER, Sanderson MJ. Mechanical stimulation induces intercelllular calcium signaling in bovine aortic endothelial cells. Am J Physiol 1993; 264:H2094-H2102.
32. Fernandes-de Lima VM, Scheller D, Tegtmeier F, Hanke W, Schlue WR. Self-sustained spreading depressions in the chicken retina and short-term neuronal-glial interactions within the gray matter neuropil.

33. Flagg-Newton J, Loewenstein WR. Experimental depression of junctional membrane permeability in mammalian cell culture. A study with tracer molecules in the 300 to 800 Dalton range. J Membr Biol 1979; 50:65-100.
34. Moreno AP, Fishman GI, Beyer EC, Spray DC. Heterotypic gap junction channels formed of connexins 43 and 45 in mammalian cells conserve the properties of hemichannels. Soc Neurosci Abstr, 1993; 19:1262.
35. Aslanidi KB, Boitsova LJ, Chailakhyan LM, Kublik LN, Marachova II, Potopova TV, Vinogradova TA. Energetic cooperation via ion-permeable junctions in mixed animal cell cultures. FASEB 1991; 283:295-297.
36. Ledbetter ML, Lubin M.. Transfer of potassium. A new measure of cell-cell coupling. J Cell Biol 1979; 80:150-165.
37. Loewenstein WR. Junctional intercellular communication: the cell-to-cell membrane channel. Physiol Rev 1981; 61,829-913.
38. Brink PR, Dewey MM. Nexal membrane permeability to anions. J Gen Physiol 1978; 72:69-78.
39. Neyton J, Trautmann A. Single-channel currents of an intercellular junction. Nature 1985; 317:331-335
40. Manivannan K., Ramanan SV, Mathias RT, Brink PR. Multichannel recordings from membranes which contain gap junctions. Biophys J 1992; 61:216-227.
41. Dermietzel R., Spray, DC. Gap junctions in the brain: Where, what type, how many, and why? Trends in Neuroscience 1993; 16:185-192.
42. Bennett MVL. Physiology of electrotonic junctions. Ann NY Acad Sci 1966; 137:509-539.
43. Socolar SJ, Politoff AL. Uncoupling cell junctions in glandular epithelium by depolarizing current. Science 1971; 172:492-494.
44. Peracchia C. The calmodulin hypothesis- six years later. In: Hertzberg EL, Johnson RG, eds. Gap Junctions New York: Alan R Liss, 1988:267-282.
45. Turin L., Warner AE. Intracellular pH in early *Xenopus* embryos: its effect on current flow between blastomeres. J Physiol (Lond). 1980; 300:489-504.
46. Spray DC, Harris AL, Bennett MVL. Gap junctional conductance is a simple and sensitive function of intracellular pH. Science 1981; 211:712-715.
47. Spray DC, Bennett MVL. Physiology and pharmacology of gap junctions. Ann Rev Physiol 1985; 47:281-303.
48. Spray DC, Stern JH, Harris AL, Bennett MVL. Comparison of sensitivities of gap junctional conductance to H and Ca ions. Proc Natl Acad Sci USA, 1982; 79:441-445.
49. Arellano RO, Ramon F., Rivera A, Zampighi GA. Lowering of pH does not directly affect the junctional resistance of crayfish lateral axons. J Membr Biol 1986; 94:293-299.
50. Campos de Carvalho AC, Verselis VK, White RL, Bennett MVL and Spray DC Calmodulin and the control of intercellular communication. Submitted.
51. Burt JM. Block of intercellular communication: interaction of intracellular H^+ and Ca^{2+}. Amer J Phsyiol 1987; 253:C607-612.
52. White RL, Doeller JE, Verselis VK, Wittenberg BA. Gap-junctional conductance between pairs of ventricular myocytes is modulated synergistically by H+ and Ca^{2+}. J Gen Physiol 1990; 95,1061-1075.
53. Lazrak A, Peracchia C. Gap junction sensitivity to physiological internal calcium regardless of pH in Novikoff hepatoma cells. Biophys J 1993; 65:2002-2012.
54. Dunina-Barkovskaya A, Plonsky I, Muravyova O, Margolis L, Chailakyan L. CO_2 effect and pH-dependency of mouse hepatocyte junctional conductance. Abstr. presented at Workshop on cell networks, Palanga, Lithuania, 1992.
55. Liu S, Taffet S, Stoner L, DelMar M, Vallano ML, Jalife J. A structural basis for the unequal sensitivity of the major cardiac and liver gap junctions to intracellular acidification. The carboxyl tail length. Biophys J 1993; 64:1422-1433.
56. Kolb HA, Somogyi R. Biochemical and biophysical analysis of cell-to-cell channels and regulation of gap junctional permeabil ity. Rev Physiol Biochem Pharmacol 1991; 118:2-47.

57. Campos de Carvalho AC, Waltzmann M, Spray DC. Gating properties of isolated liver gap junction membranes incorporated into lipid bilayers. Brazil. J Biol Med 1992; 25:81-92.
58. Burt JM, Spray DC. Single channel events and gating behavior of the cardiac gap junction channel. Proc Natl Acad Sci USA 1988; 85:3431-3434.
59. Corr PB, Snyder DW, Cain ME, Crafford Jr. WA, Gross R, Sobel BE. Electrophysiological effects of amphiphiles on canine Purkinje fibers; implications for dysrhythmia secondary to ischemia. Circ Res 1981; 49, 354-363.
60. Takens-Kwak BR, Jongsma HJ, Rook MB, van Ginneken ACG. Mechanism of heptanol-induced uncoupling of cardiac gap junctions; a perforated-patch clamp study. Am J Physiol 1992; 262,1531-1538.
61. Spray DC, Burt JM. Structure-activity relations of the cardiac gap junction channel. Amer J Physiol 1990; 258:C195-C207.
62. Burt JM, Spray DC. Volatile anesthetics reversibly reduce gap junctional conductance between cardiac myocytes. Circ Res 1989; 65:829-837.
63. Hirschi KK, Minnich BN, Moore LK, Burt JM. Oleic acid differentially affects gap junction-mediated communication in heart and vascular smooth muscle cells. Amer J Physiol 1993; 265:C1517-C1526.
64. Burt JM. Modulation of gap junction channel activity by the lipid environment. In: Peracchia C, ed. Biophysics of Gap Junction Channels. Boca Raton: CRC Press, 1991:75-93.
65. Burt JM. Uncoupling of cardiac cells by doxyls stearic acids: specificity and mechanism of action. Am J Physiol 1989; 256:C913-C924.
66. Hennemann H, Suchyna T, Lichtenberg-Frate H, Jungbluth S, Dahl E, Schwartz J, Nicholson BJ, Willeke K. Molecular cloning and functional expression of mouse connexin40, a second gap junction gene preferentially expressed in lung. J Cell Biol 1992; 117:1299-1310.
67. Massey KD, Minnich BN, Burt JM. Arachidonic acid and lipoxygenase metabolites uncouple neonatal rat cardiac myocyte pairs. Am J Physiol 1992; 263:C494-C501.
68. Agawal R, Daniel EE. Control of gap-junction formation in canine trachea by arachidonic acid metabolites. Am J Physiol 1986; 250,C495-C505.
69. Fluri GS, Rudisuli A, Willi M, Rohr S, Weingart R. Effects of arachidonic acid on the gap junction of neonatal rat heart cells. Pflugers Arch. 1990; 417:149-156.
69a. Bastiaanse EM, Jongsma HJ, et al. Heptanol-induced decrease in cardiac gap junctional conductance is mediated by a decrease in the fluidity of membranous cholesterol-rich domains. J Membr Biol 1993; 136:135-145.
70. Mazet F. Filipin and digitonin studies of membrane cholesterol in frog atrial fibers with unusual gap junction configurations. J. Mol. and Cell Card. 1987; 19:1121-1128.
71. Spray DC, Saez JC, Brosius D, Bennett, MVL, Hertzberg EL. Isolated liver gap junctions: Gating of transjunctional currents is similar to that in intact pairs of hepatocytes. Proc Natl Acad Sci USA 1986; 83:5494-5497.
72. Evans WH. A biochemical dissection of the functional polarity of the plasma membrane of the hepatocyte. Biochim Biophys. Acta 1980; 604:27-64.
73. Aylsworth, Welsch CW, Kabara JJ, Troski JE. Effects of fatty acids on gap junctional communication; possible role in tumor promotion by dietary fat. Lipids 1987; 6,445-454.
74. Meyer RA, Lampe PD, Malewicz B, Bauman W, Johnson RG. Low-density lipoprotein and apolipoprotein B enhance gap junction assembly and cell-cell communication. Exp. Cell. Res. 1991; 196:72-81.
75. Auerbach AA, Bennett MVL. A rectifying electrotonic synapse in the central nervous system of a vertebrate. J.Gen.Physiol. 1969; 53:211-237.
76. Spray DC, Harris AL. and Bennett MVL. Voltage dependence of junctional conductance in early amphibian embryos. Science 1979; 204:432-434.

77. Spray DC, Harris AL, Bennett MVL. Equilibrium properties of a voltage dependent junctional conductance. J Gen Physiol 1981; 77:75-94.
78. Harris AL, Spray DC, Bennett MVL. Kinetic properties of a voltage dependent junctional conductance. J Gen Physiol 1981; 77:95-120.
79. Spray DC, White RL, Campos de Carvalho A, Harris AL, Bennett MVL. Gating of gap junctional conductance. Biophys J 1984; 45:219-230.
80. Willecke K, Heynkes R, Dahl E, Stutenkemper R, Hennemann H, Jungbluth S, Suchyna T, Nicholson B. Mouse connexin37: cloning and functional expression of a gap junction gene highly expressed in lung. J Cell Biol 1991; 114:1049-1057.
81. Chanson M, Chandross K, Rook MB, Kessler JA, Spray DC. Gating characteristics of a steeply voltage dependent gap junction channel in rat Schwann cells. J Gen Physiol 1993; 102:925-946.
82. Moreno AP, Laing JG, Beyer EC, Spray DC. Properties of gap junction channels formed of connexin45 endogenously expressed in human hepatoma (SKHep1) cells. Submitted to Amer J Physiol
83. Lal R and Arnsdorf MF. Voltage-dependent gating and single-channel conductance of adult mammalian atrial gap junctions. Circ Res 1992;71:737-743
84. Anumonwo JM, Wang HZ, Trabka Janik E, et al. Gap junctional channels in adult mammalian sinus nodal cells. Immunolocalization and electrophysiology. Circ Res 1992;71:229-239
85. Wang H-Z, Li J, Lemanski LF, Veenstra RD Gating of mammalian cardiac gap junction channels by transjunctional voltage. Biophys J 1992; 63:139-151.
86. Rook MB, Jongsma HJ, van Ginneken ACG. Properties of single gap-junctional channels between isolated neonatal rat heart cells. Am J Physiol 1988; 225,H770-H782.
87. Moreno AP, Campos de Carvalho AC, Christ G, Spray, DC. Gap junctions between human *corpus cavernosum* smooth muscle cells: Gating properties and unitary conductance. Amer J Physiol 1993; 33:C80-C92.
88. DePalo LR, Spray DC, Hertzberg EL, Castaldi MC, Kotlikoff MI, Kessler JA. Connexin43 expression and functional gap junction formation in cultured human airway smooth muscle cells. Submitted to Amer J Physiol
89. Perez Armendariz M, Romano M, et al. Characterization of gap junctions between pairs of Leydig cells from any testis. Amer J Physiol 1994, in press.
90. Eckert R, Dunina-Barkovskaya A, Hulser D. Biophysical characterization of gap-junction channels in HeLa cells. Plugers Arch. 1993; 424:335-342.
91. Moreno AP, Eghbali B, Spray DC. Connexin32 gap junction channels in stably transfected cells. Equilibrium and kinetic properties. Biophys J 1991; 60:1267-1277
92. Spray DC, Moreno AP, Eghbali B, Fishman GI, Chanson M. Gating of gap junction channels as revealed in cells stably transfected with wild type and mutant connexin cDNAs. Biophys J 1992; 62:51-53.
93. Fishman GI, Moreno AP, Spray DC, Leinwand LA. Functional analysis of human cardiac gap junction channel mutants. Proc Natl Acad Sci U S A 1991; 88:3525-3529.
94. Beyer EC. Molecular cloning and developmental expression of two chick embryo gap junction proteins. J Biol Chem 1990; 265:14439-14443.
95. Veenstra RD, Wang HZ, Westphale EM, Beyer EC. Multiple connexins confer distinct regulatory and conductance properties of gap junctions in developing heart. Circ Res 1992; 71:1277-1283.
96. Nicholson BJ, Suchyna T, Xu LX, Hammernick P, Cao FL, Fourtner C, Barrio L, Bennett MVL. Divergent properties of different connexins in *Xenopus* oocytes. In: Hall JE, Zampighi GA, Davis RM, eds. Progress in Cell Research, Vol. 3. Elsevier, 1993:3-13.
97. Wilders R, Jongsma HJ. Limitation of the dual voltage clamp method in assaying conductance and kinetics of gap junction channels. Biophys J 1992; 63:942-953.
98. Ebihara L Beyer EC, Swenson KI, Paul DL, Goodenough DA.. Cloning and expression of a *Xenopus* embryonic gap-junction protein. Science 1989; 243, 1194-1195.

99. Bennett MVL, Verselis V, White RL, Spray, DC. Gap junctional conductance: Gating. In: Hertzberg EL, Johnson RG, eds. Gap Junctions. AR Liss, 1988:287-304.
100. Barrio LC, Suchyna T, Bargiello T, Xu LX, Roginski RS, Bennett MVL, Nicholson BJ. Gap junctions formed by connexins26 and 32 alone and in combination are differently affected by applied voltage. Proc Natl Acad Sci USA 1991; 88:8410-8414.
101. Rubin JB, Verselis VK, Bennett MV, Bargiello TA. Molecular analysis of voltage dependence of heterotypic gap junctions formed by connexins 26 and 32. Biophys J 1992; 62:183-93
102. Yox D, Rosinski C, Xu LX, Nicholson BJ. Switching of charged amino acids in membrane spanning segments of connexin26 reverses the polarity of voltage gating of their gap junctional channels. Mol Biol Cell 1993; 4(Suppl).328a.
103. Moreno AP, Rook MB, Fishman GI, Spray DC. Gap junction channels: Distinct voltage-sensitive and insensitive conductance states. Biophys J 1994; in press.
104. Moreno AP, Fishman GI, Spray DC. Kinetic analysis of single gap junction chanels: use of communication-deficient cells (SKHep1). Biophys J 1992; 61:A505.
105. Chen Y-H, DeHaan RL. Multiple channel conductance states and voltage regulation of embryonic chick cardiac gap junctions. J Membr Biol 1992; 127, 95-111.
106. Weingart R, Bukauskas FF. Gap junction channels of insects exhibit a residual conductance. Pflug Arch. *1993;* 424:192-194.
107. Rook MB, van Ginneken ACG, de Jonge B, El Aoumari A, Gros D, Jongsma HJ. Differences in gap junction channels between cardiac myocytes, fibroblasts and heterologous pairs. Am J Physiol 1992; 263:C959-C977.
108. Obaid, A.L., Socolar, S.J. and Rose, B. Cell-to-cell channels with two independently regulated gates in series: Analysis of junctional conductance modulation by membrane potential, calcium and pH. J Membr Biol 1983; 73,69-89.
109. Verselis VK, Bennett MVL, Bargiello TA. A voltage-dependent gap junction channel in Drosophila melanogaster. Biophys J 1992; 59:114-122.
110. Churchill D, Caveney S. Double whole-cell patch-clamp of gap junctions in insect epidermal cell pairs: single channel conductance, voltage dependence, and spontaneous uncoupling. Progress in Cell Research 1993; 3:239-245.
111. Chanson M, Roy C, Spray DC. Voltage-dependent gap junctional conductance in hepatopancreatic cells of *Procambarus clarkii.* Amer J Physiol 1994; 35:C569-C577.
112. Chanson M, Eghbali B, Spray DC. Voltage-dependence and unitary conductance of novel gap junctions channels in rat pheochromocytoma (PC12) cells. Biophys J 1992; 61:A506.
113. Saez JC, Spray DC, Mairn A, Hertzberg EL, Greengard P, Bennett MVL. cAMP increases junctional conductance and stimulates phosphorylation of the 27kDa principal gap junction polypeptide. Proc Natl Acad Sci USA 1986; 83:2473-2477.
114. Saez JC, Nairn AC, Spray DC, Hertzberg EL, Greengard P, Bennett MVL. The main hepatocyte gap junction protein (MP27) is phosphorylated by cAMP and Ca^{2+} dependent kinases. Eur J Biochem 1990; 192:263-273.
115. Takeda A, Saheki S, Shimazu T, Takeuchi N. Phosphorylation of the 27 kDa gap junction protein by protein kinase c in vitro and in rat hepatocytes. J Biochem (Tokyo) 1989; 106:723-731.
116. Musil LS, Beyer EC, Goodenough DA. Expression of the gap junction protein connexin43 in embryonic chick lens: molecular cloning, ultrastructural localization, and post-translational phosphorylation. J Membr Biol 1990; 116:163-175
117. Swenson KI, Piwnica-Worms H, McNamee H, Paul D. Tyrosine phosphorylation of the gap-junction protein connexin43 is required for the pp 60 v-*src*-induced inhibition of communication. Cell Regulation 1990; 1, 989-1002.
118. Garcia AM, Nicholson BJ. Dependence of cell transformation by v-src on intercellular coupling. J Cell Biol 1991; 115:192a.
119. Saez JC, Nairn A, Spray DC, Hertzberg EL. Rat connexin43: Regulation by phosphorylation in heart. In Gap Junctions (J.E. Hall,m G.A. Zampighi, R.M. Davis, Eds), Progress in Cell Research, Vol. 3, Elsevier, Amsterdam, 1993; pp.211-217.

120. Moreno AP, Saez JC, Fishman GI, Spray DC. Human connexin43 gap junction channels: Regulation of unitary conductances by phosphorylation of the channel protein. Circ Research 1994; 74.
121. Moreno AP, Fishman GI, and Spray DC. Phosphorylation shifts unitary conductance and modifies voltage dependent kinetics of human connexin43 gap junction channels. Biophys J 1992; 62:51-53.
122. Moore LK, Moreno AP, Fishman GI, Spray DC. Human connexin43 (hCx43) phosphorylation site mutants: Unitary conductance and voltage sensitivity of channels in stable transfectants. Biophys J , 1994; 66:A260.
123. Takens-Kwak BR, Saez JC, Fishman GI, Wilders R, Hertzberg EL, Spray DC, Jongsma, HJ cGMP-dependent phosphorylation of connexin43: Influence of gap junction channel conductance and kinetics. Submitted to Amer J Physiol.
124. Britz-Cunningham SH, Shah MM, Martinez A-M, Fletcher WH. Mutations in the connexin43 (a1) gene are associated with defects in heart development. Mol Biol Cell 1993; 4 (suppl):329a.
125. Bergoffen J, Scherer SS, Wang S, Scott MO, Bone LJ, Paul DL, Chen K, Lensch MW, Chance PF, Fischbeck KH. Connexin mutations in x-linked Charcot-Marie-Tooth disease. Science 1994, 1993; 262: 2039-2042.
126. Dermietzel R, Kessler JA, Hertzberg EL, Spray DC. Gap junctions between cultured astrocytes: Immunocytochemical, molecular and electrophysiological analysis. J Neuroscience 1991; 11:1421-1432.
127. Giaume C, Fromaget C, Aoumari AE, Cordier J, Glowinski J, Gros D. Gap junctions in cultured astrocytes, single-channel currents and characterization of channel-forming protein. Neuron 1991; 6,133-143.
128. Moreno AP, Eghbali B, Spray DC. Connexin32 gap junction channels in stably transfected cells: Unitary conductance. Biophys J 1991; 60:1254-1266.
129. Rudisuli A, Weingart R. Electrical properties of gap-junction channels in guinea-pig ventricular cell pairs revealed by exposure to heptanol. Pflugers Arch 1989; 415:12-21.
130. Spray DC, Chanson MA, Moreno AP, Dermietzel R, Meda P. Distinctive types of gap junction channels connect WB cells, a clonal cell line derived from rat liver. Amer J Physiol 1991; 260 (Cell Physiol. 29):C:513-C527.
131. Vink MJ, Christ GJ, Hertzberg EL, Spray DC. Gap junctions in rabbit femoral and mesenteric arterial smooth muscle in culture: Unitary and macroscopic junctional conductance and presence of connexin43. Circulation, 1993; 88:I-174.
132. Waltzmann MN, Bai S, Spray DC. Stable transfection and functional expression of a gap junction protein, connexin37 (Cx37), in a communication deficient cell line. Biophys J , 1994; 66:A260.
133. Spray DC, Campos de Carvalho AC, Moreno AP, Scemes E, Leidigh C, Kim NH, Pappas GD, Bennett MVL. Gap junction channels in marine embryos: Comparison of properties in late blastulae of squid (*Loligo paelei*) and skate (*Raja erinacea*). Biol Bull 1990; 179:225-226.
134. Moreno AP, Spray DC. Sea urchin embryos: Suitability for exogenous expression of gap junction channels. Biol Bull 1990; 179:225.
135. McMahon DG, Knapp AG, and Dowling JE. Horizontal cell gap junctions: Single channel conductance and modulation by dopamine. Proc Natl Acad Sci USA 1989; 86:7639-7643.
136. Chow I, Young SH. Opening of single gap junction channels during formation of electrical coupling between embryonic muscle cells. Devel Biol 1987; 122:332-337.
137. Spray DC, Moreno AP, Roy C, Saez JC, Burt JM, Hertzberg E, Campos de Carvalho AC, Dermietzel R, Hatcher V, Wittner M, Tanowitz H. Gap junctions between endothelial cells: Unitary conductance, identity of the channel protein, and response to second messengers and Trypanosome infection. Physiologist 1989; 32:212.
138. Spray DC, Moreno AP, Kessler JA, Dermietzel R. Characterization of gap junctions between cultured leptomeningeal cells. Brain Research 1991; 598:1-14.
139. Moore LK, Beyer EC, Burt JM Characterization of gap junction channels in A7r5 vascular smooth muscle cells. Amer J Physiol 1991; 260:C975-C981.

CHAPTER 12

MOLECULAR ORGANIZATION OF THE PLAQUE STRUCTURE OF CADHERIN-BASED CELL-TO-CELL ADHERENS JUNCTIONS AND ITS POSSIBLE INVOLVEMENT IN TUMORIGENESIS AND METASTASIS

Shoichiro Tsukita
Akira Nagafuchi
Sachiko Tsukita
Shigenobu Yonemura

1. INTRODUCTION

The adherens junction (AJ) represents a specialized class of cell-to-cell and cell-to-substrate associations:[1] the adhesion molecules are cadherins[2] and integrins,[3] respectively. This type of junction is characterized by the well developed undercoat on the cytoplasmic surface of the junctional membrane ('plaque' structure) through which actin filaments are tightly associated. AJ is commonly believed to be directly involved in tissue morphogenesis. This feature is, most likely, related to the tight association between AJ and the force-generating actin filament bundles. In addition to these mechanical properties of AJ, the notion has become widespread that AJ plays crucial roles in the control of cell growth and tissue morphogenesis through transmembrane signaling.[4,5]

Recently, the information of the adhesion molecule-based signaling in AJ is gradually increasing, and evidence has been accumulating that AJ is

Molecular Mechanisms of Epithelial Cell Junctions: From Development to Disease
Edited by Sandra Citi, M.D., Ph.D.;

directly involved in tumorigenesis and metastasis[6] (see also chapter 13 in this volume). Here we review our recent findings on the molecular organization of the plaque structure of the cell-to-cell AJ and its possible involvement in tumorigenesis and metastasis.

2. DEVELOPMENT OF THE ISOLATION PROCEDURE FOR CELL-TO-CELL AJ

As a model of the cell-to-substrate AJ, the membrane fraction prepared from smooth muscle cells was used, leading us to conclude that some unique proteins such as vinculin and talin localize in the plaque structure of cell-to-substrate AJ.[7,8] Talin was detected only in the cell-to-substrate AJ, while vinculin was localized in cell-to-cell as well as cell-to-substrate AJ. Some efforts were made to isolate the cell-to-cell AJ, mainly from cardiac muscle cells[9-11] However, the purity of the cell-to-cell AJ fraction from cardiac muscle seemed to be rather low. For example, on electrophoresis of this fraction, vinculin did not appear to dominate. Therefore, to analyze the molecular organization of the cell-to-cell AJ systematically, it was first necessary to develop a new procedure for the isolation of this type of junction.

In 1989, we attempted to isolate the cell-to-cell AJ . Our criteria for the isolation of the cell-to-cell AJ was as follows. (a) In the fraction obtained, AJ should be highly enriched at the electron microscopic level. The prominent actin bundles should not be contained in this fraction. (b) On electrophoresis of the fraction, vinculin should be clearly identified as one of the major constituents. On the basis of these criteria, we searched for the 'nonmuscle' organ appropriate for the isolation of the cell-to-cell AJ, and then we found that liver was the most appropriate for this purpose.

Using rat liver, we successfully developed a new isolation procedure for the cell-to-cell AJ.[12] From the bile canaliculi-enriched fraction obtained by homogenization of the liver and sucrose gradient centrifugation, the fraction rich in AJ was recovered by detergent treatment followed by sucrose gradient centrifugation (Fig. 12.1). In the fraction obtained, AJ was highly enriched both at light and electron microscopic levels. These isolated AJ were characterized by a prominent plaque and associated short actin filaments. Furthermore, one- and two-dimensional electrophoresis and immunological analyses revealed

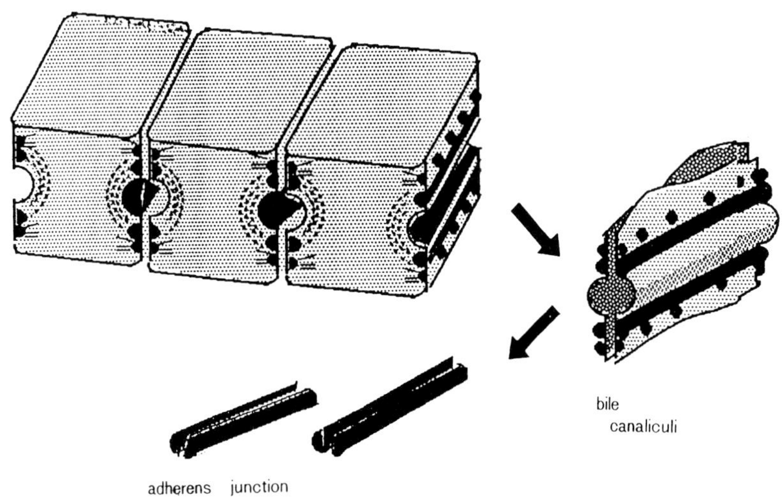

Fig. 12.1. A scheme for the isolation of adherens junctions from rat liver. The isolation procedure includes two steps: isolation of bile canaliculi and isolation of the belts of adherens junctions.

that vinculin was highly enriched in this fraction. The constituents of the plaque of AJ were selectively extracted from the isolated AJ. Upon SDS electrophoresis of this extract, 10 polypeptides including vinculin, α-actinin, and actin were dominant. Given our criteria for the isolation of AJ described above, we concluded that we succeeded in isolating the cell-to-cell AJ from rat liver and in opening a way to systematically analyze the molecular organization of the cell-to-cell AJ.

3. MOLECULAR ORGANIZATION OF THE CELL-TO-CELL AJ

Intensive studies, which mainly use immunological methods, revealed that some unique proteins, such as vinculin and α-actinin, are localized in the plaque structure of the cell-to-cell AJ. However, our knowledge of the molecular organization of the plaque structure was still fragmentary. The development of the isolation procedure for the cell-to-cell AJ enabled us to systematically analyze the molecular organization of the cell-to-cell AJ, and new information has been accumulated. The molecular organization of the cell-to-cell AJ is not as simple as was expected: its complexity appears to be due to the fact that it contains two distinct systems that are evolutionally combined; actin-based and cadherin-based systems.

As summarized in Fig. 12.2, and as discussed in detail previously,[13] we proposed that two types of molecular linkage between cadherin molecules and actin filaments exist in the plaque structure of the cell-to-cell AJ, and that many plaque-constitutive proteins responsible for these molecular linkages are integrated into the densely packed plaque structure in situ.

4. CONCENTRATION OF SRC FAMILY TYROSINE KINASES IN THE CELL-TO-CELL AJ

It was widely accepted that tyrosine phosphorylation is one of the important signals in AJ. The relationship between AJ and tyrosine phosphorylation was first illuminated by Rohrschneider,[14] who showed that the plaque of both cell-to-cell and cell-to-substrate AJ of Rous sarcoma virus-transformed cells contained the src gene product, pp60^{v-src}, which is a tyrosine kinase. Then, using the isolated cell-to-cell AJ, we first analyzed the AJ-specific tyrosine phosphorylation.[15]

When various types of rat adult tissue were pretreated with sodium orthovanadate, a potent inhibitor of tyrosine phosphatase, immunofluorescence microscopy showed that anti-phosphotyrosine polyclonal antibody specifically stained the plaque of the cell-to-cell AJ (Fig. 12.3). This indicated that the tyrosine kinase activity is elevated at the plaque of

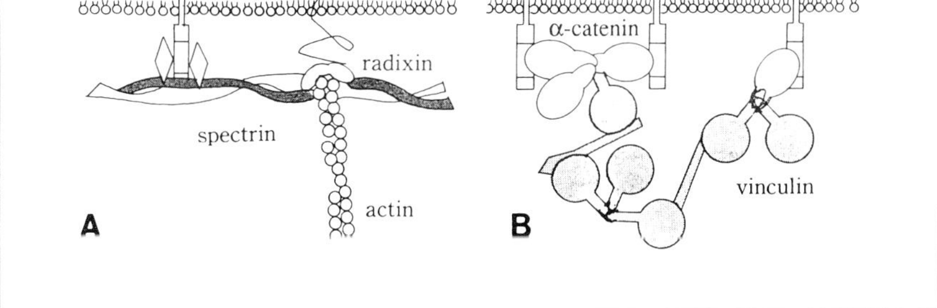

Fig. 12.2. Hypothetical molecular architecture of the actin-based (A) and cadherin-based (B) systems in the plaque structure of the adherens junctions.

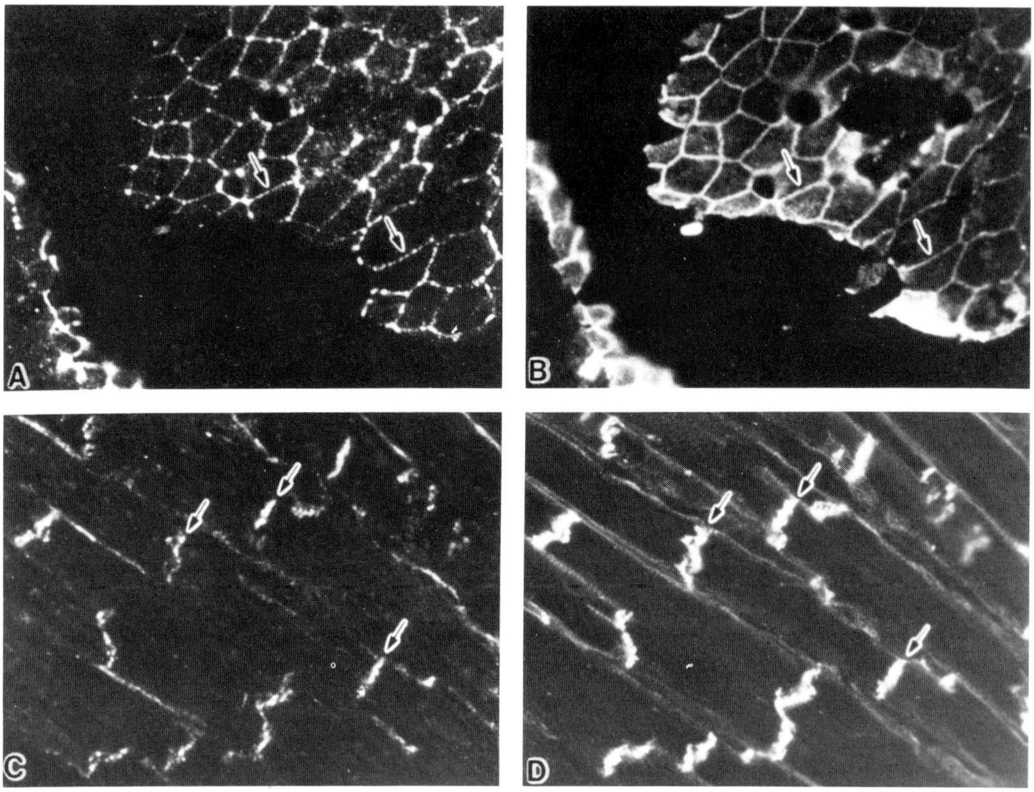

Fig. 12.3. Immunostaining for phosphotyrosine on frozen sections of rat adult intestine (A, B) and cardiac muscle (C, D) after pretreatment with sodium orthovanadate (see ref.15). The frozen sections were stained doubly with anti-phosphotyrosine pAb (A, C) and anti-tenuin mAb (B, D), both of which specifically stained the cell-to-cell AJ. x1000.

the cell-to-cell AJ of adult tissues. To identify tyrosine kinases responsible for the high level of tyrosine phosphorylation at AJ, we performed in vitro phosphorylation experiments with the isolated AJ and immunoblotting analyses with specific antibodies for tyrosine kinases. As a result, three proto-oncogenic tyrosine kinases of src family, c-yes, c-src, and lyn kinases, were identified as major tyrosine kinases in the cell-to-cell AJ of hepatocytes. Furthermore, it was immunofluorescently shown that at least two of these kinases, c-yes and c-src kinases, were enriched at the cell-to-cell AJ of various types of cells including hepatocytes. Based on these findings, we concluded that in various types of cells, specific proto-oncogenic tyrosine kinase of src-family (c-yes and c-src) are enriched to work as signal mediators in the cell-to-cell AJ where the level of tyrosine phosphorylation is elevated.

This conclusion gives us a hint regarding the possible molecular mechanism of cell transformation by v-src products. It is widely accepted that the ability of v-src kinase to bind the plasma membrane is required for its transformation activity and that v-src kinase is highly concentrated in AJ of transformed cells; concomitantly, AJ was seen to be modified or destroyed. Therefore, it is interesting to imagine that v-src kinase accumulated in AJ may override the c-yes and c-src kinase-dependent signal transduction system in normal AJ to transform cells.

5. RADIXIN AND ITS RELATIONSHIP WITH METASTASIS AND TUMOR SUPPRESSION

The manner of association (side-to-membrane or end-to-membrane) of actin filaments with the plasma membrane in cell-to-cell AJ remains unknown. However, it is clear that at least in the cell-to-cell AJ of cardiac muscle cells (intercalated discs), all actin filaments are bound to the plaque in an end-to-membrane fashion by their barbed ends.[16] Therefore, many investigators have attempted to identify the barbed-end capping protein located at the plaque of AJ. Vinculin was originally reported to have a barbed end capping activity.[17] However, it was known that this activity was attributed to contaminants and that highly purified vinculin did not cap the barbed ends of actin filaments.[18,19]

The development of an isolation procedure for cell-to-cell AJ enabled systematic analysis of its molecular architecture. Among the major constituents of the plaque of the isolated AJ, only an 82 kD protein, radixin, showed the ability to cap the barbed ends of actin filaments.[20] Sequence analysis of the radixin cDNA demonstrated that mouse radixin shares 74.9% sequence identity with mouse ezrin,[21] which was identified as a constituent of the microvilli of intestinal epithelial cells.[22-24] Furthermore, cDNA encoding another radixin/ezrin-like protein was isolated from mouse and its sequence analysis revealed that this protein is a mouse homolog of human moesin.[25] Moesin was originally reported to be an extracellular protein, and recently corrected to be an ezrin-like intracellular protein.[26,27] Comparison of the nucleotide sequences of these cDNAs pointed to a gene family consisting of ezrin, radixin, and moesin (85 kD, 82 kD, and 75 kD in SDS-PAGE, respectively). We tentatively referred to this family as 'ERM family'[13,25] (Fig. 12.4). So far ezrin (cytovillin) was reported to be localized in the microvilli, and radixin was found to be concentrated not only in the plaque of AJ but also in the cleavage furrow during cytokinesis.[28,29] In addition to these sites, recent close immunolocalization work clarified that at least one ERM family member is concentrated at the focal contacts (cell-to-substrate AJ) and ruffling membranes.[25] The most prominent feature shared by these sites is the tight association of actin filaments with plasma membranes. Although the direct interaction of ezrin and moesin with actin filaments in vitro remains to be studied, this localization of the ERM family members suggest that these members play a crucial role in the association of the barbed ends of actin filaments with the plasma membrane in general. Actually, recent experiments with antisense oligonucleotides complementary to ERM sequences revealed that ERM family members play crucial roles at least in cell-cell and cell-substrate adhesion and microvilli formation probably through the regulation of actin filament/plasma membrane interactions.[30]

Surprisingly, these ERM family members were recently found to have a close relationship to the tumor suppressor protein of neurofibromatosis 2 (NF2), one of the two major human genetic disorders both of which display autosomal dominant inheritance and involve tumors of the nervous system.[31,32] Genetic linkage studies of NF2 had revealed that this familial tumor is caused by inactivation of a tumor suppressor gene in chromosome 22q12, and recently, a candidate for the NF2 tumor

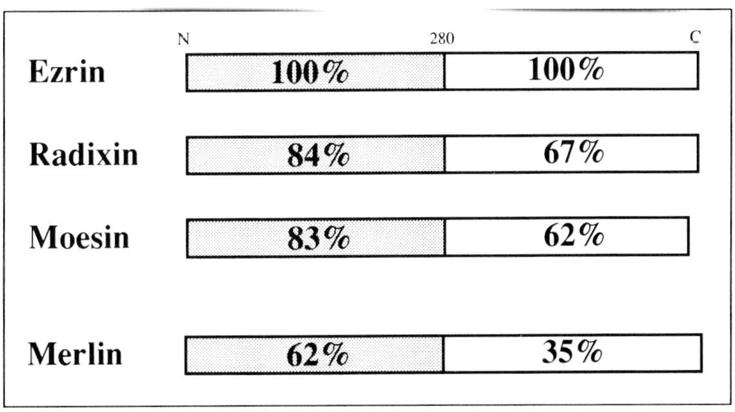

Fig. 12.4. ERM family and merlin. The structures of mouse ezrin, mouse radixin, mouse moesin, and human merlin are compared. These proteins have a homologous domain (shaded area) in their amino-terminal half. The identify of the N-terminal and C-terminal halves of each protein with those of ezrin was calculated and shown in each box.

suppressor gene was cloned. The candidate gene was shown to encode a cytoplasmic protein with a striking similarity to ERM family members, and was named 'merlin' (moesin-ezrin-radixin-like protein) (Fig. 12.4). Clusters of amino acids identical in merlin and the ERM family is concentrated in the amino-terminal half of each molecule (about 62% identity). Therefore, it is tempting to speculate that merlin, and probably other ERM family members, act as tumor suppressors by affecting the actin filament/plasma membrane association.

As reported in the sequence study of ezrin cDNA, this ERM family is included in the band 4.1 superfamily. The band 4.1 protein is one of the major constituents of the undercoat of erythrocyte membranes.[33] In erythrocytes the band 4.1 protein acts as a key protein in the association of short actin filaments with a plasma membrane protein (glycophorin C), together with spectrin. The N-terminal half of ezrin/radixin/moesin has homology with that of band 4.1 protein (about 30% identity). This conserved domain is found in all band 4.1 superfamily members. Recent PCR analysis revealed that the band 4.1 superfamily can be subclassified into at least six gene families; band 4.1 protein, ERM, talin, protein-tyrosine-phosphataseH1, NBL4, and merlin families.[34] The N-terminal half of the band 4.1 protein is responsible for its specific binding to glycophorin C (also called glycoconnectin), one of the major integral membrane proteins in erythrocytes, suggesting that the band 4.1 superfamily members are also directly associated with glycophorin-like integral membrane proteins at their N-terminal half domains. By immunoprecipitation experiments, we have recently found that ERM family members are directly associated with a broadly-distributed cell surface glycoprotein called CD44,[35] which was reported to play crucial roles in metastasis.[36-38] Therefore, further analyses of ERM family members will lead to a better understanding for the molecular mechanisms of tumor suppression and metastasis through merlin and CD44, respectively.

6. α-CATENIN AND ITS RELATIONSHIP WITH TUMOR INVASION

Cadherins are a family of transmembrane glycoproteins responsible for Ca^{2+}-dependent cell-cell adhesion.[2,4] The family includes E-cadherin/uvomorulin, N-cadherin/A-CAM P-cadherin, and L-CAM, and it was recently found that desmosomal glycoproteins are also members of this family, although their overall structure is slightly different from that of the original cadherins.[39,40] The molecular mechanism of cadherin-mediated cell adhesion was studied in two aspects. One was concerned with the role of the extracellular domain, the region that is required for interaction between cadherin molecules.

The other aspects sought to clarify the role of the cytoplasmic domain. Interestingly, the domain was the most highly conserved in amino acid sequences in this molecular family, suggesting that it plays an important role.[41] In fact, evidence was accumulated suggesting that the cytoplasmic domain of cadherin molecules is crucial for their cell-cell binding function. The importance of the cytoplasmic domain was first demonstrated by observations that cadherin molecules in which C-terminal regions of the cytoplasmic domain were deleted could not function as cell adhesion molecules, although they were expressed on the cell surface.[42,43] The loss of the cell-binding function in such mutant molecules coincided with the loss of two other properties of cadherins. First, many of the cadherin molecules in a cell are resistant to nonionic detergent extraction, and the detergent-insoluble fraction of cadherins is tightly associated with cortical actin bundles. The nonfunctional, carboxyl terminus-truncated cadherins, however, lost the ability to bind to the cytoskeleton. Second, cadherins are associated with some cytoplasmic proteins, such as α- (102 kD), β- (88 kD) and γ- (80 kD) catenin.[43] These proteins, however, could not bind to the above nonfunctional, truncated cadherins. These observations led us to speculate that cadherins form a molecular complex with the actin-based cytoskeleton, via catenins or other proteins at their cytoplasmic domain, and that this molecular association is essential for them to act as cell adhesion receptors. For further understanding of the mechanism of cadherin-mediated cell adhesion, it was therefore required to elucidate the structure and functions of catenins.

Cadherins are concentrated in cell-to-cell AJ. Therefore, we expected that E-cadherin-associated protein, catenins, are also concentrated in the isolated AJ, and that this AJ fraction offers an ideal source for isolating and characterizing catenins. Then, taking advantage of the isolated AJ, we purified α-catenin, and generated an antiserum to this protein; using this antiserum as a probe, we isolated cDNAs encoding α-catenin. The deduced amino acid sequence revealed that α-catenin resembles vinculin (Fig. 12.5).[44,45]

Using this cDNA encoding mouse α-catenin, we produced a recombinant α-catenin in *E .coli*, and with this recombinant as an antigen we obtained a mAb called α-18 which clearly recognized 'human' α-catenin. Using this mAb, in collaboration with us, Hirohashi's group examined the expression level of α-catenin in a human lung cancer cell line, PC9. PC9 cells had been known to strongly express normal E-cadherin molecules on their cell surface, but to show reduced cadherin-mediated aggregation activity, resulting in poorly-differentiated round-shaped cells. Immunoprecipitation, immunoblot and PCR analyses showed that PC9 cells lack the expression of intact α-catenin due to the homologous deletion of its gene.[46,47] Furthermore, transfection of the cDNA encoding α-catenin into PC9 cells brought about a drastic restoration of cell aggregation activity, resulting in cultures of a well-differentiated epithelial cell.[48] Most recently, using mAb α18, the loss of α-catenin expression was also observed in a human prostate cancer cell line, PC3, which showed a high expression level of E-cadherin but lacked cell-cell adhesion ability.[49] Therefore, we were led to conclude that, at least in PC9 and PC3 cells, α-catenin is a key protein not only to exert the cadherin-dependent cell adhesion but also to 'suppress' dedifferentiation phenomena. The expression of α-catenin in human adenocarcinomas was then examined, especially those of the scirrhous-type, in which cancer cells are characteristically detached from each other to form single cells that infiltrate the connective tissue in a scattered manner.[50,51] These scattered cancer cells were reported occasionally to express E-cadherin at a normal level, leading us to speculate that —like PC9 cells— these scattered cancer cells also lack α-catenin.

The scirrhous-type adenocarcinomas are most commonly found in the stomach and the breast. Interestingly, close immunohistochemical analysis showed apparently complete loss of α-catenin expression in gastric and breast scirrhous adenocarcinomas in about 55% and 75% of the cases studied, respectively.[50] Most recently, the human α-catenin gene was assigned to chromosome 5q31, for which a loss of heterozygocity (LOH) had frequently been reported:[52] The APC gene, a tumor suppressor gene in the familial adenomatous polyposis, is also located in this chromosome region.[53] And all these findings led us to suggest that the loss of α-catenin may be a key step for the phenotype of the scirrhous-type adenocarcinomas, i.e. that α-catenin is a suppressor of tumor cell invasion

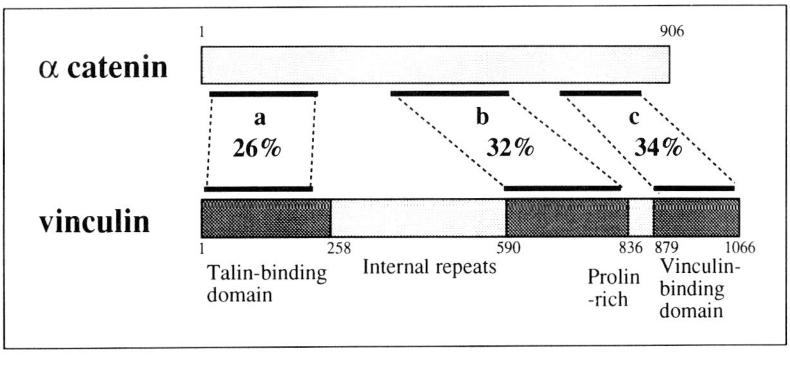

Fig. 12.5. Schematic structures of mouse α–catenin and vinculin. Vinculin contains five domains. Boundaries of each domain are shown by the number of boundary residues. Three regions (a, b, c) are conserved between α–catenin and vinculin.

as well as of cell dedifferentiation. Therefore, α-catenin is a good candidate for the "suppressor" in scirrhous-type adenocarcinoma.

7. ZO-1 AND THE DLG PRODUCT, A DROSOPHILA TUMOR SUPPRESSOR MOLECULE

Using the isolated adherens junction fraction, we have identified a 220 kD plaque protein, which was colocalized with the detergent-resistant form of cadherins in nonepithelial cells such as fibroblasts and cardiac muscle cells, at least at the immunofluorescence microscopic level.[54] In cardiac muscle cells, this protein appeared also at the electron microscopic level at so-called intercalated discs, in the immediate vicinity of the plasma membrane in the plaque of adherens junctions. Furthermore, this 220 kD protein was tightly associated with spectrin molecules. Therefore, we thought that this 220 kD protein would be a good candidate for binding directly or indirectly, cadherins to the actin-based cytoskeleton and thereby regulating cadherin function.

To further characterize this protein, we cloned a full-length cDNA encoding the 220 kD protein and sequenced it.[55] Unexpectedly, a homology search and immunological analyses showed that the mouse 220 kD protein is very similar to rat protein ZO-1, which is reportedly concentrated at tight junctions[56,57] and only a part of which had been sequenced.[58] Localization analyses of cadherins and the 220 kD/ZO-1 protein in various types of tissues further revealed that the 220 kD/ZO-1 protein precisely colocalizes with N- and P-cadherins in nonepithelial cells, and that only in epithelial cells the 220 kD/ZO-1 protein and E-cadherins to segregate into tight and adherens junctions, respectively. Furthermore, by transfecting mouse L cells with cadherin cDNAs we demonstrated that cadherins, directly or indirectly, interact with the 220 kD protein in order to colocalize. This study has cast a new light on the relationship between adherens and tight junctions, as well as on the molecular mechanisms of tight junction formation. Although it was not mentioned in original paper of Itoh et al,[55] close comparison of the amino acid ZO-1 sequence with the sequences of other proteins revealed that the amino-terminal half of this protein displays significant similarity to the product of the lethal (1) discs large-1(dlg) gene in *Drosophila*[6,59] (Fig. 12.6) (see also chapters 1 and 5 in this volume). In *Drosophila*, so far seven tumor suppressor genes have been identified by isolating recessive lethal mutations in which cell proliferation in imaginal discs continue beyond the normal limits.[60] Sequence analysis of the cDNA of dlg, one of the *Drosophila* tumor suppressor genes, revealed that its product contains some putative functional domains: a filamentous domain, an SH3 domain, and the guanylate kinase domain.[61] As shown in Figure 12.6, four regions were conserved between the amino-terminal half of ZO-1 and the dlg

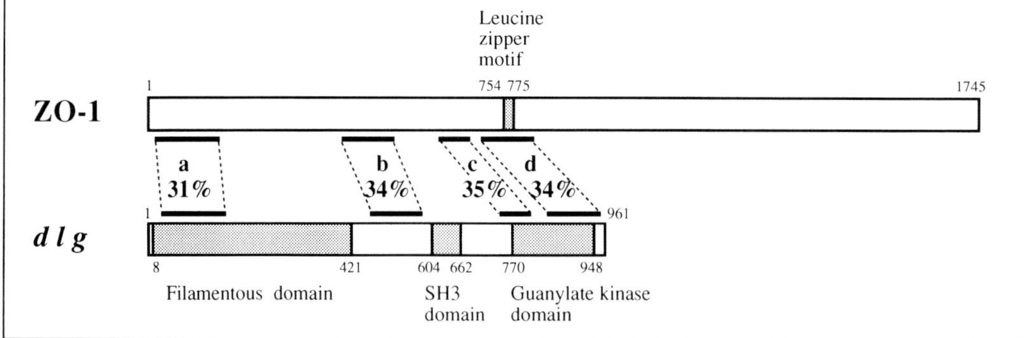

Fig. 12.6. Schematic structures of mouse ZO-1 and Drosophila dlg gene product. The dlg gene product contains three putative functional domains. Boundaries of each domain are shown by the number of boundary residues. Four regions (a, b, c, d) are conserved between the N-terminal half of ZO-1 and the dlg gene product.

gene product, and three of them are included in the filamentous and guanylate kinase domains of the dlg gene product. Interestingly, in *Drosophila* epithelial cells, the dlg gene product was reported to be localized in the undercoat of the septate junctions,[61] which are thought to be the counterpart of the vertebrate tight junctions. From the above, we speculate that ZO-1 can act as a tumor suppressor in vertebrate cells. This speculation, of course, must be evaluated by further functional analysis of ZO-1.

8. CONCLUSIONS

The results discussed above favor the idea that AJ plaque proteins play crucial roles in suppression of the malignant phenotypes of tumors such as overproliferation, transformation, dedifferentiation, invasion, and metastasis. Furthermore, as also discussed above, some proto-oncogenic tyrosine kinases such as c-yes and c-src kinases are concentrated at the cadherin-based AJ plaque, and a unique tyrosine kinase called FAK was reported to be involved in the signal transduction in the integrin-based AJ plaque.[5] Therefore, the AJ plaque structure appears to have an intimate relationship with both tumor suppressor gene and oncogene products. Further studies of the molecular organization of the cell-to-cell AJ using isolated AJ will lead us to better understanding of the role of the submembranous junctional plaque proteins not only in normal cells but also in cancer cells.

REFERENCES

1. Geiger B. Membrane-cytoskeleton interaction. Biochim Biophys Acta 1983; 737:305-341.
2. Takeichi M. The cadherins: cell-cell adhesion molecules controlling animal morphogenesis Development 1988; 102:639-655.
3. Hynes RO. Integrins: a family of cell surface receptors. Cell 1987; 48:549-554.
4. Takeichi M. Cadherin: cell adhesion receptors as a morphogenetic regulator. Science 1991; 251:1451-1455.
5. Juliano RL, Haskill S. Signal transduction from the extracellular matrix. J Cell Biol 1993; 120:577-586.
6. Tsukita Sh, Itoh M., Nagafuchi A, Yonemura S, Tsukita Sa. Submembranous junctional plaque proteins include potential tumor suppressor molecules. J Cell Biol 1993; 123:1049-1053.
7. Geiger B. A 130k protein from chicken gizzard: its localization at the termini of microfilament bundles in cultured chicken cells. Cell 1979; 18:193-205.
8. Burridge K, Connell L. A new protein of adherens plaques and ruffling membranes. J Cell Biol 1983; 97:359-367.
9. Kensler RW, Goodenough DA. Isolation of mouse myocardial gap junctions. J Cell Biol 1980; 86:755-764.
10. Colaco CALS, Evans WH. Partial purification of an intercalated disc-containing cardiac plasma membrane fraction. Biochim Biophys Acta 1982; 684:40-60.
11. Volk T, Geiger B. A 135-kD membrane protein of intercellular adherens junctions. EMBO J 1984; 3:2249-2260.
12. Tsukita Sh, Tsukita Sa. Isolation of cell-to-cell adherens junctions from rat liver. J Cell Biol 1989; 108, 31-41.
13. Tsukita S, Tsukita S, Nagafuchi A, Yonemura S. Molecular linkage between cadherins and actin filaments in cell-to-cell adherens junctions. Curr Opin Cell Biol 1992; 4:834-839.
14. Rohrschneider LR. Adhesion plaques of Rous sarcoma virus-transformed cells contain the src gene product. Proc Natl Acad Sci USA. 1980; 77:3514-3518.
15. Tsukita Sa, Oishi K, Akiyama T, Yamanashi Y, Yamamoto T, Tsukita Sh. Specific proto-oncogenic tyrosine kinases of src family are enriched in cell-to-cell adherens junctions where the level of tyrosine phoshorylation is elevated. J Cell Biol 1991; 113.867-879.
16. Ishikawa H. Identification and distribution of intracellular filaments. In: Hatano S, Ishikawa H, Sato H, eds. Cell Motility: Molecules and Organization. Tokyo: University of Tokyo Press, 1979:417-444.
17. Wilkins JA, Lin S. High affinity interaction of vinculin with actin in vitro. Cell 1982; 28:83-90.
18. Evans RR, Robson RM, Stromer MH. Properties of smooth muscle vinculin. J Biol Chem 1984; 259:3916-3924.

19. Wilkins JA, Lin S. A re-examination of the interaction of vinculin with actin. J Cell Biol 1986; 102:1085-1092.
20. Tsukita Sa, Hieda Y, Tsukita Sh. A new 82 kD-barbed end capping protein (radixin) localized in the cell-to-cell adherens junction: purification and characterization. J Cell Biol 1989; 108:2369-2382.
21. Funayama N, Nagafuchi A, Sato N, Tsukita Sa, Tsukita Sh. Radixin is a novel member of the band 4.1 family. J Cell Biol 1991; 115:1039-1048.
22. Bretscher A. Purification of an 80:000-dalton protein that is a component of the isolated microvillus cytoskeleton, and its localization in nonmuscle cells. J Cell Biol 1983; 97:425-432.
23. Gould KL, Bretcher A, Esch FS, Hunter T. cDNA cloning and sequencing of the protein-tyrosine kinase substrate, ezrin, reveals homology to band 4.1. EMBO J 1989; 8:4133-4142.
24. Turunen O, Winqvist R, Pakkanen R, Grzeschik KH, Wahlstrom T, Vaheri A. Cytovillin, a microvillar Mr 75:000 protein. cDNA sequence, prokaryotic expression, and chromosomal localization. J Biol Chem 1989; 264:16727-16732.
25. Sato N, Funayama N, Nagafuchi A, Yonemura S, Tsukita Sa, Tsukita Sh. A gene family consisting of ezrin, radixin, and moesin. Its specific localization at actin filaments/plasma membrane association sites. J Cell Sci 1992; 103:131-143.
26. Lankes W, Griesmacher A, Grunwald J, Schwarts-Albiez R, Keller R. A heparin-binding protein involved in inhibition of smooth-muscle cell proliferation. Biochem J 1988; 251:831-842.
27. LankesW, Farthmayr H. Moesin: a member of the protein 4.1-talin-ezrin family of proteins. Proc Natl Acad Sci USA 1991; 88: 8297-8301.
28. Sato N, Yonemura S, Obinata T, Tsukita Sa, Tsukita Sh. Radixin, a barbed end-capping actinomodulating protein, is concentrated at the cleavage furrow during cytokinesis. J Cell Biol 1991; 113, 321-330.
29. Yonemura S, Nagafuchi A, Sato N, Tsukita Sh. Concentration of an integral membrane protein CD43 (leukosialin, sialophorin), in the cleavage furrow through the interaction of its cytoplasmic domain with actin-based cytoskeletons. J Cell Biol 1993; 120:437-449.
30. Takeuchi K, Sato N, Kasahara H, Funayama N, Nagafuchi A, Yonemura S, Tsukita Sa, Tsukita Sh. Perturbation of cell adhesion and microvilli formation by antisense oligonucleotides to ERM family members. Submitted for publication.
31. Trofatter JA, MacCollin MM, Rutter JL, Murrell JR, Duyao MP, Parry DM, Eldridge R, Kley N, Menon AG, Pulaski K, Haase VH, Ambrose CM, Munroe D, Bove C, Haines JL, Martuza RL, MacDonald ME, Seizinger BR, Short MP, Buckler AJ, Gusella JF. A novel moesin-, ezrin-, radixin-like gene is a candidate for the neurofibromatosis 2 tumor suppressor. Cell 1993; 72:791-800.
32. Rouleau GA, Merel P, Lutchman M, Sanson M, Zucman J, Marineau C, Hoang-Xuan K, Demczuk S, Desmaze C, Plougastel B, Pulst SM, Lenoir G, Bijlsma E, Fashold R, Dumanski J, De Jong P, Parry D, Eldrige R, Aurias A, Delattre O, Thomas G. Alteration in a new gene encoding a putative membrane-organizing protein causes neurofibromatosis type 2. Nature 1993; 363, 515-521.
33. Bennett V. The spectrin-actin junction of erythrocyte membrane skeletons. Biochim Biopys Acta 1989; 988:107-121.
34. Takeuchi K, Kawashima A, Nagafuchi A, Tsukita S. Structural diversity of band 4.1 superfamily. Submitted for publication.
35. Tsukita S, Oishi K, Sato N, Sagara J, Kawai A, Tsukita S. CD44 as a plasma membrane attachment site of ERM family members. Submitted for publication.
36. Günthert U, Hofmann M, Rudy M, Reber S, Zöller M, Haussmann I, Matzku S, Wenzel A, Ponta H, Herrlich P. A new variant of glycoprotein CD44 confers metastatic potential to rat carcinoma cells. Cell 1991 65:13-24.

37. Arch R, Wirth K, Hofmann M, Ponta H, Matzku S, Herrlich P, Zöller M. Participation in normal immune responses of a metastasis-induced splice variant of CD44. Science 1992; 257:682-685.
38. Koopman G, Heider K-H, Horst E, Adolf GR, VanDenBerg F, Ponta H, Herrlich P, Pals ST. Activated human lymphocytes and aggressive non-Hodgkin's lymphomas express a homologue of the rat metastasis-associated variant of CD44. J Exp Med 1993; 177:897-904.
39. Koch PJ , Walsh MJ, Schmelz M, Goldschmidt MD, Zimbelmann R, Franke WW. Identification of desmoglein, a constitutive desmosomal glycoprotein, as a member of the cadherin family of cell adhesion molecules. Eur J Cell Biol 1990; 53:1-12.
40. Holton JL, Kenny TP, Legan PK, Collins JE, Keen JN, Sharma R, Garrod DR. Desmosomal glycoprotein 2 and 3 (desmocollins) show N-terminal similarity to calcium-dependent cell-cell adhesion molecules. J Cell Sci 1990; 97:239-246.
41. Hatta K, Nose A, Nagafuchi A, Takeichi M. Cloning and expression of cDNA encoding a neural calcium-dependent cell adhesion molecule: its identity in the cadherin gene family. J Cell Biol 1988; 106:873-881.
42. Nagafuchi A, Takeichi M. Cell binding function of E-cadherin is regulated by the cytoplasmic domain. EMBO J 1988; 7:3679-3684.
43. Ozawa M, Baribault H, Kemler R. The cytoplasmic domain of the cell adhesion molecule uvomorulin associates with three independent proteins structurally related in different species. EMBO J 1989; 8:1711-1717.
44. Nagafuchi A, Takeichi M, Tsukita Sh. The 102 kD cadherin-associated protein: similarity to vinculin and posttranscriptional regulation of expression. Cell 1991; 65:1-20.
45. Herrenknecht K, Ozawa M, Eckerskorn C, Lottspeich F, Lenter M, Kemler R. The uvomorulin-anchorage protein alpha catenin is a vinculin homologue. Proc Natl Acad Sci USA 1991; 88:9156-9160.
46. Shimoyama Y, Nagafuchi A, Fujita S, Gotoh M, Takeichi M, Tsukita Sh, Hirohashi S. Cadherin dysfunction in a human cancer cell line: possible involvement of loss of α-catenin expression in reduced cell-cell adhesiveness. Cancer Res 1992; 52:1-5.
47. Oda T, Kanai Y, Shimoyama Y, Nagafuchi A, Tsukita Sh, Hirohashi S. Cloning of the human α-catenin cDNA and its aberrant mRNA in a human cancer cell line. Biochem Biophys Res Commun 1993; 193:897-904.
48. Hirano S, Kimoto N, Shimoyama Y, Hirohashi S, Takeichi M. Identification of a neural α-catenin as a key regulator of cadherin function and multicellular organization. Cell 1992; 70:293-301.
49. Morton RA, Ewing CM, Nagafuchi A, Tsukita Sh, Isaacs WB. Reduction of E-cadherin levels and deletion of the α catenin gene in human prostate cancer cells. Cancer Res 1993; 53:3585-3590.
50. Ochiai A, Akimoto S, Shimoyama Y, Nagafuchi A, Tsukita Sh, Hirohashi S. Japn.J.Cancer Res 1994; 85:266-273.
51. Kadowaki T, Shiozaki H, Inoue M, Tamura S, Oka H, Doki Y, Iihara K, Matsui S, Iwazawa T, Nagafuchi A, Tsukita Sh, Mori T. E-cadherin and α-catenin expression in human esophageal cancer. Cancer Res 1993; 54:291-296.
52. Furukawa Y, Nakatsuru S, Nagafuchi A, Tsukita Sh, Muto T, Nakamura Y. and Horii A. Structure, expression and chromosomal assignement of the human α-catenin gene. Cytogenetics and Cell Genetics 1993; 65:74-78.
53. Kinzler KW, Nilbert MC, Vogelstein B, Bryan TM, Levy DB, Smith KJ, Preisinger AC, Hamilton, SR, Hedge P, Markham A, Carlson M, Joslyn G, Groden J, White R, Miki Y, Miyoshi Y, Nishisho I, Nakamura Y. Identification of a gene located at chromosome 5q21 that is mutated in colrorectal cancers. Science 1991; 251:1366-1370.
54. Itoh M, Yonemura S, Nagafuchi A, Tsukita Sa, Tsukita Sh. A 220-kD undercoat-constitutive protein: its specific localization at cadherin-based cell-cell adhseion sites. J Cell Biol 1991; 115:1449-1462.

55. Itoh M, Nagafuchi A, Yonemura S, Kitani-Yasuda T, Tsukita Sa, Tsukita Sh. The 220-kD protein colocalizing with cadherins in non-epithelial cells is identical to ZO-1, a tight junction-associated protein in epithelial cells: cDNA cloning and immunoelectron microscopy. J Cell Biol 1993; 121:491-502.
56. Anderson JM, Stevenson BR, Jesaitis LA, Goodenough DA, Mooseker MS. Characterization of ZO-1, a protein component of the tight junction from mouse liver and Madin-Darby canine kidney cells. J Cell Biol 1988 106:1141-1149.
57. Stevenson BR, Siliciano JD., Mooseker MS, Goodenough DA. Identification of ZO-1: a high molecular weight polypeptide associated with the tight junction (zonula occludens) in a variety of epithelia. J Cell Biol 1986; 103:755-766.
58. Willott E, Balda MS, Heintzelman M, Jameson B, Anderson JM. Localization and differential expression of two isoforms of the tight junction protein ZO-1. Am J Physiol. 1992; 262:C1119-C1124.
59. Willott E, Balda MS, Fanning AS, Jameson B, Van Itallie C, Anderson JM. The tight junction protein ZO-1 is homologous to the *Drosophila* discs-large tumor suppressor protein of septate junctions. Proc Natl Acad Sci USA 1993; 90: 7834-7838.
60. Gateff EA, Mechler BM. Tumor-suppressor genes of *Drosophila* melanogaster. CRC Crit Rev Oncogene 1989; 1: 221-245.
61. Woods DF, Bryant PJ. The discs-large tumor suppressor gene of *Drosophila* encodes a guanylate kinase homolog localized at septate junctions. Cell 1991; 66:451-464.

CHAPTER 13

The Role of E-Cadherin Mediated Cell Adhesion in Tumor Metastasis

Jürgen Behrens
Walter Birchmeier

1. INTRODUCTION: CELL ADHESION PHENOMENA IN CANCER PROGRESSION

In the last decade, the molecular analysis of tumor development was mainly directed toward the identification of components that influence the growth behavior of the cells. A variety of oncogenes encoding growth factors and their receptors were identified. However, it is clear from both clinical and experimental experience that the fatal cause of a cancer disease is not primarily the unrestricted growth of tumor cells but their capacity to metastasize, i.e. to colonize and destroy foreign tissue. In order to accomplish this, tumor cells acquire various phenotypic changes which are not dependent on the loss of growth control.[1] Examples are the enhanced motility of the cells or the production of proteases that are able to digest extracellular matrix. Most prominently, metastasizing cells exhibit changes in the adhesive interactions with the extracellular matrix, with neighboring tumor cells or with normal cells of the various body compartments encountered upon the metastatic journey. Much of our knowledge of the molecular basis of cell adhesion comes from work in cell biology and from developmental studies which led to the identification of extracellular adhesion proteins and of adhesion receptors located on the cell membrane. Many of these components which are proposed to function in developmental processes have also been assigned a role in metastasis formation; this is in line with the old concept that tumor

Molecular Mechanisms of Epithelial Cell Junctions: From Development to Disease
Edited by Sandra Citi, M.D., Ph.D.;

cells recapitulate certain stages in embryonal development.

Metastasizing cells exhibit a complex pattern of adhesion phenomena. Besides the alterations in homotypic cell-to-cell interaction in the primary tumor, a multitude of other changes in adhesive interactions take place that are characteristic for individual steps of metastasis formation. During invasion of carcinomas, cells which escape from the tumor have to cope with the basement membrane which requires digestion of and interaction with extracellular matrix components. Specific adhesion receptors on the metastatic cells are likely to play a role in this process. For instance, it has been shown that peptides containing the RGD sequence for recognition by integrin receptors block the penetration of tumor cells through basement membrane components in vitro or lung colony formation by B16 melanoma cells in vivo.[2,3] Apparently, these treatments interfered with the contact formation between cell surface receptors of the tumor cells and components such as collagen and laminin of the basement membrane of the target organ. Support for this hypothesis comes from experiments in which gene transfer of an $\alpha 2$ integrin cDNA into low metastatic rhabdomyosarcoma cells increased the metastatic potential.[4] This was most likely due to the formation of a functional $\alpha 2\beta 1$ receptor which mediates adhesion to collagen and laminin. Various other correlative and functional studies have demonstrated cell substrate adhesion as being important for the promotion of metastasis formation.[5] On the other hand, overexpression of cell matrix receptors such as $\alpha 5\beta 1$ (which interacts with fibronectin) has been shown to reduce tumorigenicity and motility of the cells. Furthermore, reduced adhesion to fibronectin and loss of $\alpha 5\beta 1$ integrin has been shown in virally transformed cells.[6,7] This reveals a dual role of cell matrix adhesion in tumorigenesis: on one hand cell substrate adhesion is essential for cell locomotion, on the other hand it can impose constraints on motility and growth and thus must be minimized under certain conditions.[8,9]

After entry into blood or lymph vessels, cells are transported to their target organs where they interact with endothelial cells lining the vessels. This contact formation is probably essential in determining the organ preference of metastasizing cells, and is reminiscent of the homing phenomena seen with normal leukocytes during colonization of tissues. Thus, while the migration of tumor cells through the mesenchymal tissue is dictated by cell-substrate interactions, the exit of cells from blood and lymph vessels initially requires cell-to-cell interactions. Here, the tumor cells could make use of the repertoire of cell surface receptors that direct leukocyte traffic under physiological circumstances, such as members of the immunoglobulin superfamily of adhesion receptors[10] and selectins.[11]

2. DISTURBANCE OF INTERCELLULAR ADHESION IN INVASIVE CARCINOMA CELLS

A distinct feature of normal epithelial cells is their tight association with each other, which results in the characteristic cobblestone-like appearance of epithelial tissues. It can be envisaged that the individual members of such a tissue exhibit only little cell motility, although movement of whole sheets of cells are observed in several epithelia upon tissue regeneration. It is obvious that the maintenance of cell contacts and thus reduced cell motility of epithelia is physiologically meaningful, given their protective functional role as diffusion barriers. In carcinomas, however, unrestricted cell movement is essential for the invasiveness of the cells and is counteracted by the various intercellular contacts made by the epithelial cells. Pathologists have observed for some decades that invasive carcinomas often show reduced epithelial differentiation, and that this is particularly prominent at invasion fronts where the cells break into surrounding mesenchymal tissue.[12,13] The invading cells can lose their epithelial appearance (they become amorphic, spindle-shaped, fibroblastoid) and in the electron microscope a reduced number of desmosomes are often observed. Based on these observations, carcinomas can be subdivided by morphological and functional criteria: well differentiated carcinomas retain epithelial tissue structure and are generally weakly invasive, and poorly differentiated carcinomas are characterized by a more amorphous tissue structure and are more highly invasive. Consequently, the state of differentiation is linked to the degree of invasiveness and can determine cancer prognosis. Early experiments involving separation of cells by physical forces revealed reduced cell-cell adhesion in carcinoma cells as compared to normal epithelial cells[14,15] which pointed to a functional disturbance of cell-cell contacts in the tumor cells.

It has been hypothesized that during progression to the invasive state, carcinoma cells undergo a process similar to the epithelial-mesenchymal conversion which occurs during development.[16] For example, at the blastula stage only a single cell type exists which resembles epithelial cells. During gastrulation, the three germinal cell layers are formed. Whereas ectoderm and endoderm remain largely epithelial in character, the newly induced mesoderm is the first example in ontogenesis of a conversion of epithelial cells to mesenchyme. Morphologically and functionally, epithelial and mesenchymal cells are entirely different; epithelial cells express intercellular junctions and are polar and largely nonmobile, whereas mesenchymal cells are generally loosely associated, nonpolarized, and surrounded by extracellular matrix. The adhesive contacts of mesenchymal cells to the environment are mediated primarily by cell-substrate adhesion molecules (see ref. 17 for a review). The conversion of epithelia to mesenchyme during mesoderm formation has recently been studied extensively and the inductive signals have been elucidated.[18,19] The phenotypic resemblance of carcinoma cells becoming invasive to epithelial cells turning into mesenchyme suggests that the molecular mechanisms underlying these two processes are also similar. Cell biological studies have revealed several molecular components capable of inducing epithelial-mesenchymal conversions in vitro. For instance, treatment of epithelial cells with scatter factor/hepatocyte growth factor, which is a motility and growth factor for various cell types, leads to the dissociation of the cells, to increased cell motility, to a change from an epithelial to a fibroblastoid shape, to invasiveness into collagen gels, but not to a loss of E-cadherin expression.[20,21] Similar results were observed in dissociation experiments of rat bladder carcinoma cells with acidic fibroblast growth factor.[22] Transformation of MDCK epithelial cells with a temperature-sensitive v-src gene also leads to morphology changes and invasiveness at the permissive temperature.[23] Similarly, expression of an estrogen receptor-fos hybrid gene in breast epithelial cells results in a hormone-inducible, rapid loss of differentiation of the cells.[24]

3. ROLE OF E-CADHERIN IN CANCER INVASION

A detailed description of the structure and composition of the various cell junctions is given elsewhere in this volume. For our discussion it is sufficient to point out the principal molecular parameters that determine the cells ability to form specific cell contact organelles. A number of studies have clearly established a central role for the cell-cell adhesion molecule E-cadherin (also called uvomorulin, L-CAM, cell-CAM 120/80, Arc-1), which is a 120 kD transmembrane glycoprotein, of which an extracellular 80 kD tryptic fragment can be released in the presence of Ca^{2+}.[25,26] In early mouse development E-cadherin first functions as an adhesion component during compaction of blastomeres; at later stages, it is confined to all epithelia originating from ecto-, meso- and endodermal tissue[25,27,28](see also chapter 4 in this volume). In the epithelium of the small intestine, E-cadherin is enriched in the adherens junctions; in other epithelia, it is uniformly present at the lateral cell surfaces.[29,30] Recently, E-cadherin expression has also been found in subsets of sensory neurons and in embryonic brain.[31,32] The cDNAs of mouse, chicken and human E-cadherin have been characterized; they code for a signal peptide at the amino terminus, a large extracellular domain with four repeated domains important in Ca^{2+} binding, a single transmembrane sequence, and a 15 kD cytoplasmic domain.[33-37] The cytoplasmic domain of E-cadherin is associated with a group of proteins, named catenins, which make contact with the microfilament network through unknown linkage proteins.[38] The cDNAs of two catenins have recently been characterized: α-catenin shows structural similarity to the junction-associated protein vinculin[39,40] and β–catenin is homologous to the *Drosophila* segment polarity gene product *armadillo*.[41-43] γ-catenin is related or identical to plakoglobin.[44,45](see also chapters 1, 8 and 12 in this volume). E-cadherin is a member of the now well established protein family of Ca^{2+} dependent cell adhesion molecules; close relatives are N-cadherin (expressed in neural and muscle cells) and P-cadherin (originally identified in mouse placenta but also found in a restricted set of human epithelia;[25,28]). Further members of the cadherin family are the muscle-specific M-cadherin,[46] the desmosomal proteins desmoglein and desmocollin (see ref. 47 for a review), as well as a series of recently discovered cadherins, of which the function is presently being analyzed (cf. [25,48,49]). E-, N-, P- (the classical cadherins) and other cadherins exhibit unique spatiotemporal expressing patterns during embryogenesis, and cells which express a particular cadherin "sort out" in vitro, suggesting that these adhesion molecules play an important role during morphogenesis.[25]

E-cadherin has been originally found as a cell-cell adhesion component through cell dissociating monoclonal antibodies directed against the protein.[26,29,50,51] It was noted already in these early studies that antibody treatment not only affected the integrity of the adherens junctions (to which E-cadherin had been localized in various epithelia) but also destroyed tight and gap junctions as well as desmosomes. This finding indicated that E-cadherin must rank high in the hierarchy of junctional components and might be a central determinant of junction formation. In fact, transfection of the E-cadherin cDNA into fibroblasts results in the formation of functional Ca^{2+} dependent contacts between the cells and induces an increase in adherens junctions and gap junctions.[33,36] Moreover, certain marker molecules such as the Na^+, K^+-ATPase are accumulated in the basolateral domain of the cells, indicating the ability of E-cadherin to induce an epithelial-like cell polarity.[52] From these results one can assume that a major function of E-cadherin is to provide the initial cell-cell contacts which then allows a self assembly process of other junctional components. However, one might also speculate that interaction of two E-cadherin molecules on neighboring cells initiates a signal that triggers a cascade of biochemical events. A similar role is currently discussed for the integrin family of substrate adhesion proteins.[17]

It is obvious from the above considerations that the E-cadherin/catenin-based adhesion system is a likely target for physiological or pathological processes that affect epithelial differentiation. Thus, changes in function, expression or structure of one of these components could lead to junctional disassembly and, consequently, to epithelial dedifferentiation. Experimentally, it could be shown that disturbance of E-cadherin function with anti-E-cadherin antibodies induces invasiveness in otherwise noninvasive epithelial cells.[53,54] Furthermore, this treatment produced phenotypic changes similar to those observed with epithelial cells which had lost E-cadherin expression after transformation with certain oncogenic retroviruses.[53] These results were the first indication that the loss of adhesive function of E-cadherin is a critical step in the promotion of epithelial cells to a dedifferentiated and invasive, i.e. more malignant phenotype. This correlation was confirmed by examining various human cell lines derived from bladder, breast, lung and pancreas carcinomas: Cell lines with an epitheloid morphology were generally noninvasive and expressed E-cadherin, whereas cell lines with a fibroblastoid morphology were invasive and had often lost E-cadherin expression.[55] Similarly, sublines of the mouse ovarian carcinoma cell line OV2944 with low capacity for spontaneous metastasis expressed E-cadherin, whereas a highly metastatic subline showed very low E-cadherin levels.[56] Fibroblastoid, highly invasive, vimentin-expressing breast cancer cell lines did also not express E-cadherin.[57] Esophageal carcinoma cell lines were examined in organotypic raft cultures:[58] Cell clones with positive E-cadherin expression showed an epitheloid shape and were non-invasive, whereas E-cadherin-negative cell lines were fibroblastic and invasive. Interestingly, down-regulation of E-cadherin expression was reported in experimental tumors derived from homogenously E-cadherin-positive transformed MDCK cells indicating host influences on E-cadherin synthesis and/or stability.[59] Invasiveness of dedifferentiated breast carcinoma cells was prevented by transfection with E-cadherin cDNA.[55] In similar experiments, invasiveness of transformed MDCK cells and of mouse mammary gland cells into chick heart fragments, and of L-cells (fibroblasts) into collagen gels was abolished by forced expression of the E-cadherin cDNA.[54,60] In the former case, the transfected cells produced differentiated tumors upon injection into nude mice. Furthermore, downregulation of E-cadherin via antisense RNA techniques induced invasiveness in otherwise noninvasive cells. Interestingly, a 50% reduction in E-cadherin protein sufficed to produce the effect, indicating a threshold level of E-cadherin expression above which invasion is suppressed. A study on E-cadherin expression in mouse epidermal carcinoma cell lines of varying malignancy also revealed a potential role for E-cadherin in the control of tumorigenicity. After transfection of E-cadherin cDNA in an E-cadherin-deficient, moderately differentiated squamous skin carcinoma cell line, the latency period and the growth of tumors produced by the transfected cells was reduced.[61] However, transfection of E-cadherin cDNA in spindle cell carcinoma cells which are poorly differentiated, did not inhibit tumorigenesis indicating that additional components are missing in these cells.[62] Interestingly, the fat tumor suppressor gene of *Drosophila* has been identified as a member of the cadherin gene family.[63]

4. E-CADHERIN EXPRESSION IN "REAL" TUMORS

The data from the in vitro experiments suggest a prime role for E-cadherin in preventing tumor invasion but can be criticized because of the somewhat artificial cell culture assay systems used. It is therefore of considerable importance that a large amount of correlative studies have meanwhile revealed a strong link between E-cadherin expression in certain carcinomas and clinical parameters, such as differentiation state and prognosis. Most investigations only report E-cadherin expression as analyzed by immunological methods (or sometimes by in situ hybridization), rarely was E-cadherin examined by functional means or by analysis of possible mutations in the gene. In most cases other junctional components were not analyzed, mainly due to the fact that specific reagents were not available. We review here a selection of these studies, which have given us many informations on the role of E-cadherin in the maintenance of the differentiated phenotype of carcinomas.

The progressive loss of the capacity of ***squamous cell carcinomas of the skin*** to differentiate has been extensively studied in a mouse skin carcinogenesis model. These tumors can be initiated by single treatment with chemical carcinogens; progression is induced by multiple applications of tumor promoters leading to squamous cell carcinomas of various grades. In the extreme, undifferentiated (spindle cell) carcinomas are produced which are highly aggressive and invasive.[64] In experimentally induced mouse skin carcinomas and derived cell lines a gradual loss of E-cadherin from dysplastic benign tumors to high grade advanced squamous cell carcinomas (SCC IV) correlated with a redistribution or absence of the cell polarity marker Na^+, K^+-ATPase.[65]

To determine whether the generation of the spindle-phenotype in mouse skin carcinomas is due to mutation of dominant or recessive genes, Balmain and collaborators recently produced hybrid cell lines from the extremely dedifferentiated Car B squamous cell carcinomas cells and the epithelioid C5N and other differentiated cells. Remarkably, all hybrids were epitheloid, indicating that the spindle-phenotype is recessive and due to loss of information which determines epitheloid morphology.[66] E-cadherin and desmoplakin were fully expressed and located in the hybrids as in the epitheloid parents;

keratin expression was also seen but no stable networks were formed. The fusion also led to suppression of malignancy as the hybrids showed longer latency of tumor growth in mice. The authors suggest that the generation of the spindle-phenotype is due to loss of gene(s) which control the complex epidermal differentiation program.

Non-melanoma skin cancer is the most common malignant neoplasia in the US; basal cell carcinoma (BCC) and squamous cell carcinoma are the most common subforms. Basal cell carcinoma arises from basal cell layers of the epidermis, and the tumors are therefore often undifferentiated. Squamous cell carcinoma of the skin arises from keratinocytes of the epidermis; they are graded as well, moderately, and poorly differentiated, or as spindle cell neoplasm. E-cadherin expression has recently been examined in 31 human basal cell carcinomas.[67] A characteristic feature of BCC is the low metastatic potential; however, local invasiveness is observed in some of the cases and these patients exhibit a higher rate of recurrency. E-cadherin was found to be preserved in all superficial BCC in which tumor nests are attached to the undersurface of the epidermis with little penetration into the dermis. Nodular BCC also showed preserved E-cadherin expression; in these, closed tumor masses of various sizes are embedded in the dermis. However, in most cases of infiltrating BCC, E-cadherin expression was clearly reduced suggesting that this property may be related to infiltrating growth.[67] To our knowledge, studies or the expression of E-cadherin in human squamous cell carcinomas of the skin have not yet been reported.

Squamous cell carcinomas of head and neck originate in the mucosa which lines the inner surfaces of the cavities of the aerodigestive tract, and 80-90% of the tumors are squamous cell carcinomas, followed by adenocarcinomas and others. According to the WHO, squamous cell carcinomas of head and neck can be subdivided by grade of epithelial differentiation into well-, moderately-, and poorly-differentiated types.[68] It has been suggested that these grades of differentiation are associated with patients survival: lower differentiation means shorter survival. Little is known about the multistep process of tumorigenesis in SCCs of the head and neck; human papilloma and Epstein-Barr virus involvement has been reported for larynx and nasopharynx carcinomas, respectively.[69,70] Recently, amplification, over-expression, and mutation of a number of oncogenes have been described in head and neck cancer.[71]

Tissue sections of 32 squamous cell carcinomas of the head and neck were recently investigated for the expression of E-cadherin.[72] We found that E-cadherin expression, as studied by immunofluorescence and in situ hybridization, is inversely correlated both with the loss of differentiation of the tumors and with lymph node metastasis. The well-differentiated SCCs expressed E-cadherin, often as strongly as the normal stratified epithelium, the moderately differentiated SCCs expressed intermediate amounts of E-cadherin or were heterogeneous, whereas the poorly differentiated SCCs were all E-cadherin-negative. Furthermore, seven of eight infiltrated lymph nodes of SCCs were E-cadherin-negative. These data indicate that the loss of the cell adhesion molecule E-cadherin plays an important role in the progression of head and neck squamous cell carcinomas, i.e., that down-regulation of expression is associated with dedifferentiation and invasion of the tumor cells in vivo.[72] A further study confirmed these findings and also correlated down-regulation of E-cadherin expression in head and neck carcinomas with an unfavorable clinical outcome.[72a] After treatment of patients by surgery and/or radiotherapy, recurrent disease or metastasis of head and neck carcinomas occurred in 50% of the cases. Significantly, the tumors of the patients who survived 30 months without further evidence of disease, had higher membranous E-cadherin expression than those from patients who developed incurable regional reccurrencies and/or distant metastases. No correlation of E-cadherin expression with lymph node metastases and survival of head and neck carcinomas was found in a third study.[73]

Fahraeus and colleagues investigated the role of components encoded by the Epstein-Barr virus (EBV) genome in the progression of squamous cell carcinomas. EBV is regularly associated with poorly differentiated (anaplastic) nasopharyngeal carcinomas which express the EBV nuclear antigens and also the membrane-associated protein LMP. It could be shown by cDNA transfection experiments that LMP specifically transforms human epithelial cells and inhibits epithelial differentiation. Concomitantly, a decreased expression of E-cadherin is observed and the cells become invasive for collagen gels.[74] The mechanisms by which a membrane protein such as LMP influences epithelial morphogenesis, needs to be investigated further.

Ductal and lobular breast carcinomas represent the majority of infiltrating breast tumors, the latter being less frequent.[75] Moll and colleagues have recently examined E-cadherin expression on cryostat and paraffin sections of 85 human primary infiltrating breast carcinomas.[76] A different picture emerged from the analysis of ductal and lobular breast carcinomas. Thus, 41 well- and moderately-differentiated infiltrating ductal carcinomas generally showed strong linear staining at the cell borders at similar levels as in luminal cells of normal mammary glands. Twenty-six poorly differentiated, more highly malignant ductal carcinomas were also positive for E-cadherin, although a higher proportion showed reduced staining which was heterogeneous and dotted over the cell borders. In contrast, 16 infiltrating lobular carcinomas which were either of the dispersed or solid type, did not express E-cadherin, while 2 cases showed weak staining. In situ lesions of lobular carcinomas (3 cases) were also E-cadherin-negative. The data indicate that loss of E-cadherin expression is an early event in the formation of the lobular but not the ductal type of breast carcinomas.[76] This finding suggests a different mode of invasion in these two breast cancer types. At the invasion front of ductal breast carcinomas, E-cadherin expression is often retained and invasive tumor cells still form true cohesive, though often thin and thread-like epithelial sheets. In contrast, lobular breast carcinomas typically invade the connective tissue in a more diffuse manner which may be reflected by the higher rate of recurrency of the patients.[76] Similar results were obtained by Gamallo et al:[77] Grade 1 breast carcinomas showed higher E-cadherin immunoreactivity than grade 2 and 3 tumors. Lobular breast carcinomas (7 cases) were all E-cadherin-negative. Oka et al.[78] examined 120 breast carcinomas for E-cadherin expression: 47% preserved E-cadherin and were mostly non-differentiated and non-invasive, whereas 53% showed reduced or negative expression and were more highly invasive and poorly differentiated. Reduced E-cadherin expression also correlated with extensive lymph node metastasis (in 74% of the cases) and distant metastasis (in 86% of the cases).

Colorectal adenocarcinomas (90-95% of colon cancers) start out as polypoid lesions which grow into the lumen or into the bowel wall. E-cadherin expression in colon cancer has been examined by several laboratories with respect to grade of differentiation, Dukes stage, and occurrence of metastasis. Dorudi et al.[79] investigated paraffin sections of 72 colorectal carcinomas: Out of 44 well- and moderately-differentiated tumors, 36 were positive for E-cadherin, whereas 24 of 28 poorly-differentiated carcinomas were E-cadherin negative. 32 of 36 Dukes A and B tumors were E-cadherin positive while 29 of 36 Dukes C1 and C2 tumors were negative. Furthermore, 20 of 32 lymph node metastases were E-cadherin negative as were 7 of 8 liver metastases. Kinsella et al[80] examined 40 patients with colorectal cancer: 12 had lymph node involvement and 9 of these showed reduced E-cadherin expression. A decrease of E-cadherin expression with decreased differentiation was found in further studies.[81-84] Mercurio and collaborators recently observed that 3 out of 5 poorly differentiated human colon carcinoma cell lines expressed E-cadherin but had lost the ability for Ca^{2+} dependent cell adhesion. These cell lines did not express α-catenin. Stable expression of α-catenin by cDNA transfection in one of the poorly differentiated cell lines fully restored cell-cell adhesion.[85] These data give an example of defective cell-cell adhesion in carcinoma cells which is not caused by change in E-cadherin expression but by a non-functional catenin.

In *gastric carcinomas*, factors of clinical significance are, besides the site and appearance of the tumor, the degree of local invasiveness and the histology. Patients with tumors of the intestinal type (showing glandular formations) generally have a better prognosis than those with tumors of the diffuse type (showing spots and clumps of tumor cells distributed within the stroma). Scirrous gastric carcinomas are invariably fatal.[86]

Mayer et al[87] recently examined E-cadherin expression in 60 gastric carcinomas and found reduced levels at a high frequency. The most significant correlation was observed between low E-cadherin expression and the degree of cellular dedifferentiation and disintegration of tissue architecture. Thus, none of the diffuse types of gastric carcinomas expressed E-cadherin. In the intermediate forms, glandular areas were mostly positive but diffuse ones were negative. Signet ring cell carcinomas were also E-cadherin negative. Furthermore, patients with low E-cadherin-expressing tumors (< 50% positive cells/lesion) appeared to have a poorer prognosis than patients with strong E-cadherin-positive carcinomas (> 50% positive cells/lesion). Similar results were found by other groups.[88-90] Shimoyama and Hirohashi[91] reported lack of E-cadherin expression only in a minor fraction of scattered gastric carcinomas.

Prostate carcinomas can be subdivided by the degree of tissue differentiation into well, moderately, and poorly differentiated lesions, whereby patients with poorly differentiated tumors carry a much worse prognosis than those with better differentiated ones. The poorer prognosis is related to the tendency of such tumors to metastasize to regional lymph nodes and distant sites. Prostate tumor development has been shown to be associated with several genetic changes defined by loss of heterozygosity such as on chromosomes 8p, 10q, and 16q.[92,93] Since the E-cadherin gene has been mapped to a region of 16q which is frequently lost in prostate cancer, it is tempting to speculate that it is the target of the mutation. Schalken, Isaacs W.B. and their colleagues have intensively studied the possible role of E-cadherin expression in both human and rat prostate carcinomas. Several transplantable rat cell lines and the corresponding tumors (from the Dunning R-3327 system) were examined. All non-invasive cell lines which formed tumors with some degree of differentiation, expressed E-cadherin mRNA, whereas poorly differentiated lines and their anaplastic tumors were E-cadherin deficient.[94] In a newly arisen, highly metastatic cell line, the decrease of E-cadherin expression occurred concomitantly with the acquisition of metastatic capability, indicating that down-regulation of E-cadherin expression is a prerequisite for invasion to occur. Ninety-two human prostate carcinomas and eight metastases were also analyzed, and a remarkable correlation between abnormal E-cadherin staining and increasing grade of the tumor was found.[95]

Besides the cases discussed here, E-cadherin expression has also been studied in various other tumors including lung, liver, bladder, and thyroid carcinomas (reviewed in ref. 96). A picture emerging from these studies is that loss of E-cadherin is a frequent, albeit not general phenomenon and correlates with decreased adhesiveness and differentiation of the tumors. We assume that these findings reflect the functional relevance of loss of E-cadherin for the invasive capability of carcinoma cells.

5. MOLECULAR MECHANISMS AFFECTING E-CADHERIN SYNTHESIS OR FUNCTION

In carcinomas, the misregulation of E-cadherin-based cell adhesion could occur on various levels, including mutations of the E-cadherin gene or of its promoter, alterations in the E-cadherin transcription rate or mRNA stability, or—in a similar fashion—disturbances of the E-cadherin associated catenins (Fig. 13.1). Alterations of the E-cadherin gene in carcinomas, such as mutations or deletions, have only rarely been observed until now. In one investigation, alterations were detected in E-cadherin mRNAs from two gastric carcinoma cell lines. In one case a 12bp in frame deletion, in the other case insertions of sequences were detected. These mutations were most likely due to abberant

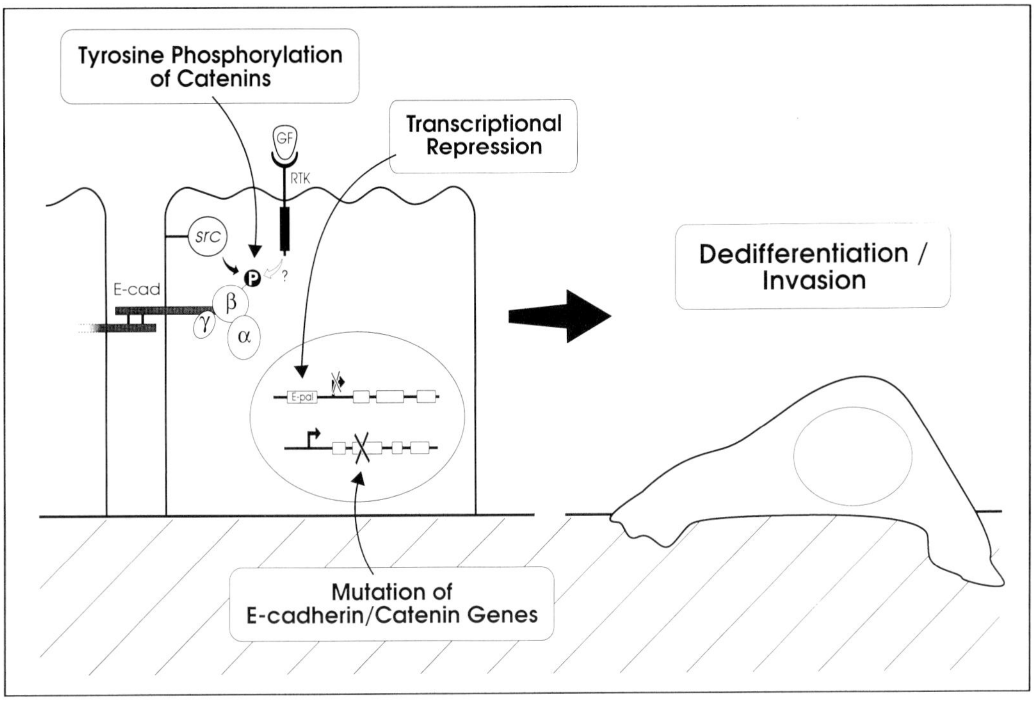

Fig. 13.1. Molecular mechanisms perturbing E-cadherin mediated cell adhesion and leading to dedifferentiation and invasion of epithelial cells. On the left, epithelial cells are shown which adhere to each other via the E-cadherin/catenin complex. Perturbation of the adhesive function of the complex can lead to dedifferentiation and invasion of carcinoma cells, as depicted on the right. Possible molecular mechanisms underlying these changes are (1) mutations of the genes for E-cadherin and the catenins, (2) transcriptional repression of the promoters of these genes, e.g. of the E-cadherin promoter through negative regulation by the E-Pal element, and (3) biochemical modifications of the individual components, such as tyrosine phosphorylation of β-catenin induced by the pp60$^{v\text{-}src}$ kinase. This could lead to a conformational change of the extracellular domain of E-cadherin thereby disrupting the contacts between two opposing molecules. Since dissociation of epithelial cells is frequently observed after growth factor (GF) treatment, the possibility of a similar modification by growth factor receptor tyrosine kinases (RTK) is indicated.

RNA splicing because DNA mutations were found at the exon-intron boundaries in these cell lines.[97] The lack of E-cadherin expression in most other carcinomas and cell lines must be due to alterations in the transcriptional regulation. Thus, a variation in functionally important transcription factors which regulate the E-cadherin promoter (or regulate the promoters of genes of other adherens junction proteins) might lead to changes in the invasiveness of carcinoma cells (Fig. 13.1). We and others have recently identified the E-cadherin promoter in order to study the mechanisms that regulate epithelial-specific expression of E-cadherin and possibly down-regulation of expression during dedifferentiation of carcinomas.[98,99] We found that a -178 to +92 bp upstream fragment of the E-cadherin gene induced specific expression of a CAT reporter gene in epithelial cells, whereas in non-epithelial cells this fragment was inactive, indicating the presence of cell type-specific regulatory elements in this region. Mutations of a proximal GC-rich region (at -25 to -58) and a CCAAT-box (at -65) resulted in loss of transcriptional activity in transient transfection experiments with epithelial cells. Mutations or deletions of a palindromic element termed E-Pal (at -86) also reduced promoter activity in epithelial cells but surprisingly, resulted in increased activity in nonepithelial and dedifferentiated carcinoma cells. This increase was dependent on the integrity of both the CCAAT-box and the GC-rich region, as judged from the activities of appropriate multiple mutants, suggesting that E-Pal suppresses the positive activity of these elements in nonepithelial cells.[100] These results indicate that the E-cadherin promoter is composed of positive and negative regulatory cis-elements, the combination of which generates epithelial specific activity. Sequence analysis as well as in vitro binding studies revealed that the E-Pal element is a potential target for helix-loop-helix transcription factors. Furthermore, we found that the GC-rich region functionally interacts with the transcription factor AP-2 at two independent sites, and that the CCAAT-binding factor of the E-cadherin promoter might be a relative of CP1 or C/EBP.

One conclusion from these results is that regulation of epithelial-specific expression of E-cadherin is largely due to specific suppression of promoter activity in nonepithelial cells rather than specific activation in epithelial cells. Apparently, the E-cadherin gene is a candidate for such a negative regulation, because the first modulation of E-cadherin expression during development occurs upon formation of the mesoderm and the neural tube which involves the loss of the protein in these tissues.[101,102] In early development, the E-cadherin gene can be considered as a kind of a housekeeping gene, being expressed already at the onset of activation of the embryonal genome in the fertilized oocyte.[103] This earliest upregulation of E-cadherin synthesis may be achieved through general positive transcription factors involved in the induction of a battery of early embryonal genes, and no factors specific for the E-cadherin promoter might be needed at this stage. In fact, our data indicate that the factors interacting with the CCAAT-box and the GC-rich region are strong, ubiquitous activators. The expression of E-cadherin is then maintained in the blastocyst stage where the protein is indispensable for compaction and for endoderm and ectoderm formation. When the mesoderm and the neural tube develop, E-cadherin is downregulated and replaced by other adhesion molecules.[101,102,104] Obviously, specific promoter repression through the E-Pal element could account for this transcriptional inactivation. In much later stages of development, E-cadherin synthesis is resumed in some mesenchymal tissues, e.g. during epithelial cell formation in the kidney and also in certain regions of the brain.[31,32,105] According to our model, loss or inactivation of the putative E-Pal repressor may then occur and transcription again be resumed by the ubiquitous factors.[100]

The negative regulatory role of E-Pal could be demonstrated in poorly differentiated carcinoma cells similarly as in fibroblasts and apparently, identical factors were found to bind to the regulatory elements in both cell types in vitro. In vivo footprinting data indicate that the CCAAT-box and the GC-rich region are occupied by nuclear factors in differentiated but not in poorly differentiated carcinoma cells (G. Hennig and J. Behrens, unpublished observations). These analogies in normal mesenchymal and malignant epithelial cells suggest that inactivation of E-cadherin expression in the embryo and in tumors might be achieved through similar regulatory pathways, which is in line with the above-mentioned concept of tumors recapitulating certain development steps in an uncontrolled fashion.

Several cases of carcinomas have been described which show apparently normal levels of membrane located E-cadherin protein but have clearly lost epithelial tissue organization and acquired a scattered phenotype (see above). In these tumors the adhesive function of E-cadherin rather than its synthesis must have been modulated. A possible target for such a regulation are the catenins which are associated with the cytoplasmic part of E-cadherin and are essential for E-cadherin function. The adhesive properties of PC9 lung cancer cells, which apparently harbor a null mutation in the α-catenin gene, could be restored by transfection with α-catenin cDNA.[106] Transformation of epithelial cells with a temperature sensitive src gene resulted in a reversible loss of differentiation, in unchanged E-cadherin expression, but tyrosine residues of β-catenin were phosphorylated at the permissive temperature.[23] It has been hypothesized that tyrosine phosphorylation of β-catenin might result in the disturbance of the adhesion complexes and lead to a non-adhesive state (Fig. 13.1).[23,107,108] If one assumes that E-cadherin plays a crucial role in the establishment and maintenance of epithelial cell contacts then any processes leading to altered adhesiveness are likely to function via disturbances of the E-cadherin based adhesion system. There are several examples of experimental modulation of cell adhesion that might involve transient deregulation of E-cadherin function. For instance, activation of the c-fos gene product in mammary epithelial cells carrying a fos-estrogen receptor hybrid leads to a change in morphology and a rapid loss of polarity.[24] After long-term activation of fos, the cells seize expression of E-cadherin. Various growth factors have cell dissociating activity and bind and activate tyrosine kinase receptors, thus possibly also interfering with the E-cadherin adhesion system by phosphorylation reactions (Fig. 13.1). It has long been known that tumor promoters such as TPA destroy the integrity of epithelial monolayers in vitro[109] indicating a role of protein kinase C in the regulation of E-cadherin activity. In line with this is the finding that inhibitors of protein kinases prevent the disrupture of cell junctions after exposure of the cells to reduced Ca^{2+} concentrations.[110] Taken together, these results suggest an important role of biochemical modifications in the regulation of the E-cadherin/catenin complex and will stimulate research on potential perturbations of these processes during tumor development.

6. OUTLOOK

In the past few years the potential role of E-cadherin as an invasion suppressor molecule has attracted the attention of many basic researchers, pathologists and clinicians. Future work will be directed towards the experimental prove for such a role in vivo, e.g. in experimental metastasis models or in transgenic animals. Obviously, developmental studies aimed at the elucidation of mechanisms that regulate the epithelial phenotype will also contribute to our understanding of malignant abberations of epithelial cells during carcinogenesis. The prognostic value of E-cadherin expression as a marker for human tumor invasiveness will have to be determined by more extensive clinical studies. Furthermore, understanding the regulation of the E-cadherin based adhesion systems will yield valuable information about possible disturbances in tumors. A consequence of this work could be the search for new drugs that reactivate E-cadherin transcription or function. For instance, the E-cadherin promoter/reporter constructs would be suitable substrates for drug screening procedures in which substances are tested for their ability to stimulate promoter activity. It will also be important to focus more on the role of catenins as modifiers of cadherin function and in the signal transduction between cells.

ACKNOWLEDGMENT

We are grateful to Jörg Hülsken for the preparation of the computer graphics. Work in our laboratory was supported by the Deutsche Krebshilfe, the Deutsche Forschungsgemeinschaft (DFG), the Wilhelm Sander-Stiftung, and the Bundesministerium für Forschung and Technologie (BMFT).

REFERENCES

1. Liotta LA, Steeg PS, Stetler-Stevenson WG. Cancer metastasis and angiogenesis: an imbalance of positive and negative regulation. Cell 1991; 64:327-336.
2. Gehlsen KR, Argraves WS, Pierschbacher MD, Ruoslahti E. Inhibition of in vitro tumor cell invasion by Arg-Gly-Asp-containing synthetic peptides. J Cell Biol 1988; 106:925-930.
3. Humphries MJ, Olden K, Yamada K. A synthetic peptide from fibronectin inhibits experimental metastasis of murine melanoma cells. Science 1986; 233:467-470.
4. Chan BMC, Matsuura N, Takada Y, Zetter B, Hemler ME. In vitro and in vivo consequences of VLA-2 expression on rhabdomyosarcoma cells. Science 1991; 251:1600-1602.
5. Zetter BR. Adhesion molecules in tumor metastasis. Sem Cancer Biol 1993; 4:219-229.
6. Giancotti FG, Ruoslahti E. Elevated levels of the $\alpha_5\beta_1$ fibronectin receptor suppress the transformed phenotype of Chinese hamster ovary cells. Cell 1990; 60:849-859.

7. Plantefaber LC, Hynes RO. Changes in integrin receptors on oncogenically transformed cells. Cell 1989; 56:281-290.
8. Ruoslahti E Integrins. J Clin Invest1991; 87:1-5.
9. Hynes RO, Lander AD. Contact and adhesive specificities in the associations, migrations, and targeting of cells and axons. Cell 1992; 68:303-322.
10. Rice GE, Bevilacqua MP. An inducible endothelial cell surface glycoprotein mediates melanoma cell adhesion. Science 1989; 246 1303-1306.
11. McEver RP. Leukocyte endothelial cell interactions. Curr Op Biol 1992; 4:840-849.
12. Gabbert H, Wagner R, Moll R, Gerharz CD. Tumor dedifferentiation: an important step in tumor invasion. Exp Metastasis 1985; 3:257-279.
13. Weinstein RS, Merck FB, Alroy J. The structure and function of intercellular junctions in cancer. Adv Cancer Res 1976; 23:23-89.
14. Coman DR. Decreased mutual adhesiveness, a property of cells from squamous cell carcinomas. Cancer Res 1944; 1:625-629.
15. McCutcheon M, Coman DR, Moore F. Studies on invasiveness of cancer. Cancer 1948; 1:460-467.
16. Valles AM, Boyer B, Thiery JP. In: Goldberg ID, ed. Cell Motility Factors. Birkhäuser Verlag, Basel 1991:17-34 .
17. Hynes RO. Integrins: Versatility, modulation, and signaling in cell adhesion. Cell 1992; 69:11-25.
18. Gurdon JB. Embryonic induction-molecular prospects. Development 1987; 99:285-306.
19. Jessell TM, Melton DA. Diffusible factors in vertebrate embryonic induction. Cell 1992; 68:257-270.
20. Weidner KM, Behrens J, Vandekerckhove J, Birchmeier W Scatter factor: molecular characteristics and effect on the invasiveness of epithelial cells. J Cell Biol 1990; 111:2097-2108.
21. Stoker M, Gherardi E, Perryman M, Gray J Scatter factor is a fibroblast-derived modulator of epithelial motility. Nature 1987; 327:239-242.
22. Jouanneau J, Gavrilowic J, Caruelle D, Jaye M, Moens G, Caruelle J-P, Thiery JP. Secreted or non-secreted forms of acid fibroblast growth factor produced by transfected epithelial cells influence cell morphology, motility and invasive potential. Proc Natl Acad Sci USA 1991; 88:2893-2997.
23. Behrens J, Vakaet L, Friis R, Winterhager E, Van Roy F, Mareel MM, Birchmeier W Loss of epithelial differentiation and gain of invasiveness correlates with tyrosine phosphorylation of the E-cadherin/β-catenin complex in cells transformed with a temperature-sensitive v-*src* gene. J Cell Biol 1993; 120:757-766.
24. Reichmann E, Schwarz H, Deiner EM, Leitner I, Eilers M, Berger J, Busslinger M, Beug H. Activation of an inducible c-fos ER fusion protein causes loss of epithelial polarity and triggers epithelial-fibroblastoid cell conversion. Cell 1992; 71:1-20.
25. Takeichi M. Cadherin cell adhesion receptors as morphology regulators. Science 1991; 251:1451-1455.
26. Hyafil F, Babinet C, Jacob F. Cell-cell interactions in early embryogenesis: A molecular approach to the role of calcium. Cell 1981; 26:447-454.
27. Edelman GM. Morphoregulatory molecules. Biochemistry 1988; 27:3533-3543.
28. Kemler R, Ozawa M, Ringwald M. Calcium-dependent cell adhesion molecules. Curr Opin Cell Biol 1989; 1:892-897.
29. Behrens J, Birchmeier W, Goodman SL, Imhof BA. Dissociation of Madin-Darby canine kidney epithelial cells by the monoclonal antibody anti-Arc 1: Mechanistic aspects and identification of the antigens as a component related to uvomorulin. J Cell Biol 1985; 101:1397-1315.
30. Boller K, Vestweber D, Kemler R. Cell-adhesion molecule uvomorulin is localized in the intermediate junctions of adult intestinal epithelial cells. J Cell Biol 1985; 100:327-332.
31. Shimamura K, Takahashi T, Takeichi M. E-cadherin expression in a particular subset of sensory neurons. Dev Biol 1992; 152:242-254.

32. Shimamura K, Takeichi M. Local and transient expression of E-cadherin involved in mouse embryonic brain morphogenesis. Development 1992; 116:1011-1019.
33. Nagafuchi A, Shirayoshi Y, Okazaki K, Yasuda K, Takeichi M. Transformation of cell adhesion properties by exogenously introduced E-cadherin cDNA. Nature 1987; 329:341-343.
34. Ringwald M, Schuh R, Vestweber D, Eistetter H, Lottspeich F, Engel J, Dölz R, Jähnig F, Epplen J, Mayer S, Müller C, Kemler R. The structure of the cell adhesion molecule uvomorulin. Insights into the molecular mechanism of Ca^{2+} dependent cell adhesion. EMBO J 1987; 6:3647-3653.
35. Gallin WJ, Sorkin BC, Edelman GM, Cunningham BA. Sequence analysis of a cDNA clone encoding the liver cell adhesion molecule, L-CAM. Proc Natl Acad Sci USA 1987; 84:2808-2812.
36. Mege R-M, Matsuzaki F, Gallin WJ, Goldberg JI, Cunningham BA, Edelman GM. Construction of epithelioid sheets by transfection of mouse sarcoma cells with cDNAs for chicken cell adhesion molecules. Proc Natl Acad Sci USA 1988; 85:7274-7278.
37. Bussemakers MJG, van Bokhoven A, Mees SGM, Kemler R, Schalken JA. Molecular cloning and characterization of the human E-cadherin cDNA. Mol Biol Rep 1993; 17:123-128.
38. Ozawa M, Baribault H, Kemler R. The cytoplasmic domain of the cell adhesion molecule uvomorulin associates with three independent proteins structurally related in different species. EMBO J 1989; 8:1711-1717.
39. Herrenknecht K, Ozawa M, Eckerskorn CH, Lottspeich F, Lenter M, Kemler R. The uvomorulin-anchorage protein α-catenin is a vinculin homologue. Proc Natl Acad Sci USA 1991; 88:9156-9160.
40. Nagafuchi A, Takeichi M, Tsukita S. The 102 kD cadherin-associated protein: Similarity to vinculin and posttranslation regulation of expression. Cell 1991; 65:849-857.
41. McCrea PD, Turck CW, Gumbiner B. A homology of the armadillo protein in *Drosophila* (plakoglobin) associated with E-cadherin. Science 1991; 254:1359-1361.
42. Butz S, Stappert J, Weissig H, Kemler R. Plakoglobin and β-catenin: dinstinct but closely related. Science 1992; 257:1142-1144.
43. Pfeifer M, McCrea PD, Green KJ, Wieschaus E, Gumbiner BM. The vertebrate adhesive junction protein β-catenin and plakoglobin and the *Drosophila* segment polarity gene armadillo form a multigene family with similar properties. J Cell Biol 1992; 118:681-691.
44. Knudsen KA Wheelock MJ. Plakoglobin or an 83 kD homologue distinct from β-catenin interacts with E-cadherin and N-cadherin. J Cell Biol 1992; 118:671-679.
45. Piepenhagen PA, Nelson WJ. Defining E-cadherin-associated protein complexes in epithelial cells: Plakoglobin, β- and γ-catenin are distinct components. J Cell Sci 1993; 104:751-762.
46. Donalies M, Cramer M, Ringwald M, Starzinsky-Powitz A. Expression of M-cadherin a member of the cadherin multigen family, correlates with differentiation of skeletal-muscle cells. Proc Natl Acad Sci USA 1991; 88:8024-8028.
47. Buxton RS, Magee AI. Structure and interactions of desmosomal and other cadherins. Semin Cell Biol 1992; 3:157-167.
48. Kemler R Classical cadherins. Sem Cell Biol 1992; 3:149-155.
49. Sano K, Tamihara H, Heimark RL, Obata S, Davidson M, St. John T, Taketani S, Suzuki S. Protocadherins: A large family of cadherin-related molecules in central nervous system. EMBO J 1993; 12:2249-2256.
50. Imhof BA, Vollmers HP, Goodman SL, Birchmeier W. Cell-cell interaction and polarity of epithelial cells. Specific perturbation using a monoclonal antibody. Cell 1983; 35:667-675
51. Vestweber D, Kemler R. Rabbit antiserum against a purified surface glycoprotein decompacts mouse preimplantation embryos and reacts with specific adult tissues. Exp Cell Res 1984; 152:169-178.
52. McNeill H, Ozawa M, Kemler R, Nelson WJ. Novel function of the cell adhesion molecule uvomorulin as an inducer of cell surface polarity. Cell 1990; 62:309-316.

53. Behrens J, Mareel MM, Van Roy FM, Birchmeier W. Dissecting tumor cell invasion: epithelial cells aquire invasive properties after the loss of uvomorulin-mediated cell-cell adhesion. J Cell Biol 1989; 108:2435-2447.
54. Chen W, Öbrink B. Cell contacts mediated by E-cadherin restrict invasive behavior of L-cells. J Cell Biol 1991; 114:319-327.
55. Frixen U, Behrens J, Sachs M, Eberle G, Voss B, Warda A, Löchner D, Birchmeier W. E-cadherin mediated cell-cell adhesion prevents invasiveness of human carcinoma cell lines. J Cell Biol 1990; 111:173-185.
56. Hashimoto M, Niwa O, Nitta Y, Takeichi M, Yokoro K. Unstable expression of E-cadherin adhesion molecules in metastatic ovarian tumor cells. Jpn J Cancer Res 1989; 80:459-463.
57. Sommers CL, Thompson EW, Torri JA, Kemler R, Gelman EP, Byers SW. Cell adhesion molecule uvomorulin expression in human breast cancer cell lines: Relationship to morphology and invasive capacities. Cell Growth and Differ 1991; 2:365-372.
58. Doki Y, Shiozaki H, Tahara H, Inoue M, Oka H, Iihara K, Kadowaki T, Takeichi M, Mori T. Correlation between E-cadherin expression and invasiveness in vitro in human esophageal cancer cell line. Cancer Res 1993; 53:3421-3426.
59. Mareel MM, Behrens J, Birchmeier W, DeBryne GK, Vleminckx K, Hoogewus A, Fiers WC, Van Roy FM. Downregulation of E-cadherin expression in Madin Darby Canine Kidney (MDCK) cells inside tumors. Int J Cancer 1991; 47:922-928.
60. Vleminckx K, Vakaet L, Mareel M, Fiers W, Van Roy F. Genetic manipulation of E-cadherin expression by epithelial tumour cells reveals an invasion suppressor role. Cell 1991; 66:107-119.
61. Navarro P Gomez M, Pizarro A, Gamallo C, Quintanilla M, Cano A. A role for the E-cadherin cell-cell adhesion molecule during tumor progression of mouse epidermal carcinogenesis. J Cell Biol 1991; 115:517-533.
62. Navarro P, Lozano E, Cano A. Expression of E- or P-cadherin is not sufficient to modify the morphology and the tumorigenic behavior of murine spindle carcinoma cells. J Cell Sci 1993; 105:923-934.
63. Mahoney PA, Weber U, Onofrechuk P, Biessmann H, Bryant PJ, Goodman CS. The fat tumor suppressor gene in *Drosophila* encodes a novel member of the cadherin gene superfamily. Cell 1991; 67:853-868.
64. Buchmann A, Ruggeri B, Klein-Szanto AJP, Balmain A. Progression of squamous carcinoma cells to spindle carcinomas of mouse skin is associated with an inbalance of H-ras alleles on chromosome 7. Cancer Res 1991; 51:4097-4101.
65. Ruggeri B, Caamano J, Slaga TJ, Conti CJ, Nelson WJ, Klein-Szanto AJP. Alteration in the expression of uvomorulin and Na^+ K^+-adenosine triphosphatase during mouse skin tumor progression. Am J Pathol 1992; 140:1179-1185.
66. Stoler AB, Stenback F, Balmain A. The conversion of mouse skin squamous cell car cinomas to spindle cell carcinomas is a recessive event. J Cell Biol 1993; 122:1103-1117.
67. Pizarro A, Benito N, Navarro P, Palacios P, Cano A, Quintanilla M, Contreras F, Gamallo C. E-cadherin expression in basal cell carcinomas. Br J Cancer 1993; 1994; 63:157-162.
68. Wahi PN, Cohen B, Luthra UK, Torloni H. Histological typing of oral and oropharyngeal tumours. Geneva: WHO 1971.
69. Bryan RL, Bevan IS, Crocker J, Young LS. Detection of HPV6 and 11 in tumors of the upper respiratory tract using the polymerase chain reaction. Clin Otolaryngol 1990; 15:177-180.
70. Fahraeus R, Rymo L, Thim JS, Klein G. Morphological transformation of human keratinocytes expressing the LMP gene of Epstein-Barr virus. Nature 1990; 345:447-449.
71. Field JK. Oncogenes and tumour-suppressor genes in squamous cell carcinoma of the head and neck. Eur J Cancer 1992; 28B 67-76.
72. Schipper JH, Frixen UH, Behrens J, Unger A, Jahnke K, Birchmeier W. E-cadherin expression in squamous cell carcinomas of head and neck: Inverse correlation with tumor dedifferentiation and lymph node metastasis. Cancer Res 1991; 51:6328-6337.

72a. Mattijssen V, Peters HM, Schalkwijk L, Manni JJ, Van't Hof-Grootenboer B, De Mulder PHM, Ruiter DJ. E-cadherin expression in head and neck squamous cell carcinoma is associated with clinical outcome. Int J Cancer 1993; 55:580-585.
73. Bowie G, Caslin AW, Roland NJ, Field JK Jones AS, Kinsella AR Expression of the cell-cell adhesion molecule E-cadherin in squamous cell carcinoma of the head and neck. Clin Otolaryngol 1993; 18:196-201.
74. Fahraeus R, Weiching C, Trivedi P, Klein G, Öbrink B. Decreased expression of E-cadherin and increased invasive capacity in EBV-LMP-transfected human epithelial and murine adenocarcinoma cells. Int J Cancer 1992; 52:834-838.
75. Henderson IC, Harris JR, Kinne DW, Hellman S. Cancer of the breast. In: De Vita VT, Hellman S, Rosenberg SA, eds. Cancer principles and practice of oncology 3rd ed. Philadelphia: JB Lippincott Co, 1989:1197-1268.
76. Moll R, Mitze M, Frixen UH, Birchmeier W. Differential loss of E-cadherin expression in infiltrating ductal and lobular breast carcinomas. Am J Pathol, 1993; 143:1731-1742.
77. Gamallo C, Palacios J, Suarez A, Pizarro A, Navarro P, Quintanilla M, Cano A. Correlation of E-cadherin expression with differentiation grade and histological type in breast carcinoma. Am J Pathol 1993; 142:987-993.
78. Oka H, Shiozaki H, Kobayashi K, Inoue M, Tahara H, Kobayashi T, Takatsuka Y, Matsuyoshi N, Hirano S, Takeichi M, Mori T. Expression of E-cadherin cell adhesion molecules in human breast cancer tissues and its relationship to metastasis. Cancer Res 1993; 53:1696-1701.
79. Dorudi S, Sheffield JP, Poulsom R, Northover JMA, Hart IR. E-cadherin expression in colorectal cancer. An immunochemical and *in situ* hybridization study. Am J Pathol 1993; 142:981-986.
80. Kinsella AR, Green B, Lepts GC, Hill CL, Bowie G, Taylor BA. The role of the cell adhesion molecule E cadherin in large bowel tumor cell invasion and metastasis. Br J Cancer 1993; 67:904-909.
81. Van der Wurff AAM, Ten Kate J, Van der Linden EPM, Dinjens WNM Arends J-W, Bosman FT L-CAM expression in normal, premalignant and malignant colon mucosa. J Pathol 1992; 168:287-291.
82. Nigam AK, Savage FJ, Boulos PB, Stamp GWH, Liu D, Pignatelli M. Loss of cell-cell and cell-matrix adhesion molecules in colorectal cancer. Br J Cancer 1993 ; 68:507-514.
83. Van Aken J, Cuvelier CA, De Wever N, Gao Y, Mareel MM, Roels HJ. Immunohistochemical analysis of E-cadherin expression in human colorectal tumors. Pathology Res and Pract 1993; 189:975-978.
84. Mareel M, Bracke M, Van Roy F, Vakaet L. Expression of E-cadherin in embryogenetic ingression and cancer invasion. Int J Dev Biol 1993 ; 37:227-235.
85. Breen E, Clarke A, Steele GD, Mercurio AM. Poorly differentiated colon carcinoma cell lines deficient in α-catenin expression express high levels of surface E-cadherin but lack Ca^{2+}-dependent cell-cell adhesion. Cell Adhesion and Communication 1994; in press.
86. Mc Donald JS, Steele G, Gunderson LL. Cancer of the stomach. In: De Vita VT Hellman S, Rosenberg SA, eds. Cancer principles and practice of oncology 3rd ed. Philadelphia: JB Lippincott Co, 1989:65-799 .
87. Mayer B, Johnson JP, Leitl F, Jauch KW, Heiss MM, Schildberg FW, Birchmeier W, Funke I. E-cadherin expression in primary and metastatic gastric cancer: Down-regulation correlates with cellular dedifferentiation and glandular dedifferentiation. Cancer Res 1993; 53:1690-1695.
88. Shiozaki H, Tahara H, Oka H, Miyata M, Kobayashi K, Tamura S, Iihara K, Doki Y, Hirano S, Takeichi M, Mori M. Expression of immunoreactive E-cadherin adhesion molecules in human cancers. Am J Pathol 1991; 139:17-23.
89. Matssuura K, Kawanishi J, Fujii S, Imamura M, Hirano S, Takeichi M, Niitsu Y. Altered expression of E-cadherin in gastric cancer tissues and carcinomatous fluid. Br J Cancer 1992; 66:1122-1130.

90. Oka H, Shiozaki H, Kobayashi K, Tahara H, Tamura S, Miyata M, Doki Y, Iihara K, Matsuyoshi N, Hirano S. Immunochemical evaluation of E-cadherin adhesion molecule expression in human gastric cancer. Virchows Arch A Pathol Anat Histpathol 1992; 421:149-156.
91. Shimoyama Y, Hirohashi S. Cadherin intercellular adhesion molecule in hepatocellular carcinomas: Loss of E-cadherin expression in an undifferentiated carcinoma. Cancer Letters 1991; 57:131-135.
92. Carter BS, Ewing CM, Ward WS, Treiger BF, Aalders TW, Schalken JA, Epstein JI, Isaacs WB. Allelic loss of chromosomes 16q and 10q in human prostate cancer. Proc Natl Acad Sci USA 1990; 87:8751-8755.
93. Bergheim URS, Kumini K, Collins VP, Ekman P. Deletion mapping of chromosome 8, 10, and 16 in human prostate carcinomas. Genes, Chromosomes, and Cancer 1991; 3:215-220.
94. Bussemakers MJ, van Morselaar RJ, Giroldi LA, Ichikawa T, Issacs JT, Takeichi M, Debruyne F M, Schalken JA. Decreased expression of E-cadherin in the progression of rat prostatic cancer. Cancer Res 1992; 52:2916-2922.
95. Umbas R, Schalken JA, Aalders TW, Carter BS, Karthaus HFM, Schaafsma HE, Debruyne FMJ, Isaacs WB. Expression of the cellular adhesion molecule E-cadherin is reduced or absent in high-grade prostate cancer. Cancer Res 1992; 52:5104-5109.
96. Birchmeier W, Behrens J. Cadherin expression in carcinomas: role in the formation of cell junctions and in the prevention of invasiveness. Biochim Biophys Acta 1994; in press.
97. Oda T Kanai Y, Oyama T, Yoshiura K, Shimoyama Y, Birchmeier W, Sugimura T, Hirohashi S. E-cadherin gene mutations in human gastric carcinoma cell lines. Proc Natl Acad Sci USA 1994; 91:1858-1862.
98. Behrens J, Löwrick O, Klein-Hitpass L, Birchmeier W. The E-cadherin promoter: functional analysis of a GC-rich region and an epithelial cell-specific palindromic regulatory element. Proc Natl Acad Sci USA 1991; 88:11495-11499.
99. Ringwald M, Baribault H, Schmidt C, Kemler R. The structure of the gene coding for the mouse cell adhesion molecule uvomorulin. Nucl Acid Res 1991; 19:6533-6539.
100. Löwrick O, Hennig G, Voss B, Lüscher B, Birchmeier W, Behrens J. Epithelial-specific activity of the E-cadherin promoter results from the action of positive (CCAAT-binding AP-2 like) and negative (helix-loop-helix) transcription factors. In preparation.
101. Thiery J-P, Duband J-L, Rutishauser U, Edelman GM. Cell adhesion molecules in early chick embryogenesis. Proc Natl Acad Sci USA 1982; 79:6737-6741.
102. Takeichi M. The cadherins: cell-cell adhesion molecules controlling animal morphogenesis. Development 1988; 102:639-655.
103. Vestweber D, Gossler A, Boller K, Kemler R. Expression and distribution of cell adhe sion molecule uvomorulin in mouse preimplantation embryos. Dev Biol 1987; 124:451-456.
104. Nose A, Takeichi M. A novel cadherin cell adhesion molecule. Its expression pattern associated with implantation and organogenesis. J Cell Biol 1986; 103:2649-2658.
105. Vestweber D, Kemler R, Ekblom P. Cell adhesion molecule uvomorulin during kidney development. Dev Biol 1985; 112:213-221.
106. Hirano S, Kimoto N, Shimoyama Y, Hirohashi S, Takeichi M. Identification of a neural α-catenin as a key regulator of cadherin function and multicellular organization. Cell 1992; 70:293-301.
107. Matsuyoshi N, Hamaguchi M, Taniguchi S, Nagafuchi A, Tsukita S, Takeichi M. Cadherin mediated cell-cell adhesion is perturbed by v-*src* tyrosine phosphorylation in metastatic fibroblasts. J Cell Biol 1992; 118:703-714.
108. Hamaguchi M, Onishi Y, Gotoh B, Takeichi M, Nagai Y. $p60^{v\text{-}src}$ causes tyrosine phosphorylation and inactivation of the N-cadherin-catenin cell adhesion system. EMBO J 1993 ; 12:307-314.
109. Ojakian GK. Tumor promoter-induced changes in the permeability of epithelial tight junctions. Cell 1981; 63:23:95-103.
110. Citi S. Protein kinase inhibitors prevent junction dissociation induced by low extracellular calcium in MDCK epithelial cells. J Cell Biol 1992; 117:169-178.

CHAPTER 14

Molecular Mechanisms of Pemphigus Diseases

Masayuki Amagai

1. INTRODUCTION

Many of you have experienced blisters several times in your life. Blisters are induced by many different causes, like friction blisters by mechanical trauma, burns by heat injuries, viral vesicles by infection of herpes zoster or herpes simplex. These blisters are accompanied by degeneration or necrosis of keratinocytes. However very rarely blisters are produced in more specific way, by loss of cell adhesion of keratinocytes without killing epidermal cells, termed acantholysis, such as in pemphigus. Extensive studies have been accumulated last decades in the field of investigative dermatology to understand the mechanism of blister formation in pemphigus. In 1960s antibodies against skin were demonstrated in sera of pemphigus patients by indirect immunofluorescent staining, and pemphigus was characterized as an autoimmune disease. In 1980s pemphigus autoantigens, target molecules of these autoantibodies, have been studied with immunochemical methods, such as immunoprecipitation and immunoblotting. And in 1990s cDNAs for pemphigus autoantigens were isolated and demonstrated that pemphigus antigens are of the cadherin type of cell adhesion molecules. This finding has fascinated many researchers outside of dermatology from the biological point of view, who investigate mechanisms of cell to cell adhesion of epithelial cells.

2. PEMPHIGUS IS A DISEASE OF CELL ADHESION OF KERATINOCYTES CAUSED BY AUTOANTIBODIES

Pemphigus is defined as a group of chronic blistering diseases, which histologically show acantholysis (loss of cell to cell adhesion of keratinocytes)

Molecular Mechanisms of Epithelial Cell Junctions: From Development to Disease
Edited by Sandra Citi, M.D., Ph.D.; © 1994, R.G. Landes Company.

and blister formation within the epidermis.[1] Pemphigus is divided into two major types, pemphigus vulgaris (PV) and pemphigus foliaceus (PF).

Clinically PV is characterized by erosions of mucous membranes and flaccid blisters of the skin (Fig. 14.1A). In typical cases of PV, oral lesions precede the cutaneous lesions by several months, and a generalized bullous phase with clear loose bulla or painful erosions occurs on the face, trunk, and extremities. Blisters arise on any stratified squamous epithelia, including esophagus or vagina. Normal appearing skin may be induced to blister by the application of gentle pressure with fingers (Nikolsky sign). Before the advent of corticosteroid therapy PV was almost invariably fatal because of lesions on large areas of skin with resultant loss of epidermal barrier function, leading to loss of body fluid or secondary bacterial infection. Pemphigus vegetans is a variant of PV and patients develop papillomatous and vegetative lesions on affected skin.

PF is clinically characterized by small flaccid blisters, but since the lesions are very

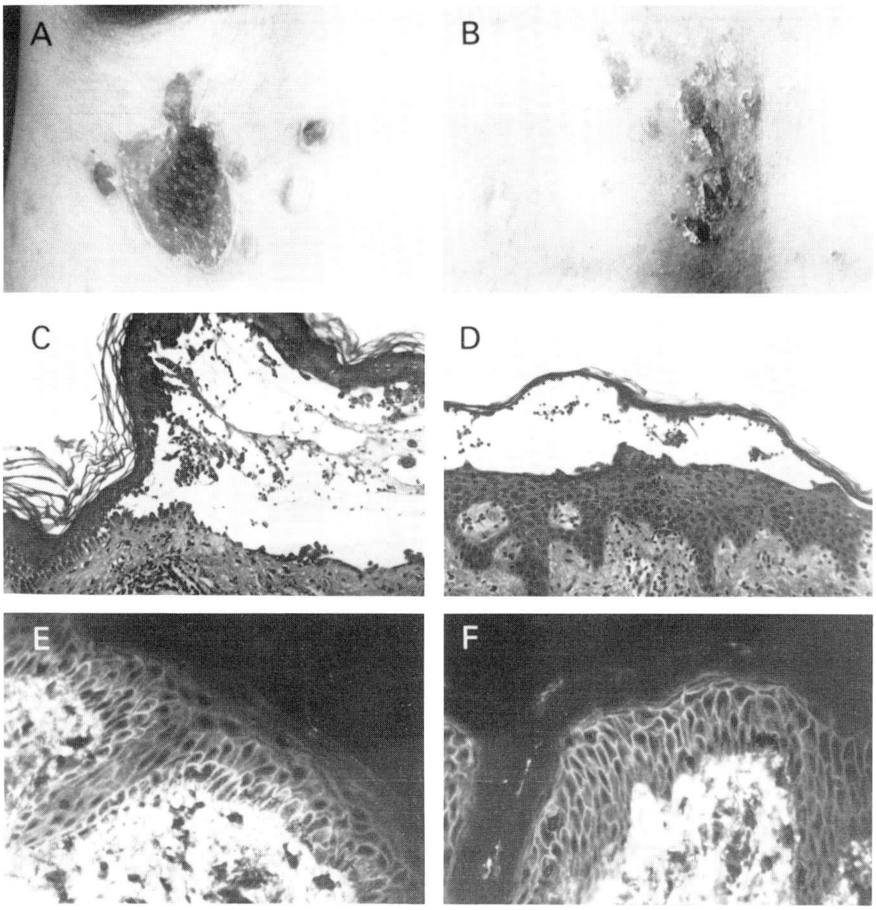

Fig. 14.1. Clinical, histological, and immunohistochemical features of pemphigus. Skin lesions of PV patients are characterized by wide-spread erosions and flaccid blisters (A) and those of PF patients are characterized by scales, erythema, and erosions (B). Histologically, both types of pemphigus show loss of cell adhesion of keratinocytes in the epidermis. In PV the splits occurs just above the basal layer (C), and in PF the loss of cell adhesion occurs at granular layer, in the upper part of the epidermis (D). Autoantibodies in patients' sera stain cell surface of keratinocytes in normal human skin. Some PV sera stain more strongly in the basal and parabasal layers (E) and some PF sera stain more prominently the cell surface in the upper epidermis (F).

superficial they rupture easily leaving erosions surrounded by erythema, that is often covered with crusting and scales (Fig. 14.1B). The lesions develop typically on the face and upper trunk, but may become generalized. PF patients do not develop mucous membrane lesions or large skin erosions, and in general PF has a much better prognosis than does PV. Nikolsky sign is positive. There are two PF variants. One is Brazilian PF or fogo selvagem,[2,3] which is an endemic form of PF found in hot humid wooded areas with poor population in central and western Brazil. An infectious etiology, such as a virus transferred by arthropods, is assumed but not yet proven. The other is pemphigus erythematosus, which is a localized variant of PF, combined with lupus erythematosus.

Histologically, these lesions in both PV and PF indicate intraepidermal blister formation with loss of cell-to-cell adhesion of keratinocytes (Fig. 14.1C and 14.1D). In PV a split occurs deep in the epidermis, just above the basal cell layer, termed suprabasilar acantholysis. Despite the loss of cell-cell adhesion, the basal epidermal cells remain tightly attached to the basement membrane zone, even when all cell-cell contact of the apical and lateral faces of the cell is lost. This distinctive histology pattern has been compared to a "row of tombstones". In contrast, in PF the intraepidermal split is in the granular layer, the upper part of the living epidermis. This difference of the depth of the split may explain the difference of their prognoses.

What causes the loss of cell adhesion of keratinocytes in pemphigus? In both PV and PF, autoantibodies of the IgG isotype are bound in vivo in a characteristic pattern around the plasma membranes of affected epithelia and detected in the sera of patients,[4,5] indicating that autoimmune process is involved in the pathogenesis of pemphigus. Some PV sera tend to stain keratinocyte cell surface more strongly in the basal and parabasal layer (Fig. 14.1E) and some PF sera stain more intensely in the upper part of the epidermis (Fig. 14.1F). However, most sera from both types of pemphigus stain throughout the epidermis and their fluorescence staining patterns are indistinguishable. Furthermore, recently, in situ hybridization with cDNAs for PV antigen (PVA) and PF antigen (PFA) demonstrated that the expression of PVA mRNA was detected in the lower part of the epidermis, and PFA mRNA was detected throughout the epidermis with weak expression in the basal and parabasal layers.[6] Therefore, the expression pattern of PVA and PFA in the epidermis might explain the localization of the split in PV and PF.

From a number of observations, the circulating autoantibodies in pemphigus patients have been proven to be pathogenic. First, disease activity in patients often correlates with anti-cell surface antibody titer, as determined by indirect immunofluorescence.[7] Second, neonates of mothers with PV may have transient disease due to maternal IgG that crosses the placenta.[8] As maternal antibody is catabolized, disease subsides. Third, IgG from patients with pemphigus alone, without complement or inflammatory cells, can induce loss of cell-cell adhesion, with the same histology as seen in patients, in skin organ culture.[9,10] Lastly, passive transfer of IgG from patients sera to neonatal mice results in loss of cell-cell adhesion and blisters with typical histological findings[11] (Fig. 14.2). The fab' fragment of IgG from Brazilian PF patients was shown to be sufficient to cause blisters in neonatal mice.[12]

Then, what are target molecules these pathogenic autoantibodies attack? The target molecules, pemphigus antigens, have been characterized by immunochemical methods. PVA and PFA have been identified as a 130 kD glycoprotein and a 160 kD glycoprotein, respectively, by immunoprecipitation and immunoblotting.[13]

For PFA immunoblot analysis with SDS extracts of normal human epidermis or bovine muzzle desmosomal preparation showed that about one-third of PF sera recognized a 160 kD polypeptide.[14-18] The PF sera, which failed to bind the 160-kD band, presumably could no longer bind PFA epitopes that were destroyed during the SDS-PAGE procedure. Immunoprecipitation of nonionic-detergent extracts of epidermis demonstrated that all PF sera bind a complex of 260-, 160- and

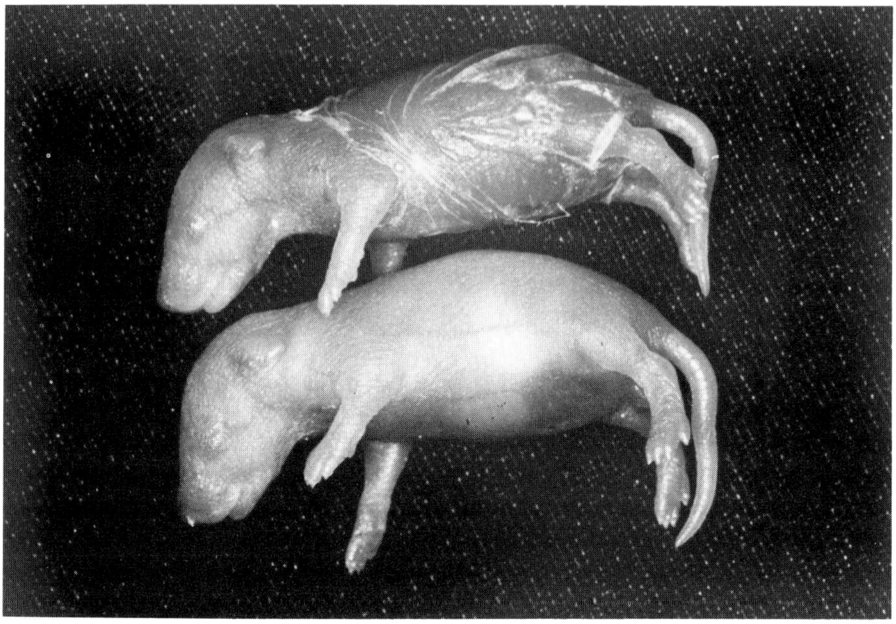

Fig. 14.2. Neonatal mouse disease model for pemphigus. Passive transfer of IgG prepared from pemphigus patients' sera into neonatal mice induces blister formation with typical histological findings of pemphigus. The top and bottom mice were injected with PV IgG and normal IgG, respectively. Note that the top mouse has blisters on the trunk and in contrast the bottom mouse has normal appearing skin (18 hours after the injection).

85-kD polypeptides.[19] It was later demonstrated by sequential immunoprecipitation that the 260-kD band is a complex form of the 160-kD and 85-kD bands and the 85-kD band is plakoglobin.[20] Plakoglobin is a plaque protein found in both desmosomes and adherence junctions, and does not contain epitopes for PF sera. This 160-kD protein is the PFA and was subsequently identified to be desmoglein I (Dsg1), a desmosomal transmembrane glycoprotein, by immunochemical evidences with anti-desmoglein I monoclonal and polyclonal antibodies and PF sera.[21,22]

PVA was first characterized by immunoprecipitation analysis of nonionic detergent extract of mouse and human cultured keratinocytes.[14,23] In these studies all PV sera tested immunoprecipitated a 130-kD glycoprotein with a 85-kD polypeptide. Under nonreducing conditions these two molecules were linked as a 210-kD molecule by disulfide bond formation. This 85-kD molecule was also shown to be plakoglobin.[20] By immunoblot analysis all PV sera tested also recognized the 130-kD molecule in normal human epidermal extract with modified extraction procedure.[16] The 130-kD PVA was not found in desmosome preparation as a major component, and the real nature of the molecule had been unknown until the isolation of cDNA.

3. PEMPHIGUS ANTIGEN IS A CADHERIN TYPE OF CELL ADHESION MOLECULE BOTH IN PV AND PF

PFA was identified as desmoglein I as discussed above at immunochemical level, and cDNAs for bovine and human desmoglein were isolated by several groups.[24-27] The deduced amino acid sequence showed significant homology with cadherins and indicated that desmoglein I belongs to the cadherin supergene family (Fig. 14.3) (see also chapter 8 in this volume).

To isolate a cDNA for PVA, the only probe available was PV patients' sera which can react PVA. However, because human sera contain numerous non-specific IgGs which give false-positive clones, the specificity of PV IgG was increased by affinity-purifying PV sera on immunoblot of the 130 kD PVA. This affinity-purified anti-PVA specific IgG

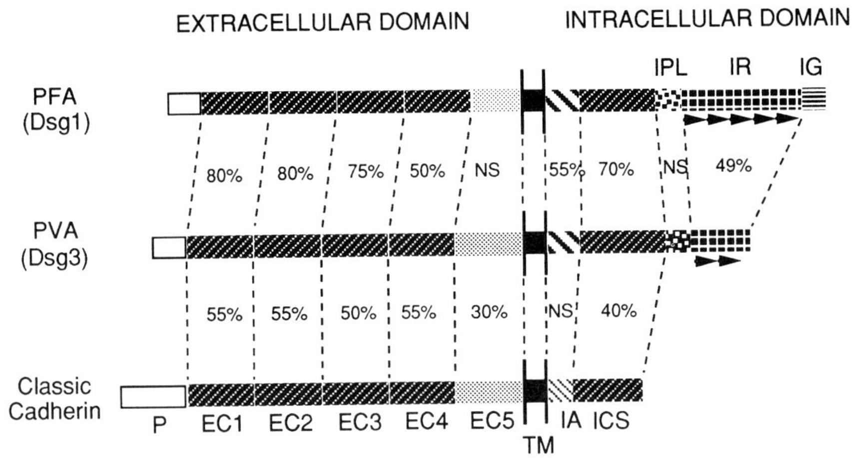

Fig. 14.3. Molecular structure of PVA, PFA and classic cadherins. PVA and PFA have a basic structure of classic cadherins. Figures indicate % homology for each domain between the molecules. NS stands for not significant. An Arrow head indicates a 29 amino acid repeating unit, which is found only in the desmoglein subfamily. P, pro-sequence; EC, extracellular domain; TM, transmembrane region; IA, intracellular anchor domain; ICS, intracellular cadherin type segment; IPL, proline-rich linker; IR, intracellular repeat; IC, intracellular glycine rich domain.

was used to screen a λgt11 expression cDNA library derived from normal human cultured keratinocyte and cDNA for PVA was isolated.[28] The deduced amino acid sequence of PVA showed significant homology only with members of the cadherin supergene family, most markedly with desmoglein I, PFA (Fig. 14.3). These findings indicate that autoantigens for both PV and PF are cadherin type of cell adhesion molecules and that pemphigus is an anti-cadherin autoimmune disease.

PVA and PFA have many common molecular features, and both have their own unique 29 residue repeats in the cytoplasmic domain, which are not found in classic cadherins. Thus, PVA and PFA form a "desmoglein" subfamily in the cadherin supergene family. According to a new nomenclature,[29] the genes (proteins) for PVA and PFA are termed *DSG3* (Dsg3) and *DSG1* (Dsg1), respectively. *DSG1* and *DSG3* are expressed only in stratified squamous epithelium.[28,30,31] cDNA for *DSG2* was isolated from a human colon carcinoma cDNA library[32] and expressed in basal cells of stratified squamous epithelia and also in simple epithelial cells.[31]

More interestingly, *DSG1*, *DSG2*, and *DSG3* were assigned to chromosome 18 using PCR amplification of those gene specific-DNA fragments in a somatic cell hybrid panel.[33-35] Just recently our group mapped *DSG1* and *DSG3* on the band q12 of human chromosome 18 by fluorescence *in situ* hybridization. Furthermore, both genes were localized on a 320-kb genomic fragment separated by pulsed-field gel electrophoresis, suggesting the possibility of a cluster for the desmoglein gene family on chromosome 18.[36] Type 1 desmocollin (*DSC1*), type 3 desmocollin (*DSC3*), N-cadherin were also mapped on human chromosome 18,[37,38] and *DSC3* were regionally mapped to human chromosome 18q12 (our unpublished data). It will be of interest to determine whether there is a big gene cluster for desmosomal cadherins, including yet identified ones, on chromosome 18.

4. LOCALIZATION OF PVA (DSG3) WITHIN KERATINOCYTE DESMOSOMES

Previous localization studies of PVA with PV sera have been contradictory. Immunofluorescence studies of mouse keratinocytes in culture have suggested that PV antibodies might bind desmosomes.[39] However, by immunoperoxidase staining of PV patients' skin, IgG was distributed diffusely along the keratinocyte cell membrane.[40] In addition, this diffuse localization of PVA on plasma membrane was supported by indirect immunoferritin on trypsinized guinea pig keratinocytes,[41] by immunoperoxidase staining of skin from neonatal mice injected with PV IgG,[42] and by immunogold staining of bovine tongue with a rabbit antibody raised against a presumptive PVA from bovine desmosomes.[43] However, contrary results of ultrastructural localization, which demonstrated PV IgG binding predominantly in desmosome, were obtained by immunoferritin detection of PV IgG on patients' skin[44] and by immunogold localization of PV IgG binding on cultured squamous cell carcinoma keratinocytes.[45]

By preembedding immunogold electron microscopic studies, only slight labeling could be detected in desmosomes, presumably because of difficulty in gold penetration on intact desmosomes. Therefore, Karpati et al used the unique property of cadherins to resist trypsinization in the presence of calcium (0.01% trypsin in 1 mM calcium) and studied keratinocytes whose desmosomes were made more accessible to gold-labeled probes by this treatment. In addition, for increased specificity human IgG was affinity purified from PV patients' sera and rabbit antibodies were raised with bacterial fusion proteins containing the extracellular portions of PVA, and used for immunolocalization.[46] In this study there was extensive labeling of the extracellular part of desmosomes but not of the interdesmosomal cell membrane. Furthermore, neonatal mice injected with PV IgG affinity-purified on the PVA fusion protein showed desmosomal separation (as discussed below) with the IgG localized to desmosomal cores. These results indicate that PVA is organized and concentrated within the desmosome and PVA has a property of desmoglein family also in terms of localization.

5. IMMUNOGENIC AND PATHOGENIC DOMAINS OF PEMPHIGUS ANTIGENS

In pemphigus, autoantibodies attack a cadherin type of cell adhesion molecule and cause loss of cell adhesion of keratinocytes. In that case, do autoantibodies in patients with pemphigus have any preference recognition sites within the molecule? And autoantibodies against which part of the molecule do play a pathogenic role in inducing loss of cell adhesion? Using bacterial fusion proteins which contain different part of the extracellular domain of PVA or PFA, major immunogenic domains and pathogenic domains were determined as an initial step to understand pathophysiological mechanism of blister formation in pemphigus.[47-49]

For PVA,[47] seven fusion proteins with β-galactosidase containing EC1-2, EC3-5, and each individual extracellular domains, EC1 to EC5, were produced and subjected to immunoblot analysis with 23 PV sera, 15 PF and 5 normal control sera. The results demonstrated that major immunogenic domains of PVA are located on the EC1, EC2 and EC4 domains, which were recognized by up to 60% of PV sera (Fig. 14.4). None of PF nor normal sera reacted with any of these fusion proteins.

To determine pathogenic domains of PVA, IgG was affinity-purified from PV sera on fusion proteins representing either the amino (EC1-2) or the carboxy (EC3-5) terminus of the extracellular PVA, and injected into neonatal mice,[47] which is an established disease model for pemphigus.[11] PV IgG affinity-purified on the EC1-2 fusion protein caused suprabasilar acantholysis, the characteristic histological finding of PV, but PV IgG affinity-purified on the EC3-5 fusion protein did not (Fig. 14.4). These results indicate that at least one pathogenic epitope, which is sufficient to induce loss of cell adhesion of keratinocytes in neonatal mice, is located on the amino-terminal region of PVA. It is interesting to know that the importance of the amino-terminal region of classic cadherins in homophilic adhesion has been suggested by studies using chimeric molecules between E- and P-cadherin, amino acid substitutions by site-directed mutagenesis, and epitope mapping of monoclonal antibodies that inhibit homophilic binding.[50]

However, PV IgG affinity-purified on this EC1-2 fusion protein could induce only microscopic blisters in neonatal mice but not gross blisters, even though PV IgG was purified from as much as 50 ml of a PV serum and injected into one neonatal mouse. When whole IgG was prepared by ammonium sulfate precipitation and injected into neonatal mice, 1.5 ml serum is enough to cause gross blisters. Therefore, the failure to cause gross blisters with affinity-purified PV IgG would not be simply due to a low amount of injected PV IgG. Furthermore, PV IgG depleted of reactivity against the EC1-2 fusion protein was still capable to cause gross blisters. These findings indicate that these bacterial fusion proteins represent only a part of epitopes and miss most of pathogenic epitopes, presumably conformational ones.

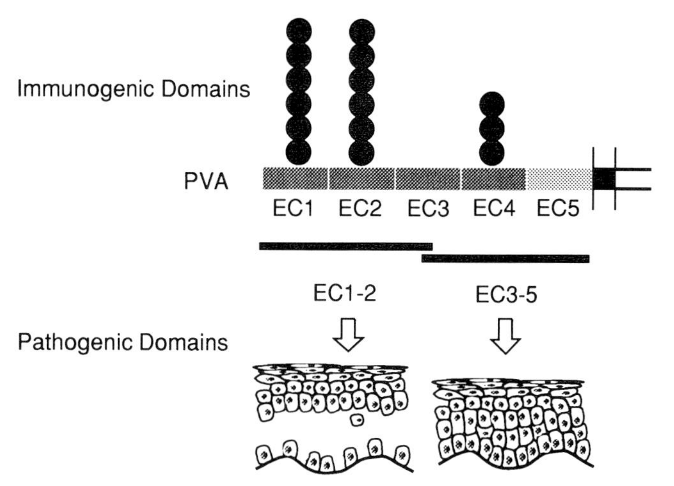

Fig. 14.4. Immunogenic and pathogenic domains for PVA analyzed with bacterial fusion proteins. A closed circle indicates approximately 10% of PV sera which reacted with a fusion protein for each domain. Major immunogenic regions of the extracellular domain of PVA are located on EC1, EC2, and EC4 domains. PV IgG affinity-purified on the EC1-2 fusion protein, but not IgG affinity-purified on the EC3-5 fusion protein, caused suprabasilar acantholysis, the typical histological finding of PV.

As for PFA(Dsg1), bacterial fusion proteins representing three different parts of the extracellular domain of *human* PFA were produced and examined by immunoblot analysis,[48] even though immunoblot is not an ideal method to detect PFA because only one third of PF sera recognize PFA in epidermal extract.[14,15] Seven of 13 Brazilian PF sera reacted the carboxy-terminal region of the extracellular domain, corresponding roughly to EC4-5 domains, and only 3 of 13 Brazilian PF sera recognized the amino-terminal region, corresponding EC1-2 domains. More recently another set of bacterial fusion protein representing EC1-2, EC3-5, EC1-5, and IC (intracellular domain) of *bovine* Dsg1 were also produced and used for immunoblot analysis with PF sera,[49] because PF sera detect Dsg1 more frequently in *bovine* desmosome preparation than in *human* epidermal extracts.[16,18] Approximately half of 20 PF sera tested reacted with the Dsg1 in bovine desmosome preparation, and almost half of these Dsg1 positive PF sera recognized EC1-2, EC1-5, and IC fusion proteins, although none of them recognized the EC3-5 fusion protein. More interestingly, it is known that approximately half of PV sera recognize the Dsg1 in addition to PVA(Dsg3).[51-53] Most of those PV sera containing antibodies against Dsg1 reacted only with the IC, but not EC1-2, EC3-5 or EC1-5 fusion proteins of *bovine* Dsg1, suggesting IgG against Dsg1 in PV sera does not play a pathogenic role. Pathogenic domains of PFA has not been characterized at molecular level yet.

6. THE EXTRACELLULAR DOMAIN OF PVA (DSG3) MEDIATES WEAK BUT HOMOPHYLIC BINDING

Hemophilic type of binding of classic cadherins[54,55] was directly demonstrated by introducing the cadherins into fibroblasts which are deficient in cadherin expression but expressing catenins, anchoring molecules for classic cadherins.[56,57] Transfected fibroblasts with the cadherin acquired strong Ca^{2+}-dependent aggregating activity by expressing the cadherin derived from the introduced cDNA and were morphologically transformed to form colonies in which cells were tightly connected to each other. One of the unique features of classic cadherins as cell adhesion molecules is that they mediate physiological interaction with the involvement of cytoplasmic anchoring molecules and cytoskeleton network and show strong adhesion with morphological change of the cells. Classic cadherins do not mediate adhesion if the cytoplasmic domain containing binding sites for catenins is severely truncated, even though the extracellular domain is properly expressed on the cell surface.[58,59] In contrast, Ca^{2+} independent cell adhesion molecules, such as molecules in immunoglobulin superfamily or receptor-ligand interaction, mediate pure molecular interaction and show loose adhesion without morphological changes.

The cytoplasmic domain of PVA, like Dsg1, binds only to plakoglobin, but does not bind to α - or β-catenin.[20] Plakoglobin could be identical to γ-catenin,[60] but could be a distinct protein.[61] Thus, the cytoplasmic domain of PVA might not function in L cells to mediate cell-cell adhesion. Furthermore, when Dsg1 was transfected to a mouse fibroblast, 3T3-L1, devoid of desmosomes, Dsg1 was not expressed stably on the cell surface and did not function properly as a cell adhesion molecule.[32] Therefore, a chimeric molecule (PVEC) with the extracellular domain of PVA and the cytoplasmic domain of E-cadherin was introduced into a mouse fibroblast, L cells, to determine whether the extracellular domain of PVA mediates homophilic binding as do classic cadherins.[62]

Immunofluorescence and flow cytometry analyses of chimeric PVEC stable transfectants with PV sera indicated that the PVA extracellular domain of this chimeric molecule was expressed in proper orientation on the cell surface. Immunoprecipitation analysis of these transfectants with PV sera showed that the cytoplasmic domain of this chimeric molecule bound the catenins, which were coprecipitated with the chimeric molecule. However, surprisingly, these L cell transfectants showed only slight aggregation compared to E-cadherin L cell transfectants. This weak aggregation was, however, specific and homophilic, because the PVEC transfectants preferentially adhered to cells expressing the chimeric molecule, but not to cells expressing E-cadherin or to control cells transfected with neomycin resistant gene alone. These results indicate that the extracellular domain of PVA mediates weak homophilic adhesion and is not interchangeable in function with the extracellular domain of E-cadherin. PVA might require other desmosomal components, such as desmocollin or desmoplakin, to function maximally as a cell adhesion molecule.

7. NEW PEMPHIGUS ENTITIES, PARANEOPLASTIC PEMPHIGUS AND IGA PEMPHIGUS

There are two recently described pemphigus variants, paraneoplastic pemphigus and IgA pemphigus.

Paraneoplastic pemphigus (PNP)[63] is thought to be caused by autoimmune process and patients with PNP have painful mucosal erosions and polymorphic skin eruptions, including blistering and erosive lesions, in the context of an occult or known neoplasm, such as lymphomas, thymomas or leukemia. The mucosal and cutaneous blisters show a characteristic combination of suprabasilar acantholysis similar to PV, and keratinocyte necrosis that may resemble erythema multiforme or a lichenoid inflammatory eruption. All patients have tissue-bound and circulating autoantibodies that bind to the cell surface of not only stratified squamous epithelia but also simple, columnar, and transitional epithelia. These autoantibodies immunoprecipitate a complex of four high molecular weight proteins from keratinocytes, 250-, 230-, 210- and 190-kD. 250- and 210-kD bands were identified to be desmoplakin I and II.[64] The 230-kD band is thought to be the 230-kD bullous pemphigoid antigen, a component of hemidesmosome and homologous to desmoplakin.[63] The autoantibodies from these patients were demonstrated to be pathogenic by passive transfer into neonatal mice, where they induce acantholysis of skin and esophageal mucosa. More recently another immunoprecipitatable 170-kD protein was suspected to be a pathogenic autoantigen because this protein was considered to be a transmembrane protein by biochemical characterization.[65] The pathophysiologic mechanism of PNP is yet to be clarified, and identification of this 170 kD autoantigen at molecular level might open new avenues in the field of cell adhesion mechanism of keratinocytes.

Recently, IgA autoantibodies against keratinocyte cell surface were described in patients suffering from recalcitrant vesiculopustular eruptions with neutrophilic infiltration. This entity is proposed to be called intercellular IgA vesiculopustular dermatosis (IAVPD) or so-called IgA pemphigus.[66-68] The unique point of this entity is pemphigus-like deposition of IgA but not IgG and the presence of circulating anti-cell surface IgA autoantibodies. IAVPD is divided into two subtypes, a subcorneal pustular dermatosis (SPD) type and an intraepidermal neutrophilic (IEN) type. Circulating IgA from SPD type recognized 115- and 105-kD proteins on immunoblot using bovine desmosome preparation, which comigrated with desmocollin I and II.[68] The presence of IgA against PFA(Dsg1) was also reported in the literature.[67,69] Pathogenic role of IgA autoantibodies in this entity is yet to be proven. Desmocollin I and II are also known to be recognized by a subset of pemphigus sera, particularly by Brazilian PF sera.[70]

8. PERSPECTIVES

By cDNA cloning, the target molecule in this antibody-mediated autoimmune disease, pemphigus, turned out to be a cadherin type of cell adhesion molecule. It is plausible that autoantibodies in pemphigus interfere with the function of pemphigus antigen, inhibit cell-cell adhesion of keratinocytes and initiate blister formation. However, why autoantibodies against a single molecule can destroy cell adhesions mediated by other molecules, such as E- or P-cadherin or desmocollins? Pemphigus IgG is known to stimulate plasminogen activator, and resulting plasmin might play a role in destroying other cell adhesion molecules.[10] Plasmin can be one explanation. However, Langerhans cells or infiltrating lymphocytes migrate between keratinocytes, and keratinocytes themselves move upwards along differentiation. The state of adhesion must be changed dynamically minute by minute. Furthermore, evidence that phosphorylation of the cytoplasmic domains of classic cadherins and catenins regulates the adhesion function of cadherins have been accumulating[71-73] (see also chapters 12 and 13 in this volume). Therefore, it is of interest to speculate that an on-and-off mechanism would exist for desmosomal assembly-disassembly in keratinocytes, presumably mediated by phosphorylation, and that the binding of pemphigus IgG to desmogleins could misinduce some signal which causes the retraction of cell adhesion molecules from cell surface. The cytoplasmic domain of desmocollins was found to have desmosomal assembly elements,[74] and desmogleins might have desmosome disassembly element, which could be stimulated by pemphigus antibodies.

From the clinical point of view it is invaluable to produce recombinant proteins with the proper conformation of the authentic pemphigus antigens. Bacterial fusion proteins could not generate the conformation of PVA as discussed above. So higher level expression systems, such as baculovirus expression or mammalian expression (which does post-translational modification) is the way to go. For diagnostic purposes, we currently rely on immunofluorescence staining of normal human skin with patients' sera, however, this is not an ideal technique because each skin section has slightly different amounts of antigen and also this is labor-intensive and time consuming. If recombinant proteins of pemphigus antigen with proper conformation were available, diagnostic tests such as ELISA would be beneficial and could be performed at any hospital with minimal labor. As for treatment, conventional therapies are administration of corticosteroid or immunosupressant, which have severe side effects. If the recombinant protein expressing all the pathogenic epitopes were available, antigen-specific plasmapheresis would be possible as a new therapeutic strategy, which could absorb out specifically pathogenic IgG from patients' circulation using a affinity column with the recombinant proteins.

Acknowledgment

I would like to thank Drs. Takashi Hashimoto and Takeji Nishikawa for reviewing this manuscript.

References

1. Stanley JR. Pemphigus. In: T. B. Fitzpatrick, A. Z. Eisen, K. Wolff, I. M. Freedberg, and K. F. Austen, eds. Dermatology in General Medicine. 4th. New York: MacGraw-Hill, 1993:606-614.
2. Diaz LA, Sampaio S, Rivitti EA, Martins CR, Cuhna PR, Lombardi C, Almeida FA, Castro RM, Macca ML, Lavrado C et al. Endemic pemphigus foliaceus-"Fogo Selvagem": I. Clinical features and immunopathology. J Am Acad Dermatol 1988; 20:657-669.
3. Diaz LA, Sampaio S, Rivitti EA, Martins CR, Cuhna PR, Lombardi C, Almeida FA, Castro RM, Macca ML, Lavrado C et al. Endemic pemphigus foliaceus (fogo selvagem): II. Current and historical epidemiological aspects. J Invest Dermatol 1989; 92:4-12.
4. Beutner EH, and Jordon RE. Demonstration of skin antibodies in sera of pemphigus vulgaris patients by indirect immunofluorescent staining. Proc Soc Exp Biol Med 1964; 117:505-510.
5. Stanley JR. Pemphigus and pemphigoid as paradigms of organ-specific, autoantibody-mediated diseases. J Clin Invest 1989; 83:1443-1448.
6. Arnemann J, Sullivan KH, Magee AI, King IA, Buxton RS. Stratification-related expression of isoforms of the desmosomal cadherins in human epidermis. J Cell Sci 1993; 104:741-750.
7. Sams W, and Jordon RE. Correlation of pemphigoid and pemphigus antibody titres with activity of disease. Br J Dermatol 1971; 84:7-13.
8. Merlob P, Metzker A, Hazaz B, Rogovin H, Reisner SH. Neonatal pemphigus vulgaris. Pediatrics 1986; 78:1102-1105.
9. Schiltz JR, and Michel B. Production of epidermal acantholysis in normal human skin in vitro by the IgG fraction from pemphigus serum. J Invest Dermatol 1976; 67:254-260.
10. Hashimoto K, Shafran KM, Webber PS, Lazarus GS, Singer KH. Anti-cell surface pemphigus autoantibody stimulates plasminogen activator activity of human epidermal cells. J Exp Med 1983; 157:259-272.
11. Anhalt GJ, Labib RS, Voorhees JJ, Beals TF, Diaz LA. Induction of pemphigus in neonatal mice by passive transfer of IgG from patients with the disease. N Engl J Med 1982; 306:1189-1196.
12. Rock B, Labib RS, and Diaz LA. Monovalent Fab' immunoglobulin fragments from endemic pemphigus foliaceus autoantibodies reproduce the human disease in neonatal Balb/c mice. J Clin Invest 1990; 85:296-299.
13. Stanley JR. Cell adhesion molecules as targets of autoantibodies in pemphigus and pemphigoid, bullous diseases due to defective epidermal cell adhesion. Adv Immunol 1993; 53:291-325.
14. Stanley JR, Koulu L, Thivolet C. Distinction between epidermal antigens binding pemphigus vulgaris and pemphigus foliaceus autoantibodies. J Clin Invest 1984; 74:313-320.
15. Stanley JR, Klaus KV, Sampaio SA. Antigenic specificity of fogo selvagem autoantibodies is similar to North American pemphigus foliaceus and distinct from pemphigus vulgaris autoantibodies. J Invest Dermatol 1986; 87:197-201.
16. Hashimoto T, Ogawa MM, Konohana A, Nishikawa T. Detection of pemphigus vulgaris and pemphigus foliaceus antigens by immunoblot analysis using different antigen sources. J Invest Dermatol 1990; 94:327-331.
17. Hashimoto T, Konohana A, and Nishikawa T. Immunoblot assay as an aid to the diagnoses of unclassified cases of pemphigus. Arch Dermatol 1991; 127:843-847.
18. Ogawa MM, Hashimoto T, Konohana A, Castro RM, Nishikawa T. Immunoblot analyses of Brazilian pemphigus foliaceus antigen using different antigen sources. Arch Dermatol Res 1990; 282:84-88.
19. Eyre RW, and Stanley JR. Human autoantibodies against a desmosomal protein complex with a calcium-sensitive epitope are characteristic of pemphigus foliaceus patients. J Exp Med 1987; 165:1719-1724.

20. Korman NJ, Eyre RW, Klaus KV, Stanley JR. Demonstration of an adhering-junction molecule (plakoglobin) in the autoantigens of pemphigus foliaceus and pemphigus vulgaris. N Engl J Med 1989; 321:631-635.
21. Koulu L, Kusumi A, Steinberg MS, Klaus-Kovtun V, Stanley JR. Human autoantibodies against a desmosomal core protein in pemphigus foliaceus. J Exp Med 1984; 160:1509-1518.
22. Stanley JR, Koulu L, Klaus KV, Steinberg MS. A monoclonal antibody to the desmosomal glycoprotein desmoglein I binds the same polypeptide as human autoantibodies in pemphigus foliaceus. J Immunol 1986; 136:1227-1230.
23. Stanley JR, Yaar M, Hawley NP, Katz SI. Pemphigus antibodies identify a cell surface glycoprotein synthesized by human and mouse keratinocytes. J Clin Invest 1982; 70:281-288.
24. Koch PJ, Walsh MJ, Schmelz M, Goldschmidt MD, Zimbelmann R, Franke WW. Identification of desmoglein, a constitutive desmosomal glycoprotein, as a member of the cadherin family of cell adhesion molecules. Eur J Cell Biol 1990; 53:1-12.
25. Goodwin L, Hill JE, Raynor K, Raszi L, Manabe M, Cowin P. Desmoglein shows extensive homology to the cadherin family of cell adhesion molecules. Biochem Biophys Res Commun 1990; 173:1224-1230.
26. Nilles LA, Parry D, Powers EE, Angst BD, Wagner RM, Green KJ. Structural analysis and expression of human desmoglein: a cadherin-like component of the desmosome. J Cell Sci 1991; 99:809-821.
27. Wheeler GN, Parker AE, Thomas CL, Ataliotis P, Poynter D, Arnemann J, Rutman AJ, Pidsley SC, Watt FM, Rees DA. Desmosomal glycoprotein DGI, a component of intercellular desmosome junctions, is related to the cadherin family of cell adhesion molecules. Proc Natl Acad Sci USA 1991; 88:4796-4800.
28. Amagai M, Klaus KV, Stanley JR. Autoantibodies against a novel epithelial cadherin in pemphigus vulgaris, a disease of cell adhesion. Cell 1991; 67:869-877.
29. Buxton RS, Cowin P, Franke WW, Garrod DR, Green KJ, King IA, Koch PJ, Magee AJ, Rees DA, Stanley JR. Nomenclature of the desmosomal cadherins. J Cell Biol 1993; 121:481-483.
30. Koch PJ, Goldschmidt MD, Zimbelmann R, Troyanovsky R, Franke WW. Complexity and expression patterns of the desmosomal cadherins. Proc Natl Acad Sci U S A 1992; 89:353-357.
31. Buxton RS, and Magee AI. Structure and interactions of desmosomal and other cadherins. Semin Cell Biol 1992; 3:157-167.
32. Koch PJ, Goldschmidt MD, Walsh MJ, Zimbelmann R, Franke WW. Complete amino acid sequence of the epidermal desmoglein precursor polypeptide and identification of a second type of desmoglein gene. Eur J Cell Biol 1991; 55:200-208.
33. Arnemann J, Spurr NK, Buxton RS. The human gene (DSG3) coding for the pemphigus vulgaris antigen is, like the genes coding for the other two known desmogleins, assigned to chromosome 18. Hum Genet 1992; 89:347-350.
34. Arnemann J, Spurr NK, Magee AI, Buxton RS. The human gene (DSG2) coding for HDGC, a second member of the desmoglein subfamily of the desmosomal cadherins, is, like DSG1 coding for desmoglein DG1, assigned to chromosome 18. Genomics 1992; 13:484-486.
35. Arnemann J, Spurr NK, Wheeler GN, Parker AE, Buxton RS. Chromosomal assignment of the human genes coding for the major proteins of the desmosome junction, desmoglein DGI (DSG), desmocollins DGII/III (DSC), desmoplakins DPI/II (DSP) and plakoglobin DPIII (JUP). Genomics 1991; 10:640-645.
36. Wang Y, Amagai M, Minoshima S et al. The human genes for desmogleins (DSG1 and DSG3) are located in a small region on chromosome 18q12. Genomics 1944; 20:492-495.
37. Walsh FS, Barton CH, Putt W, Moore SE, Kelsell D, Spurr N, Goodfellow PN. The N-cadherin gene maps to human chromosome 18 and is not linked to the E-cadherin gene. J Neurochem 1990; 55:805-812.

38. King IA, Arnemann J, Spurr NK, Buxton RS. Cloning of the cDNA (DSC1) coding for human type 1 desmocollin and its assignment to chromosome 18. Genomics 1993; 18:185-194.
39. Jones J, Yokoo KM, Goldman RD. Further analysis of pemphigus autoantibodies and their use in studies on the heterogeneity, structure, and function of desmosomes. J Cell Biol 1986; 102:1109-1117.
40. Wolff K, and Schreiner E. Ultrastructural localization of pemphigus auto-antibodies within the epidermis. Nature 1971; 229:59-60.
41. Takigawa M, Imamura S, Ofuji S. Surface distribution of pemphigus antibody-binding substance(s) on isolated guinea pig epidermal cells. An immunoferritin electron microscopic study. J Invest Dermatol 1978; 71:182-185.
42. Takahashi Y, Patel HP, Labib RS, Diaz LA, Anhalt GJ. Experimentally induced pemphigus vulgaris in neonatal BALB/c mice: a time-course study of clinical, immunologic, ultrastructural, and cytochemical changes. J Invest Dermatol 1985; 84:41-46.
43. Jones J, Yokoo KM, Goldman RD. A cell surface desmosome-associated component: identification of a tissue-specific cell adhesion molecule. Proc Natl Acad Sci USA 1986; 83:7282-7286.
44. Iwatsuki K, Takigawa M, Imaizumi S, Yamada M. In vivo binding site of pemphigus vulgaris antibodies and their fate during acantholysis. J Am Acad Dermatol 1989; 20:578-582.
45. Akiyama M, Hashimoto T, Sugiura M, Mishikawa T. Ultrastructural localization of pemphigus vulgaris and pemphigus foliaceus antigens in cultured human squamous carcinoma cells. Br J Dermatol 1991; 125:233-237.
46. Karpati S, Amagai M, Prussick R, Cehrs K, Stanley JR. Pemphigus vulgaris antigen, a desmoglein type of cadherin, is located within keratinocyte desmosomes. J Cell Biol 1993; 122:409-415.
47. Amagai M, Karpati S, Prussick R, Kalus-Kovtun V, Stanley JR. Autoantibodies against the amino-terminal cadherin-like binding domain of pemphigus vulgaris antigen are pathogenic. J Clin Invest 1992; 90:919-926.
48. Allen EM, Giudice GJ, Diaz LA. Subclass reactivity of pemphigus foliaceus antiantibodies with recombinant human desmoglein. J Invest Dermatol 1993; 100:685-691.
49. Dmochowski M, Hashimoto T, Amagai M et al. The extracellular aminoterminal domain of bovine desmoglein I (Dsg1) is recognized exclusively by pemphigus foliaceus sera, whereas its intracellular domain is recognized by sera of various types of pemphigus. J Invest Dematol: in press.
50. Nose A, Tsuji K, Takeichi M. Localization of specificity determining sites in cadherin cell adhesion molecules. Cell 1990; 61:147-155.
51. Eyre RW, Stanley JR. Identification of pemphigus vulgaris antigen extracted from normal human epidermis and comparison with pemphigus foliaceus antigen. J Clin Invest 1988; 81:807-812.
52. Labib RS, Camargo S, Futamura S, Martins CR, Rock B, Anhalt GJ, Diaz LA. Pemphigus foliaceus antigen: characterization of a keratinocyte envelope associated pool and preparation of a soluble immunoreactive fragment. J Invest Dermatol 1989; 93:272-279.
53. Calvanico NJ, Martins CR, Diaz LA. Characterization of pemphigus foliaceus antigen from human epidermis. J Invest Dermatol 1991; 96:815-821.
54. Takeichi M. Cadherins: a molecular family important in selective cell-cell adhesion. Annu Rev Biochem 1990; 59:237-252.
55. Takeichi M. Cadherin cell adhesion receptors as a morphogenetic regulator. Science 1991; 251:1451-1455.
56. Nagafuchi A, Shirayoshi Y, Okazaki K, Yasuda K, Takeichi M. Transformation of cell adhesion properties by exogenously introduced E-cadherin cDNA. Nature 1987; 329:341-343.
57. Nose A, Nagafuchi A, Takeichi M. Expressed recombinant cadherins mediate cell sorting in model systems. Cell 1988; 54:993-1001.

58. Nagafuchi A, Takeichi M. Cell binding function of E-cadherin is regulated by the cytoplasmic domain. EMBO J 1988; 7:3679-3684.
59. Ozawa M, Ringwald M, Kemler R. Uvomorulin-catenin complex formation is regulated by a specific domain in the cytoplasmic region of the cell adhesion molecule. Proc Natl Acad Sci U S A 1990; 87:4246-4250.
60. Knudsen KA, Wheelock MJ. Plakoglobin, or an 83-kD homologue distinct from beta-catenin, interacts with E-cadherin and N-cadherin. J Cell Biol 1992; 118:671-679.
61. Piepenhagen PA, Nelson WJ. Defining E-cadherin-associated protein complexes in epithelial cells: plakoglobin, beta-and-gamma-catenin are distinct components. J Cell Sci 1993; 104:751-762.
62. Amagai M, Karpati S, Klaus-Kovtun V et al. The extracellular domain of pemphigus vulgaris antigen (Desmoglein 3) mediates weak homophilic adhesion. J Invest Dermatol 1994; 102:402-408.
63. Anhalt GJ, Kim SC, Stanley JR, Korman NJ, Jabs DA, Kory M, Izumi H, Ratrie H, Mutusim D, Ariss-AL et al. Paraneoplastic pemphigus. An autoimmune mucocutaneous disease associated with neoplasia. N Engl J Med 1990; 323:1729-1735.
64. Oursler JR, Labib RS, Ariss AL, Burke T, O'Keefe EJ, Anhalt GJ. Human autoantibodies against desmoplakins in paraneoplastic pemphigus. J Clin Invest 1992; 89:1775-1782.
65. Anhalt GJ, Hilu J, Takehara K, Lee A, Casciola-Rosen L. Identification of a putative transmembrane autoantigen in paraneoplastic pemphigus. J Invest Dermatol 1993; 100:508a.
66. Huff JC, Golitz LE, Kunke KS. Intra-epidermal neutrophilic IgA dermatosis. N Engl J Med 1985; 313:1643-1645.
67. Beutner EH, Chorzelski TP, Wilson RM, Kumar V, Michel B, Helm F, Jablonska S. IgA pemphigus foliaceus. Report of two cases and a review of the literature. J Am Acad Dermatol 1989; 20:89-97.
68. Ebihara T, Hashimoto T, Iwatsuki K, Takigawa M, Ando M, Ohkawara A, Nishikawa T. Autoantigens for IgA anti-intercellular antibodies of intercellular IgA vesiculopustular dermatosis. J Invest Dermatol 1991; 97:742-745.
69. Chorzelski TP, Beutner EH, Kowalewski C, Olszewska M, Maciejowska E, Seferowicz E, Kumar V, Jablonska S. IgA pemphigus foliaceus with a clinical presentation of pemphigus herpetiformis. J Am Acad Dermatol 1991; 24:839-844.
70. Dmochowski M, Hashimoto T, Garrod DR, Nishikawa T. Desmocollins I and II are recognized by certain sera from patients with various types of pemphigus, particularly Brazilian pemphigus foliaceus. J Invest Dermatol 1993; 100:380-384.
71. Matsuyoshi N, Hamaguchi M, Taniguchi S, Nagafuchi A, Tsukita S., Takeichi M. Cadherin-mediated cell-cell adhesion is perturbed by v-src tyrosine phosphorylation in metastatic fibroblasts. J Cell Biol 1992; 118:703-714.
72. Behrens J, Vakaet L, Friis R, Winterhager E, Van Roy F, Mareel MM, Birchmeier W. Loss of epithelial differentiation and gain of invasiveness correlates with tyrosine phosphorylation of the E-cadherin/beta-catenin complex in cells transformed with a temperature-sensitive v-SRC gene. J Cell Biol 1993; 120:757-766.
73. Hamaguchi M, Matsuyoshi N, Ohnishi Y, Gotoh B, Takeichi M, Nagai Y. p60v-src causes tyrosine phosphorylation and inactivation of the N-cadherin-catenin cell adhesion system. EMBO J 1993; 12:307-314.
74. Troyanovsky SM, Eshkind LG, Troyanovsky RB, Leube RE, Franke WW. Contributions of cytoplasmic domains of desmosomal cadherins to desmosome assembly and intermediate filament anchorage. Cell 1993; 72:561-574.

CHAPTER 15

Molecular Mechanisms of Keratin Gene Disorders and Other Bullous Diseases of the Skin

Pierre A. Coulombe
Elaine Fuchs

1. INTRODUCTION

In recent years, remarkable progress has been made in our understanding of the genesis and maintenance of complex tissues such as stratified squamous epithelia. Some of this progress has led to an understanding of the molecular bases underlying blistering diseases affecting primarily the skin. Skin blisters arise when the function of an important component of the structural framework of the tissue is altered, leading to a loss of tissue cohesion and/or integrity. The isolation and characterization of these defective components and the comparison with their normal counterparts, at the gene and protein levels, provide an effective approach to determine the genetic bases of these diseases. In this article we review the progress recently made in our understanding of the molecular basis underlying selected human skin blistering disorders. The diseases included (listed in Table 15.1) were selected because they allow us to review the major components that are responsible for the structural integrity and functional cohesiveness of skin tissue. A particular emphasis is placed on two genetic skin diseases caused by mutation in keratins: epidermolysis bullosa simplex and epidermolytic hyperkeratosis. Due to a number of recent discoveries, these diseases are presently among the best understood of blistering disorders.

Molecular Mechanisms of Epithelial Cell Junctions: From Development to Disease
Edited by Sandra Citi, M.D., Ph.D.;

Table 15.1. Localization of tissue cleavage in selected skin blistering diseases

SKIN TISSUE COMPARTMENT		GENODERMATOSES	AUTOIMMUNE DISORDERS
EPIDERMIS	SUPRABASAL LAYERS	Epidermolytic Hyperkeratosis Darier-White Disease	Pemphigus Foliaceus Pemphigus Vulgaris
	BASAL LAYER	EB Simplex	
DERMO-EPIDERMAL INTERFACE	LAMINA LUCIDA	EB Junctional	Bullous Pemphigoid
DERMIS	ANCHORING FIBRILS	EB Dystrophica	EB Acquisita

2 . FUNCTIONAL ORGANIZATION OF THE EPIDERMIS

The epidermis is the prototype of a stratified squamous epithelium. It is the product of a specific program of terminal differentiation, in which a delicate balance must be maintained between cell growth and cell differentiation.[1] The epidermis shows a typical structural and functional polarization in its vertical axis, with distinct compartments organized as cell layers. Mitotic activity is restricted to a population of cells attached to the underlying connective tissue, i.e., basal cells. Upon a signal or set of signals (yet to be identified), basal cells become committed to a program of terminal differentiation.[1,2] This commitment triggers a complex series of molecular events, the exact chronological order of which is not fully known. These events include an exit from the cell cycle, the upward migration to the suprabasal compartment, and the induction of a specific set of genes encoding differentiation-specific products. Activation of the expression of several of these differentiation-specific genes occurs in a defined sequence, which is necessary for differentiating suprabasal cells to progressively acquire the structural and functional characteristics of the ultimate product of the pathway. In the epidermis and other cornifying epithelia, the product of the terminal differentiation pathway is a squame, i.e., a highly flattened cell without a nucleus or any other recognizable organelle, and containing mainly structural proteins extensively cross-linked via ε-(γ-glutamyl)lysine isopeptide bonds catalyzed by epidermal transglutaminases.[3] Finally, these squames are literally cemented to one another via a complex, ceramide lipid-rich substance secreted at the late stage of differentiation.[4] The cornified layers of the upper epidermis allow the skin to play important roles as a protective envelope and barrier.[5]

2.1. Keratins are Major Constitutents in Epidermis and Other Stratified Epithelia

The most abundant proteins in all epidermal cells are the keratins, which are intermediate filament-forming (IF) proteins.[6] More than 30 different keratins have been characterized up until now, each encoded by at least one gene.[7,8] Keratins have been classified in two major types (I and II) based on homology at the DNA sequence level.[9] Type I keratin proteins are smaller (MW 40-56.5 Kd) and acidic, whereas the type II tend to be larger (MW 53-67 Kd) and relatively basic-neutral.[10,11] Keratin filaments are obligatory heteropolymers comprised of type I and type II sequences in a 1:1 molar ratio.[12-14] As will be described in greater details below, the first step in keratin filament assembly consists in the formation of a coiled-coil heterodimer involving one type I and one type II keratin chains.[15-17] A several micrometer-long 10 nm filament contains between 12 to 16 heterodimers in cross-section, and of thousands of them when including the longitudinal dimension.[18] Because of this strict heteromeric assembly requirement, a member of each subtype of keratin sequences is expressed in epithelial cells.[11] A proper balance between type I and type II keratins is achieved partly by their coordinate regulation at the transcriptional level,[19] but also by regulation at the post-translational level, since unpaired keratins display a very short intracellular half-life and are thought to be targeted for rapid degradation.[20,21]

Work in many laboratories established that many keratins are expressed as specific type I-type II "pairs" in a differentiation-specific fashion in the various types of epithelia in situ. In all stratified squamous epithelia, therefore, the type II keratin K5 (58 kD) and the type I keratin K14 (50 kD) are expressed in the mitocally-active cells of the basal layer, where together they form 15-20% of total proteins.[22] As or soon after basal cells commit themselves to terminal differentiation, a down-regulation of the synthesis of K5 and K14 occurs,[19] concomitant with the induction of different sets of keratins, whose identity is specific to the program of terminal differentiation being executed.[7,11] In epidermis and other cornifying epithelia, these suprabasally-expressed keratins are mainly the type II K1 (67 kD) and the type I K10 (56.5 kD).[23] Expression of these differentiation-specific keratins is so abundant that together with K5 and K14, these keratins form 85 % of total proteins in terminally differentiated epidermal cells.[23] Other stratified squamous epithelia characterized by distinct programs of terminal differentiation express different type I and type II keratins in their differentiating cell layers (Table 15.2). The "switch" in keratin gene expression that occurs as progenitor basal cells become committed to differentiation constitutes a highly conserved phenomenon among mammalian species, but the underlying molecular basis remains poorly understood.[7,24]

2.2. Keratin Filaments Are Integrated within a Supracellular Network in Stratified Epithelia

Keratin filaments are organized into a network spanning the entire cytoplasm of stratified squamous epithelial cells (Fig. 15.1A). As judged by electron microscopy, 10 nm filaments interact with various subcellular components to form an intricate network. Thus filaments are anchored at or near the surface of the nucleus (Fig. 15.1 A and 15.1B) and

Table 15.2. Pairwise expression of selected type I - type II keratins

TYPE I KERATINS	MW (kDa)	pI	TYPE II KERATINS	MW (kDa)	pI	EPITHELIAL DISTRIBUTION	REFERENCES
K8	52.5	6.1	K18	45	5.7	all simple epithelial cells	10
K5	58	7.4	K14	50	5.3	progenitor basal cells in stratified squamous epithelia	10,22,23
K1	67	7.8	K10	56.5	5.3	differentiating suprabasal layers layers in cornifying epithelia	23
K3	63	7.5	K12	55	4.9	differentiating suprabasal layers in cornea	145
K4	59	7.3	K13	54	5.1	differentiating suprabasal layers in internal "wet" epithelia	146
K6	56	7.8	K16	48	5.1	no defined pattern; suprabasal layers of "complex"epithelia undergoing hyperproliferation	147

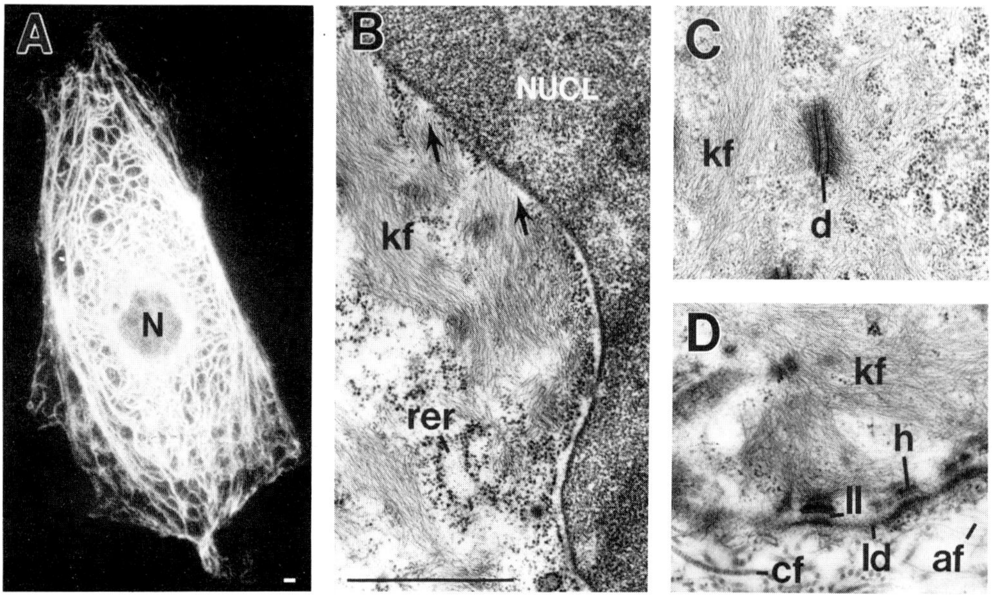

Fig. 15.1. Architecture and organization of the keratin filament network in epithelial cells. A, A cultured PtK2 (kidney) epithelial cell was processed for the immunofluorescence labeling of its keratin filament network with previously described monoclonal antibody L2A1[130] followed by Texas Red-conjugated goat anti-mouse IgG[56]. Keratin filaments span the entire cytoplasm, from the surface of the nucleus (N) to the outer cell membrane. B-D, Adult mouse tongue was snap-frozen in liquid nitrogen-cooled isopentane. Sections (20 μm) were cut, thawed in the presence of buffered 2% glutaraldehyde, dehydrated in graded ethanols and propylene oxide, and embedded in epoxy resin (J. Folmer and P.A. Coulombe, unpublished data). Keratin filaments (kf) are seen to anchor directly at the outer nuclear (N) membrane (arrows in frame B), at the level of desmosomes (d) (C), and at the level of hemidesmosomes (h) (D) in a basal epithelial cell. Note that the various cell membranes can not be distinguished in this preparation due to lack of osmification. Other abbreviations: rer, rough endoplasmic reticulum, ll, lamina lucida; ld, lamina densa; af, anchoring fibril, cf, collagen fiber. Scale bars = 1 μm.

they extend towards the periphery where they interact with large macromolecular complexes at cell membranes. In epithelia such as epidermis, tongue, esophagus and cornea, such membrane complexes consist mostly of desmosomes, which mediate cell-cell attachment, and hemidesmosomes, which mediate attachment of basal cells to the underlying extracellular matrix (Fig. 15.1C and 15.1D). Desmosomes and hemi-desmosomes are multiproteins assemblies spanning the plasma membrane; some of these components are oriented such that they are accessible from the cytoplasmic side, and in fact physically interact with intermediate filaments.[25,26] Albeit their related ultrastructure, these cell adhesion-mediating membrane complexes are immunologically and biochemically distinct.[25-27] The desmosome and hemidesmosome components mediating the anchorage of keratin filaments have yet to be identified, and efforts to that end are currently in progress in several laboratories (see chapters 8 and 9 in this volume).

The radial disposition of the keratin filament bundles around the nucleus, combined with their anchorage at desmosomal plaques (and hemidesmosomal plaques in the case of basal keratinocytes) at the periphery of keratinocytes (Fig. 15.1), results in a configuration in which the filaments of a given cell are physically connected with those of adjacent cells, thereby creating a supracellular structural framework within the stratified epithelial tissue. Moreover, this supramolecular network is continuous with the underlying connective tissue matrix through a set of interactions involving distinct macromolecular complexes. Hence, integral components of hemidesmosomes located on their exoplasmic face interact with protein components of the basal lamina, itself "attached" to the underlying extracellular matrix by anchoring fibrils composed on collagen type VII (see below). Thus, because of its intracellular architecture and its anchorage at multiple points of cell-cell and cell-matrix contacts, the keratin filament network appears virtually uninterrupted in the living layers of the epidermis. While the supracellular nature of this structural network certainly contributes to confer to the epidermis its remarkable physical resilience, the interdependence of its multiple components is such that alteration of any of them usually has repercussions for the entire tissue. In fact, as we shall review in this paper, this is exactly what happens in several diseases affecting primarily the skin.

It should be noted that keratin filament-associated proteins are not restricted to cell or nuclear surface components. There are in fact a number of known proteins that have been isolated on the basis of their physical interaction with detergent-extracted keratin filaments; because of this selection bias, they tend to be mediating aspects of filament-filament interactions and will not be considered in the present text (for a review see ref. 28).

2.3. Functional Organization of the Basement Membrane Zone in Skin

Through its network of keratin filaments anchored at sites of cell-cell and cell-matrix junctions, the epidermis is firmly attached to the upper dermis through a complex set of interactions involving cell surface components of basal cells and several specialized extracellular polymers.[29-31] The elements involved in these interactions can be grouped into four distinct compartments on the basis of ultrastructural criteria: together they form a unit known as the basement membrane zone. These compartments can be briefly described as follows:

- Integral membrane proteins and cell surface macromolecules of basal keratinocytes. Among the best known are the integrins, a large family of cell-surface molecules existing as heterodimers of α and β chains, which bind various extracellular matrix ligands and thus play a major role in cell adhesion.[32] Specific integrins are found concentrated within hemidesmosomal plaques (e.g. $\alpha_6\beta_4$;[33]). Several aspects of integrin structure and function are detailed elsewhere in this book (see chapter 16);
- Anchoring filaments, which are thin fibers of 2-8 nm diameter spanning the entire lamina lucida (25-50 nm) between the plasma membrane of basal keratinocytes (especially at the level of hemidesmosomes) and the lamina densa.[30] These fibers contain an heterotrimeric protein complex with 175 kD, 170 kD, and 125 kD subunits produced by epidermal basal cells.[34] This complex has been characterized independently by several laboratories, and has thus been given different names, such as GB300 antigen or nicein,[35] kalinin,[36] and possibly, epiligrin.[37] Treatment of cultured epidermal keratinocytes with antibodies directed against component(s) of this heterotrimer causes their detachment from the underlying matrix, documenting their important role in keratinocyte adhesion;[36]
- The lamina densa portion of the basal lamina. The ultrastructure of the dermo-epidermal interface strongly suggests that the lamina densa functions as a structural scaffold at the level of which occur, on one hand, the attachment of basal keratinocytes via the anchoring filaments, and on the other hand, the attachment of the epidermis-basal lamina complex to the upper dermis via the anchoring fibrils. The lamina densa of the dermo-epidermal interface is a complex polymer containing type IV collagen, laminin, nidogen/entactin, and heparan sulfate proteoglycans as main constituents.[30]
- The anchoring fibrils are fibrous aggregates of ~800 nm length and 20-75 nm diameter that are composed primarily of type VII collagen, a non-fibrous type of collagen composed of three $\alpha1$ (VII) chains (reviewed in ref. 38; Fig. 15.1, frame D). The building block of these fibrils is a antiparallel dimer involving two type VII collagen triple helices (each ~450 nm long) whose distal carboxyl

termini are linked by disulfide bonds. The amino terminus of each triple-helical collagen molecule contains a large globular domain, termed NC1, which interacts with type IV collagen and laminin within the lamina densa, as well as with anchoring plaques, which are electron-dense, discrete structures located below the lamina densa in the upper dermis.[39] Remarkably, several of these anchoring fibrils have both their NC1 domains attached to the lamina densa, and the intervening fibrillar domains loop into the dermis where they wrap around dermal fibrillar collagen fibers (type I and type III). Others such fibrils are attached to the basal lamina and extend into the dermis to terminate at the so-called anchoring plaques.[39] Such an organization of these fibrils as a network intermingled with the conventional fibrous collagens ensures, under normal conditions, a firm anchorage of the epidermis—lamina densa complex to the dermis.[31,38]

Additional vital components of the basement membrane zone will be identified as state-of-the-art biochemical and molecular genetics techniques are increasingly applied by researchers active in that field.

3. MOLECULAR MECHANISMS UNDERLYING SKIN BLISTERING DISORDERS

As briefly outlined in the preceding section, a number of distinct but interconnected functional elements that exist at the various levels of epidermis and upper dermis are collectively providing a complex and highly resilient structural framework allowing the skin to play important roles as a protective envelope and as a barrier. It comes as no surprise, therefore, that the known human skin blistering diseases have been discriminated on the basis of the exact location of the plane of tissue cleavage, which in turn is determined by the nature of the defective component. Thus there are diseases in which the plane of cleavage initiating skin blister formation consistently occurs within either (i) the epidermis, (ii) the upper dermis, or (iii) at the dermo-epidermal interface (reviewed in refs. 5 and 40). The unique clinical features and prognosis characterizing these various skin blistering diseases are, to a large extent, determined by the nature and the subtissular location of the defective component.

There are two major types of mechanisms which can lead to blistering at any one of these three distinct levels: (i) inherimutations in genes encoding specific types of structural proteins and resulting in a functionally defective product, collectively referred to as genodermatoses; and (ii) presence of circulating autoantibodies directed against, and whose binding affect the function of, specific structural components within the skin tissue framework; these are referred to as autoimmune skin disorders.[5,40] In this section we will provide selected examples of genodermatoses and autoimmune disorders associated with blistering at each of these three levels of skin (Table 15.1) and discuss them briefly. In the next section, we will expand on epidermolysis bullosa simplex and epidermolytic hyperkeratosis, two distinct genodermatoses characterized by epidermal cell lysis and intraepidermal blistering, and which are caused by mutations in specific keratin genes.

3.1. Intraepidermal Blistering Disorders

The plane of cleavage in intraepidermal blistering disorders can be located either within the basal cell layer, at the basal-suprabasal layer junction, and even at higher levels within the epidermis, e.g., within the spinous and granular layer.

3.1.1 Blistering within the basal layer of epidermis

The classic example of a blistering disorder in which the primary defect localizes to the basal layer of epidermis is epidermolysis bullosa (EB) simplex. EB simplex refers to a groups of bullous genodermatoses usually inherited in an autosomal dominant fashion and affecting ~1:25,000 individuals, in which basal cells lyse in response to minor mechanical trauma, thereby causing the epidermis to blister.[41-43] There are several clinical subtypes of EBS, ranging from very mild to very severe (Table 15.3), but in most cases the disease manifests itself at birth or during early infancy (for a recent clinical review see ref. 43). Electron microscopic studies of clinically non-affected skin in EBS patients revealed the presence of anomalies in the structure or the organization of keratin filament bundles in the basal cells of epidermis, while suprabasal cells appeared mostly normal (refs. 45 and 45; Fig. 15.2). When performed in freshly blistered skin, the same analyses indicated that the plane of tissue cleavage is consistently located between the nucleus and the lowermost portion of basal cells,[46,47] suggesting that these had lost their ability to withstand mechanical stress.[47]

It is interesting to consider for a moment the manner with which the molecular basis of EBS diseases was discovered. Although Ingrid Anton-Lampbrecht had proposed in 1983,[46] based on electron microscopic observations, that EBS may have as its genetic basis a defect in keratin filaments, most of the subsequent research efforts directed towards elucidating the etiology of EBS focused on the possibility that cytolytic enzymes or cell surface defects might be responsible for basal cell disruption and skin blistering (e.g. refs. 48-53). In fact, it is only in recent years that researchers began to uncover genetic evidence which complemented Anton-Lampbrecht's ultrastructural findings. Thus, there was the demonstration, by Albers and Fuchs (refs. 54 and 55; see also ref. 56), that deletion mutations affecting the α-helical rod domain of a keratin sequence often behaved in a dominant-negative fashion when transfected into cultured epithelial cells, that is, affected the structural behavior of the entire pool of wild-type proteins. Thereafter, Kitajima et al.[57] described keratin-containing spherical inclusions in the peripheral cytoplasm of epidermal keratinocytes isolated

Table 15.3. Clinical and histological features of the major EBS clinical subtypes

PARAMETER	DOWLING MEARA	KOEBNER	WEBERCOCKAYNE
Mode of Inheritance	Autosomal Dominant	Autosomal Dominant	Autosomal Dominant
Onset	Birth	Birth/ Early Infancy	Childhood and Later
Lethality	Yes	No	No
Distribution of Blisters	Generalized	Generalized	Localized: Hands/Feet
Keratin Filament Network in Epidermal Basal Cells	Disrupted	Disrupted	Mildly disrupted to not detectable
Tonofilament Clumping in Epidermal Basal Cells	Yes	No	No
Involvement of Internal Stratified Squamous Epithelia	Yes	No	No
Blister Healing without Scarring	Yes	Yes	Yes
Additional Distinguishing Feature	Herpetiform Grouping of Blisters	Intraepidermal Bullae	Intraepidermal Bullae

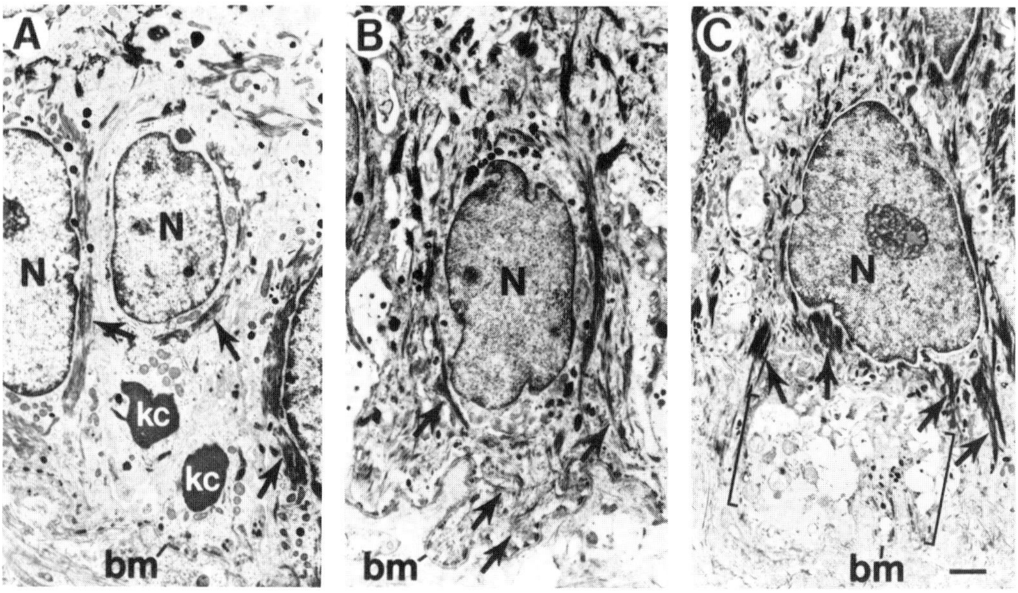

Fig. 15.2. Ultrastructure of epidermal basal cells in epidermolysis bullosa simplex patients. Back skin samples from an unaffected region of a Dowling Meara patient, A, a Weber-Cockayne patient, C, and a healthy individual, B, were processed for routine electron microscopy.[47,61] In the subnuclear cytoplasm of basal cells, the Dowling Meara skin sample (frame A) contains large keratin clumps (kc), a typifying feature of this EBS subtype, whereas the Weber-Cockayne sample, C, display extensive vacuolization and disorganization of keratin filaments (arrows). In contrast, normal basal cells, B, display a extensive array of keratin filament bundles (arrows) spanning the entire cytoplasm. N, nucleus; bm, basement membrane. Scale bar = 1 μm. Reproduced from ref. 60 with permission.

and cultured from EBS patients. These authors noted that these inclusions were similar to those seen when human keratinocytes are transfected with mutant keratin genes. The critical finding was the creation of an EBS-like phenotype in transgenic mice when mutant versions of a basal layer-specific keratin were expressed in a tissue-specific fashion in transgenic animals.[47,58] Of particular interest, these transgenic mouse studies demonstrated that the wide range of clinical manifestations of EBS diseases could be reproduced by expressing various "types" of mutations in the K14 sequence in a tissue-specific fashion.[47] Thus, mice engineered to express a K14 mutant which severely disrupted keratin filament assembly displayed a phenotype most consistent with Dowling Meara-EBS, which is the most severe form of the disease[58] (Table 15.3). Alternatively, mice expressing K14 mutants affecting filament assembly in a much milder fashion showed a milder skin blistering phenotype, limited to the hands and feet, a clinical picture consistent with Weber Cockayne-EBS[47] (Table 15.3). These transgenic experiments, therefore, not only provided the key evidence indicating that EBS is caused by a defective K5-K14 filament network in basal cells, they also demonstrated that the full range of human EBS diseases could be accounted for on the basis of different mutations in a single gene (for reviews see refs. 59 and 60). Subsequent studies conducted on patients with Dowling Meara-EBS and Koebner-EBS demonstrated the presence of point mutations in either the type II K5 or the type I K14 genes[61-65] (see Table 15.4), expressed at high levels in the basal layer of stratified squamous epithelia such as epidermis. Point mutations have since then been evidenced in the K5 gene of several kindreds with Weber-Cockayne-EBS,[66-68] in a rare instance of recessively inherited EBS,[69] and in additional incidences of Dowling Meara- and Koebner-EBS (listed in Table 15.4).

That such mutations in the K5 or K14 gene are functionally responsible for the EBS phenotype is supported by two lines of evidence. First, chromosomal linkage analysis performed by various groups[62,66,69-71] (also, several of the refs. listed in Table 15.4) have mapped the genetic defect in several different EBS kindreds to either the long arm of chromosome 17, at a location known to contain a cluster of type I keratin genes (including K14),[72] or the long arm of chromosome 12, at a location where a cluster of type II keratin genes had been previously mapped (including K5).[73] Second, genetic engineering of the specific point mutations in the cloned K14 cDNA or K5 cDNA, followed by testing in vivo and in vitro, have verified that the Dowling Meara mutations can cause the tonofilament clumping and shortened keratin filaments that are the morphological hallmarks of this disease.[61,74] In contrast, Weber-Cockayne mutations have more subtle effects, and may especially affect lateral interactions between IF subunits in the context of the filament.[71] In Section 4 below, we review the effects of such point mutations on the ability of keratins to assemble into 10 nm filaments, and correlate the position and nature of these mutations along the keratin sequence with the clinical severity of the disease in the patients in which they were discovered.

In EBS, thus, the defective component is the intracellular keratin filament network, blistering generally necessitates mechanical trauma, and results from cytolysis. Note that there are two additional forms of epidermolysis bullosa diseases, which are clinically, pathologically, and genetically distinct from EBS. In EB junctional, the plane of skin tissue cleavage is located at the dermoepidermal interface, while in EB dystrophica, it is located below the lamina densa in the upper dermis.[43,75] These two EB variants do not involve the lysis of epidermal cells, as their name would imply. However, as in EB simplex, mechanical trauma generally precedes blister formation, onset is at birth or during early infancy, and several clinical subtypes have been distinguished based on distribution and severity of blisters. Since EB junctional and dystrophica are both subepidermal blistering disorders, they will be discussed later in this text.

3.1.2 Blistering above the basal cell layer of epidermis

There are several blistering diseases for which skin tissue cleavage occurs above the basal layer of epidermis. In some cases, such as pemphigus vulgaris and darier's disease, the plane of cleavage tends to be localized at the interface between the basal layer and the first layer of suprabasal cells, while in other cases, such as pemphigus foliaceus and epidermolytic hyperkeratosis, the plane of cleavage truly localizes within the suprabasal compartment.[5,40] Of these diseases, only epidermolytic hyperkeratosis involves cytolysis as the primary mechanism initiating blister formation, and as in EBS, this generally ensues mechanical trauma. For the other three diseases mentioned here, lysis of epidermal cells is not required for initiating blister formation; rather, the major histological finding is acantholysis, i.e., an apparent loss in the cells' ability to remain attached to one another.

Epidermolytic hyperkeratosis (EHK; also called bullous congenital ichtyosiform erythroderma or BCIE) is a relatively rare genetic disorder characterized by an autosomal dominant inheritance. In EHK, cell lysis occurs in the suprabasal layers of the epidermis, and blistering is induced by mechanical trauma.[46,76,77] There is considerable heterogeneity in the clinical presentation of the EHK, and several subtypes have been distinguished based upon a number of criteria.[77,78] Electron microscopy of the non-blistered epidermis of EHK patients revealed a general loss of cytoarchitecture in suprabasal cells, and obvious alterations in their keratin filament network, including tonofilament clumping and formation of a shell of tonofilaments around distorded nuclei.[46] The chronic cell lysis occurring in the suprabasal compartment of the tissue leads to an enhanced basal cell proliferation rate, resulting in massive hyperkeratosis and a "scaly skin" phenotype, which are hallmarks of this genodermatose.[76-78]

Again, as it had been done for EBS, the essential clinical features of human EHK, as well as its typifying ultrastructural features in the suprabasal cells of epidermis, were reproduced in transgenic mice by targeting the expression of an assembly-defective mutant of keratin K10 to the suprabasal layers of the epidermis.[79] Subsequent studies revealed the presence of point mutations in either the K1 or the K10 genes in several independent incidence of EHK[80-87] (see Table 15.4). More recently, a point mutation was discovered in the K9 gene of the affected members of a large kindred with epidermolytic palmoplantar hyperkeratosis (EPPK), which, as its name suggest, is a form of EHK in which blistering is restricted to the palms and soles.[88] The type I keratin K9 is expressed only in the differentiating layers of the unusually thick palmar and plantar epidermis,[23,89] explaining the distribution of blisters in EPPK. The defective structural component responsible for the EHK phenotype is thus the intracellular keratin filament network of the differentiating suprabasal cells of the epidermis. This makes the comparison EHK and EBS from a keratin biology standpoint quite interesting, and this exercise is done below.

Darier-White's disease (and the related Hailey–Hailey disease) is a relatively rare genodermatose inherited in an autosomal dominant fashion, in which papule-like lesions affect the skin and other mucous membranes.[90] In contrast to EBS and EHK, the onset of the disease is usually at the end of the first decade or during the second decade of life. Histologically, the seminal changes in the epidermis of Darier-White's patients are acantholysis and a typical pattern of dyskeratosis. Electron microscopy of lesional epidermis revealed a marked loss of desmosomes, and frequent "detachment" of keratin filament bundles from desmosomal plaques.[91,92] Intracellular levels of various desmosomal components such as desmoplakins, desmogleins and desmocollins are unusually high in suprabasal keratinocytes of active skin lesions.[93] Given the striking pattern of acantholysis, and consistent with the remarkable changes in desmosome number, ultrastructure, and/or intracellular localization, the defective structural component(s) responsible for this disease appear to influence intercellular adhesion between keratinocytes in the suprabasal layers of the epidermis. Adhesion of basal cells to the underlying basal lamina, however, is not functionally altered,[90] suggesting that hemidesmosomes and other pertaining basal cell components are not affected. The molecular basis underlying Darier-White's disease has yet to be discovered, although recent mapping data provided evidence for its linkage to chromosome 12q23-q24.1 in several kindreds with this disease.[94-96] Note that this location is more distal from the centromere than the cluster of type II keratin genes (12q11-q13; 73). Additional studies will be required to determine the identity of the gene involved in these Darier kindreds.

As described by Stanley[29,97] (see also chapter 14 in this volume), pemphigus refers to a group of blistering diseases of skin and other mucous membranes which are characterized histologically by acantholysis in the suprabasal compartment of the affected epithelia, and immunopathologically by the detection of in vivo bound and circulating IgGs directed against keratinocyte cell surface components. The incidence of pemphigus diseases is unknown, and the mean age of onset is 50-60 years old, although the range is broad and includes children.[97] There are two major types of pemphigus :*v ulgaris (PV)*, in which the plane of cleavage occurs at the interface between the basal and suprabasal compartments (as in Darier-White's disease) and *foliaceus (PF)*, in which the plane of cleavage occurs in the granular layers of epidermis.[29,98] This distinction between PV and PF was at first mystifying, since the use of either PV or PF patient's serum for the indirect immunostaining of normal human epidermis resulted in a keratinocyte surface staining over its entire thickness.[99] However, the subsequent use of PV and PF patients' sera to identify, purify and characterize the cell surface antigens targeted by the autoantibodies provided a likely explanation for the different planes of epidermal cleavage associated with each disease. Both these antigens belong to the desmoglein (dsg) family of cadherin-type adhesions molecules.[100] The major PV antigen is a 130 kD glycoprotein, called dsg3, while the PF antigen is a 160 kD glycoprotein called dsg1.[99-102] dsg1 and dsg3 are part of macromolecular complexes which include plakoglobin at sites of desmosomes and possibly, adherens junctions.[29,103-105] The pathogenicity of PV (and PF) autoantibodies has been demonstrated in vivo by their injection in neonatal mice, which results in the formation of blisters that histologically mimic the corresponding type of pemphigus.[106] Autoantibody binding could thus interfere, either directly or indirectly, with the function of their target antigens at sites of cell adhesion complexes, perturbing cell-cell interaction and leading to blister formation. The molecular basis underlying the development of this autoimmune response in pemphigus patients is not yet known. These patients have a markedly increased frequency of the HLA-DR4 and DRw6 haplotypes in the class II major histocompatibility complex, which might be involved in pemphigus antigen presentation to the immune system.[97]

3.2 Subepidermal blistering

3.2.1 Blistering at the dermo-epidermal interface

In EB junctional, skin tissue cleavage occurs exactly within the lamina lucida, i.e., in the electron microscopy-defined narrow space between the lamina densa and the plasma membrane of the basal keratinocytes.[107] Inheritance of this genodermatose is autosomal recessive, and two major clinical subtypes have been distinguished based on the prognosis for survival, distribution and severity of skin blisters, and involvement of other epithelial tissues.[43,75] Ultrastructural studies conducted on the skin of patients suffering from Herlitz's EB junctional, a severe form, have shown changes in the number and/or the structure of hemidesmosomes in basal cells,[108] and an absence of anchoring filaments in the lamina lucida. This latter ultrastructural finding is further supported by immunological data. Indeed, several laboratories reported a partial or complete loss of reactivity of epidermal keratinocytes from EB Junctional patients to antisera recognizing component(s) of anchoring filaments (in particular, the heteromeric triplet of proteins known as either nicein or kalinin; see refs. 34 and 109). Future studies will be required to identify the mutations underlying this form of EB. Given that both anchoring filaments and the adhesion molecules at the surface of basal keratinocytes appear to be oligomers of multiple proteins,[30] it is conceivable that mutations in a number of different genes could elicit an EB junctional phenotype.

A similar type of blistering occurs in bullous pemphigoid (BP), an autoimmune disorder usually occurring in the elderly, and characterized immunopathologically by the presence of immune deposits in the epidermal basement membrane zone.[29,110] The direct detection of C3 or IgGs in the skin of BP patients, or alternatively, the use of BP patient's serum for the indirect immunostaining of sections of normal individual's skin, both result in a characteristic linear signal along the basement membrane zone. Electron microscopic examination of freshly blistered skin in BP patients reveals a loss of anchoring filaments and hemidesmosomes at that level.[98] Immunochemical studies using serum from several BP patients have indicated that the dominant target antigen is a 230 kD protein which, by immunoelectron microscopy, localizes specifically to hemidesmosomes.[111-113] The cDNA encoding the BP230 antigen has been cloned[103] and subsequent analyses revealed that it bears a significant sequence homology (and predicted protein domains) to desmoplakin I, a desmosomal protein.[114] BP patients' sera often bind additional antigens,[115] the most prominent of which is another hemidesmosomal component of 180 kD molecular mass.[116] Injection of BP IgG autoantibodies into rabbit cornea results in their deposition at the basement membrane zone and causes subepithelial blister formation,[117] suggesting that this binding event is involved in the pathogenesis of this disease. It is still not clear, however, whether this is a direct effect on hemidesmosome function, or whether it requires complement fixation and the subsequent cascade of inflammatory events.[110]

3.2.3 *Blistering below the basal lamina*

There is a form of epidermolysis bullosa characterized by a plane of skin cleavage localizing in the upper papillary dermis, immediately below the basement membrane zone. Sub-lamina densa blisters are associated with atrophic or hypertrophic scarring, and can lead to severe mutilations and loss-of-function. Accordingly this form of EB has been called *Dystrophica*.[31,43] As is the case for EB simplex and junctional, several clinical subtypes of EB dystrophica are recognized based principally on the morphology and

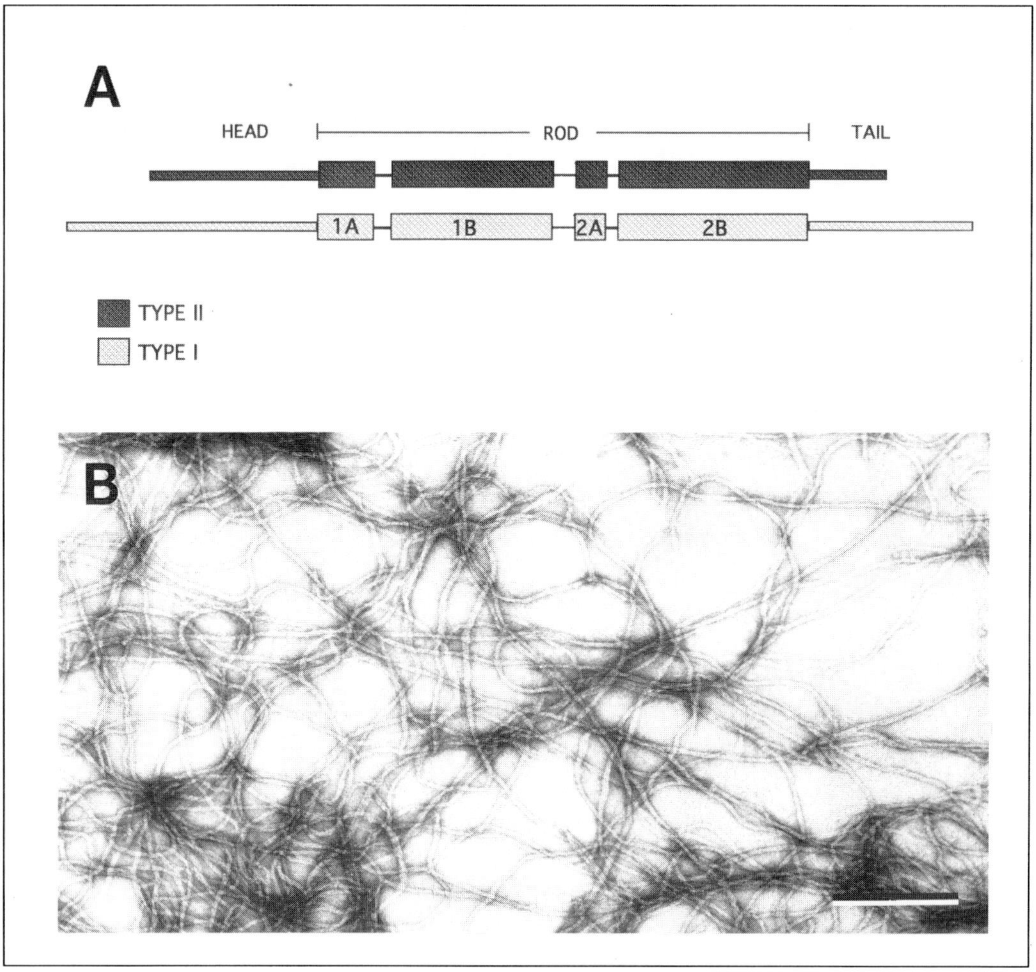

Fig. 15.3. Secondary structure and in vitro assembly properties of human type II keratin K5 and type I keratin K14. A, as in all intermediate filament (IF) proteins, keratins feature a central "rod" domain of ≈310 amino acids characterized by an heptad repeat of apolar residues leading to an α-helical structure (larger boxes). This repeat and the α-helicity are interrupted by linker segments at three conserved locations within the rod, generating subdomains 1A, 1B, 2A and 2B. The rod domain is flanked by non-helical sequences at both ends (head and tail), whose length and amino acid composition vary among keratins and IF proteins in general. The length of these non-helical sequences is the primary determinant of IF protein size, and is greater in type II than in type I keratins.[7,123] B, electron micrograph of negatively-stained filaments reconstituted in vitro from purified recombinant human K5 and K14 proteins[15]. Note that these filaments are several micrometers-long and regular, and display an average diameter of 10.5 nm when negatively-stained with 1% uranyle acetate. Scale bar = 200 nm.

distribution of blisters.[43] EB dystrophica may be genetically inherited in an autosomal or recessive fashion, and disease onset occurs at birth or during early infancy. The dominant forms are relatively mild compared to the recessive forms, which may be lethal. Ultrastructural studies have shown that in virtually all forms of EB dystrophica, ranging from localized and mild to severe and generalized, there are ultrastructural anomalies in the anchoring fibrils of the papillary dermis.[30,31,38] These anomalies range from the accumulation of type VII collagen epitopes within epidermal basal cells, which suggests an impairment of its folding and/or secretion, to the complete absence of these fibrils underneath the lamina densa. Linkage analyses and direct DNA sequencing performed on several kindreds covering many clinical subtypes of EB dystrophica demonstrated that the genetic basis underlying this disorder resides in mutations in the collagen type VII gene ([118,151] and refs. therein). Therefore, in analogy to EBS (see below) and other collagen diseases,[119] it appears that the full clinical range of EB dystrophica can be accounted on the basis of mutations in a single gene (two partner genes in the cases of EBS) and the severity of the phenotype may well depend on the location and nature of the point mutation along the collagen type VII sequence.[31,119]

The *Acquisita* subtype is an adult-onset form of EB disease which has a phenotype very similar to EB dystrophica, including a sub-lamina densa plane of skin cleavage.[120] Interestingly, EB acquisita is associated with the presence of autoantibodies to collagen type VII,[121] the building block of anchoring fibrils, thereby reinforcing the notion that the loss of collagen type VII function is the cause of the trauma-induced cleavage at that level.

4 . THE MOLECULAR GENETICS OF EBS AND EHK

Epidermolysis bullosa simplex and dpidermolytic hyperkeratosis are two genetic disorders caused by mutations in distinct keratin genes in which the epidermis, primarily, blisters when subjected to trivial trauma. In this final section, we discuss the relationship between point mutation in keratin proteins, the keratin gene affected, and the clinical manifestations of these diseases. Before doing so, however, we need to introduce some fundamentals of keratin filament assembly.

4.1 OVERVIEW OF KERATIN FILAMENT ASSEMBLY AND PROPERTIES

Intermediate filament proteins, including keratins, are defined by their ability to form 10 nm filaments. At least in vitro, 10 nm filament formation is a remarkable self-assembly process, which can be driven solely by the primary sequence of IF proteins[12,14] (Fig. 15.3). Whether in vivo, accessory proteins or factors are required for 10 nm filament formation is unknown.[122] Electron microcopy analyses, including STEM, conducted on various types of native or in vitro-reconstituted IFs together suggest that they are comprised of three to four protofibrils, each of ~4.5 nm diameter, and each containing 8 protein monomers per cross-section.[18] As further considered below, thousands of monomers are involved in the making of a several micrometer-long filament.

All IF proteins feature a central, 310 amino acid-long, α-helical domain flanked by non-helical head and tail domains whose length and composition vary extensively (reviewed in refs. 6 and 123) (Fig. 15.3). The α-helical domain features an heptad repeat (abcdefg) in which every first (a) and fourth (d) residues tend to be hydrophobic, and this repeat is interrupted at three conserved locations by short spacer sequences, defining subdomains 1A, 1B, 2A, and 2B[6,123] (Fig. 15.3). Also characteristic of IF sequences is the occurrence, at both the amino-terminal (beginning) and carboxy-terminal (end) portions of their rod domain, of short sequences (15-20 residues) which are highly conserved across IF sequence types and throughout evolution.[6,7,124] As discussed below, substitutions at these positions often have devastating consequences for 10 nm filament structure and function, and have been found to cause EBS and EHK, depending of which keratin gene is affected.[24]

Because of the heptad repeat in their α-helical domain, all IF proteins spontaneously form coiled-coil dimers.[125] The α-helical domain from two monomeric chains associate in parallel and in register to form a ~44-46 nm-long rod-like dimer.[18,123,126] For keratins, type I and type II rod sequences are absolutely required to form a coiled-coil dimer productive for 10 nm filament assembly.[15-17] Heterodimers of type I-type II IF chains rapidly associate laterally with an antiparallel orientation to form tetramers, the next intermediate in the 10 nm filament assembly pathway.[15] The exact axial alignment of the antiparallel dimers within the tetramer has been much debated in recent years, and evidence for various alignments was obtained using different approaches and different IF proteins as prototypes (e.g. refs. 15, 125 and 128. and references sited therein). As most IFs feature an axial repeat ("beading") with a ~21-22 nm periodicity under some preparative conditions for electron microscopy, this implies the existence of a stagger between the 44-46 nm long dimer molecules in the polymer context. If the stagger does not take place at the tetramer level, it must therefore occur at a later stage of assembly. The 21-22 nm repeat also implies that the highly-conserved ends of the rod domain overlap by 1-3 nm (discussed in ref. 126).

Tetramers must obviously polymerize laterally and longitudinally to form full-width 10 nm filaments. However, the mechanisms involved are not well understood. The tetramer is the only assembly intermediate which is relatively easy to obtain in high yields in vitro.[127] Intriguingly, data accumulated so far suggest that the cytoplasmic pool of soluble IF subunits consists of tetramers.[129,130] An important property of keratin and other intermediate filaments that was recently uncovered is their dynamic nature in vivo. This was determined from a careful monitoring of the behavior of exogenously added IF proteins in the context of transfection assays[54,55] or microinjection experiments,[131,132] and fluorescence recovery after photobleaching of pre-labeled IF networks in living cultured cells.[133] An exchange of IF assembly subunits between the polymerized pool (filaments) and a soluble pool in the cytoplasm of interphase cells (tetramers?) has been postulated to explain the dynamic behavior of these filaments,[15,132] and could be at least partly mediated by post-translational modifications (e.g. phosphorylation; reviewed in ref. 134).

4.2 Molecular Basis Underlying the Similarities and Differences Between EBS and EHK

It is quite interesting to compare the EBS and EHK diseases from a keratin biology standpoint, for it provides a biological rationale for the similarities and differences in the physiopathology of these two disorders.

Consider, for instance, the following similarities between the two diseases:

- trauma-induced cytolysis, initiating blister formation (see above);
- an autosomal dominant mode of transmission (rare cases of recessive inheritance have been reported for EBS; see ref. 69);
- occurrence of distinct clinical subtypes said to be either "generalized' (lesions over most of the body surface) or "localized" (lesions mostly at extremities such as hands and feet).[43] Clinical "subtypes" of EBS and EHK have been defined because of marked differences in clinical presentation, including distribution and severity of blisters, prognosis, and other important manifestations;
- anomalies in the architecture of the keratin filament network in a subset of epidermal cells. The degree of anomaly typically varies significantly among the various clinical subtypes of these two diseases, and often correlate with their severity. Thus, in the most severe subtypes, there is even "clumping of tonofilaments"[46] (Fig. 15.2). "Clumps" refer to unpolymerized keratins in the form of aggregates;
- Such anomalies in the structure of keratin filaments and/or of the network they form result in a significant loss of physical strength in the affected cells of the epidermis, which become unable to cope with their normal load of mechanical stress.

To a large extent, the striking resemblance in the pathophysiology of these two diseases has its roots in the fact that both are caused by point mutations in specific keratin sequences. These point mutations, as reviewed below, do not impair the ability of mutant keratins to form, upon interaction with a suitable pairing partner, the assembly subunit which gets incorporated within 10 nm filaments.[61,74,135] It is in fact beyond this assembly step that mutant-containing subunits interferes with the formation of 10 nm filaments. As a result, mutant-containing filaments can be shortened and/or unraveled (loss of compactness), depending on the interaction(s) that are affected.[61,71,74] This type of behavior is a prerequisite for a mutant to function in a dominant negative fashion,[136] and is responsible for the autosomal dominant mode of transmission of EBS and EHK (with rare exceptions). Depending on the location of the codon affected by a point mutation within the keratin primary sequence, and depending as well on the nature of the amino acid substitution, the effect(s) on the assembly and structure of keratin filament assembly will be variable. Thus some mutations perturb so much the normal course of interactions between assembly subunits that these can not form genuine 10 nm filaments in vivo: such mutations lead to the accumulation of keratins as amorphous cytoplasmic aggregates known as clumps.[58] In some other cases, the interactions between the assembly subunits are perturbed in a much more subtle fashion, e.g., within the context of the filaments themselves. In such cases, the ultrastructural features of these filaments in vitro[71,74] or in vivo[45,47] are only modestly perturbed, and greater amounts of stress need be applied before such filaments are disrupted. Therefore, the location and type of point mutations in keratin sequences are the major factors determining the various clinical subtypes of these diseases, and these various subtypes correspond in fact to differential sensitivities to mechanical trauma.

On the other hand, consider the following distinctions between EBS and EHK:

- distinct plane of cleavage within the epidermis: in EBS, cytolysis affects progenitor basal cells, while differentiating suprabasal cells are generally normal;[46,107] in EHK, cytolysis affects only suprabasal cells, and these are progressively more affected as they progress though the program of terminal differentiation;[79]
- hyperkeratosis, which is a hallmark of EHK and is absent in EBS skin. It is thought that in EHK, the chronic cell lysis occurring in the suprabasal compartment somehow results in an increased proliferation rate in basal keratinocytes. This hyperplasia, coupled to structural aberrations in the differentiating layers of the tissue, probably account for the marked hyperkeratosis;[79]
- involvement of stratified epithelia other than the epidermis: the oral mucosa, tongue, esophagus, and cornea can show erosions or blistering in EBS, while they do not in EHK. In EBS, nails may be lost, but will regrow and have a normal structure, while in EHK, the nail plates can be abnormal.[43,78]

These distinctions can be accounted for by the notion that different sets of keratin genes, that are expressed in different subsets of epidermal cells, are altered in EBS and EHK, respectively. All mutations that have been discovered in EBS patients were in either the K5 or K14 keratin gene, which are specifically expressed in the basal layer of the epidermis and all other stratified squamous epithelia (Table 15.4). Whenever keratin mutant-expressing basal cells are spared from trauma-induced blistering, they will, upon commitment to terminally differentiate, downregulate the synthesis of basal-specific keratins (including the mutant allele) and upregulate tremendously the synthesis of keratins K1 and K10. The vast molar excess of K1/K10 generated relative to the persisting amounts of K5/K14 is likely to "rescue" the assembly defect in differenting suprabasal cells, explaining why keratin filament anomalies are confined mostly to the basal compartment.[47,58,59] In contrast, all mutations thus far discovered in EHK patients affect either the K1 or the K10 genes, whose levels increase immensely as postmitotic epidermal cells progress through the program of terminal differentiation. In this particular case, there is no possible "rescue" due to a subsequent induction of the expression of wild-type keratin sequences; in contrast, there should be a progressive worsening of the keratin filaments anomalies as epidermal cells differentiate, and that is exactly what occurs.[79] Finally, many more epithelial tissues can be affected in EBS compared to EHK patients, because the pattern of K1 and K10 expression, being restricted to cornifying epithelia, is much more restricted than that of K5 and K14 (Table 15.2).

4.3 Distribution of Point Mutations Along Keratin Sequences in EBS and EHK, and Their Relationship to Disease Severity

Table 15.4 lists the keratin mutations that have been reported so far in forty-four independent incidences of EBS and EHK (incl. EPPK) and Figure 15.4 depicts their distribution along a generic model of the secondary structure of keratin proteins. From the mutations documented so far, the following trends can be identified:

- Approximately 80% of the mutations (35 out of 44) occur in α-helical subdomains of keratin sequences;
- Of the 35 mutations in α-helical subdomains, a disproportionately high number occur in the highly conserved sequence motifs located at the amino-terminus of subdomain 1A and at the carboxy-terminus of subdomain 2B;
- mutations at a single arginine residue, located in the very beginning of subdomain 1A, account for a remarkable ~40% (17 out of 44) of all mutations discovered in EBS and EHK/ EPPK (Table 15.4). This particular Arg codon, "CGC", is conserved to a greater than 90 % extent among all known IF sequences (independently of type and species; see ref. 61). Moreover, only the type I keratin genes are mutated at this position (codon 125 in K14 and codon 156 in K10).
- Excluding those occurring in the K14-Arg_{125} and K10-Arg_{156} residues, mutations do not appear to affect preferentially the type I (12 out of 27) or the type II (15 out of 27) keratin genes;
- There are nine point mutations affecting residues outside α-helical subdomains (see Table 15.4). Five are located in the linker domain between coil 1 and coil 2 ("L12" in Fig. 15.4), whose length and sequence vary between intermediate filament proteins. The other four are located in the portion of the keratin head domain closest to subdomain 1A, involve type II keratin genes, and affect residues which are highly conserved among them;[66,81,86]
- There exists a relationship between the position and nature of the point mutations along the keratin sequence and the severity of the clinical phenotype.[61,71,74] This important notion is further discussed below.

It is interesting to comment on the distribution of these point mutations along keratin protein sequences in light of extensive studies which exploited experimental mutagenesis to determine the sequence requirements for 10 nm filament assembly in vitro and filament network formation in vitro (reviewed in ref. 6 and 24). The type I keratin K14, in particular, and the type II keratin K5 sequences have been subjected to such mutagenesis studies. Thus, the tail and most (> 90%) of the head domain of K14, as well as the entire tail domain of K5, can all be deleted and the resulting mutants will together assemble into 10 nm filaments in vitro under slightly modified, but still "physiological" conditions.[137] Accordingly, it is not

Fig. 15.4. (Opposite) Distribution of point mutations along the generic keratin protein sequence. The secondary structure shared by keratin proteins K1, K5, K9, K10 and K14 is depicted. The rod domain comprises α-helical subdomains 1A and 1B (coil 1), and 2A and 2B (coil 2), which are separated by linkers segment L1, L'2 and L2. The rod domain is flanked by non-helical amino-terminal ("head") and carboxy-terminal ("tail") sequences, whose length vary among these keratins. Each filled circle represent a point mutation documented in either epidermolysis bullosa simplex or epidermolytic hyperkeratosis, as listed on Table 15.4 above. Note that a total of 27 mutations locate to the very beginning of subdomain 1A, and that 17 of them affect a highly conserved arginine (Arg) residue in the type I keratin K10 And K14. The position of the mutations is approximative.

Table 15.4. Keratin point mutations discovered in patients with epidermolysis bullosa simplex and epidermolytic hyperkeratosis

Disease	keratin gene[1,2]	Mutation	Domain location[3]	References
EBS-DM	K14	R125C (n = 3)	rod (1A)	59,61,64
EBS-DM	K14	R125H (n = 5)	rod (1A)	61,64
EBS-recessive	K14	E144A	rod (1A)	69
EBS-DM	K5	E475Q	rod (2B)	63
EBS-K	K14	L384P	rod (2B)	62
EBS-K	K5	L463P	rod (2B)	65
EBS-K	K14	M272R	linker L12	68
EBS-WC	K5	M327T	linker L12	71
EBS-WC	K5	N329K	linker L12	71
EBS-WC	K5	R331C	linker L12	67
EBS-WC	K14	V270M	linker L12	67
EBS-WC	K5	I161S (n = 2)	head	66
EHK	K10	R156H (n = 5)	rod (1A)	80,82,83
EHK	K10	R156C (n = 2)	rod (1A)	82,85
EHK	K10	R156P	rod (1A)	87
EHK	K10	R156S	rod (1A)	87
EHK	K10	M150R	rod (1A)	85
EHK	K10	N154H	rod (1A)	82
EHK	K10	L161S	rod (1A)	83
EHK	K10	Y160D	rod (1A)	82
EHK	K10	Y160N	rod (1A)	84
EHK	K1	S185P	rod (1A)	87
EHK	K1	N187P	rod (1A)	87
EHK	K1	N187S	rod (1A)	86
EHK	K1	S192P	rod (1A)	86
EHK	K10	K439E	rod (2B)	85
EHK	K10	L442Q	rod (2B)	82
EHK	K1	Y481C	rod (2B)	85
EHK	K1	E489Q	rod (2B)	83
EHK	K1	L160P	head	81
EHK	K1	V154G	head	86
EPPK	K9	N160Y	rod (1A)	88

[1]Point mutations in keratin coding sequences have so far been reported in 44 incidences of EBS/EHK/EPPK. 19 of these point mutations affect either the K5 of K14 gene and thus cause EBS; 23 of these mutations affect either the K1 or K10 gene and cause EHK; one point mutation affect the K9 gene in the diseased members of a large EPPK pedigree.

[2]Codon numbering is derived from the following published sequences: K14 (148); K10 (149); K5 (21); K1 (150); K9 (89).

[3]Figure 15.4 below depicts the localization of each of these point mutations along the generic keratin protein sequence.

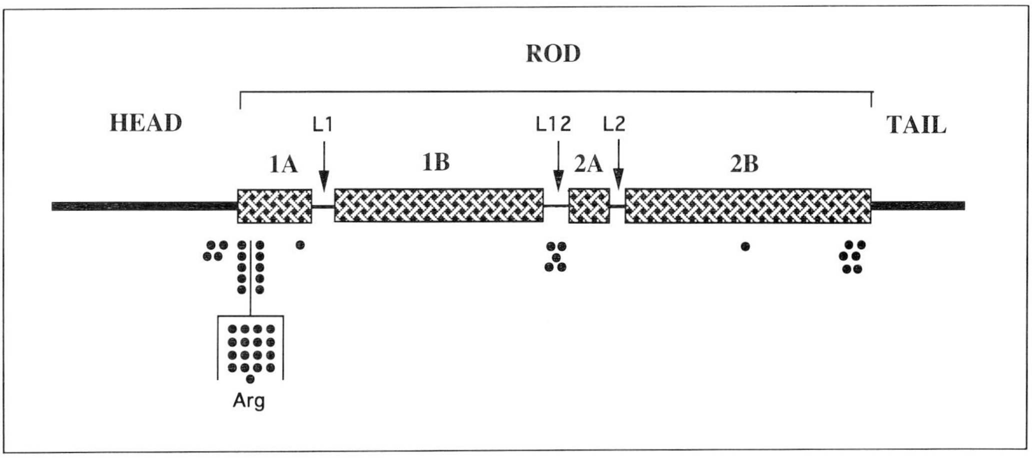

surprising that at least so far, no point mutation has been found in these end domains in either the K5/K14 or the K1/K10 pair. Wilson et al.[137] found, on the other hand, that the head domain of the type II K5 was essential for this process, and point mutations have been found in the head domain of this keratin in two instances of WC-EBS,[66] and in the head domain of K1 in two instances of EHK.[81,86]

In contrast to the non-helical end domains, rod domain sequences are absolutely required for normal 10 nm filament assembly to proceed. Thus, when the end portions of the human K14 rod domain were either deleted[54,56] or point-mutagenized,[135] the consequences on 10 nm filament assembly and network formation were severe. Similar observations were made when other intermediate filament sequences were subjected to similar manipulations (reviewed in ref. 138). As noted above, the end portions of the rod domain sequence contain the most conserved amino acid residues among intermediate filament proteins. In a study in which single proline substitutions were engineered throughout the rod domain of K14, Letai et al.[135] found that these substitutions were significantly better tolerated if engineered in the middle portion of the rod domain (e.g., subdomains 1B, 2A, and linker sequences; see Fig. 15.4) as opposed to the highly-conserved end portions of this domain. Remarkably, ~72 % of all point mutations found in EBS /EHK / EPPK patients affect one among the first 20 or the last 20 amino acid residues of the rod domain (Fig. 15.4). Thus, the experimental mutagenesis studies conducted on keratin and other intermediate filament sequences had enormous implications for epidermal blistering diseases, in addition to their significance for the assembly and organization of 10 nm filaments.

As noted above, there exists a relationship between the position and nature of the point mutations along keratin sequences and the severity of the clinical phenotype. This can be best demonstrated in EBS diseases, for which three major clinical subtypes have been clearly delineated based on the type and distribution of blisters and clinical course of the disease in affected individuals.[43] Thus, the Dowling-Meara subtype is the most severe form of EBS and is unique by the frequent involvement of internal stratified squamous epithelia and the presence of tonofilament clumps in epidermal basal cells (Fig. 15.2 and Table 15.3). Point mutations have been discovered in nine incidences of Dowling Meara-EBS, and they all affect highly conserved amino acid residues in the proximal portion of subdomain 1A or the distal portion of subdomain 2B (Table 15.4). The Weber-Cockayne subtype is the mildest form of EBS: it affects only the epidermis, blisters occur primarily on hands and feet, and ultrastructurally the keratin filament network of epidermal basal cells is barely disrupted (Fig. 15.2 and Table 15.3). Accordingly, the point mutations found in such patients so far affect residues located outside of the α-helical sequences: they occur in either the linker L12 sequence (n = 4) or in the head domain (n =2) (Table 15.4). Finally, the Koebner subtype of EBS is of intermediate severity (Table 15.3) and two of the three point mutations known to cause this form of the disease are located in the middle portion of the rod domain (Table 15.4). A direct relationship between location of point mutation and severity of the clinical phenotype also appears to exist for EHK (see refs. 82, 85 for further discussion).

This relationship has been directly tested in studies comparing the ability of various EBS point mutations to disrupt K5-K14 filament assembly in vitro, and in some cases, the keratin filament network by transient transfection of cultured epidermal cells. As expected, the Dowling-Meara point mutations were the most potent disruptors in the context of in vitro 10 nm filament assembly assays using bacterially-expressed, purified human keratins. These mutations often lead to shortened and unraveled 10 nm filaments.[61,74] This was especially so for the point mutations occurring so frequently in codon 125 of the K14 sequence, i.e., R125C and R125H.[74] In contrast, the point mutations discovered in Koebner-EBS and in Weber-Cockayne-EBS caused significantly milder disruptions when tested in vitro.[71,74] For instance, the L12 point mutation found in two independent incidence of WC-EBS do not to affect significantly the polymer elongation phase during filament assembly, but appears to "destabilize" the lateral interactions between protofibrils such that the filament are significantly "unraveled" compared to those assembled from wild-type K14 and K5.[71] Whenever tested in parallel, the phenotypes caused by the expression of some these mutants following transient transfection of cultured epidermal cells were consistent with the in vitro assembly data.[74] Characterization of the effects of additional EBS point mutations, as well as EHK point mutations in the K1 and K10 proteins,[82] should test further the validity of the relationship between their nature and location and the clinical presentation of the disorder.

4.4 Implications for the Function of Keratin Filaments in Epidermis

The recent studies in which dominant negative mutants of keratin sequences were expressed in transgenic mice have provided compelling evidence for the function of keratin filaments in the epidermis in situ.[47,58,79] Thus, in transgenic mice expressing dominant negative mutants of either K14 or K10 in a tissue-specific fashion, blistering of the epidermis arose as a result of cytolysis in the basal and suprabasal compartments, respectively, in response to mild mechanical trauma. Transgenic epidermal cells expressing mutant keratins featured a disruption of their keratin filament network, along with alterations in cytoarchitecture as well as in nuclear shape. The severity of the blistering phenotype correlated with the extent with which the keratin filament network was perturbed, and was found to be a function of the type of mutation engineered, i.e., its ability to disrupt 10 nm filament assembly in vitro, and the levels at which is was expressed relative to the mouse endogenous keratin homologs. The blistering phenotype is specific to the expression of dominant negative mutants, since overexpression of wild-type K14[139,140] K5,[141] K18,[142] and K19[143] in a tissue-specific fashion reportedly does not perturb the structure and function of epithelial cells expressing them. These observations are most consistent with a *loss-of-function phenotype,*[24,47,79] and point to two distinct roles for epidermal keratins: (i) to impart mechanical strength without which epidermal cells cannot cope with their normal load of mechanical stress (a uniquely important function in skin) and rupture readily, and (ii) to provide a scaffold responsible for at least part of the cytoarchitecture. While a definitive validation of this interpretation requires the generation and characterization of the corresponding gene knockouts in mouse, there is already an emerging body of evidence pointing to a general role of IFs in imparting mechanical strength to cells and tissues (reviewed in ref. 59; see also ref. 144). The multiplicity of keratins and other IF protein sequences, and their highly complex regulation, strongly suggest that additional functions await discovery.

Acknowledgments

The authors would like to thank Ms. Janet Folmer (J.H.U.) for providing the electron micrographs shown in Figures 15.1 and 15.2.

Author's Note

Point mutations in the γ2–chain (100 kD subunit) of kalinin/nicein/epiligrin have recently been discovered in patients suffering from the junctional form of epidermolysis bullosa disease.[152,152]

References

1. Fuchs E. Epidermal differentiation: The bare essentials. J Cell Biol 1990; 111:2807-2814.
2. Watt FM. Terminal differentiation of epidermal keratinocytes. Curr Opin Cell Biol 1989; 1:1107-1115.
3. Rice RH, Green H. Presence in human epidermal cells of a soluble protein precursor of the cross-linked envelope: Activation of the cross-linking by calcium ions. Cell 1979; 18:681-688.
4. Swatzendruber DC, Wertz PW, Madison KC, Downing DT. Evidence that the corneocyte has a chemically-bound lipid envelope. J Invest Dermatol 1987; 88:709-713.

5. Fritsch PO, Elias PM, Hintner HH. Mechanisms of vesicle formation and classification. In: Fitzpatirck TB, Eisen AZ, Wolff K, Freedberg IW, Austen KF, eds. Dermatology in general Medicine, Vol. I, 4th Edit. New York: McGraw-Hill, 1993:577-584.
6. Fuchs E, Weber K. Intermediate filaments: Structure, dynamics, function and disease. Ann Rev Biochem 1994, Ann Rev Biochem 1994; 63:345-382.
7. Fuchs E. Keratins as biochemical markers of epithelial differentiation. Trends Genet 1988; 4:277-281.
8. Steinert PM, Roop DR. Molecular and cellular biology of intermediate filaments. Ann Rev Biochem 1988; 57:593-625.
9. Fuchs E, Coppock S, Green H, Cleveland D. Two distinct classes of keratin genes and their evolutionary significance. Cell 1981; 27:75-84.
10. Moll R, Franke WW, Schiller D, Geiger B, Krepler R. The catalog of human cytokeratins: Patterns of expression in normal epithelia, tumors and cultured cells. Cell 1982; 31:11-24.
11. Sun TT, Eichner R., Schermer A, Cooper D, Nelson WG, Weiss RA. Classification, expression, and possible mechanisms of evolution of mammalian epithelial keratins: A unifying model. In: A Levine, W Topp, G van de Woude, JD Watson, eds. Cancer Cells, Vol. 1: The transformed phenotype. Cold Spring Harbor Press, 1984:169-176.
12. Steinert PM, Idler WW, Zimmerman SB. Self-assembly of bovine epidermal keratin filaments in vitro. J Mol Biol 1976; 108:547-567.
13. Franke WW, Schiller DL, Hatzfeld M, Winter S. Protein complexes of intermediate-sized filaments: melting of cytokeratin complexes in urea reveals different polypeptide separation characteristics. Proc Natl Acad Sci (USA) 1983; 80:7113-7117.
14. Eichner R, Sun TT, Aebi U. The role of keratin subfamilies and keratin pairs in the formation of human epidermal keratin filaments. J Cell Biol 1986; 102:1767-1777.
15. Coulombe PA, Fuchs E. Elucidating the early stages of keratin filament assembly. J Cell Biol 1990; 111:153-169.
16. Hatzfeld M, Weber K. The coiled-coil of in vitro assembled keratin filaments is a heterodimer of type I and type II keratins: use of site-specific mutagenesis and recombinant protein expression. J Cell Biol 1990; 110:1199-1210.
17. Steinert PM. The two-chain coiled-coil molecule of native epidermal keratin intermediate filaments is a type I-type II heterodimer. J Biol Chem 1990; 265:8766-8774.
18. Aebi U, Haner M, Troncoso J, Eichner R, Engel A. Unifying principles in intermediate filament (IF) structure and assembly. Protoplasma 1988; 145:73-81.
19. Stellmach VM, Leask A, Fuchs E. Retinoid-mediated transcriptional regulation of keratin genes in human epidermal and squamous cell carcinoma cells. PNAS (USA) 1991; 88:4582-4586.
20. Kulesh DA, Cecena G, Darmon YM, Vasseur M, Oshima RG. Posttranslational regulation of keratins: Degradation of mouse and human keratins 8 and 18. Mol Cell Biol 1989; 9:1553-1565.
21. Lersch R, Stellmach V, Stocks C, Giudice G, Fuchs E. Isolation, sequence and expression of a human keratin K5 gene: transcriptional regulation of keratins and insights into pairwise control. Mol Cell Biol 1989; 9:3685-3697.
22. Nelson W, Sun TT. The 50- and 58-kD keratins classes as molecular markers for stratified squamous epithelia: cell culture studies. J Cell Biol 1983; 97:244-251.
23. Fuchs E, Green H. Changes in keratin gene expression during terminal differentiation of the keratinocyte. Cell 1980; 19:1033-1042.
24. Coulombe PA. The cellular and molecular biology of keratins: Beginning a new era. Curr Opin Cell Biol 1993; 5:17-29.
25. Schwarz MA, Owaribe K, Kartenbeck J, Franke WW. Desmosomes and hemi-desmosomes: Constitutive molecular components. Annu Rev Cell Biol 1990; 6:461-491.
26. Jones JCR, Green KJ. Intermediate Filament-plasma membrane interactions. Curr Opin Cell Biol 1991; 3:127-132.

27. Jones JCR, Yokoo KM, Goldman RD. Is the hemidesmososme a half-desmosome ? An immunological comparison of mammalian desmosomes and hemidesmosomes. Cell Motil Cytoskel 1986; 6:560-569.
28. Foisner R, Wiche G. Intermediate filament-associated proteins. Curr Opin Cell Biol 1991; 3:75-81.
29. Stanley JR. Cell adhesions molecules as targets of autoantibodies in pemphigus and pemphigoid, bullous diseases due to a defective epidermal cell adhesion. Adv Immunol 1992; 53:291-325.
30. Burgeson RE. Basement membranes. In: Fitzpatrick TB, Eisen AZ, Wolff K, Freedberg IW, Austen KF, eds. Dermatology in general Medicine, Vol. I, 4th Edit. New York: McGraw-Hill, 1993:329-338.
31. Uitto J, Christiano AM. Inherited Epidermolysis Bullosa: Clinical features, molecular genetics, and pathoetiologic mechanisms. Dermatol Clin 1993; 11:549-563.
32. Watt FM, Hertle MD. Keratinocytes integrins. In: IM Leigh, EB Lane, and FM Watt, eds. The keratinocyte handbook. Cambridge: Cambridge University Press, 1994: Submitted for publication.
33. Jones JCR, Kurpakus MA, Cooper HM, Quaranta V. A function for the integrin $\alpha 6\beta 4$ in the hemidesmosome. Cell Regul 1991; 2:427-438.
34. Marinkovich MP, Verrando P, Keene DR, Meneguzzi G, Lunstrum GP, Ortonne JP, Burgeson RE. Basement membrane proteins Kalinin and Nicein are structurally and immunologically identical. Lab Invest 1993; 69:295-299.
35. Verrando P, Blanchet-Bardon C, Pisani A, Thomas L, Cambazard F, Eady RAJ, Schofield O, Ortonne JP. Monoclonal antibody GB3 defines a widespread defect of several basement membranes and a keratinocyte dysfunction in patients with lethal junctional epidermolysis bullosa. Biochim Biophys Acta 1988; 942:45-46.
36. Rouselle P, Lunstrum GP, Keene DR, Burgeson RE. Kalinin: An epithelium-specific basement membrane-adhesion molecule that is a component of anchoring filaments. J Cell Biol 1991; 114:567-76.
37. Carter WG, Ryan MC, Gahr PJ. Epiligrin, a new adhesion ligand for integrin $\alpha 3\beta 1$ in epithelial basement membranes. Cell 1991; 65:599-610.
38. Burgeson RE. Type VII collagen, anchoring fibrils, and epidermolysis bullosa. J Invest Dermatol 1993; 101:252-255.
39. Keene DR, Sakai LY, Lunstrum GP, Morris NP, Burgeson RE. Type VII collagen forms an extended network of anchoring fibrils. J Cell Biol 1987; 104:611-621.
40. Crosby DL, Diaz LA. Bullous Diseases: Introduction. Dermatol Clin 1993; 11:373-378.
41. Pearson RW. Dermatology in general medicine: The mechanobullous diseases (epidermolysis bullosa), 1st ed. New York: McGraw-Hill, 1971.
42. Gedde Dahl T, Anton-Lamprecht I. Epidermolysis bullosa. In: Emory AEH, Rimoin DL eds. Principles and practice in medical genetics. New York: Livingstone, 1981:672-687.
43. Fine JD, Bauer EA, Briggaman RA, Carter DM, Eady RAJ, Esterly NB, Holbrook KA, Hurwitz S, Johnson L, Lin A, Pearson R, Sybert VP. Revised laboratory and clinical criteria for subtypes of inherited Epidermolysis Bullosa. J Am Acad Dermatol 1991; 24:119-135.
44. Anton-Lamprecht I, Snyder UW. Epidermolysis bullosa herpetiformis Dowling Meara: report of a case and pathogenesis. Dermatol 1982; 164:221-226.
45. Haneke E, Anton-Lamprecht I. Ultrastructure of blister formation in epidermolysis bullosa hereditaria: V. Epidermolysis bullosa simplex localisata tye Weber-Cockayne. J Invest Dermatol 1982; 78:219-223.
46. Anton-Lamprecht I. Genetically-induced abnormalities of epidermal differentiation and ultrastructure in ichtyoses and epidermolysis: pathogenesis, heterogeneity, fetal manifestations, and prenatal diagnosis. J Invest Dermatol 1983; 81 (Suppl. 1): 149s-156s.
47. Coulombe PA, Hutton ME, Vassar R, Fuchs E. A function for keratins and a common thread among Epidermolysis Bullosa Simplex diseases. J Cell Biol 1991; 115:1661-1674.

48. Savolainen ER, Kero M, Pihlajaniemi T, Kivirikko KI. Deficiency of galactosylhydroxylysyl glycosyltransferase, an enzyme of collagen synthesis, in a family with dominant epidermolysis bullosa simplex. N Engl J Med 1981; 304-309.
49. Sanchez G, Seltzer JL, Eisen AZ, Stapler P, Bauer EA. Generalized dominant epidermolysis bullosa simplex: Decreased activity of a gelatinolytic protease in cultured fibroblasts as a phenotypic marker. J Invest Dermatol 1983; 81:576-584.
50. Takamori K, Ikeda S, Naito K, Ogawa H. Proteases are responsible for blister formation in recessive dystrophic epidermolysis bullosa and epidermolysis bullosa simplex. Brit J Dermatol 112, 1985; 121:1985-1993.
51. Fine JD, Griffith RD. A specific defect in glycosylation of epidermal cell membranes: definition in skin from patients with epidermolysis bullosa simplex. Arch Dermatol 1985; 121:1292-1299.
52. Winberg JO, Real D, Gedde-Dahl T. Gelatinase expression in generalized epidermolysis bullosa simplex fibroblasts. J Invest Dermatol 1986; 87:326-333.
53. Tidman MJ, Eady RA, Leigh IM, Macdonald, DM. Keratin expression in epidermolysis bullosa (Dowling Meara). Stockholm: Acta Derm Venereol, 1988; 68:15-20.
54. Albers K, Fuchs E. The expression of mutant epidermal keratin cDNAs transfected in simple epithelia and squamous carcinoma lines. J Cell Biol 1987; 105:791-806.
55. Albers K, Fuchs E. Expression of mutant epidermal keratin cDNAs in epithelial cells reveals possible mechanisms for initiation and assembly of intermediate filaments. J Cell Biol 1989; 108:1477-1493.
56. Coulombe PA, Chan YM, Albers K, Fuchs E. Deletions in epidermal keratins leading to alterations in filament organization in vivo and intermediate filament assembly in vitro. J Cell Biol 1990; 111:3049-3064.
57. Kitajima Y, Inoue, S, Yaoita H. Abnormal organization of keratin intermediate filaments in cultured keratinocytes of epidermolysis bullosa simplex. Arch Dermatol Res 1989; 281:5-10.
58. Vassar R, Coulombe PA, Degenstein L, Albers K, Fuchs E. Mutant keratin expression in transgenic mice causes marked abnormalities resembling a human genetic skin disease. Cell 1991; 64:365-380.
59. Fuchs E, Coulombe PA. Of mice and men: Genetic skin diseases of keratin. Cell 1992; 69:899-902.
60. Coulombe PA, Fuchs E. Epidermolysis Bullosa Simplex. Semin Dermatol 1993; 12:173-190.
61. Coulombe PA, Hutton ME, Letai A, Hebert A, Paller AS, Fuchs E. Point mutations in human keratin 14 genes of Epidermolysis Bullosa Simplex patients: Genetic and functional analyses. Cell 1991; 66:1301-1311.
62. Bonifas JM, Rothman Al, Epstein EH. Epidermolysis Bullosa Simplex: Evidence in two families for keratin gene abnormalities. Science 1991; 254:1202-1205.
63. Lane EB, Rugg EL, Nausaria H, Leigh IM, Heagerty HHM, Ishida-Yamamoto A, Eady RAJ: A mutation in the conserved helix termination peptide of keratin 5 in hereditary skin blistering. Nature 1992; 356:244-246.
64. Stephens K, Sybert VP, Wijsman EM, Ehrlich P, Spencer A. A keratin 14 mutational hot spot for epidermolysis bullosa simplex, Dowling Meara: Implications for diagnosis. J Invest Dermatol 1993; 101:240-243.
65. Dong W, Ryynanen M, Uitto J. Identification of a Leucine-to-Proline mutation in the keratin 5 gene in a family with the generalize koebner type of epidermolysis bullosa simplex. Hum Mutat 1993; 2:94-102.
66. Chan YM, Yu QC, Fine JD, Fuchs E. The genetic basis of Weber-Cockayne epidermolysis bullosa simplex. Proc Natl Acad Sci (USA) 1993; 90:7414-7418.
67. Rugg EL, Morley SM, Smith FJD, Boxer M, Tidman MJ, Navsaria H, Leigh IM, Lane EB. Missing links: Weber-Cockayne keratin mutations implicate the L12 linker domain in effective cytoskeletal function. Nature Genetics 1993; 5:294-300.
68. Humphries MM, Sheils DM, Farrar GJ, Kumar-Singh R, Kenna PF, Mansergh FC, Jordan SA, Young M, Humphries P. A mutation (Met->Arg) in the type I keratin K14 gene responsible for autosomal dominant epidermolysis bullosa simplex. Hum Mutat 1993; 2:37-42.

69. Hovnavian A, Pollack E, Hilal L, Rochat A, Prost C, Barrandon Y, Goosens M. A missense mutation in the rod domain of keratin 14 associated with recessive epidermolysis bullosa simplex. Nature Genetics 1993; 3:327-332.
70. McKenna KE, Hughes AE, Bingham EA, Nevin NC. Linkage of epidermolysis bullosa simplex to keratin gene loci. J Med Genet 1992; 29:568-570.
71. Chan YM, Yu QC, Christiano A, Uitto J, Kucherlapati RS, Leblanc-Strasecki J, Fuchs E. Mutations in the non-helical linker segment L1-2 of keratin 5 in patients with Weber-Cockayne epidermolysis bullosa simplex. J Cell Sci 1994, 107:765-774.
72. Rosenberg M, RayChaudhury A, Shows TB, Le Beau MM, Fuchs E. A group of type I keratin genes on human chromosome 17: Characterization and expression. Mol Cell Biol 1988; 8:722-736.
73. Rosenberg M, Fuchs E, Le Beau MM, Eddy RL, Shows TB. Three epidermal and one simple epithelial type II keratin genes map to human chromosome 12. Cytogenet Cell Genet 1992; 57:33-38.
74. Letai A, Coulombe PA, McCormick MB, Yu QC, Hutton E, Fuchs E. Disease severity correlates with position of keratin point mutation in patients with Epidermolysis Bullosa Simplex. Proc Natl Acad Sci (USA) 1993; 90:3197-3201.
75. Bauer EA, Briggaman RA. Hereditary epidermolysis bullosa. In: Fitzpatrick TB, Eisen AZ, Wolff K, Freedberg IW, Austen KF, eds. Dermatology in General Medicine, Vol. I, 4th Edit. New York: McGraw-Hill, 1993: 329-338.
76. Ackerman AB. Histopathologic concept of Epidermolytic Hyperkeratosis. Arch Dermatol 1970; 102:253-259.
77. Williams ML, Elias PM. Genetically transmitted, generalized disorders of cornification: The ichtyoses. Dermatol Clin 1987; 5:155-178.
78. Baden H. Keratinizing disorders. In: Alper JC, ed. Genetic disorders of the skin. St Louis: Mosby Year Book, 1991:170-194.
79. Fuchs, Esteves RA, Coulombe PA. Transgenic mice expressing a mutant keratin 10 gene reveal the likely genetic basis Epidermolytic Hyperkeratosis. Proc Natl Acad Sci (USA) 1992; 89:6906-6910.
80. Cheng J, Syder AS, Yu QC, Letai A, Paller AS, Fuchs E. The genetic basis for Epidermolytic Hyperkeratosis: A disorder of Differentiation-specific epidermal keratin genes. Cell 1992; 70:811-819.
81. Chipev CC, Korge BP, Markova N, Bale SJ, DiGiannova JJ, Compton JG, Steinert PM. A leucine-proline mutation in the H1 subdomain of keratin 1 causes Epidermolytic Hyperkeratosis. Cell 1992; 70:821-828.
82. Chipev CC, Yang JM, , DiGiannova JJ, Steinert PM, Marekov L, Compton JG, Bale SJ. Preferential sites in keratin 10 that are mutated in Epidermolytic Hyperkeratosis. Am J Hum Genet 1994; 54:179-190.
83. Rothnagel JA, Dominey AM, Dempsey LD, Longley MA, Greenhalgh DA, Gagne TA, Huber M, Frenk E, Hohl D, Roop DR. Mutations in the rod domain of keratins 1 and 10 in epidermolytic hyperkeratosis. Science 1992; 257:1128-1130.
84. Rothnagel JA, Longley MA, Holder RA, Kuster W, Roop DR. Prenatal diagnosis of epidermolytic hyperkeratosis by direct gene sequencing. J Invest Dermatol 1994; 102:13-16.
85. Syder AJ, Yu QC, Paller AS, Guidice G, Pearson R, Fuchs E. Genetic mutations in the K1 and K10 genes of patients with Epidermolytic Hyperkeratosis: Correlation between location and disease severity. J Clin Invest 1994, 93:1533-1542.
86. Yang JM, Chipec CC, DiGiovanna JJ, Bale SJ, Marekov LN, Steinert PM, Compton JG. Mutations in the H1 and 1A domains of keratin 1 gene in epidermolytic hyperkeratosis. J Invest Dermatol 1994; 102:17-23.
87. McLean WHI, Eady RAJ, Dopping-Hepenstal PJC, McMillan JR, Leigh IM, Navsaria HA, Higgins C, Harper JI, Paige DG, Morley SM, Lane EB. Mutations in the rod 1A domain of keratin 1 and 10 in bullous congenital ichtyosiform erythroderma (BCIE). J Invest Dermatol 1994; 102:24-30.

88. Torchard D, Blanchet-Bardon C, Serova O, Langbein L, Narod S, Janin N, Goguel AF, Bernheim A, Franke WW, Lenoir GM, Feunteum J. Epidermolytic palmoplantar keratoderma co-segregates with a keratin 9 mutation in a pedigree with breast and ovarian cancer. Nature Genetics 1994; 6:106-110.
89. Langbein L, Heid HW, Moll I, Franke WW. Molecular characterization of the body-site specific human epidermal cytokeratin 9: cDNA cloning, amino acid sequence, and tissue specificity of gene expression. Differentiation 1994. Submitted for publication.
90. Philips SB, Baden HP. Darier-White disease (keratosis follicularis) and other miscellaneous hyperkeratotic disorders. In: Fitzpatrick TB, Eisen AZ, Wolff K, Freedberg IW, Austen KF, eds. Dermatology in general Medicine, Vol. I, 4th Edit. New York: McGraw-Hill, 1993:547-552.
91. Caulfield JB, Wilgram GF. An electron microscopic study of dyskeratosis and acantholysis in Darier's disease. J Invest Dermatol 1963; 41:57-67.
92. El-Gothamy Z, Kamel MM. Ultrastructural observations in Darier's disease. Am J Dermatopathol 1988; 10:306-310.
93. Burge SM, Garrod DR. An immunohistochemical study of desmosomes in Darier's disease and Hailey-Hailey's disease. Br J Dermatol 1991; 124:242-251.
94. Bashir R, Munro CS, Mason S, Stephenson A, Rees JL, Strachan T. Localization of a gene for Darier's disease. Hum Molec Genet 1993; 11:1937-1939.
95. Craddock N, Dawson E, Burge S, Parfitt L, Mant B, Roberts Q, Daniels J, Gill M, McGuffin P, Powell J, Owen M. The gene for Darier's disease maps to chromosome 12q23-q24.1. Hum Molec Genet 1993; 11:1941-1943.
96. Parfitt E, Burge S, Craddock N, Roberts E, McLean WHI, Weissenbach J, McGuffin P, Owen M. The gene for Darier's disease maps between D1278 and D1279. Hum Molec Genet 1994; 3:35-38.
97. Stanley JR. Pemphigus. In: Fitzpatrick TB, Eisen AZ, Wolff K, Freedberg IW, Austen KF, eds. Dermatology in general Medicine, Vol. I, 4th Edit. New York: McGraw-Hill, 1993:606-614.
98. Lever WF. Pemphigus and pemphigoid. J Am Acad Dermatol 1979; 1:2-32.
99. Stanley JR, Koulu L. Thivolet C. Distinction between epidermal antigens binding pemphigus vulgaris and pemphigus foliaceus autoantibodies. J Clin Invest 1984; 74:313-320.
100. Buxton RS, Cowin P, Franke WW, Garrod DR, Green KJ, King IA, Koch PJ, Magee AI, Rees DA, Stanley JR, Steinberg MS. Nomenclature of the desmosomal cadherins. J Cell Biol 1993; 121:481-483.
101. Eyre RW, Stanley JR. Identification of pemphigus vulgaris antigen extracted from normal human epidermis and comparison with pemphigus foliaceus antigen. J Clin Invest 1988; 81:807-814.
102. Hashimoto T, Ogawa MM, Konohana A, Nishikawa T. Detection of pemphigus vulgaris and pemphigus foliaceus antigens by immunoblot analysis using different antigen sources. J Invest Dermatol 1990; 94:327-335.
103. Stanley JR, Tanaka T, Mueller S, Klaus-Kovtun V, Roop DR. Isolation of complementary DNA for bullous pemphigoid antigen by use of patients' autoantibodies. J Clin Invest 1988; 82:1864-1870.
104. Korman NJ, Eyre RW, Klaus-Kovtun V, Stanley JR. Demonstration of an adhering-junction molecule (plakoglobin) in the autoantigens of pemphigus vulgaris and pemphigus foliaceus. N Engl J Med 1989; 321:631-635.
105. Kárpáti S, Amagai M, Prissuck R, Cerhs K, Stanley JR. Pemphigus vulgaris antigen, a desmoglein type of cadherin, is localized within keratinocyte desmosomes. J Cell Biol 1993; 122:409-415.
106. Anhalt GJ, Labib RS, Voorhees JJ, Beals TF, Diaz LA. Induction of pemphigus in neonatal mice by passive transfer of IgG from patients with the disease. N Engl J Med 1982; 306:1189-1193.
107. Pearson RW. Studies on the pathogenesis of epidermolysis bullosa. J Invest Dermatol 1962; 39:551-575.
108. Chapman SJ, Leigh IM, Tidman MJ, Eady RAJ: Abnormal expression of hemidesmosome-like structures by junctional epidermolysis bullosa keratinocytes in vitro. Br J Dermatol 1990; 123:137-144.

109. Verrando P, Pisani A, Ortonne JP. The new basement membrane antigen recognized by the monoclonal antibody GB3 is a large size weight glycoprotein: modulation of its expression by retinoic acid. Lab Invest 1991; 64:85-92.
110. Stanley JR. Bullous pemphigoid, cicatricial pemphigoid, and chronic bullous diseases of childhood. In: Fitzpatrick TB, Eisen AZ, Wolff K, Freedberg IW, Austen KF, eds. Dermatology in general Medicine, Vol. I, 4th Edit. New York: McGraw-Hill, 1993:615-625.
111. Stanley JR, Hawley-Nelson P, Yuspa SH, Shevach EM, Katz SI. Characterization of bullous pemphigoid antigen: A unique basement membrane protein of stratified squamous epithelia. Cell 1981; 24:897-905.
112. Klatte DH, Kurpakus MA, Grelling KA, Jones JCR: Immunochemical characterization of three components of the hemidesmosomes and their expression in cultured epithelial cells. J Cell Biol 1989; 109:3377-3390.
113. Robledo MA, Kim SC, Korman NJ, Stanley JR, Labib RS, Futamura S, Anhalt GJ. Studies of the relationship of the 230-kD and 180-kD bullous pemphigoid antigens. J Invest Dermatol 1990; 94:793-797.
114. Green KJ, Virata ML, Elgart GW, Stanley JR, Parry DAD: Comparative structural analysis of desmoplakin, bullous pemphigoid antigen, and plectin: members of a new gene family involved in organization of IFs. Int J Biol Macromol 1992; 14:145-153.
115. Labib RS, Anhalt GJ, Patel HP, Mutasim DF, Diaz LA. Molecular heterogeneity of the bullous pemphigoid antigens as detected by immunoblotting. J Immunol 1986; 136:1231-1235.
116. Diaz L, Ratrie H, Saunders WS, Futamura S, Squiquera HL, Anhalt GA, Giudice GJ. Isolation of a human epidermal cDNA corresponding to the 180 kD autoantigen recognized by bullous pemphigoid and herpes gestationis sera. J Clin Invest 1990; 86:1088-1094.
117. Anhalt GJ, Bahn CF, Labib RS, Voorhees JJ, Sugar A, Diaz LA. Pathogenic effects of bullous pemphigoid autoantibodies on rabbit corneal epithelium. J Clin Invest 1981; 68:1097-1104.
118. Hillal L, Rochat A, Duquesnoy P, Blanchet-Bardon C, Wechsler J, Martin N, Christiano AM, Barrandon Y, Uitto J, Goosens M, Hovnavian A. A homozygous insertion-deletion in the type VII collage gene (col7A1) in Hallopeau-Siemens dystrophic epidermolysis bullosa. Nature Genetics 1993; 5:287-293.
119. Prockop DJ. Mutations in collagen genes as a cause of connective tissue diseases. N Engl J Med 1992; 326:540-546.
120. Woodley DT, Gammon WR, Briggaman RA. Epidermolysis Bullosa Acquisita: Autoimmunity to type VII collagen. In: Fitzpatrick TB, Eisen AZ, Wolff K, Freedberg IW, Austen KF, eds. Dermatology in general Medicine, Vol. I, 4th Edit. New York: McGraw-Hill, 1993:630-635.
121. Woodley DT, Burgeson RE, Lunstrum G, Bruckner-Tuderman L, Reese MJ, Briggaman RA. Epidermolysis bullosa acquisita antigen is the carboxyl terminus of type VII procollagen. J Clin Invest 1988; 81:683-687.
122. Klymkowsky MW, Maynell LA, Nislow C. Cytokeratin phosphorylation, cytokeratin filament severing and the solubilization of the maternal mRNA Vg1. J Cell Biol 1991; 114:787-797.
123. Parry DAD: Primary and secondary structure of IF protein chains and modes of molecular aggregation. In: RD Goldman and PM Steinert, eds. Cellular and molecular biology of intermediate filaments. Plenum Press Corp, 1990:175-204.
124. Conway JF, Parry DAD. Intermediate filament structure. 3: Analysis of sequence homologies. Int. J. Biol. Macromol. 1988; 10:79-98.
125. McLachlan AD. Coiled-coil formation and sequence regularities in the alpha-helical regions of keratins. J Mol Biol 1978; 124:297-304.
126. Heins S, Wong PC, Muller S, Goldie K, Cleveland DW, Aebi U. The rod domain of NF-L determines neurofilament architecture, whereas the end domains specify filament assembly and network formation. J Cell Biol 1993; 6:1517-1533.

127. Geisler N, Schünemann J, Weber K. Chemical cross-linking indicates a staggered and antiparallel protofilament of desmin intermediate filaments and characterizes one higher-level complex between protofilaments. Eur J Biochem 1992; 841-852.
128. Steinert PM, Marekov LN, Fraser RDB, Parry DAD: Keratin intermediate filament structure. Cross-linking studies yield quantitative information on molecular dimensions and mechanisms of assembly. J Mol Biol 1993; 230:436-452.
129. Soellner P, Quinlan RA, Franke WW. Identification of a distinct soluble subunit of an intermediate filament protein: tetrameric vimentin from living cells. Proc Natl Acad Sci (USA) 1985; 82:7929-7933.
130. Chou CF, Riopel CM, Rott LS, Omary MB. A significant soluble fraction in simple epithelial cells: Lack of an apparent phosphorylation and glycosylation role in keratin solubility. J Cell Sci 1993; 105:433-444.
131. Vikstrom KL, Borisy GG, Goldman RD. Dynamic aspects of intermediate filament networks in BHK-21 cells. Proc Natl Acad Sci (USA) 1991; 86:549-553.
132. Miller RK, Khuon S, Goldman RD. Dynamics of keratin assembly: Exogenous type I keratin rapidly associates with type II keratin in vivo. J Cell Biol 1992; 122:123-135.
133. Vikstrom KL, Lim SS, Goldman RD, Borisy GG. Steady-state dynamics of intermediate filament network. J Cell Biol 1992; 118:121-130.
134. Eriksson JE, Opal P, Goldman RD. Intermediate filament dynamics. Curr Opin Cell Biol 1992; 4:99-104.
135. Letai A, Coulombe PA, Fuchs E. Do the ends justify the mean? Proline mutations at the ends of the keratin coiled-coil rod segment are more disruptive that internal mutations. J Cell Biol 1992; 116:1181-1195.
136. Herskowitz I. Functional inactivation of genes by dominant negative mutations. Nature 1987; 329:219-222.
137. Wilson AK, Coulombe PA, Fuchs E. The roles of K5 and K14 head, tail and R/KLLEGE domains in keratin filament assembly in vitro. J Cell Biol 1992; 119:401-414.
138. Albers K, Fuchs E. The molecular biology of intermediate filament proteins. Int Rev Cytol 1992, 134:243-271.
139. Vassar R, Rosenberg M,Ross S, Tyner A, Fuchs E. Tissue-specific and differentiation-specific expression of a human K14 gene in transgenic mice. Proc Natl Acad Sci (USA) 1989; 86:1563-1567.
140. Leask A, Rosenberg M, Vassar R, Fuchs E. Regulation of a human epidermal keratin gene: sequences and nuclear factors involved in keratinocyte-specific transcription. Genes Dev. 1990; 4:1985-1998.
141. Byrne C, Fuchs E. Probing keratinocyte and differentiation specificity of the human K5 promoter in vitro and in transgenic mice. Mol Cell Biol 1993; 13:3176-3190.
142. Abe M, Oshima RG. A single human keratin 18 gene is expressed in diverse epithelial cells of transgenic mice. J Cell Biol 1990; 111:1197-1206.
143. Bader BL, Franke WW. Cell-type specific and efficient synthesis of human cytokeratin 19 in transgenic mouse. Differentiation 1990; 45:109-118.
144. Baribault H, Price J, Miyai K, Oshima RG. Midgestional lethality in mice lacking keratin 8. Genes Dev 1993; 7:1191-1202.
145. Schermer A, Jester JV, Hardy V, Milano D, Sun TT. Transient synthesis of K6 and K16 keratins in regenerating rabbit corneal epithelium: keratin markers for an alternative pathway of keratinocyte differentiation. Differentiation 1989; 42:103-110.
146. Van Muijen GNP, Ruiter DJ, Franke WW, Achtstaetter T, Haasnoot WHB, Ponec M, Warnaar SO. Cell type heterogeneity of cytokeratin expression in complex epithelia and carcinomas as demonstrated by antibodies specific for cytokeratins nos. 4 and 13. Exp Cell Res 1986; 162:97-113.
147. Weiss RA, Eichner R, Sun TT. Monoclonal antibody analysis of keratin expression in epidermal diseases: a 48- and a 56-kD keratin as molecular markers of hyperproliferative keratinocytes. J Cell Biol 1984; 98:1397-1406.
148. Marchuk D, McCrohon S, Fuchs E. Remarkable conservation of structure among intermediate filament genes. Cell 1984; 39:491-499.

149. Rieger M, Franke WW. Identification of an orthologous mammalian cytokeratin gene. J Mol Biol 1988; 204:841-856.
150. Steinert PM, Parry DAD, Idler WW, Johnson LD, Steven AC, Roop DR. Amino acid sequences of mouse and human epidermal type II keratins of Mr 67,000 provides a systematic basis for the structural and functional diversity of the end domains of keratin intermediate filament subunits. J Biol Chem 1985; 260:7142-7149.
151. Christiano AM, Greenspan DS, Hoffman GG, Zhang X, Tamai Y, Lin AN, Dietz HC, Hovnavian A, Uitto J. A missense mutation in type VII collagen in two affected siblings with recessive dystrophic epidermolysis bullosa. Nature Genetics 1993; 4:62-66.
152. Pulkkinen L, Christiano AM, et al. Mutations in the γ2 chain gene (LAMC2) of kalinin/laminin5 in the junctional forms of epidermolysis bullosa. Nature Genetics 1994; 6:293-298.
153. Aberdam D, Galliano MF, et al. Herlitz's junctional epidermolysis bullosa is linked to mutations in the gene (LAMC2) for the γ2 subunit of nicein/kalinin (laminin-5). Nature Gentics 1994; 6:299-304.

CHAPTER 16

Role of Integrins in Cell-Substratum Attachment and Polarity of Normal and Neoplastic Epithelial Cells

Pier Carlo Marchisio
Michele De Luca

1. INTRODUCTION

Epithelial cells may be distinguished from other cell types by the presence of discrete membrane domains that mark their morphological and functional polarization. The definition of such membrane domains goes side by side with the formation of specific junctions and the targeted exposure of membrane molecules by means of sorting mechanisms operating within individual cells.

The mechanisms that are involved in defining and maintaining cell polarity have been widely reviewed particularly in simple columnar epithelia[1-3] and will not be discussed here. Rather, in this chapter, we would like to give a brief account of the role of adhesion mechanisms in defining the initial phenomena leading to the formation of a polarization axis and define the concept of adhesion-dependent polarity. In accordance with previous reports,[4] we would like to analyze the potential role of different adhesion molecules in changing the cytoskeletal organization and hence the shape of cells. Moreover, we intend to deal with the changes of adhesion-dependent polarity in epithelial diseases that involve hyperproliferation or neoplastic transformation.

Most of this chapter will be centered on the epidermis where the contact of basal keratinocytes with the underlying basal lamina and their reciprocal

Molecular Mechanisms of Epithelial Cell Junctions: From Development to Disease
Edited by Sandra Citi, M.D., Ph.D.;

cell-cell interactions are essential for the assembly, maturation, maintenance and correct function of the epidermis.[5] In this respect the epidermis is not different from any epithelium in which tissue polarity depends on topographically correct interactions of defined membrane domains.

The epidermis is a complex squamos epithelium that forms the external layer of the skin. Within this epithelium only the most basal keratinocytes, that are in contact with the basement membrane, undergo proliferation. A fraction of these basal cells continue to sustain DNA synthesis and mitosis while others leave the basement membrane and begin migrating to the external surface. During this terminal differentiation keratinocytes undergo progressive changes that end up with the formation of dead squamous elements that are eventually shed.[6] Among the above changes, keratinocytes change their adhesive properties and regulate the expression of molecules involved in cellular interactions.[5]

Several classes of molecules (for review see[7]) are involved in cell-cell and cell-matrix interaction and play a major role in the regulation of epithelial cell proliferation, mobility and differentiation. Among these molecules, a superfamily of transmembrane heterodimer surface receptors named integrins (for review see refs. 8-12) have been associated with the assembly of epithelial[13] and endothelial cells[14] as well as to a steadily growing set of other interactive phenomena. Integrins are then associated not only with the recognition of extracellular matrix components but also with the genesis of cell-cell interactions (for review see ref. 15). We propose that integrins may be involved not only in purely mechanical functions but may also play a direct role in controlling the functional polarity as well as the proliferation and differentiation properties of epidermal keratinocytes and of their transformed counterparts.

2. INTEGRINS

2.1 GENERAL FEATURES

Integrins belong to a class of adhesion receptors which were discovered about a decade ago[16-20] (reviewed in refs. 8-12). They were so called, and this name is now generally accepted, because they were both integral membrane proteins and could play a major role in the integration of functions between the extracellular matrix and the cytoskeleton.[8] Originally, integrins were exclusively considered as receptors for the extracellular matrix[16] and the human prototype integrin was identified in a molecular complex binding a specific domain of fibronectin.[18,19] Later, a new human integrin was discovered[21] whose specific function was to bind a different ligand, i.e. vitronectin, which, similarly to fibronectin, was both a soluble plasma protein and a component of extracellular matrix. On the basis of biochemical and structural similarities, it was found that one older member of the integrin family had already been discovered and characterized: this was the platelet major fibrinogen receptor or the glycoprotein gpIIb/IIIa complex[22,23] (for reviews see also refs. 24 and 25) now also called αIIb/β3.

The main criterion for defining integrins is that they are non-covalently associated glycoprotein heterodimers composed of one α and β chain (e.g. ref. 8). The original classification of three subfamilies based on a shared β chain[8] is now challenged by the discovery of new β chains and by the fact that a few α chains may associate with more than one β chain giving rise to receptors endowed with different binding properties.

One further original assumption, namely the strict ligand specificity of each integrin, was formally disproved by the discovery that several integrins may bind to different matrix proteins[26] and that the ligand specificity of individual integrins may be regulated by exogenous factors.[27]

Moreover, the definition of integrins as molecules involved in matrix recognition is fading because the involvement of integrins in the formation of intercellular bonds[14,28-31] implies that integrins have a much wider spectrum of adhesive functions than previously expected.

Finally, integrins are no longer held to be as simple adhesion devices but are considered receptors involved in transducing intracellular signals (reviewed by refs. 12 and 32–33).

3. ROLE AND REGULATION OF SOME INTEGRIN FUNCTIONS

3.1 Tissue assembly and polarity

Information on the role of integrins in the maintenance of animal tissue organization may be extrapolated from in vitro data. It is very important to gain more understanding of the function of integrins in assembled tissues in vivo as well and during histogenesis in the embryo.

One major step forward has been the discovery that integrins are located at cell-to-cell boundaries not only transiently as in lymphocytes contacting endothelial cells (reviews in refs. 34 and 35) but also in homogeneous cell populations like those forming the epidermis[13,28-31,36,37] or the vascular endothelium.[14] This novel location suggests that integrins may integrate and support the role of CAMs, cadherins or other cell adhesion molecules[38-40] in keeping cells together in a tissue organization.

A few integrins have been described at intercellular boundaries. In human endothelial cells (and also in in situ endothelia), two β1 integrins (α2β1 and α5β1) line the boundaries of confluent cells.[14] In addition, αv, but not β3, is also intercellular. The β chain associated with intercellular αv, is not yet known because the alternative complex αvβ1 is apparently missing from or poorly expressed in endothelial cells and no information is yet available about the possible presence of other heterodimers sharing αv, e.g. αvβ5[41] and αvβ6.[42] In any case the distribution of intercellular integrins seems to overlap that of CD31/PECAM,[43,44] endoCAM,[45] endoglin,[46,47] the Ca^{2+}-dependent cell-cell adhesion molecule recognized by the MoAb Ec6C10[48] or VE-cadherin.[49] A major question is whether intercellular integrins could entertain homophylic bonds like CAMs do, i.e. reciprocally interact in adjacent cells, or whether integrins could recognize an intercellular matrix ligand or a counter-receptor exposed on adjacent cells.

4. INTEGRINS AND BASEMENT MEMBRANE PROTEINS IN NORMAL HUMAN EPIDERMIS AND EPIDERMAL KERATINOCYTES

In human epidermis, two β1 integrins (α2β1 and α3β1) discretely line the apicolateral domain mostly in basal cells but also in some suprabasal cells.[13,28,29,31] These two integrins (see above) share the capacity of binding laminin and collagen. However, known matrix molecules are missing from intercellular spaces in intact epidermis and the possibility remains that either homophylic interactions may occur or a hypothetical counter-receptor may be exposed on adjacent membranes. Recently it has been proposed that α2β1 and α3β1 may entertain reciprocal heterophylic bonds[50] No obvious basal location of β1 integrins has been detected in intact human epidermis.[29] However, the new integrin heterodimer α9β1 has been found to be expressed pericellularly in the basal cells of squamous epithelia.[51]

The intercellular location of integrins in such highly polarized tissues, like epithelia and endothelia, raise the question whether integrins are involved in defining cell polarity. In epidermal cells, as well as in many other epithelial cells, the exclusion of β1 integrins and the restricted location of α6β4 to the basal domain in contact with the basement membrane suggests that the sorting of integrins to defined membrane domains is involved in building up a polarity axis. In endothelial cells, that lack α6β4, the role of leading, but not exclusive, basal integrin may be played by αvβ3 that has been also shown to be involved in binding to von Willebrand factor and to fibrinogen.[52-55] This indicates that a domain-dependent exposure of integrins may be a general occurrence in defining cell polarity.

In normal human keratinocytes in culture α6β4 is exclusively found at the basal surface in areas that are poor in microfilaments.[13] Such discontinuous location produces a typical "leopard skin" pattern (Fig. 16.1) that is related to the formation of hemidesmosomes.[56-58] In this same system, α2β1 and α3β1 are found at lateral boundaries

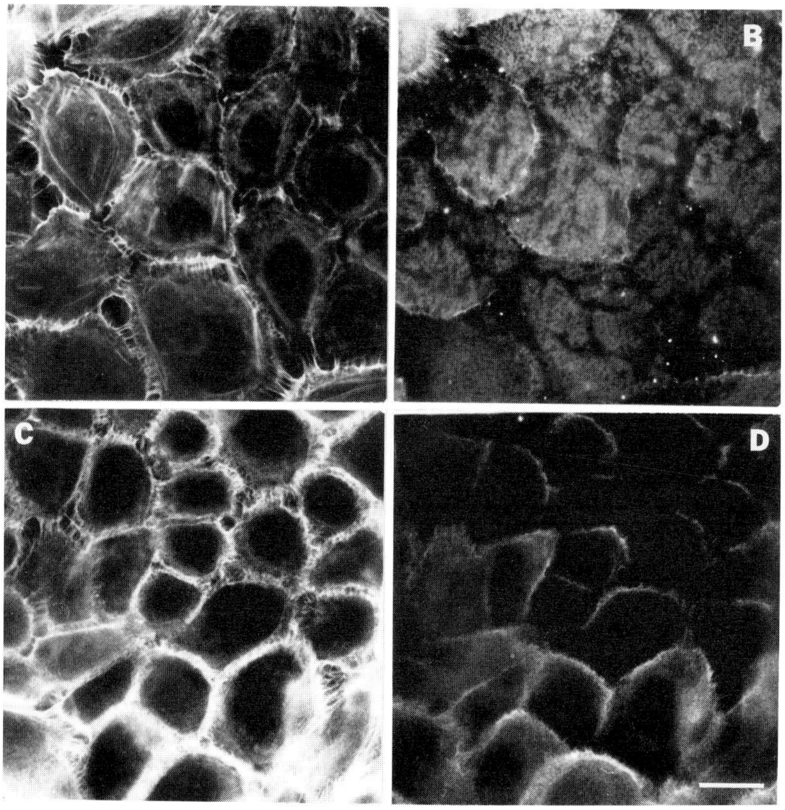

Fig. 16.1. This figure shows the "leopard skin" pattern of integrin β4 (b) in a colony of normal human keratinocytes which is also stained for F-actin by fluorescein-tagged phalloidin (a). The "leopard skin" basal pattern of integrin β4 is lost in high passage keratinocytes and replaced by pericellular distribution (d). The F-actin pattern of the same colony is shown in c. The change of integrin topography indicates the loss of adhesion-dependent polarity that takes place both in situ *and* in vitro *whenever hyperproliferation or altered growth regulation occur.*

in accordance to the pattern found in normal epidermis.

In epidermal cells growing in low Ca^{2+} and thus unable to enter their differentiation pathway (for review see ref. 59), basal β1 integrins have been observed.[28] Then, under conditions which do not permit epidermal maturation, like in conditions of low Ca^{2+}, keratinocytes lose their typical integrin polarity[13] and, therefore, are to be considered in a non-physiological state.

Under normal conditions we have never been able to see any obvious surface exposure of the prototype fibronectin receptor α5β1 in cultured keratinocytes as well as in sections of normal epidermis (see below).

The cellular mechanism driving the sorting of membrane adhesion receptors is not yet known unequivocally. On the basis of topography the enrichment of integrins at defined surface domains depends on the reciprocal contact of individual cells and of cells with the basement membrane. Indeed, in cultured keratinocytes, a6β4 antibodies prevent adhesion while β1 antibodies do not;[29,60] this indicates that α6β4 is functionally involved in the recognition of basement membrane components. So do αvβ3 antibodies in endothelial cells.[61] In both cell types, β1 antibodies impair cell aggregation in vitro and are presumably involved in tissue organization in vivo by cooperating with other adhesion molecules in preventing the formation of intercellular bonds.[13,14] Intercellular integrins may play specialized physiological functions in human endothelia. Evidence suggests that, at least in vitro, confluent endothelial cells exposed to β1 but not to β3 antibodies, lose their capacity of restraining macromolecules in the luminal compartment.[14] This indicates that integrins may also be involved in the control of intercellular permeability in intact blood vessels in situ but this may occur also in other epithelial tissues that separate two discrete compartments.

A major step toward the understanding of the domain-specific exposure of integrins has been recently made by Giancotti et al.[62] who showed that the β4 subunit is synthesized by keratinocytes and first exposed homogeneously over the whole cellular surface. Then, because of selective proteolysis of its cytodomain operated by calpain, α6β4 is removed from the domains that are not in contact with the basement membrane.

The composition of the epithelial basement membrane is probably crucial in driving the basal positioning of α6β4 in epidermal keratinocytes and in other epithelia whose adhesion depends on α6β4. In human keratinocyte colonies cultured under conditions that allow full epidermal differentiation we have reported[63] that (i) the basement membrane protein kalinin is codistributed with laminin and collagen type IV as well as with the bullous pemphigoid antigen in footprints deposited by growing and migrating colonies; (ii) the integrin heterodimer α6β4 is codistributed with the same molecules suggesting its spatial association with basement membrane components; (iii) the distribution pattern of α6β4 and kalinin underneath individual cells is identical and is characterized by a typical "leopard skin" pattern complementary to the distribution of submembraneous F-actin microfilament network. We suggest that α6β4 displays not only an adhesive role for keratinocytes in view of its reported association to hemidesmosomes[56-58] but may also be involved in organizing the molecules of the epithelial extracellular matrix, including those forming the basement membrane zone and involved in hemidesmosome assembly, a function proposed for other integrins in other cellular systems (for review see ref. 64).

The coherence between the basal exposure of α6β4 and the presence of kalinin in the texture of the basement membrane is indirectly supported by new data showing that when α6β4 is missing from epithelial cells, like in the thyroid gland (Orlandi et al, submitted) and the kidney tubular system (in preparation), also kalinin is not a component of the corresponding basement membrane. Overall, the data suggest that kalinin either is the ligand or is part of the ligand complex of α6β4.

In summary, integrins may no longer be considered as simple matrix receptors but are also involved in defining the initiation of epithelial polarity and in regulating the assembly of cells in cooperation with other membrane adhesion molecules and some components of the basement membrane.

5. ADHESION-DEPENDENT POLARITY OF KERATINOCYTES IS REGULATED BOTH IN VIVO AND IN VITRO

The sorting of different integrins to different membrane domains is a property that is established in normal keratinocytes undergoing controlled growth both in vitro and in vivo. In fact, in the fetal epidermis, where cells proliferate at a higher rate, both β4 and β1 integrins are distributed pericellularly and thus fail to be sorted.[65] A very similar phenomenon, namely the loss of integrin polarity, occurs also at the borders of an epidermal wound where keratinocytes temporarily resume migration and proliferation properties over a provisional matrix that are similar to those of immature epidermis.[66]

To check whether the establishment of adhesion-dependent polarity could depend on the rate of proliferation we tested high-passage in vitro keratinocytes that had lost proliferation control.[67] As reported above, keratinocyte adhesion to basal lamina and lateral interactions among basal epidermal cells are mediated, beside other molecules, by integrin receptors that are sorted to defined membrane domains. The hemidesmosome-associated integrin α6β4 was sharply localized to the basal surface of basal cells while α2β1 and α3β1 were enriched laterally. This integrin sorting pattern was perfectly reproducible in vitro by cultured keratinocytes and took place progressively in primary or secondary culture in the presence of 1.8 mM Ca^{2+}. The polarized topography of integrins was gradually lost with higher passage numbers and, between the 5th and 7th passage, there was a complete pericellular redistribution of these integrins (Fig. 16.1). Along with the decreased basal adhesive value of α6β4 a marked increase in the number of focal contacts was observed in high passage keratinocyte colonies along with the lack of hemidesmosome formation. A similar loss of polarized topography of integrins occurred in low Ca^{2+} culture conditions that prevent the maturation of keratinocytes.

A novel unexpected set of data was obtained by this set of experiments. In fact increasing the number of culture passages beyond the fifth induced the appearance of the fibronectin receptor α5β1 on the surface of keratinocytes, particularly at intercellular junctions and in some focal contacts. The receptor α5β1 was not detectably exposed by low passage cells. We propose that forcing keratinocytes into more frequent cell cycles by continuous passaging may perturb the polarized topography of integrins and the adhesion mechanisms of keratinocytes and supported the concept that low passage keratinocytes are, in our opinion, the most reliable in vitro models for studying the physiology of epidermal cells.

6. ADHESION-DEPENDENT POLARITY OF KERATINOCYTES IS LOST IN PSORIASIS

Psoriasis (*psoriasis vulgaris*) is a hyperproliferative cutaneous disease of unknown etiology and etiopathogenesis. Alteration of keratinocyte adhesiveness to basal lamina has been proposed as the initial disturbance leading to poorly controlled proliferation. Our data obtained on high passage keratinocytes (see above) suggested that adhesion-dependent integrin polarized topography could have been impaired also in psoriasis.[68] Therefore, the expression and function of integrins in psoriatic keratinocytes were examined, both in vivo and in vitro. We found that (i) in psoriatic keratinocytes, observed in skin sections and in culture, the integrin heterodimers α2β1, α3β1 and α6β4 lost their polarized distribution on the plasma membrane; (ii) the role of these integrins in mediating keratinocyte adhesion in vitro was altered; (iii) psoriatic keratinocytes formed focal contacts containing both β1 and β4 integrins. As pointed out in a previous paragraph, in normal adult keratinocytes the α5β1 fibronectin receptor is poorly expressed and diffusely distributed on the basal keratinocyte plasma membrane and is not organized in defined adhesive structures. In contrast, psoriatic keratinocytes showed a clear fibronectin receptor staining in vivo, and organized α5β1 in typical focal contacts in vitro without any obvious increase of its expression and synthesis. Unexpectedly, these multiple alterations of integrins were also present in the uninvolved skin and in the corresponding cultured keratinocytes biopsed from psoriatic patients, suggesting a key role for altered integrin-mediated adhesion in the pathogenesis of this disease. Most integrin alterations detected in psoriasis vulgaris[68] were recently observed also in the psoriatic syndrome that is associated to chronic liver diseases.[69] However, the disturbances of adhesion-dependent polarity were limited to overt lesions and was not observed in biopsies of clinically normal skin.

7. ALTERATIONS OF ADHESION-DEPENDENT POLARITY IN CARCINOMAS

Human cancers are in large majority (80-90%) of epithelial origin. It is rather obvious that their progression and invasiveness stem from altered social behavior, i.e. from loss of correct adhesion properties. We are currently analyzing different carcinomas and in all cases we have found that primary alterations involve disturbances of adhesion-dependent polarity and/or changes of basement membrane composition.

Because of our current interests, the first carcinomas we tested were those originating from the epidermis, i.e. basal cell carcinomas (BCC) and squamous cell carcinomas (SCC)[70] where we studied the expression and topography of some integrins and basement membrane proteins by immunochemical procedures. It has been shown that the typical cell-to-cell distribution of α2β1 and α3β1 found in normal epidermis was replaced by pericellular distribution in both BCC and SCC cells. BCC and SCC also showed different patterns of expression of α6β4: while SCC showed high expression and pericellular distribution of α6β4, BCC cells did not express this integrin at all. The absence of α6 and β4 subunits from BCC extracts was confirmed by Western blotting. The molecular composition of the basement membrane was markedly different in the two types of epidermal tumors. While laminin and collagen type IV were conserved in the basement membrane zone of both tumors, kalinin was absent from BCC cells. Then, the simultaneous loss of expression of α6β4 and kalinin in BCC cells but not in SCC cells, indicates that α6β4 integrin and one of its potential ligands may be coregulated in both BCC and SCC, thus suggesting a role for this phenomenon in the pathogenesis and clinical behavior of these epidermal tumors.

Very similar results were obtained in a number of cutaneous precancerous lesions (Savoia and Marchisio, submitted). In either seborrheic keratosis or in actinic keratosis, pericellular staining for α2β1 and α3β1 was focally conserved in upper epidermal layers. In *epidermodysplasia verruciformis* and in Bowen's disease, the expression of the β1 chain was mostly pericellular but α2 and α3 were observed also basally indicating that α2β1 and α3β1 were both exposed in contact with the basement membrane. The basal pattern of α6β4 was preserved in *epidermodysplasia verruciformis* and appeared similar to that of normal epidermis; in other preneoplastic skin lesions the polarity of α6β4 was lost and replaced by pericellular distribution. Topography changes of a6 and β4 chains were maximal in seborrheic keratosis where α6β4 completely lost its polarized distribution and was mostly detected on the lateral domain of keratinocytes. Then, in preneoplastic skin disorders, as reported for cutaneous carcinomas, a progressive failure to sort integrins to defined membrane domains takes place in epidermal keratinocytes. We propose that the abnormal expression and topography of integrins in preneoplasic skin disorders may represent early markers of neoplastic transformation and be related to the progression of these diseases toward aggressive cutaneous tumors.

The data obtained in epidermal keratinocytes, that just represent one species of epithelial cells, are progressively being confirmed in other squamous epithelia (e.g. those lining the lower female genital tract or the upper alimentary tract) and their derived tumors[71] (Sacchi et al, in preparation). Similarly, other epithelial cells that form unicellular sheets and their derived neoplasms also show clear adhesion-dependent polarity and altered adhesion patterns. We are now analyzing the columnar epithelium of the gross intestine and the derived colorectal neoplasias, the mammary gland and breast tumors as well as the endometrium and endometrial carcinomas (in preparation). In all these epithelia altered control of cell proliferation, not necessarily to be considered as frankly malignant, implies loss of adhesion-dependent polarity and membrane redistribution of β4 and β1 integrins. Interesting data have recently been obtained in the thyroid gland where α3β1 replaces α6β4 as the integrin linking the cellular basal surface with the basement membrane. In all the thyroid carcinomas, but not the benign adenomas, derived from this unique epthelium that reverses its polarity according to function, α3β1, as well as E-cadherin, changes its polarity and becomes pericellular. Interestingly, these changes may be detected also in fine-needle biopsies and may represent useful markers in the differential diagnosis between thyroid adenomas and follicular carcinomas (Orlandi et al, submitted).

8. CONCLUSIONS

A major recent development in the field of integrins has been the discovery that this family of matrix receptors is involved not only in the formation of cell-matrix adhesion but also in cell-cell bonds. It is not yet clear whether this additional role of integrins implies a new molecular recognition mechanism or whether signals originated thereafter may be involved in defining the cascade of events that lead to the establishment of the axis of polarity in epithelial cells. Even if evidence is not yet available, we are inclined to suggest that the establishment of adhesion-dependent polarity is the prerequisite whereby epithelial cells initiate the cascade of events that build up their typical polarized phenotype and their own functional properties. Integrins, once they have defined their position, may generate signals that drive the sorting of specific molecules to given membrane domains, induce the formation of different types of junctional devices in their right position and even control proliferation rates. The reason why we suggest a direct role for integrins in generating metabolic signals is essentially the observation that whenever the rate of epithelial cell proliferation is deregulated adhesion-dependent polarity is lost. We also wonder whether the opposite is true. In other words is the polarized epithelial phenotype established when adhesion-dependent polarity is prevented? We are attempting experiments where the processing of the apico-laterally exposed β4 is prevented by inhibiting the proteolytic activity of calpain and find that the pattern of keratinocyte growth is altered in vitro. The same occurs when we lower the concentration of extracellular Ca^{2+} and find that keratinocytes behave in a different way and fail to sort their integrins[13,65] and thus also fail to begin their process of maturation.[59]

A direct involvement of integrins as environment sensors capable of transducing messages for the establishment and maintenance of the polarized epithelial phenotype remains highly speculative until such signals are identified.

Acknowledgments.

The authors thank all the scientists who shared the reagents that made our work possible. We also thank Ottavio Cremona, Paola Savoia and Livio Trusolino for stimulating discussions. This work was supported by grants from the Italian National Research Council, CNR Special Project ACRO "Applicazioni Cliniche della Ricerca Oncologica" and from AIRC, Associazione Italiana per la Ricerca sul Cancro.

References

1. Rodriguez-Boulan E, Nelson WJ. Morphogenesis of the polarized epithelial cell phenotype. Science 1989; 245:718-725.
2. Rodriguez-Boulan E, Powell SK. Polarity of epithelial and neuronal cells. Ann Rev Cell Biol 1992; 8:395-427.
3. Wollner DA, Nelson WJ. Establishing and maintaining epithelial cell polarity. Roles of protein sorting, delivery and retention. J Cell Sci 1992; 102:185-190.
4. Klein G, Langegger M, Timpl R, Ekblom P. Role of laminin A chain in the development of epithelial cell polarity. Cell 1988; 55:331-341.
5. Watt FM. Terminal differentiation of epidermal keratinocytes. Curr Opin Cell Biol 1989; 6:1107-1115.
6. Fuchs E. Epidermal differentiation: the bare essentials. J Cell Biol 1990; 111:2807-2814.
7. Abelda SM, Buck CA. Integrins and other cell adhesion molecules. FASEB J 1990; 4:2869-2880.
8. Hynes RO. Integrins: a family of cell surface receptors. Cell 1987; 48:549-544.
9. Buck CA, Horwitz AF. Cell surface receptors for extracellular matrix molecules. Annu Rev Cell Biol 1987; 3:179-205.
10. Ruoslahti E, Pierschbacher MD. New perspectives in cell adhesion: RGD and integrins. Science 1987; 238:491-497.
11. Ruoslahti E. Integrins. J Clin Invest 1991; 87:1-5.
12. Hynes RO. Integrins: versatility, modulation, and signaling in cell adhesion. Cell 1992; 69:11-25.
13. Marchisio PC, Bondanza S, Cremona O, Cancedda R, De Luca M. Polarized expression of integrin receptors (α6β4, α2β1, α3β1, and αvβ5) and their relationship with the cytoskeleton and basement membrane matrix in cultured human keratinocytes. J Cell Biol 1991; 112:761-773.
14. Lampugnani MG, Resnati M, Dejana E, Marchisio PC. The role of integrins in the maintenance of endothelial monolayer integrity. J Cell Biol 1991; 112:479-490.
15. Marchisio PC. Integrins and tissue organization. Adv. Neuroimmunol 1991; 1: 213-233.
16. Horwitz AK, Duggan R, Greggs C, Decker C, Buck C. The cell-substrate attachment (CSAT) antigen has properties of a receptor for laminin and fibronectin. J Cell Biol 1985; 101:2134-2144.
17. Gardner JM, Hynes RO. Interaction of fibronectin with its receptor on platelets. Cell 1985; 42:439-448.

18. Pytela R, Pierschbacher MD, Ruoslahti E. A 125/115 kDa cell surface receptor specific for vitronectin interacts with the arginine-glycine-aspartic acid adhesion sequence derived from fibronectin. Proc Natl Acad Sci USA 1985; 82:5766-5770.
19. Pytela R, Pierschbacher MD, Ruoslahti E. Identification and isolation of a 140 kD cell surface glycoprotein with properties expected of a fibronectin receptor. Cell 1985; 40:191-198.
20. Giancotti FG, Comoglio PM, Tarone G. A 135000 molecular weight plasma membrane glycoprotein involved in fibronectin-mediated cell adhesion: immunofluorescence localization in normal and RSV-transformed fibroblasts. Exp Cell Res 1986; 163:47-52.
21. Pytela R, Pierschbacher MD, Ginsberg MH, Plow EF, Ruoslahti E. Platelet membrane glycoprotein IIb/IIIa: member of a family of Arg/Gly/Asp-specific adhesion receptors. Science 1986; 231:1559-1562.
22. Jennings LK, Phillips DR. Purification of glycoproteins IIb-IIIa from human platelet plasma membrane and characterization of a calcium-dependent glycoprotein IIb-IIIa complex. J Biol Chem 1982; 257:10458-10466.
23. Plow EF, Loftus JC, Levin EG, Fair DS, Dixon D, Forsyth J, Ginsberg MH. Immunologic relationship between platelet membrane glycoprotein GpIIb-IIIa and cell surface molecules expressed by a variety of cells. Proc Natl Acad Sci USA 1986; 83:6002-6006.
24. Ginsberg MH, Loftus JC, Plow EF. Cytoadhesins, integrins and platelets. Thromb Haemost 1988; 59:1-6.
25. Phillips DA, Charo IF, Parise CV, Fitzgerald LA. The platelet membrane glycoprotein IIb-IIIa complex. Blood 1988; 71:831-843.
26. Kirchhofer D, Languino LR, Ruoslahti E, Piersbacher MD. α2β1 integrins from different cell types show different binding specificities. J Biol Chem 1990; 265:615-618.
27. Gailit J, Ruoslahti E. Regulation of the fibronectin receptor affinity by divalent cations. J Biol Chem 1988; 265:12927-12932.
28. Carter WG, Wayner EA, Bouchard TS, Kaur P. The role of integrins α2β1 and α3β1 in cell-cell and cell-substrate adhesion of human epidermal cells. J Cell Biol 1990; 110:1378-1404.
29. De Luca M, Tamura RN, Kajiji S, Bondanza S, Rossino P, Cancedda R, Marchisio PC, Quaranta V. Polarized integrin mediates keratinocyte adhesion to basal lamina. Proc Natl Acad Sci USA 1990; 87:6888-6892.
30. Kaufmann R, Frosch D, Westphal C, Weber L, Klein CE. Integrin VLA-3: ultrastructural localization at cell-cell contact sites of human cell cultures. J Cell Biol 1989; 109:1807-1815.
31. Larjava H, Peltonen J, Akiyama SK, Yamada SS, Gralnick HR, UittoJ, Yamada KM. Novel function for β1 integrins in keratinocyte cell-cell interactions. J Cell Biol 1990; 110:803-815.
32. Damsky CH, Werb Z. Signal transduction by integrin receptors for extracellular matrix: cooperative processing of extracellular information. Curr Opin Cell Biol 1992; 4:772-781.
33. Juliano RL, Haskill S. Signal transduction from the extracellular matrix. J Cell Biol 1992; 120:577-585.
34. Hemler ME. VLA proteins in the integrin family: structures, functions and their role in leukocytes. Ann Rev Immunol 1990; 8:365-400.
35. Larson RS, Springer TA. Structure and function of leukocyte integrins. Immunol Rev 1990; 114:181-217.
36. Konter U, Kellner I, Klein E, Kaufmann R, Mielke V, Sterry W. Adhesion molecule mapping in normal human skin. Arch Dermatol Res 1989; 281:454-462.
37. Staquet MJ, Levarlet B, Dezutter-Dambuyant C, Schmitt D, Thivolet J. Identification of specific human epithelial cell integrin receptors as VLA proteins. Exp Cell Res 1990;187:277 283.
38. Edelman GM. Cell adhesion molecules in the regulation of animal form and tissue pattern. Ann Rev Cell Biol 1986;2: 81-116.
39. Ekblom P, Vestweber D, Kemler R. Cell-matrix interactions and cell adhesion during development. Ann Rev Cell Biol 1986; 2: 27-47.
40. Takeichi M. The cadherins: cell-cell adhesion molecules controlling animal morphogenesis. Development 1988; 102:639-655.

41. Cheresh D, Smith JW, Cooper HM, Quaranta V. A novel vitronectin receptor integrin (αvβx) is responsible for distinct adhesive properties of carcinoma cells. Cell 1989; 57:59-69.
42. Busk MR, Pytela R, Sheppard D. Characterization of the integrin alfa v beta 6 as a fibronectin-binding protein. J Biol Chem 1992; 267:5790-5796.
43. Newman PJ, Berndt MC, Gorski J, White GC, Lyman S, Paddock C, Muller WA. PECAM-1 (CD31) Cloning and relation to adhesion molecules of the immunoglobulin gene superfamily. Science 1990; 247:1219-1222.
44. Simmons D, Walker C, Power C, Pigott R. Molecular cloning of CD31, a putative intercellular adhesion molecule closely related to carcinoembryonic antigen. J Exp Med 1990; 171:2147-2152.
45. Albelda SM, Oliver PD, Romer LH, Buck CA. EndoCAM : a novel endothelial cell-cell adhesion molecule. J Cell Biol 1990; 110:1227-1237.
46. Gougos A, Letarte M. Identification of a human endothelial cell antigen with monoclonal antibody 44G4 produced against a pre-B leukemic cell line. J Immunol 1988; 141:1925-1933.
47. Gougos A, Letarte M. Primary structure of endoglin, an RGD-containing glycoprotein of human endothelial cells. J Biol Chem 1990; 265:8361-8364.
48. Heimark RL, Degner M, Schwarz SM. Identification of a Ca2+-dependent cell-cell adhesion molecule in endothelial cells. J Cell Biol 1990; 110:1745-1756.
49. Lampugnani MG, Resnati M, Raiteri M, Pigott R, Pisacane A, Houen G, Ruco LP, Dejana E. A novel endothelial-specific membrane protein is a marker of cell-cell contacts. J Cell Biol 1992; 118:1511-1522.
50. Symington BE, Takada Y, Carter WJ. Interaction of integrins α3β1 and α2β1: potential role in keratinocyte intercellular adhesion. J Cell Biol 1993; 120:523-535.
51. Palmer EL, Ruegg C, Ferrando R, Pytela R, Sheppard D. Sequence and tissue distribution of the integrin α9 subunit, a novel partner of β1 that is widely distributed in epithelia and muscle. J Cell Biol 1993; 123:1289-1297.
52. Cheresh D. Human endothelial cells synthesize and express an Arg-Gly-Asp-directed receptor involved in attachment to fibrinogen and von Willebrand factor. Proc Natl Acad Sci USA 1987; 84:6471-6475.
53. Cheresh D, Spiro RC. Byosynthetic and functional properties of an Arg-Gly-Asp-directed receptor involved in human melanoma cell attachment to vitronectin, fibrinogen and von Willebrand factor. J Biol Chem 1987; 262:17703-17711.
54. Dejana E, Lampugnani MG, Giorgi M, Gaboli M, Federici AB, Ruggeri ZM, Marchisio PC. Von Willebrand factor promotes endothelial cell adhesion via an Arg-Gly-Asp-dependent mechanism. J Cell Biol 1989; 109:367-375.
55. Dejana E, Lampugnani MG, Giorgi M, Gaboli M, Marchisio PC. Fibrinogen induces endothelial cell adhesion and spreading via release of endogenous matrix proteins and the recruitment of more than one integrin receptor. Blood 1990; 75:1509-1517.
56. Stepp MA, Spurr-Michaud S, Tisdale A, Elwell J, Gipson IK. α6β4 integrin heterodimer is a component of hemidesmosomes. Proc Natl Acad Sci USA 1990; 87:8970-8974.
57. Sonnenberg A, Calafat J, Janssen H, Daams H, van der Raaij-Helmer LM, Falcioni R, Kennel SJ, Aplin JD, Baker J, Loizidou M, et al Integrin α6β4 complex is located in hemidesmosomes, suggesting a major role in epidermal cell-basement membrane adhesion. J Cell Biol 1991; 113:907-917.
58. Jones JC, Kurpakus MA, Cooper HM, Quaranta V. A function for the integrin alpha 6 beta 4 in the hemidesmosome. Cell Regul 1991; 2:427-438.
59. Green H. The keratinocyte as differentiated cell type. The Harvey Lectures 1980; Series 74:101-139.

60. Ryynanen J, Jaakkola S, Engvall E, Peltonen J, Uitto J. Expression of β4 integrins in human skin: comparison of epidermal distribution with β1-integrin epitopes, and modulation by calcium and vitamin D3 in cultured keratinocytes. J Invest Dermatol 1991; 97:562-567.
61. Dejana E, Colella S, Conforti G, Abbadini M, Gaboli M, Marchisio PC. Fibronectin and vitronectin regulate the organization of their respective Arg-Gly-Asp adhesion receptors in cultured human endothelial cells. J Cell Biol 1988; 107:1215-1223.
62. Giancotti FG, Stepp MA, Suzuki S, Engvall E, Ruoslahti E. Proteolytic processing of endogenous and recombinant beta 4 integrin subunit. J Cell Biol 1992; 118:951-959.
63. Marchisio PC, Cremona O, Savoia P, Pellegrini G, Ortonne JP, Verrando P, Burgeson RE, Cancedda R, De Luca M. The basement membrane protein BM-600/nicein codistributes with kalinin and the integrin α6β4 in human cultured keratinocytes. Exp Cell Res 1993; 205:205-212.
64. McDonald JA. Extracellular matrix assembly. Ann Rev Cell Biol 1988; 4:183-207.
65. Hertle MD, Adams JC, Watt FM. Integrin expression during human epidermal development in vivo and in vitro. Development 1991; 112:193-206.
66. Cavani A, Zambruno G, Marconi A, Manca V, Marchetti M, Giannetti A. Distinctive integrin expression in the newly forming epidermis during wound healing in humans. J Invest Dermatol 1993; 101:600-604.
67. De Luca M, Pellegrini G, Bondanza S, Cremona O, Savoia P, Cancedda R, Marchisio PC. The control of polarized integrin topography and the organization of adhesion-related cytoskeleton in normal human keratinocytes depend upon number of passages in culture and ionic environment. Exp Cell Res 1992; 202:142-150.
68. Pellegrini G, De Luca M, Orecchia G, Balzac F, Cremona O, Savoia P, Cancedda R, Marchisio PC. Expression, topography, and function of integrin receptors are severely altered in keratinocytes from involved and uninvolved psoriatic skin. J Clin Invest 1992; 89:1783-1795.
69. Giannelli G, Savoia P, Schiraldi O, et al Psoriatic lesions in patients with chronic liver disease are distinct from psoriasis vulgaris lesions based on integrin adhesion receptors. Hepatology, in press.
70. Savoia P, Trusolino L, Pepino E, Cremona O, Marchisio PC. Expression and topography of integrins and basement membrane proteins in epidermal carcinomas: basal but not squamous cell carcinomas display loss of α6β4 and BM-600/nicein. J Invest Dermatol 1993; 101:352-358.
71. Surico N, Priori L, Savoia P, et al. New diagnostic perspectives offered by integrin investigation. In: Genazzani AR, D'Ambrogio, Artini PG, eds. Current Investigations in Gynecology and Obstetrics. Carnforth: The Parthenon Publishing Group, 1993:509-511.

INDEX

C

D

E

F

G

U

V

W

X

Y

Z

ZO-2, 10, 40, 92-96, 99-101, 114, 126-127

Receive a FREE BOOK of your choice.

R. G. Landes Company publishes four series of books: *Medical Intelligence Unit*, *Molecular Biology Intelligence Unit*, *Neuroscience Intelligence Unit* and *Biotechnology Intelligence Unit*. The books address topics about which information is accumulating rapidly. Our publishing program conforms to the pace of research in these areas by *printing all books within 90 days of receipt of the manuscripts*.

Please help us out—Just answer the questions below.
Select your FREE BOOK from the back cover
and return this postpaid card. Thanks.

1. We want to make our books available on CD-ROM. The electronic (CD) copy of a book would only be available to an individual who purchased the book. All copyright restrictions would be waived. Text, references, figures and tables could be used in the disk-holder's publications or slide presentations without requesting permission from the copyright holder (us). Would this be valuable to you?

2. Would you pay $20 in addition to the cost of the book in order to receive an electronic copy of the book on CD-ROM?

3. All books in our series cost at least $89.95. We would like to sell them at a discount as part of a deposit account subscription system. Would you be interested in receiving any five books per year from our Intelligence Series for a single payment of $270–effectively $54.00 per book?

1.	2.	3.
Yes ________	**Yes** ________	**Yes** ________
No ________	**No** ________	**No** ________

PLEASE TEAR ALONG PERFORATIONS

TO RECEIVE YOUR FREE BOOK

Please select three titles (in order of preference) from the list on the back cover of this book. Indicate the authors name below in the spaces provided. Your request will be filled based on availability.

1.________________ 2.________________ 3. ________________

Please complete the shipping information below

Name: ________________________________

Company: ________________________________

Address: ________________________________

City: ________________ State/Province: ________________

Country: ________________ Postal Code: ________________

My computer type is Macintosh ____; IBM ____; Other ________________

Do you own ______ or plan to purchase ______ a CD-ROM drive?

NO POSTAGE
NECESSARY
IF MAILED
IN THE
UNITED STATES

BUSINESS REPLY MAIL

FIRST-CLASS MAIL PERMIT NO. 7200 AUSTIN TX

POSTAGE WILL BE PAID BY THE ADDRESSEE

R. G. LANDES COMPANY
PO BOX 4858
AUSTIN TX 78765-9837